SPOROPOLLENIN DINOFLAGELLATE CYSTS

THEIR MORPHOLOGY AND INTERPRETATION

by

William R. Evitt

Department of Geology
Stanford University

Additional copies of SPOROPOLLENIN DINOFLAGELLATE CYSTS may be obtained by sending $30.00 to: Robert T. Clarke, Treasurer, AASP Foundation, c/o Mobil Research & Development Corp., P.O. Box 819047, Dallas, Texas 75381-9047. Make check payable to AASP Foundation.

Library of Congress Catalog Card Number, 84-72457
ISBN 0-931871-00-X

Printed in the United States by Hart Graphics, Inc.
8000 Shoal Creek Blvd., Austin, Texas 78758

This volume honors the memory of

GEORGES DEFLANDRE and ALFRED EISENACK
1897-1973 1891-1982

whose studies on fossil microplankton,
during 90 collective years of active
research, laid the foundation of
our knowledge of
dinoflagellate cysts.

Frontispiece - REPRESENTATIVE MODERN AND FOSSIL DINOFLAGELLATES

A - Theca of modern freshwater *Peridinium limbatum* in oblique ventral view, showing striate growth bands between reticulate original portions of thecal plates. Round Pond, Falmouth, Massachusetts; X 675. See Chapters 1 and 9.

B - Cavate cyst of Paleogene *Deflandrea phosphoritica*, in dorsoventral view, showing open intercalary archeopyle through two wall layers that differ in form and thickness and are separated by a smaller (apical) and larger (hypocystal) pericoel. Septarienton, Denmark; X 450.

C - Probably extrathecal envelope of Late Cretaceous *Palaeoperidinium pyrophorum* in ventral view, showing reticulate and striate areas that replicate the exterior relief of the unpreserved theca (compare A). Red Bank Formation, New Jersey; X 575. See Chapter 9.

D, G - Proximate autocyst of Late Jurassic *Leptodinium mirabile* in apical view (D) and as detail of midventral surface (G), showing prominent paratabulation expressed by low parasutural septa. Open precingular archeopyle in D lies on sagittal line opposite parasulcus whose paratabulation is shown in full in G. Naknek Formation, Alaska; X 935 (D) and X 1065 (G). See Chapter 5 and compare Figure 5.12 B.

E - Chorate cyst of Early Cretaceous *Oligosphaeridium complex* in ventral view. Each trumpetlike process corresponds to a single plate of the once-surrounding theca (but some plates are unrepresented by processes). The open apical archeopyle truncates the top of the specimen; its dislocated operculum (with four apical processes) overlies the edge of the main body at 8 o'clock. Rurikfjellet Member of Janusfjellet Formation, Spitzbergen (from material courtesy of Bindra Thusu); X 425. See Chapter 11.

F - Covering of probably motile cell of Late Cretaceous *Dinogymnium acuminatum* in ventral view, showing single convoluted wall layer whose form suggests modern *Gymnodinium* or *Balechina*. Moreno Formation, California; X 550. See Chapter 12.

H - Cyst of Middle Jurassic *Nannoceratopsis gracilis* in right lateral view. Single large hypocystal paraplate covers most of visible area; parasutural grooves separating paraplates of epicyst, paracingulum, and parasulcus are visible in upper right quarter. Inferior Oolite, England; X1000. See Chapter 12.

I - Theca of modern freshwater *Ceratium furcoides* in ventral view. Thecal plates have separated along sutures following treatment with sodium hypochlorite. Sulcus (open area in line with longest apical plate) lies well to the left of ventral midline. Felt Lake, Stanford, California; X 825.

J - Cornucavate cyst of Late Cretaceous *Xenascus ceratioides* in ventral view, showing ceratioid outline defined by prominent horns. The three prominent horns, which define the ceratioid outline, and the other projections are formed by the periphragm which otherwise closely surrounds the ovoidal endocyst. Upper Goru Formation, Pakistan; X 435.

Note. Previous publication of these figures: A - Evitt and Wall (1968, possibly the first published SEMs of dinoflagellates); C - Cover of "Bioscience", October 1964; F - Evitt, Clarke, and Verdier (1967); H - Piel and Evitt (1981); I - Wall and Evitt (1975); J - Evitt (1961a).

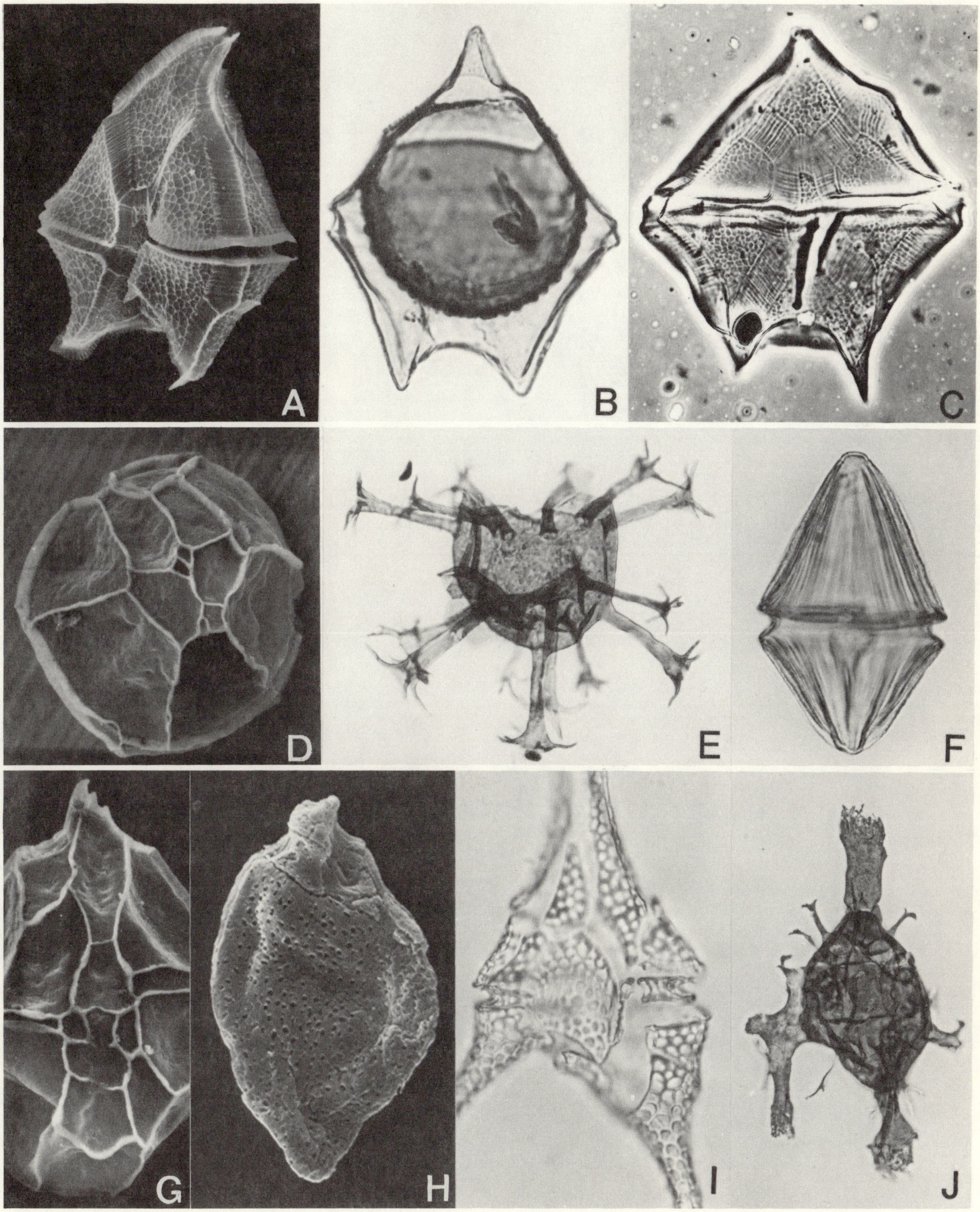
A
B
C
D
E
F
G
H
I
J

CONTENTS

Page

Part II. Morphology of the Cyst

Part III. Interpretation of the Cyst

PREFACE

This volume aims to summarize what is currently known about the morphology of sporopollenin dinoflagellate cysts and to provide guides for their interpretation. Its immediate basis is a short-course involving about 80 hours of lectures, demonstrations, and laboratory work, which I have presented to numerous small groups of professional paleontologists and to my advanced graduate students during the last dozen years. More generally it draws on experience with fossil dinoflagellates that began in 1956, shortly after I joined William S. Hoffmeister's palynology research group in the Carter Oil Company, then the site of geological research for the Standard Oil Company of New Jersey (now The Exxon Corporation).

The volume is divided into three main parts. Part I treats selected aspects of the biology and the geologic record of dinoflagellates, providing a general background for the rest of the text. Selected references to more comprehensive treatments of important topics not covered in this book are listed at the end of each of the two chapters of Part I.

Part II is a comprehensive review of the morphology of sporopollenin dinoflagellate resting cysts. Its four chapters explore the basic organization of the cyst, its wall structure and surface features, tabulation and paratabulation, and the archeopyle or excystment aperture.

Part III deals with the interpretation of sporopollenin dinoflagellate cysts. The first two chapters present a discussion of principles of interpretation and descriptions of major morphological categories. Four following chapters, organized according to these major categories, cover a broad selection of morphological features and problems of interpretation, the latter illustrated by specific cases among both fossil and modern dinoflagellates. A final chapter concerns classification schemes for sporopollenin cysts and reviews several hypotheses and speculations regarding the evolution of the dinoflagellates.

The book was written with the expectation that it will be used most extensively by paleontologists, but also with the hope that neontologists will find in it a useful synthesis of information on the sporopollenin resting cyst, a feature of the dinoflagellate life cycle which, although richly documented in the fossil record, is only now being recognized as important in many living species.

Except for highly selected topics which are included in the interest of background and general perspective, the emphasis throughout is on morphology. Other important matters, such as stratigraphic and geographic distribution, ecology and paleoecology, are not covered. Two considerations in particular have led me to this approach. In the first place, morphology must be the starting point for the paleontological study of any group of organisms. Upon it depends every identification and every ulterior application. In the second place, morphology has been the focus of my own studies and, consequently, is the area in which I am most likely to be able to make a useful contribution.

Although I have eschewed the use of "I," the book is largely a personal statement. This is especially true of those parts in which both substance and philosophy reflect peculiarities of my own experience and outlook (e.g., portions of Chapter 2 and Part III). When I write "we," I am not being rhetorical, but intend specifically "the reader and I" as we explore together the current understanding of and speculation about our subject. Although I picture this effort as another small attack on the perennially intractable problem of reducing Nature's works to the level of human understanding, I invite the reader to regard my guidance critically, with full awareness that my selectivity has tightly constrained the scope of the material being placed before him. Anyone using this work should consult other recent works—for example, the comprehensive coverage by Tappan (1980)—for discussion of many topics barely touched on here and for a more even-handed treatment of many matters on which I have stressed my own opinions.

The term "cyst" has been applied to a variety of structures produced by living dinoflagellates and differing in form, function, and composition. Except where the context clearly indicates otherwise, "dinoflagellate cyst" and "fossil dinoflagellate" throughout this book refer to the strikingly resistant organic-walled container that formed probably around a zygotic cell during a resting period of the life cycle. Preservation of fossil cysts of this type demonstrates their remarkable resistance to degradative processes. Dinoflagellate cysts of siliceous and calcareous composition fall outside the scope of this work.

Throughout the text, "modern" should be understood as "present day" (i.e., "living" in contrast to "fossil"), whereas "Holocene" (in preference to "Recent") is used in a geological sense to refer to post-Pleistocene phenomena or materials.

A word about morphological terms. Although an extensive terminology is presented in the following pages, no attempt is made to be encyclopedic. The reader who concentrates very long on dinoflagellates will soon enough become acquainted with the multitude of available terms. While it is tempting to design a term for each neatly definable morphological feature, the abundance of intergradations among slightly different features constantly alerts us to the folly of attempting too great precision when describing the morphology of organisms. While a specialized vocabulary is essential, at times virtually every specialized term can be more confusing than helpful. Technical terms should inconspicuously but effectively facilitate clear communication, not becloud or adorn it. "When in doubt, write it out" is a useful admonition that space saved at the expense of clarity is false economy. Therefore, the definitions of terms given here should be taken as applying to idealized examples.

With few exceptions, names of genera and species of fossil dinoflagellates are used in the sense of Lentin and Williams (1981) or the responsible authors of names proposed more recently. Names of modern dinoflagellates follow diverse sources, especially Balech (1974 and others), Dodge (1982), Loeblich and Loeblich (1966-1974), and Sournia (1973, 1978, 1982). All dinoflagellate names which appear in the text have been published, except for "*Lacrymodinium*" (see Index of Generic and Specific Names). In no case have I intended in this work formally to introduce a new name, new combination, or other nomenclatural novelty under the I.C.B.N.

Debate about the appropriateness or validity of generic distinctions as made currently has been avoided, but genera and species are cited copiously to illustrate particular morphological features. In citing genera, the morphology attributed to their type species has been borne in mind wherever possible in the hope that the examples will remain relevant and intelligible as other species are reattributed to new or different genera and current generic concepts are gradually modified. The line drawings that illustrate the work are intended as guides to concepts and structures, rather than as aids to identification of individual taxa. They are diagrammatic in varying degrees, and magnifications are not stated.

Completion of the main part of the manuscript in early 1983 established the cut-off point for consideration of most new publications. The text has been adjusted in places to take cognizance of two significant works (Sarjeant, 1982b; Bujak and Davies, 1983) which became available too late to be considered adequately. Citations of the collective work edited by Taylor (in press) refer only to Taylor's chapter on dinoflagellate morphology, which was read in manuscript; the other chapters in this work were not seen.

William R. Evitt
Stanford, California
March 1984

ACKNOWLEDGMENTS

I would like to acknowledge my particular debt to four parties who were remote from the circumstances that led directly to this volume but who played critical roles in my involvement with dinoflagellates. The exploration research division of Standard Oil Company (New Jersey)—and, especially, William S. Hoffmeister—by their progressive attitude and generous support enabled me to dedicate my research efforts to dinoflagellates at a time when the applicability of these microfossils to petroleum exploration was unknown and unpredictable. Trygve Braarud (University of Oslo) extended warm encouragement in a brief personal meeting which came at a critical time. Georges Deflandre (Ecole Pratique des Hautes Etudes, Paris) and Alfred Eisenack (University of Tübingen) were the source of inspiration and challenge through correspondence and personal contacts extending over many years.

Many other individuals contributed directly or indirectly to realization of the work. John E. Bennett (ARCO Oil and Gas Company) encouraged me to organize the first of the series of two-week instructional sessions on which the book is based. All together, 150 participants in those sessions have served as a responsive testing ground for many of the ideas presented here. Many aspects of the presentation have benefitted from stimulating discussions with my graduate students and colleagues, who have also restrained some of my wilder notions and have shown patient tolerance of my preoccupation with this enterprise. Discussions with F. J. R. Taylor (University of British Columbia), Barrie Dale (University of Oslo), and contributors to a seminar during one academic quarter at Stanford (Svein B. Manum, University of Oslo, and my students, David K. Goodman and Jeffrey A. Stein) were especially critical for development of ideas relating to dinoflagellate tabulation patterns.

The writing was begun in 1978 when generous financial support from the following sources made possible the extension of a brief sabbatical leave to a full year free from usual academic duties: Atlantic Richfield Co., Chevron Oil Field Research Co., Mobil Oil Corporation, Gulf Research and Development Co., and Stanford University School of Earth Sciences. A contribution from Exxon Production Research Company partially covered the cost of manuscript preparation. The help of these organizations is gratefully acknowledged.

For assistance in the terminal phases of manuscript preparation, I extend my warm thanks to Sarah Pierce Damassa (Winchester, Massachusetts), Lewis E. Stover (Exxon Production Research Co.), and F. J. R. Taylor (University of British Columbia), who read critically the entire manuscript and offered many suggestions which have improved its substance and clarity. Three of my current graduate students at Stanford—Anthony N. Bint, Joyce Lucas-Clark, and Javier Helenes—provided valuable technical input and helped crystallize my ideas in the later stages of writing. A fourth student, David I. Wharton, and Lucy E. Edwards (U.S. Geological Survey) have my very special gratitude for meticulous reviews of the final text and illustrations, and for a wealth of thoughtful suggestions which greatly improved the accuracy, style, and coherence of the work. The quality of the line illustrations reflects the skill of Perfecto Mari (School of Earth Sciences, Stanford), who prepared them from my pencil roughs. Martha Helenes assisted in preparing the indices and in checking references to taxa and citations of literature. Finally, I thank my wife, Gisela, for the most fundamental contribution of all — loving encouragement and support that made the whole undertaking both possible and worthwhile.

INTRODUCTION

Microscopic in size and occupying a position among the most primitive eukaryotic organisms, dinoflagellates (Class Dinophyceae) are algal unicells referred (along with classes Ebriophyceae, Ellobiophyceae, and Syndiniophyceae) to the division of the plant kingdom known as the Pyrrhophyta. Dinoflagellates are important constituents of the marine and freshwater biota today and are represented by abundant fossils in Upper Triassic and younger strata. The pre-Triassic record includes a single Silurian species and, possibly, some of the fossils generally treated as acritarchs.

To the neontologist dinoflagellates present an array of unusual biological features, including unique flagellar structure and nuclear organization, distinctive organelles of sophisticated design, varied metabolic regimes involving remarkable examples of symbiosis and parasitism as well as more conventional photosynthesis and phagotrophy, and complicity in economically and ecologically important outbreaks of red tides. To the paleontologist they offer a rich and extensive fossil record characterized by high morphological diversity and rapid evolution—a combination that means excellent potential for geologic dating and correlation and nearly limitless opportunities for descriptive systematics.

Important impetus has been given to the study of dinoflagellate cysts in recent years by their role in petroleum exploration, where their small size, abundance in many strata deficient in other fossils, and broad geographic distribution combine with their resistant composition and rapid evolution to make them highly useful in subsurface stratigraphic studies. Reciprocally, the wealth of observations that have accumulated over the last 25 years as a consequence of this practical application of dinoflagellate cysts has helped advance their scientific study to levels far beyond what would have been attainable in that period otherwise.

Four solid and nonliving materials produced by dinoflagellates are incorporated into structures that serve generally to enclose the cell—cellulose, sporopollenin, calcite, and silica. It is consistent with the diversity characteristic of dinoflagellates that these materials appear at different times in the life cycles of different species and are used for different purposes.

Cellulose (which, alone among the four substances, is not preserved in fossil dinoflagellates) is the principal component of the theca that encloses the metabolically active motile cell in many photosynthetic and phagotrophic species living today. However, it also occurs in living nonmotile cysts, both in the photosynthetically metabolizing stage of the gonyaulacoid genus *Pyrocystis* and the metabolically inactive stage of *Ceratium*.

Sporopollenin, the inclusive name for a group of complex biopolymers derived by oxidative polymerization of carotenoids or carotenoid esters (Brooks and Shaw, 1973), is believed to be the highly resistant organic material most important in the dinoflagellate fossil record. It appears to be the principal constituent in the walls of those dinoflagellate cysts that, heretofore, have been characterized rather awkwardly and ambiguously as "organic-walled." Following Bujak and Davies (1983), "sporopollenin" has been adopted here as a convenient and acceptably accurate replacement for "organic-walled," although the precise compositions of geologically resistant cyst walls have not been determined. Sporopollenin also occurs today in cysts of some living species that have cellulosic thecae, as well as in a subthecal membrane in some of these species (Morrill and Loeblich, 1981).

Calcite forms the cysts of certain fossil and living peridinioids (Calciodinellaceae, *Scrippsiella*), and the covering of the coccoid vegetative stage of *Thoracosphaera* (Tangen et al., 1982). The diversity and abundance of calcareous dinoflagellates in the fossil record is only just now becoming appreciated (Keupp, 1979a, 1979b, 1980, 1981).

Finally, silica forms the presumed resting cysts of fossil *Peridinites* (Lefevre, 1933; Dale, 1978) and *Jusella* (Vozzhennikova, 1963), as well as the stellate spicules first described by Ehrenberg and now known to be endoskeletal elements from gymnodinioid dinoflagellates of the genus *Actiniscus* and several related genera (Dumitrică, 1973). Silica has also been identified as one ingredient of the cyst wall in certain living freshwater species of *Peridinium* and *Ceratium* (Eren, 1969a; Chapman et al., 1982).

An especially unusual and intriguing aspect of the study of dinoflagellates is that neontologists and paleontologists, in the manner of the fabled blind men and the elephant, have been concerned almost exclusively with different stages in the dinoflagellate life cycle. Neontologists have focused their attention on the haploid motile stage, which is normally devoid of geologically preservable structures. Even the distinctive cellulosic theca of the motile cells in many of the most familiar living species is an easily degraded structure probably unknown in the fossil record. Paleontologists, in contrast, have found dinoflagellates only in the form of structures believed to represent almost exclusively the highly resistant cyst, which housed a diploid zygote during a resting period following syngamy. While neontologists have largely ignored comparable cysts of living species, paleontologists, in their turn, have often paid little attention to important aspects of motile dinoflagellates.

Fortunately, recent advances in knowledge of the dinoflagellate life cycle and better communication among specialists, facilitated by such international meetings as the Planktonic Conferences and the Penrose (1978) and Hexrose (1981) conferences on modern and fossil dinoflagellates, are leading rapidly to a more holistic understanding of dinoflagellates. Fossil dinoflagellates can now be interpreted with a better comprehension of the biologic role of the resting cyst and its relation to the metabolically more significant motile stage, and the resting cyst can now be accorded more appropriate attention in the study of both the biology and taxonomy of living dinoflagellates.

Although the existence of an encysted stage in some living species had been known for many years, paleontological studies, especially during the decade 1953-1963, brought questions about the nature and role of the dinoflagellate resting cyst into increasingly sharp focus and began to suggest some of the answers. Tappan (1980) reviews early studies of cyst, theca, and reproductive cycles in living dinoflagellates, while Sarjeant (1974) recounts the history of the study of fossil dinoflagellates and traces the development of ideas about them. In one sense, this development of ideas has been a case of the past serving as a key to the present, for hypotheses first formulated on the basis of studies of fossil dinoflagellates have been confirmed and expanded through observations on living species. In another and less fortunate sense (as will be pointed out in Chapters 1 and 2), we have been busy reinventing the wheel, for some of the observations basic to our current understanding were made once by a number of astute early investigators, and forgotten, only to be made again more recently.

Geologically preservable resting cysts, whether of organic or mineral composition, are produced by only a minority of living species distributed somewhat erratically among the genera and higher taxa of modern dinoflagellates. There is no reason to believe the situation was significantly different in the geologic past. Consequently, in this book, where our main concern is with the morphology of the cyst and its interpretation, the coverage of dinoflagellate systematics at large will be quite uneven.

The abbreviated classification of higher taxonomic categories presented in Table In.1 will be our general guide to dinoflagellate systematics. Two of the orders listed (Prorocentrales and Dinophysiales) are not certainly known either to produce preservable cysts today or to have fossil representatives. However, it has been suggested that some of the acritarchs may represent the Prorocentrales (Lister, 1970) and that a recently described Jurassic form (Helby and Stover, 1984) may represent the Dinophysiales. Jurassic dinoflagellates of the genus *Nannoceratopsis*, until recently attributed to the Dinophysiales, are now assigned to a separate order (Piel and Evitt, 1980). The ordinal arrangement shown in Table In.1 contrasts in minor ways with treatments by Taylor (1980) and Tappan (1980) as follows: 1) Contrary to Taylor, who erected a new order, Gonyaulacales, for *Gonyaulax* and its allies, we will retain these dinoflagellates in the Peridiniales. Sporopollenin cysts produced in this order will be our principal concern throughout this volume.

Table In.1 - CLASSES AND ORDERS OF PYRRHOPHYTA

Division	Pyrrhophyta Pascher 1914
Class	Dinophyceae Fritsch 1929
Order	Prorocentrales Lemmermann 1910
Order	Dinophysiales Lindeman 1928
Order	*Peridiniales Haeckel 1894
Order	Gymnodiniales Lemmermann 1910
Order	Nannoceratopsiales Piel and Evitt 1980
Class	Ebriophyceae Loeblich III 1970
Class	Ellobiophyceae Loeblich III 1970
Class	Syndiniophyceae Loeblich III 1976

*As noted in the text, Taylor's (1979) proposal to distinguish a new order Gonyaulacales from the Peridiniales is not adopted here.

2) We will construe the Gymnodiniales to include resistant organic-walled structures of two quite different types. On the one hand are cysts of the sort produced today by some members of the pseudocolonial gymnodinialean family Polykrikaceae. On the other hand are Cretaceous fossils referred to *Dinogymnium*, which we will treat as probably representing motile cells with a thick sporopollenin cover, contrary to Tappan, who regards them as cysts close to the peridinialean family Lophodiniaceae. Note that Taylor (1980) regards the Gymnodiniales as an artificial grouping distinguished by features acquired independently by dinoflagellates descended along several different lines from gonyaulacoids, peridinioids, or both. 3) We will treat Jurassic *Nannoceratopsis* as representing a distinct order, Nannoceratopsiales.

In 1885, Bütschli recorded that about 26 genera and 90-95 species of dinoflagellates had been described. Today, described living marine taxa alone include well over 100 genera and about 2000 species (Taylor, in press). In addition, about 200 living freshwater species have been described (Bourrelly, 1970), and fossil dinoflagellates today include approximately 2500 species in over 500 genera (2210 species and 413 genera treated by Lentin and Williams, 1981). The number of fossil taxa, especially, is growing rapidly.

Part I. On Dinoflagellates in General

Before concentrating on our main topic, cyst morphology, it will be well to build a certain perspective. We need to review dinoflagellate biology at least to the extent of assuring that the cyst, a structure with distinctive character and specific function, can be seen as an integral part of the dinoflagellate picture. Also, certain aspects of fossil dinoflagellates apart from their morphology merit special emphasis. Most important among these are the extraordinary limitations which the highly selective nature of the dinoflagellate fossil record imposes on inferences that can be drawn from that record. These and assorted other topics bearing on living and fossil dinoflagellates in general are covered in the two chapters of Part I.

Chapter 1. Aspects of Living Dinoflagellates

See Loeblich III (1967, 1970), Tappan (1980) and Taylor (1976a, 1976b, 1978, 1980, in press) for extensive reviews and references.

SYSTEMATIC POSITION AND BASIC CHARACTERISTICS

Living dinoflagellates exhibit a great diversity in form, habit, and habitat that belies their systematic position near the base of the phylogenetic tree of the eukaryotes (Figure 1.1). Their primitiveness is shown especially by properties of the nucleus, mitotic apparatus, and chloroplast. The nuclear structure (typically with chromosomes permanently condensed) and the mitotic apparatus (with spindles external to the nuclear membrane) are perhaps the most primitive in any eukaryote. The chloroplast structure and the pigment assortment that includes chlorophyll a and c_2, but not c_1, suggest that only the red algae may be more primitive. However, the general organization of the dinoflagellate cell and extreme specializations to be found in certain taxa hardly match the usual concept of primitive. As an example of a highly specialized organelle, consider the light-sensitive structure, with eye-like succession of lens, fluid-filled "camera", retinoid, and pigment backing, which occurs in a few species (Francis, 1967; Greuet, 1970). Less spectacular but interesting for their widespread occurrence are the vacuolelike pusules, fluid-filled bodies which occur two per cell and possibly have an excretory or assimilative function.

Chloroplasts may be present or absent, and holophytic, phagotrophic, saprophytic, symbiotic, and parasitic nutritional regimes occur. Planktonic forms inhabit the open sea, coastal and estuarine waters, and rivers and lakes—environments which, collectively, encompass extreme ranges in temperature, salinity, and other aspects of water chemistry.

Throughout this volume our concern will be centered on two distinct but intimately related structures produced by some dinoflagellates: the theca and the cyst. In those dinoflagellate species in which they occur, both these structures are, in effect, containers for the cell protoplast.

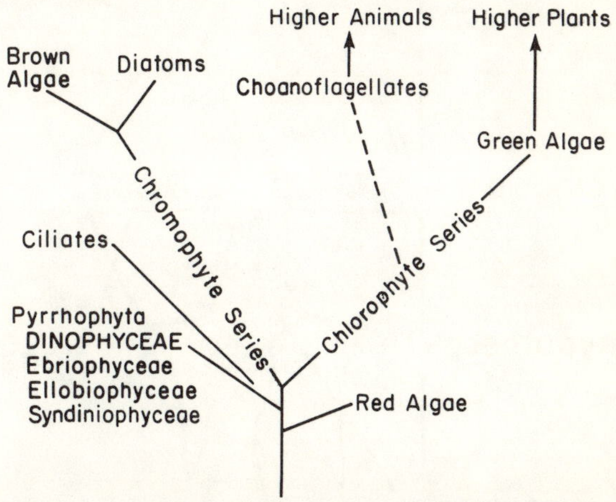

Figure 1.1 - POSSIBLE RELATIONS OF DINOFLAGELLATES
The Division Pyrrhophyta and Class Dinophyceae are shown in relation to some other major groups of organisms near the base of the phylogenetic tree of the eukaryotes (simplified from Taylor, 1978).

Figure 1.2 shows the principal features of a peridinialean theca and the terminology applied to them. Although the flagella are illustrated, these structures are organelles of the living cell and not a part of the theca.

The theca is part of the cell covering in certain living dinoflagellates during the motile stage of their life cycle. It is constructed of **thecal plates**, which are thin cellulosic units fitted together closely along linear junctions termed **sutures**. The pattern of plate arrangement constitutes the **tabulation** and is characteristic for a genus or species. Motility of the thecate cell results from the action of two flagella of characteristic form and position. They emerge from the theca through one or two **flagellar pores** on the surface arbitrarily considered ventral. One flagellum is longitudinal and has a long-period beat. The other is transverse, lies in a circular path in a plane about perpendicular to the first, has the form of a ribbon strongly undulated along one edge, and has a short-period beat. The two flagella in concert impart a

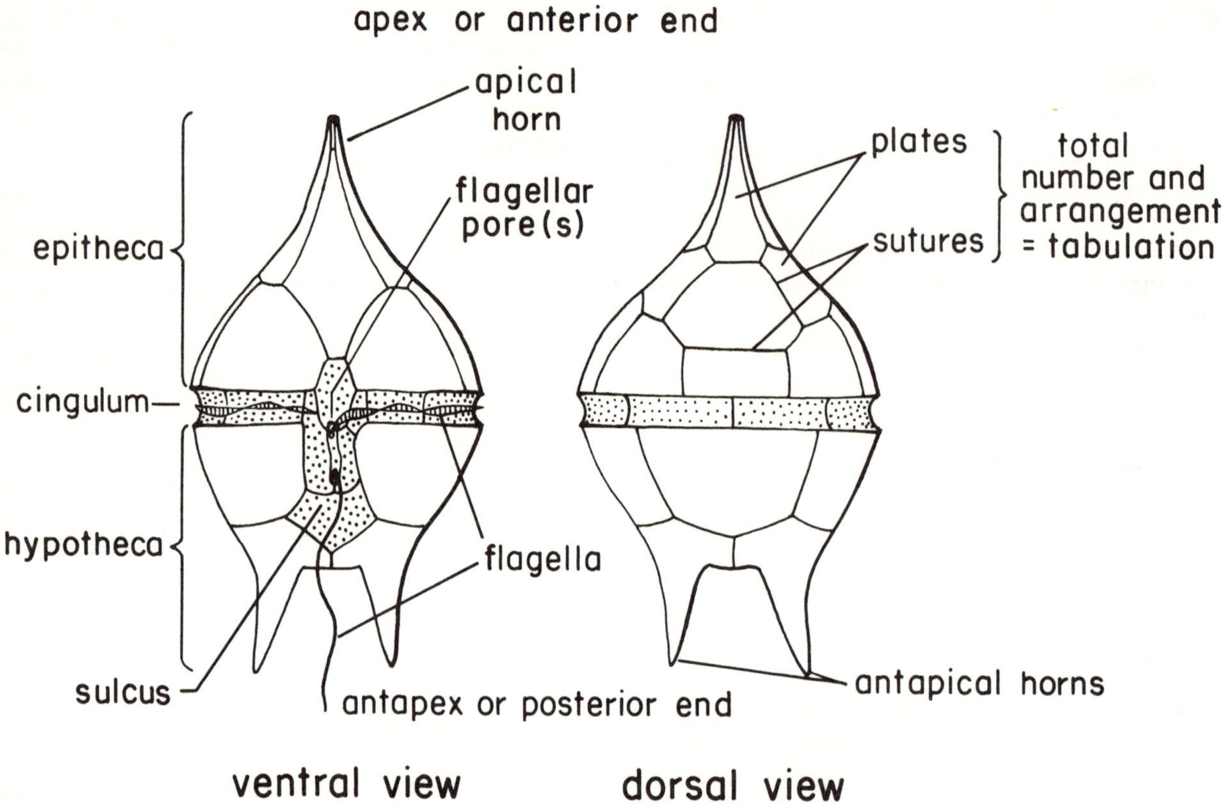

Figure 1.2 - PRINCIPAL FEATURES OF THE THECA IN A PERIDINIALEAN DINOFLAGELLATE
(For corresponding terms applicable to the cyst, see Figure 3.1)

characteristic spiral motion to the swimming cell (Jahn et al., 1965; LeBlond and Taylor, 1976). Forward rates of one meter per hour have been recorded (e.g., *Gonyaulax polyedra*), but rapid changes in direction of progress and rotation are typical (Peters, 1929b; Hand et al., 1965; Loeblich, 1967).

Characteristically (except in Prorocentrales) the transverse flagellum occupies a channel or depression known as the **cingulum** (or girdle) which divides the theca into an anterior **epitheca** and posterior **hypotheca**. The longitudinal flagellum may lie partly in an elongate ventral depression located mostly on the hypotheca and known as the **sulcus**. Corresponding principal features of the cyst of a peridinialean dinoflagellate will be considered in Chapter 3 (Figure 3.1).

The different compositions of dinoflagellate thecae and cysts, which are probably more varied in detail than has been documented by the few determinations yet made, account for their contrasting resistance to degradation in nature and in the laboratory. The theca, which is composed of thin plates of cellulose (or a cellulose-like glucan; Nevo and Sharon, 1969; Loeblich III, 1970) is a relatively transient structure, readily destroyed by bacteria and consequently unknown in the fossil state. In certain circumstances, it is attacked so rapidly and extensively that thecae of near-surface species never reach the depositional interface where more resistant cysts accumulate in large numbers (Pollingher, 1981). In contrast, sporopollenin cysts generally are highly resistant and are geologically long-lived. In the laboratory, both recent and fossil cysts can withstand severe chemical treatments, including, among others, 70 percent hydrofluoric acid, concentrated hydrochloric and nitric acids, five percent sodium hypochlorite, and acetolysis (10 minutes at boiling in about 1:8 concentrated sulfuric acid and acetic anhydride). Even after one or two hundred million years, the thin and still flexible walls of many fossil cysts retain delicate surficial relief and their apparently original three-dimensional shape. Resistance, of course, is a relative matter, and some differences observed in the resistance of modern resting cysts under both natural and laboratory conditions (Dale, 1976) may largely reflect differences in sporopollenin content. Cysts that are poorly resistant, for whatever reasons, will not concern the paleontologist.

The pellicle, a layer that lies beneath the theca in some living dinoflagellates has also been observed to contain sporopollenin in certain species. However, at present the relationship among the occurrence of a pellicle, the presence of sporopollenin in it, the acetolysis resistance of this layer, and the formation of sporopollenin cysts by the same species is unclear (Morrill and Loeblich, 1981).

Despite the impermanence of thecal cellulose in contrast to sporopollenin, an appreciation of thecal morphology is fundamental to the interpretation and identification of sporopollenin cysts. There are two principal reasons for this. First, the morphology of the theca is the basis for the classification of those living dinoflagellates in which a theca occurs, and most sporopollenin cysts are believed to have been produced by thecate species. Second, important morphological characteristics of cysts include many features that are analogous to thecal features—most of them now labeled by descriptive terms incorporating either the prefix "para-" or the stem word "-cyst."

LIFE CYCLE IN THE PERIDINIALES

The life cycle of peridinialean dinoflagellates may involve phases of vegetative and sexual reproduction, as well as periods of nonmotility when the cell is enclosed in a cyst wall. Cysts, even for a single species, may be of more than one type in terms of morphology, function, and resistance to degradative processes. Although these organisms have been the subject of extensive biological study for more than a hundred years, complete life cycles, including possible episodes of encystment, are as yet known for only a relatively few species, and discovery of the sexual phase of the cycle has been especially elusive. Several observers around the turn of the century reported phenomena or cellular features which can now be appreciated as pertaining to sexual reproduction, but Diwald (1938) appears to have been the first to describe the life cycle of a single species (*Peridiniopsis lubiniensiforme*) through both its sexual and asexual phases. Unfortunately, this landmark paper was overlooked by later workers, and dinoflagellates were commonly said to reproduce only vegetatively. Even the comprehensive synthesis on dinoflagellate biology by Chatton (1952) contains only six lines about sexual reproduction, and the section on fossil dinoflagellates in the same treatise (Deflandre, 1952a) gives no suggestion that these fossils may represent cysts.

Among the many contributions toward better understanding of life cycles, two further papers of interest are those of Braarud and Rahat. Braarud (1949) described cysts that formed within thecae in cultures of two marine dinoflagellates and were released subsequently when the thecae went to pieces. In the case of *Protogonyaulax tamarensis*, the wall of the cyst was smooth; probably it was nonpreservable, like the cysts of this species recorded by Anderson and Wall (1978). However, the spinose cysts of *Gonyaulax grindleyi* were later shown (Evitt and Davidson, 1964) to be morphologically similar to many fossil hystrichospheres and to be resistant to acetolysis. Braarud did not recognize any sexual processes in connection with the formation of these cysts, and thought they were vegetative resting stages, but it now seems more likely that they were zygotic cysts.

Rahat (1968) studied cultures of the freshwater species *Peridinium westii* (now *P. cinctum* f. *westii*), and reported a variety of observations, including 1) binary fission of the protoplast within thecate cells, 2) rupture of the theca and release of freely motile cells which subsequently formed new thecae, 3) emergence of naked protoplasts from resting cysts whose walls are now known to be resistant to acetolysis, and 4) development of some of the excysted individuals into normal, mature, thecate motile cells. Although Rahat recognized these as steps in the life cycle, he was not aware of their sequence or of the alternation between vegetative and sexual generations they represented.

More recent investigations have illuminated some of the diverse adaptations dinoflagellates have incorporated into their life cycles (see, for example: Beam and Himes, 1974; Buchanan, 1968; Chapman et al., 1981; Pfiester, 1975 and later; Walker and Steidinger, 1979; Turpin et al., 1978; von Stosch, 1964, 1972, 1973). These and similar studies increasingly suggest that the preservable cysts upon which we are focusing our attention in this volume are the resistant housings of hypnozygotes—i.e., zygotic cells that were in a resting stage. In other words, they reflect a period of nonmotility following sexual reproduction. However, since any dinoflagellate covering with substantial amounts of sporopollenin is conceivably preservable, it is possible that the fossil record may include dinoflagellate remains that were associated with other stages in the life cycle. Examples might include resistant-walled vegetative resting cysts (as opposed to zygotic resting cysts) formed in conjunction with mitotic cell division or in response to severely adverse conditions, lightly sporopollenin-charged membranes such as appear to occur beneath the theca in some living species (Loeblich III and Volcani, 1969; Morrill and Loeblich, 1981), or even the walls of motile cells or structures formed outside the theca (see Chapter 9 under *Palaeoperidinium pyrophorum* and Chapter 12 under *Dinogymnium*).

The life cycle shown schematically in Figure 1.3 and summarized in the following paragraphs approximates several that have been observed. Until more is learned about the distribution of sporopollenin among dinoflagellates, this seems to be a plausible model to explain the origin of most fossil dinoflagellate cysts as hypnozygotes.

1. During the period favorable for rapid growth and population expansion, vegetative fission dominates, yielding successive generations of motile schizonts that are haploid in chromosome count. In some species, growth is marked by the appearance of conspicuous bands of material accreted along the margins of individual plates (not shown in Step 1 of the figure). In other species, the accretionary material is less clearly differentiated from that of the original plates.

2. At certain times and for reasons as yet unknown, some of the schizonts function as gametes, pair up, and fuse to form zygotes. One or both thecae may be lost in the process. In some living species the gametes all appear similar, but sexual dimorphism also occurs (Beam and Himes, 1980). Morphological differences between smaller male and larger female gametes of the same species may be as great as differences that have been used to distinguish subgenera (von Stosch, 1964).

3. The diploid zygote constructs a new theca resembling the vegetative theca (if this has been lost), and resumes its motility as a planozygote (i.e., a motile zygote). In some species, the planozygote has paired flagella, one from each gamete. As cell size increases, bands of new thecal material may become prominent along the sutures (Frontispiece A; Peters, 1927, 1929a).

4. In several of the species studied, the zygotic theca ultimately becomes much thicker and considerably larger than the vegetative theca; its outline becomes less regular, and the protoplast acquires a distinctly granular appearance, in part from the development of many minute oil droplets. In addition, one to several distinctly reddish bodies commonly become visible within the protoplast. Finally, the activity level diminishes and, after as long as l5 days, the flagella are lost. As the cell becomes nonmotile (i.e., a hypnozygote), the protoplast shrinks, pulling away from the theca, and one or two membranes destined to be future cyst walls become visible in the space between the contracted protoplast and the theca. With the appearance of these cyst walls, or shortly thereafter, the thecal plates break apart or are destroyed by bacteria and the completed cyst is exposed.

5. The hypnozygote or resting cyst henceforth behaves as a sedimentary particle, settling through the water column as turbulence permits and coming to rest on the depositional interface.

6. Following a period of obligate dormancy, the protoplast excysts (typically through an operculate archeopyle in the case of cysts that have formed a geologically resistant wall) and the cycle closes as meiotic divisions produce new haploid cells, each of which develops flagella and a theca. Both the dormancy period and the timing of meiotic divisions vary according to species. Cysts may germinate after as little as 12 hours in the nonmotile state, but in some species it has been impossible to induce excystment until after rest periods of 3-4 months. A change in ambient temperature is probably the immediate stimulus for excystment in many species in nature and it has been effective in laboratory study of species with an extended dormancy period.

The two cell divisions of meiosis may occur in close succession during the later part of the encysted stage, or one or both of them may be delayed until after excystment. It is assumed that the first two divisions, whenever they occur, accomplish the return to the haploid state, although evidence has been cited (Beam et al., 1977) for a special abbreviated meiosis in some peridinialeans. Dale (1983, p. 111) summarized the main variations in details of sexual cycles of dinoflagellates.

While syngamy is now thought to be a normal part of the life cycle of peridinialean dinoflagellates, the hypnozygote is only one of many different zygotes produced by dinoflagellates today (von Stosch, 1972), and it is provided with a resistant wall in only certain species. In other words, the sexual phase of the life cycle may run its course without formation of a resting cyst, or the hypnozygote may lack a preservable wall. The modern genus *Ceratium* provides instructive examples of this variability.

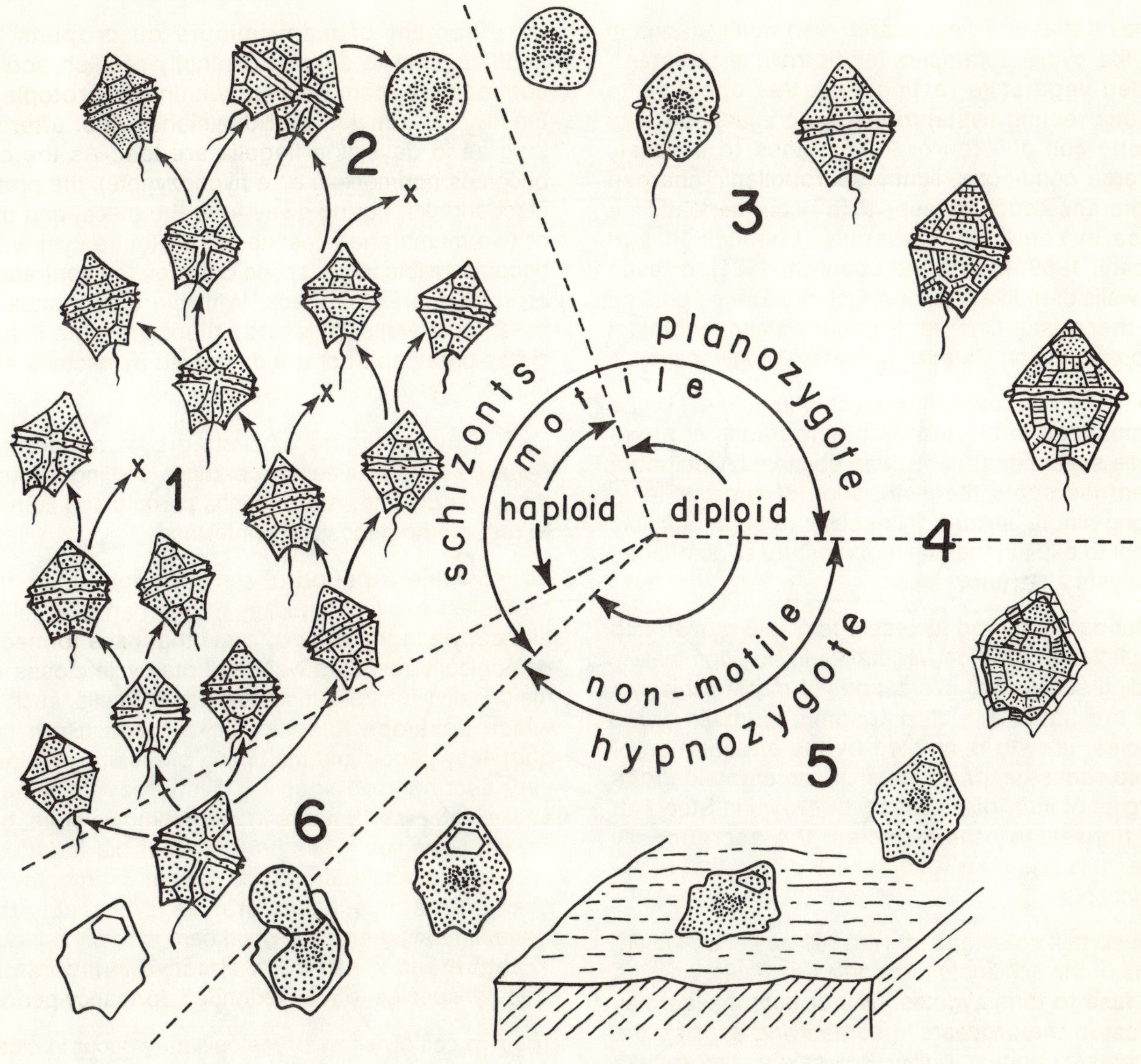

Figure 1.3 - SCHEMATIC LIFE CYCLE OF CYST-PRODUCING DINOFLAGELLATE
(Numbers refer to steps described in text; x's in Step 2 represent the unfigured second products of binary fission.)

Most living marine species of the genus appear not to produce a hypnozygote, although cysts have been reported in *C. furca*, *C. fusus*, and *C. tripos* (Jollos, 1910). The zygote of freshwater *C. cornutum* forms a resting cyst (von Stosch, 1972; Happach-Kasan, 1982) and zygotic cysts have long been suspected in freshwater *C. hirundinella* (Hall, 1925). Chapman et al. (1981) reported that vegetative cells of the latter species also appear to form persistent (but not geologically preservable) cysts, some of whose walls contain traces of silica. These authors also confirmed the observation by Entz (1925) that motile cells of *C. hirundinella* emerge from the cysts through a slitlike opening of constant position.

Cysts of *C. hirundinella* from varved sediments in Lake Zurich were viable after 6.5 years (Huber and Nipkow, 1923) but, apparently, these bodies eventually disintegrate and disappear, for Wall and Evitt (1975) found no trace of *Ceratium* cysts in the same varves, although empty sporopollenin cysts of *Peridinium cinctum* were present in good condition. Huber and Nipkow (1923) had determined that *Peridinium* cysts from Lake Zurich sediments were viable after 16.5 years, and Dale (1983) successfully germinated common types of marine sporopollenin cysts (including *Gonyaulax spinifera* and *G. grindleyi*)

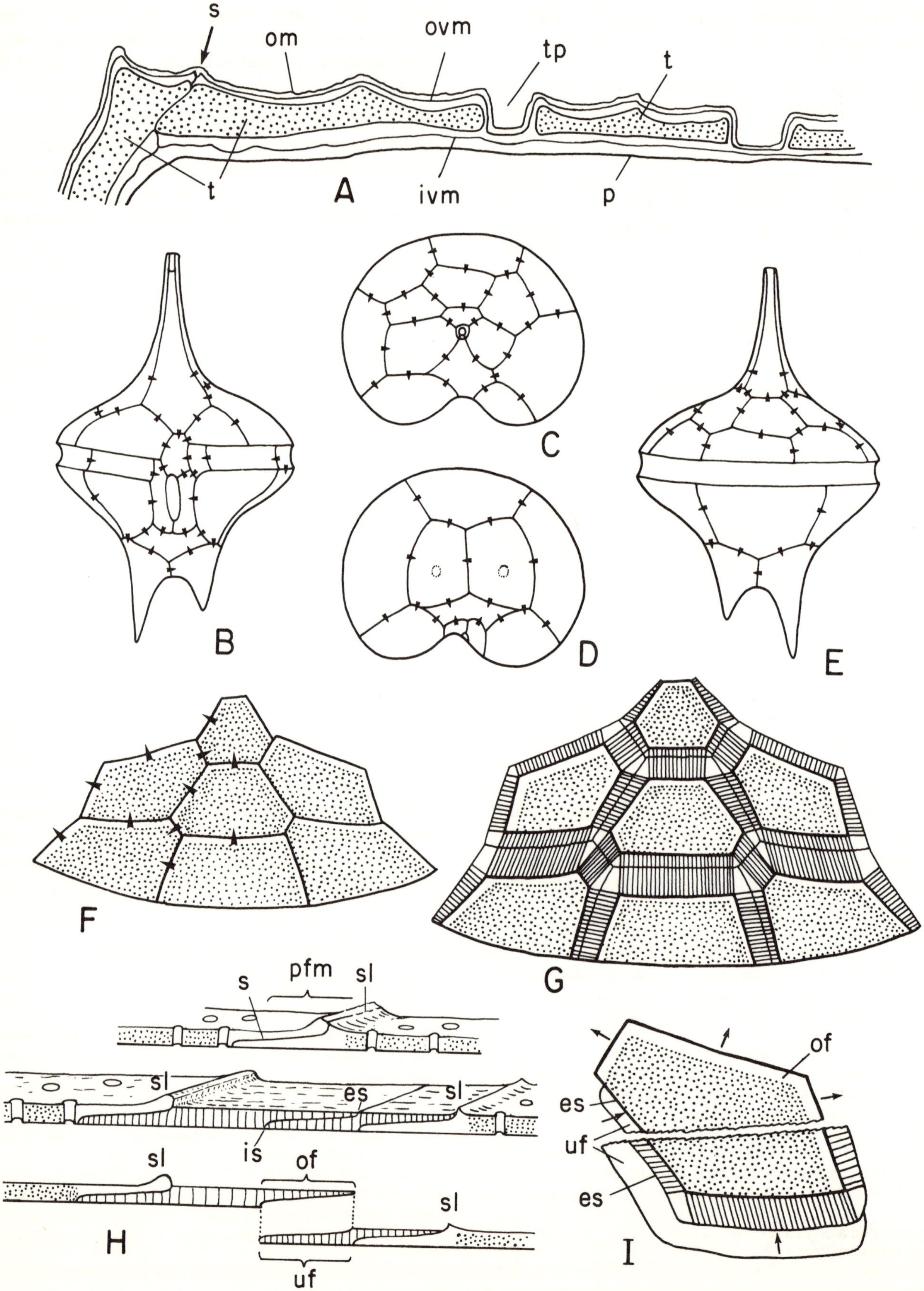
s
om
ovm
tp
t
t
A
ivm
p
B
C
D
E
F
G
pfm
s
sl
sl
es
sl
is
sl
of
sl
H
uf
of
es
uf
es
I

Figure 1.4

from Oslofjord sediments after storage for 5.5 years at temperatures of 4-7° C.

It is certainly a gross misrepresentation to characterize encystment as simply a response to onset of adverse conditions—an explanation that has frequently been offered. Most likely, the resting cyst represents an adaptive coupling between sexual reproduction, on the one hand, and survival during periods when conditions favoring prolific vegetative growth are temporarily diminished, on the other. The complexity of the adaptation is suggested by observations on *Protogonyaulax tamarensis* (= *P. excavata*) and *Peridinium cinctum* f. *westii*.

Nonpreservable (i.e., sporopollenin-free) resting cysts of toxin-producing *Protogonyaulax tamarensis* apparently function along the New England coast as "seed" populations for successive blooms through the year (Anderson and Wall, 1978). Cysts produced at the end of the spring bloom could be germinated after a relatively short time, such as would be appropriate for them, in nature, to yield the vegetative stock for a second bloom later in the same season. However, the obligate dormancy of cysts collected at the end of a late summer bloom was longer, apparently timed to carry them through the winter until the coastal waters could be expected to begin warming in the spring.

In contrast, freshwater *Peridinium cinctum* f. *westii*, which inhabits Lake Kinneret (Israel-Jordan) in tremendous numbers, encysts as the water temperature rises (Eren, 1969b). Its sporopollenin resting cysts are evidently designed for oversummering, when nutrient levels are low. They excyst when ambient lake temperatures drop with the onset of winter (Pollingher, 1981). West (1909) even reported that, of two species living in a single English pool, cysts of one (*Peridinium cinctum* var. *lemmermannii*) were adapted for overwintering, while those of the other (*P. aciculiferum*) were for oversummering.

Whatever the varied functions of dinoflagellate resting cysts may be, it is particularly striking and, as we shall see, of great importance in the interpretation of the fossil record that a majority of living dinoflagellate species thrive without ever producing them. Although Dale (1983) reported that at least 170 sporopollenin cyst morphotypes are known to be formed today and suggested that such cysts probably occur in something over 200 species, this still accounts for only about 10 percent of all known species. So far, only about a third of the known modern cysts have been correlated with their respective motile stages. The cyst-producing species are concentrated in a few major taxonomic categories, especially those modern families of the order Peridiniales that are based on the genera *Peridinium*, *Protoperidinium*, and *Gonyaulax*. Even among these families, only a minority of the species are cyst producers.

AMPHIESMA, PELLICLE, AND THECA

The **amphiesma** (Loeblich III, 1970) is the complex and unique covering of the dinoflagellate cell (Dodge and Crawford, 1970a-b; Dodge, 1971; Dürr, 1979a-c). It consists of a series of membranes and associated structures (Figure 1.4 A), including flattened vesicles arranged in a mosaic-like fashion. In many species a cellulosic plate forms within each vesicle. These plates may be so thin as to be visible only in

Figure 1.4 - AMPHIESMA AND THECAL FEATURES ASSOCIATED WITH PLATE OVERLAP AND GROWTH

A - Schematic section of amphiesma, about X 40,000, showing portions of two thecal plates and suture between them. ivm - inner vesicular membrane, om - outermost membrane, ovm - outer vesicular membrane, p - pellicle, s -suture, t - thecal plates, tp - trichocyst pore. Modified from Dodge and Crawford, 1970.

B-E - Schematic views of tabulation in *Protoperidinium* (hypothetical details of a meta penta pattern, see Chapter 4), showing systematic distribution of plate overlap (arrows). B - Ventral, C - apical, D - antapical, E - dorsal. Flagellar pores and additional platelets occupy elliptical area in B. Overlap based on Gocht and Netzel, 1974, and original studies.

F - Schematic diagram of six dorsal epithecal plates, showing overlap directions (arrows), areas occupied by trichocyst pores (stippled), and porefree margins (unstippled).

G - Same group of plates after marginal growth (striate bands and unpatterned triangles). Sutures between plates show externally as lines about two-thirds the way from original edge of overlapping plate toward original edge of adjacent underlapping plate.

H - Schematic views of the junction between two thecal plates, showing portions penetrated by trichocyst pores (stippled in sections through plates), porefree margin (pfm) of original plates (plain in sections), and growth increments along sutures (lined in sections). Upper figure shows original plates; middle figure shows plates after growth; lower figure shows sections of two plates after growth, separated along sutures. es - external suture, is - internal suture, of - overlap flange, pfm - porefree margin, s - suture, sl - sutural list, uf - underlap flange.

I - One plate extracted from the association in F and G, showing condition before (upper part of figure) and after (lower part) addition of marginal increment of plate material. Arrows point in overlap direction. Abbreviations as in H.

section under the transmission electron microscope, or they may reach several micrometers in thickness and fit closely together with plates in adjacent vesicles to form a more or less rigid theca. In many other species no cellulose is deposited within the vesicles. In such cases the large number of cavities, their small size, and close juxtaposition apparently contribute toward a firm but flexible cell surface by imparting a certain stiffness to the amphiesma.

On the basis of the presence or absence of a theca, dinoflagellates were formerly classified into two major categories, commonly referred to as **armored** or thecate and **unarmored** or naked. However, electron microscopy suggests that this distinction, while still descriptively useful, is not a fundamental one. The vesiculate amphiesma is a significant common denominator in them all, regardless of the absence, presence, or thickness of plates.

The most exterior portion of the amphiesma is an extremely thin outer membrane, which encloses the flagella as well as covering the thecal plates. It is easily destroyed and can be seen clearly only in electron micrographs of specially prepared specimens. Beneath this lies a more substantial **vesicular layer** bounded by outer and inner vesicular membranes and including flattened cavities in which thecal plates may be formed. Under the vesicular layer in some species lies another layer, the **pellicle**, which is thin, fibrous, and flexible, but relatively tough. Cellulose may be present in the pellicle or not, and a reported resistance to acetolysis suggests that sporopollenin is involved in its composition. Possibly the pellicle, at least in some species, is a precursor to the resistant wall of the resting cyst, being transformed into that more substantial structure by deposition of additional sporopollenin. Unusually thick pellicles characterize the athecate motile cells in one group of living dinoflagellates (Taylor, 1980), causing them to resemble an important group of fossils (*Dinogymnium*) whose nature is discussed in Chapter 12.

Opinions have differed (E. R. Cox, 1978) as to whether the plasmalemma, or true cell membrane, is represented by the outermost membrane of the amphiesma or by one lying still more internally than the pellicle. If the outermost membrane is taken to be the plasmalemma (and this is the current consensus), then the amphiesma is unusual among algal cell coverings in having a substantial wall structure within, instead of outside, this bounding membrane (Dodge, 1974).

The general similarity of the peripheral structures of the dinoflagellate cell to those found in the cortex of ciliates has occasioned the suggestion of close relationship (Figure 1.1; Corliss, 1975; Taylor, 1976b).

Since the plates of a dinoflagellate theca develop within vesicles separated by membranes, they are not actually fused together, and usually they can be dissociated readily following brief treatment with Javelle water (i.e., dilute sodium hypochlorite solution). This technique of "dissection" is critical for determining the details of tabulation that are basic to the taxonomy of dinoflagellates with well-developed thecae. The monograph on the major living genus *Protoperidinium* by Balech (1974) includes excellent examples of comparative morphology of thecal plates in a major living genus.

The junctions between adjacent thecal plates appear to differ depending upon plate location. Simple butt joints and tongue-and-groove contacts are reported in the sulcus and cingulum (Balech, 1974), but elsewhere an **overlap flange** on one plate extends over an **underlap flange** on its neighbor (Figure 1.4 H-I). Such overlap along an essentially oblique sutural surface was already described clearly by Schütt (1895). Subsequent observers have reported that the sense of this overlap varies systematically with position on the theca (Kofoid, 1901; Balech, 1974; Dürr and Netzel, 1974; Gocht and Netzel, 1974; Gocht, 1979). In general, the overlap gradient is away from middorsal pre- and postcingular plates, which have been termed **keystone plates** because of their central position in the overlap arrangement, and toward the midventral sulcals (Figure 1.4 B-E). Although a considerable consistency has been noted in overlap relationships among apparently closely related species, details of overlap patterns have not been investigated exhaustively. Without many more observations, the existence of even a few inconsistencies (e.g., the contrary sense of overlap between apparently homologous plates in *Ceratium* and *Gonyaulax* reported by Happach-Kasan, 1982) urges caution in using this character as a critical guide to either genetic relationships or plate homologies.

Although the theca may appear to enclose the motile cell fully, in fact it does not do so, for a small number of rather prominent openings lie between the thecal plates and a large number of minute ones penetrate them. In the first group are the flagellar pore(s) that open through the ventral surface, and two openings of unknown function that occur in only certain species, one situated at the apex and the other lying close to or at the edge of one of the anteroventral plates. The position and something of the form of these openings are occasionally suggested by markings in the cyst wall.

Much smaller and more numerous are the **trichocyst pores** that perforate the cellulosic plates of most species (Figure 1.4 A). Their loci may be hardly visible or quite conspicuous, depending upon both the pore diameter and the prominence of sculpture elements associated with them on the external thecal surface. The distribution pattern of trichocyst pores is commonly species specific. Electron microscopy has shown that ejectile organelles, the trichocysts (Bouck and Sweeny, 1966; Dodge, 1971), lie directly beneath these perforations. In a manner grossly reminiscent of coelenterate nematocysts, dinoflagellate trichocysts fire stiff or mucoid charges through the openings, but the exact function of the resulting extrusions and the manner in which they penetrate the pellicle and other membranes is unknown. Nor is there complete agreement that providing passage for trichocyst discharges is the principal function of the openings themselves (Balech, 1974, 1980). It is interesting that the exterior surfaces of some Holocene and fossil cysts bear elements of minor surface relief (e.g., minute coni or spinules) arranged in patterns closely similar to the patterns of trichocyst pores.

In thecae which have not yet been enlarged by intercalary growth, the fields of trichocyst pores on two adjoining plates extend unequally close to the intervening suture. Since each pore penetrates only one plate (Gocht and Netzel, 1974), pores on the underlapping plate occur virtually up to the suture, whereas the overlapping plate bears a porefree margin coextensive with the overlap zone (Figure 1.4 H). The porefree area may also be marked by meshes of the surface reticulum that are larger and more regular than elsewhere. Provided the overlap zone is sufficiently broad (common in *Protoperidinium* and its allies; less so in *Gonyaulax* and *Ceratium*), these features, in combination, make it possible to determine the direction and approximate extent of plate overlap in an intact theca without examining directly the overlap structure itself.

Intimately associated with overlap between thecal plates is the manner in which individual plates are enlarged by addition of material along their margins in order to accommodate the increasing volume of a growing cell (Figure 1.4 F-I). Basically, a plate is enlarged by addition of thecal material along its sutural edges as the cell volume increases. **Intercalary growth bands** (or, simply "intercalary bands") result when the incremental material is visibly different from that of the original plate (Frontispiece A, Figure 1.4 G). Such bands are prominent in many species, varying in width from specimen to specimen and from place to place in a single theca. In many species of *Protoperidinium* and similar genera, the bands are transversely striate and lack trichocyst pores, thereby contrasting sharply with the differently patterned surface of the original portions of the plates. In these taxa, material is added unequally to the edges of the two plates that meet along a suture, causing the lines of the suture as seen on the external and internal surfaces of a mature theca to run closer to one edge of the intercalary band than to the other (Figure 1.4 H). Trichocyst pores may occur in the growth bands in *Gonyaulax* (Dürr, 1979a) and *Ceratium* (Happach-Kasan, 1982)

Cyst walls of some peridinioid species show especially well the extent to which differences in patterning on the cyst wall can be correlated with original plates and secondary increments. Modern freshwater *Peridinium limbatum* provides an example closely similar to many fossil cysts. In this case (Frontispiece A, Figure 1.5 A-C) the reticulate/perforate original plates and striate intercalary bands of the theca are matched on the outermost surface of the cyst by, respectively, polygonal areas bearing many short conical projections and intervening bands of surface devoid of such projections. In the Cretaceous and Paleocene fossil, *Palaeoperidinium pyrophorum* (Frontispiece C), intercalary bands and polygonal original plates show up with even greater clarity, although (for reasons that will be explored further in Chapter 9) this species should not be taken as a typical fossil cyst.

In contrast to these examples, many peridinioid cysts lack intercalary bands, even when lines suggesting the boundaries between thecal plates are clearly shown by a variety of surficial features. In *Wilsonidium* (Figure 1.5 G), for instance, rows of tubercles and short spines mimic thecal sutures, although the cyst wall along these rows has no discontinuities comparable to the sutures between thecal plates. While such "geometrical" or "spatial" similarities between features of cyst and theca are manifestly determined by their common genetic control, it is quite clear from comparative morphological studies that, in some species, the theca itself played a critical role in determining the thecalike aspects of cyst features by acting as a sort of template during the formation of the immediately

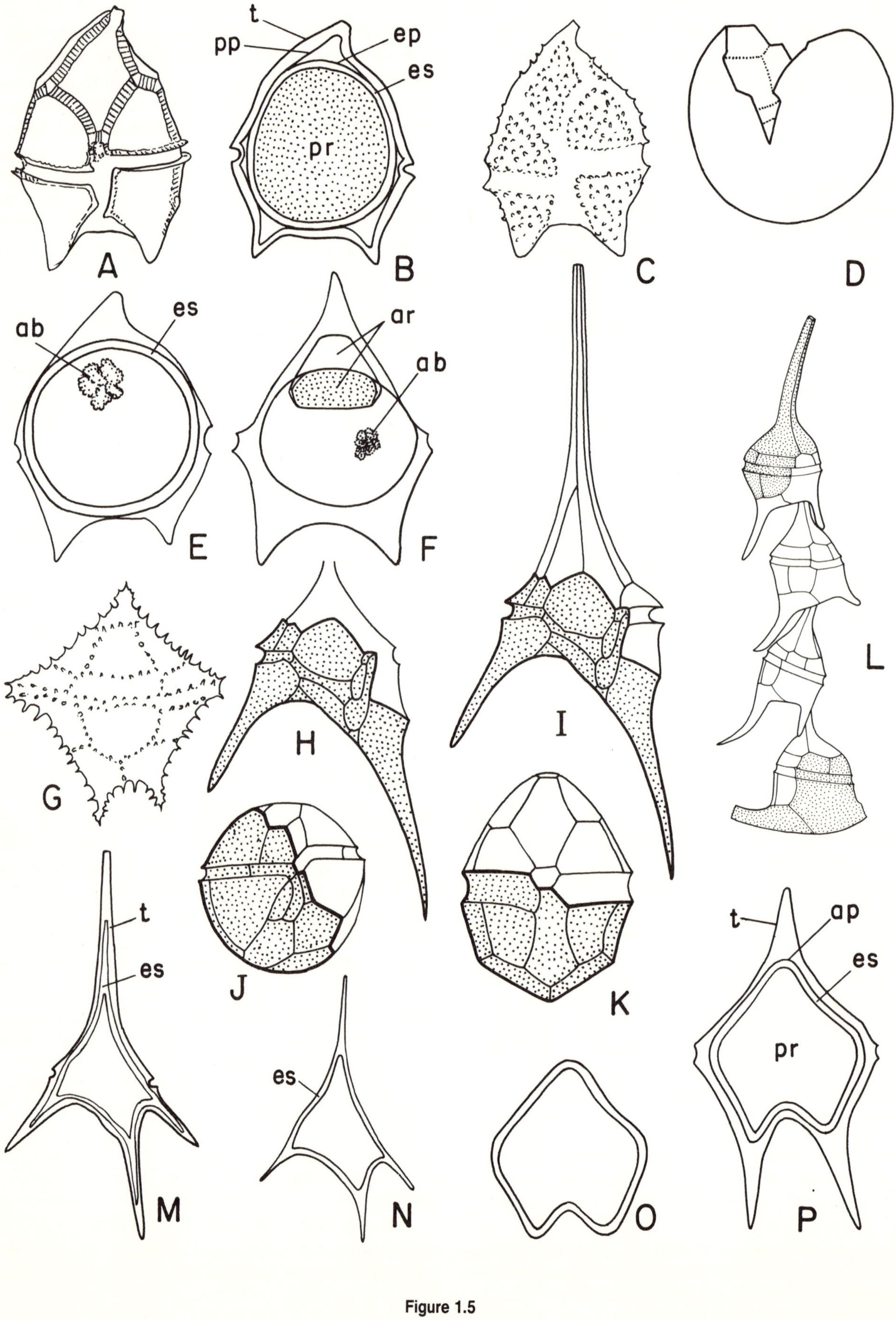

Figure 1.5

underlying cyst wall. We will explore this topic further in Chapter 4.

Thecal features in many living dinoflagellates of *Gonyaulax*- and *Ceratium*-type differ somewhat from those in *Peridinium*. The effects of incremental plate growth tend to be less conspicuous, and growth appears to be limited to those margins of a plate along which it overlaps its neighbors. The incremental material is not normally striate and it may bear trichocyst pores. In some species, it is hardly distinguishable from the original plate, but in others a contrast in surface texture or relief makes the difference quite obvious (Dürr and Netzel, 1974; Gocht, 1979a). Such intercalary bands do not appear to have any counterpart in the distribution of surface features on the cysts produced by living gonyaulacoid and ceratioid species. For the condition in fossil cysts, see discussion in Chapter 4.

VEGETATIVE REPRODUCTION AND THE MOTILE SCHIZONT

Comments under this heading will be brief, since the delicate cysts that may form during the vegetative part of the life cycle are probably rare constituents of sedimentary deposits.

Two styles of cell division appear to be common. In one, known as **desmoschisis** (Figure 1.5 H-K), the parent theca separates along a particular sequence of sutures as the protoplast divides, and each daughter cell initially retains one half of the old theca. In *Gonyaulax*, *Triadinium*, *Ceratium*, and their relatives, in which desmoschisis is the common type of vegetative division, the theca normally separates along an oblique line that passes through the area where the flagella emerge at the intersection of cingulum and sulcus. The resultant left anterior half of the theca carries the transverse flagellum and the right posterior half carries the longitudinal flagellum. Among *Peridinium* and its allies, desmoschisis is rare, but it occurs in some species (e.g., *Peridiniopsis lubiniensiforme*; Diwald, 1938, as *Glenodinium*) accompanied by division of the theca along the cingulum.

As cell division proceeds, or after it is complete, new thecal plates begin to appear. They first become visible over the exposed portion of each daughter cell and contrast strongly with the much thicker thecal remnant that was inherited from the parent and covers the rest of the cell. This conspicuous stage has been reported often, together with the inference that the daughter cell constructs only

Figure 1.5 - FEATURES OF THECAE AND CYSTS

A-E - *Peridinium limbatum*. A modern freshwater species whose theca (A) shows reticulate plates (unpatterned in figure) separated by conspicuous striate intercalary bands, while the cyst (C) has the corresponding areas marked by fields of short coni and intervening smooth strips. B - Longitudinal optical section of complete encysted specimen, showing cellulosic theca (t), two sporopollenin layers of cyst wall (pp - periphragm, ep - endophragm), and nonpreservable content (es - endospore, pr - protoplast). D - Endocyst only, agape along archeopyle suture; external surface is smooth but interior is delicately textured between distinct parasutural grooves (dotted). E - Longitudinal section of cyst only (compare B) with prominent endospore (es) and accumulation body (ab). A-C, E are vv*; D is an oblique left latv; E and D at different scales from A-C. From culture, courtesy L. Pfiester; and from Round Pond, Massachusetts, modified from Evitt and Wall, 1968.

F - Fossil cyst of *Deflandrea* cf. *D. phosphoritica*, showing preserved remnant of accumulation body (ab) and archeopyles (ar) in periphragm (polygonal) and endophragm (stippled); dv. Oligocene, Germany.

G - *Wilsonidium* sp. An Eocene peridinioid cyst whose surface ornament suggests sutural lines without traces of intercalary bands like those in the cyst of *Peridinium limbatum* (compare C); dv. Eocene, Maryland.

H-K - Diagrams of ventral surfaces of modern thecae, showing line along which theca separates during vegetative reproduction. Stippling identifies portion inherited from parent cell. H-I - Division along diagonal line of sutures in freshwater *Ceratium furcoides*; Felt Lake, Stanford, California. In I, tabulation is distinct in exposed portion of new theca, although the plates are conspicuously thinner and more delicately structured than in old portion, which will be discarded as young cell matures. At an earlier stage (H), apical horn of the new portion is incomplete and plates not yet delineated. J - Stage comparable to I, in vegetative division of marine *Triadinium sphaericum*. Based on SEM, courtesy H. Netzel. K - In freshwater *Peridiniopsis lubiniensiforme* the line of thecal separation is nearly equatorial, but proximal (left) and distal parts of cingulum remain with different portions of the parent theca. Based on Diwald, 1938.

L - Chain of four individuals of marine *Ceratium* in dv, showing variation in thecal shape. Stippling identifies the two portions of the parent theca, with characteristics of *C.* (*Tripoceratium*) *tripos* (hypothecal horns broken), whereas hypothecae of the two daughter cells conform to *C.* (*Biceratium*) *californiense*. Pacific coastal waters, California; based on Kofoid, 1909b.

M-N - *Ceratium furcoides*. Sectional views of cyst within theca (t) and of isolated cyst of this modern freshwater species; both vv. The cyst contrasts with that of *P. limbatum* (compare B) in lacking a sporopollenin wall about the endospore (es); none of the structures shown here is geologically preservable. Felt Lake, Stanford, California.

O-P - *Protoperidinium oblongum*. A modern marine species. In the sectional view of an encysted specimen (P), the theca (t) encloses a cyst composed of a single resistant wall, the autophragm (ap), lined by the endospore (es) and containing the protoplast (pr). The cyst alone is shown in O. Both dvv; Pacific coastal waters, California.

*For explanation of abbrevations of views, see page 51.

the missing half of the old theca. However, Dürr (1979a) showed that for *Gonyaulax polyedra* this is not so. The thecal remnant soon falls away, exposing a new layer of thecal plates which had formed beneath it and leaving the cell once more with a covering of uniform thickness. As the cell ages, the new theca gradually increases in thickness. In some species, the cells resulting from desmoschisis may remain in close contact, forming chains of individuals (Figure 1.5 L)—up to 70 cells reported in *Gessnerium monilatum* (Steidinger and Williams, 1970, as *Gonyaulax*).

Among the relatively few examples showing oblique division which have been studied in detail, some variation in the course of the line of separation over the hypotheca has been observed (H. Netzel, pers. comm.), leaving uncertain the reliability of this character for assessing plate homology and biologic relationship.

In the second style of vegetative cell division, termed **eleutheroschisis**, the parent theca is completely discarded at an early stage, and each of the initially naked schizonts develops its own new theca from the beginning. The protoplast may divide either before or after ecdysis, which may take place in several ways. The old or new protoplasts may emerge through an opening formed by release of one or more plates of the old theca (Boltovskoy, 1973), or by the spreading apart of several convergent sutures. Alternatively, the theca may separate essentially in halves, normally along the cingulum. Eleutheroschisis is the typical form of vegetative division in *Peridinium* and its allies.

Vegetative division has been observed to occur in cultures with frequencies ranging from about once in four days to once in nine hours (Loeblich III, 1967) and both styles of division may occur in the same species. Eleutheroschisis is preceded in some species by the formation of an ecdysial cyst, several types of which have been observed. One of these, which forms around the protoplast within the theca, possibly includes the pellicle and, if this layer contains a sufficient amount of sporopollenin, it is conceivably preservable.

Other temporary cysts resembling ecdysial cysts are produced occasionally by motile vegetative cells, apparently as hasty and short-lived responses to a sudden onset of unfavorable conditions. It is not certain that any of them contain sporopollenin or are preservable.

ACCUMULATION BODIES

The objects referred to here occur in the form of one to several small globular masses within the protoplast. They are visible in some motile cells but are most obvious in resting cysts, where their brownish to reddish color tends to be intensified (Figure 1.5 E-F). Although their exact nature and function are unknown, they have been observed to increase in size within the cell and have been termed **accumulation bodies** (D. L. Taylor, 1968; Bibby and Dodge, 1972) on the premise that they represent small masses of metabolic waste materials. Their color apparently results from a concentration of carotenoids and, since the very resistant material, sporopollenin, is regarded as a polymeric carotenoid derivative, this composition may explain both the observed resistance of accumulation bodies to acetolysis of modern encysted cells and the common occurrence of virtually identical objects inside many fossil cysts. It had once been suggested that these objects in the fossil record may represent cell nuclei (Eisenack, 1961) or symbionts (Eisenack, 1966). Accumulation bodies have also been termed "red bodies," "brown bodies," "eye spots," "oil droplets," and, in the case of fossil examples, "omphali" (Dörhöfer and Davies, 1980).

Although variable in number, form, and position, accumulation bodies have a generally characteristic appearance, even in fossil cysts. The presence of such objects within unruptured fossil specimens is probably reasonable presumptive evidence for identifying an otherwise appropriately constituted specimen as a dinoflagellate. However, reliance upon the position of an accumulation body as a critical criterion for distinguishing dorsal from ventral surfaces (Dörhöfer and Davies, 1980) seems unjustified, even within a confirmed dinoflagellate.

LIVING GENERA PARTICULARLY RELEVANT TO THE FOSSIL RECORD

Several living genera of peridinialean dinoflagellates stand out from the rest as having special importance in a consideration of dinoflagellate cysts and the fossil record. Not only do most preservable cysts produced by living dinoflagellates represent species currently referred to these genera, but also a majority of fossil cysts, although described under different generic names, have features that suggest relatively close affinities to these same taxa. These important modern genera are *Peridinium*, *Protoperidinium*, *Gonyaulax*, and *Ceratium*. In Chapter 5 we will explore in detail the tabulation patterns that distinguish them; at this point only a few introductory comments are offered.

Before 1979 *Protoperidinium* was seldom mentioned in the literature on fossil dinoflagellates, whereas *Peridinium* was cited frequently and the derivative term "peridinioid" is still in wide use. This situation has changed as a consequence of the taxonomic revision of *Peridinium* by Balech (1974). Previously *Peridinium* was used in a broad sense to encompass several hundred species of marine and freshwater dinoflagellates, and *Protoperidinium*, although long an available name, had not been applied widely even among living species. Currently *Peridinium s.s.* includes only a very moderate number of exclusively freshwater species, most of the marine species formerly referred to it having been reattributed to *Protoperidinium*. More or less similar marine species are also referred to several other genera which we need not consider here. It is likely that the application of *Peridinium* to freshwater species will be restricted further (Boltovskoy, 1979).

Distinctions among the genera that were formerly included within the scope of *Peridinium* are based on thecal features that seldom have clear counterparts in the cyst (e.g., four plates in the cingulum of *Protoperidinium*, as compared to six in *Peridinium*). Therefore, they are not easy for the student of dinoflagellate cysts to apply, and, certainly, habitat alone is an unreliable criterion for judging biological affinities. Although most fossil cysts represent the marine environment and all living species of *Peridinium s.s.* inhabit fresh water, features visible in some fossil cysts are more compatible with modern freshwater *Peridinium* than with modern marine *Protoperidinium* (Boltovskoy, 1979). In view of these uncertainties, the reader must recognize the reservation necessarily implicit in the use of a such a term as "peridinioid" or the generic names "*Peridinium*" and "*Protoperidinium*" in reference to cysts whose thecate equivalents cannot be established by direct observation or experiment. In this volume **peridinioid** most frequently will have a broad connotation, signifying general similarity to and presumed affinity with *Peridinium* and related taxa. The term will also be used in the sense of a peridinioid outline (see Chapters 3 and 9) without necessarily implying relationship. Unless the context clearly indicates otherwise, use of "*Peridinium*" without mention of "*Protoperidinium*" (or *vice versa*), implicitly includes the unmentioned alternative.

After *Peridinium*, *Gonyaulax* and *Ceratium* are the best known genera of living peridinialean dinoflagellates. Although fewer species have been referred to *Gonyaulax* than to *Ceratium*, some of them are especially significant by virtue of wide geographic distribution or involvement in production of red tides. As in the case of *Peridinium*, recent taxonomic revisions have affected the nomenclature of species referred to *Gonyaulax*, and further revisions are imminent. For instance, Taylor (1979) proposed *Protogonyaulax* as a new genus for a group of species, including several toxic ones, formerly included in *Gonyaulax*, and gross differences in cyst morphology provide plausible grounds for further revision of the species remaining under *Gonyaulax*. The thecal patterns that distinguish *Gonyaulax* from *Protogonyaulax*, unlike those separating *Peridinium* from *Protoperidinium*, are recognizable in many fossil cysts and will be important considerations in Chapter 5. There we will also analyze thecal patterns of two other living genera (*Pyrodinium* and *Triadinium*, the latter until recently known as *Goniodoma* or *Heteraulacus*) and compare them to *Gonyaulax* and *Protogonyaulax*.

In passing, it may be well to note that the continued wide use of the generic name *Gonyaulax* today is more *de facto* than *de jure*. Strictly speaking, *Gonyaulax* Diesing 1860 almost certainly is a synonym of *Spiniferites* Mantell 1850, since cysts known as *S. bentori* have been cultured to yield thecae identified as *G. spinifera*, the type species of *Gonyaulax*. Protection of the name *Gonyaulax* will eventually require either conservation under Article 14 of the I.C.B.N. or sanction of the simultaneous use of different generic names based on cysts and thecae for the respective objects.

The genus *Ceratium* includes a large number of living species. Most of them are marine, but the few freshwater species include both the much-studied *C. hirundinella* and a similar species (*C. furcoides*) that has commonly been included as a forma of it. Three or four usually prominent horns are characteristic of the genus, along with a tabulation that has been considered either quite distinctive (Kofoid, 1907) or basically *Gonyaulax*-like (Wall and Evitt, 1975). Cysts strongly suggestive of *Ceratium* are conspicuous elements of the fossil record, where they are restricted to strata of latest Jurassic and Cretaceous age (Frontispiece I, J).

Features that are diagnostic of a few other living genera have been recognized among fossil cysts, and some living species of these genera are also known to produce cysts. For example, *Pyrophacus*, with a distinctively tabulate theca, is known from equally distinctive cysts (Wall and Dale, 1971), but the total number of species is small. Also, *Cladopyxis* and *Palaeophalacroma*, although unimportant genera on the contemporary scene, are possibly related to a large number of as yet poorly understood fossil cysts which have generally been referred to *Microdinium*. These will be discussed in Chapter 5.

ZYGOTIC THECA AND ZYGOTIC CYST OR HYPNOZYGOTE

Although the zygotic theca itself is not preserved in sediments, its formation and morphology are of interest because this is the body within which the fossilizable resting cyst develops, and part of the cyst may form in intimate contact with the theca.

The zygotic theca and the vegetative theca appear to grow in similar ways, although mature stages of the former may be significantly larger than those of the latter in the same species (Pfiester, 1975 and later).

While cysts of many dinoflagellates are now known, attempts to culture individual cyst-producing species through their life cycle have met with mixed success. Approximately 50 marine species have been induced to excyst in the laboratory, producing thecate individuals whose specific identity has been confirmed by reference to conventional thecal criteria (see Wall and Dale, 1968b and later). However, encystment has not been induced successfully on a regular basis in marine species under culture. In contrast, both excystment and encystment have been studied in a number of freshwater species, with special attention given most recently to the sexual phase of the life cycle (e.g., Claparède and Lachmann, 1859; Entz, 1925; Pfiester, 1975-1977; von Stosch, 1964, 1973). Cyst morphology has not been the focus of principal interest in any of these studies.

Although different numbers of layers have been observed in the walls of cysts from different living species and a special terminology has been applied to them (von Stosch, 1973; Pfiester, 1975; Dürr, 1979c), it is not certain that these terms always have been attached to exactly homologous structures, nor is it certain which layers are represented in the still more varied walls of fossil cysts. In the examples diagrammed (Figure 1.5 B, E) a conspicuous, thick, clear layer, refringent in transmitted light and testing positively for cellulose in at least some species (von Stosch, 1973; Anderson and Wall, 1978) immediately surrounds the granular-appearing protoplast. This distinctively appearing layer has been called the **endospore** (von Stosch, 1973, in *Woloszynskia apiculata*) and is, presumably, homologous in different species. Where tested, it is readily destroyed by acetolysis and (unless its outer portion is preservable as suggested in the next paragraph) has not been observed in fossil specimens. No geologically preservable layer appears to occur more centrally than the endospore, and, in some living cyst-producing species (e.g., *Ceratium furcoides*, Figure 1.5 M-N), none exists outside it. But, in other modern examples, one or two sporopollenin layers lie outside the endospore.

The first of these appears to be just the outermost fraction of the endospore rendered preservable by incorporation of sporopollenin. The terminology that biologists have applied to the living cyst does not distinguish this layer from the endospore. In the terminology that paleontologists have developed for the preservable cyst wall, it is the **endophragm** in cysts constructed like that of *Peridinium limbatum* (Figure 1.5 B) and the **autophragm** in some cysts with only a single preservable wall (e.g., *Protoperidinium oblongum*, Figure 1.5 O-P). The external preservable layer in the cyst of *Peridinium limbatum* and many similar cysts is the biologists' **mesospore** and the paleontologists' **periphragm**. This is the layer mentioned earlier as possibly being derived from the pellicle by addition of further sporopollenin. The same layer is termed the **autophragm** in preserved cysts if it is the only layer present. The all-enclosing cell membrane and the **exospore** (= zygotic theca), being geologically unpreservable, are unrepresented in the terminology of preservable cyst structures. The complexity of the sporopollenin cyst wall will be discussed in Chapter 3, where reasons for uncertainty in the recognition of homologous layers will become apparent.

Dürr (1979c) described the cyst of *Peridinium cinctum* as revealed by electron microscope studies, with special attention to the wall layers and the membranes associated with their formation. He considered that vesicles analogous to those in which thecal plates are formed probably occur also in the early stage of mesospore and endospore formation, but that the constituent vesicles of each of these layers fuse at an early stage and their inner and outer membranes become continuous structures. The mesospore and endospore are then laid down, each between a pair of membranes, without any kind of sutural interruptions comparable to those between thecal plates.

A distinctive feature unique to the cyst is the excystment aperture or archeopyle (Chapter 6). In most cysts this opening is formed by separation of the cyst wall along a definite and preformed line (archeopyle suture), which normally results in the partial or complete isolation of a portion of the wall as an operculum. Such archeopyles (Figure 1.5 F; 1.7) are frequently conspicuous features of resting cysts, but in some species, both living and fossil, the opening for excystment appears to form in a less regular way, perhaps by general breakdown of a part of the wall.

The occurrence of cysts in thecae of modern marine species was observed long ago (Carter, 1858), as was the process of encystment in freshwater species (Claparède and Lachmann, 1859). However, familiarity with these phenomena and with the features that distinguish isolated dinoflagellate cysts grew slowly, witness the illustrations now recognizable as dinoflagellate cysts that appear along with other uncertainly identified objects in reports of some of the oceanographic expeditions around the turn of the century (e.g., Lohmann, 1904). Entz (1925) described cysts of freshwater *Ceratium*, while Braarud (1945), Nordli (1951), and Erdtman (1954) figured cysts from modern marine species of *Gonyaulax*. Chorate cysts now known to represent species of the *Gonyaulax spinifera* group were observed in the Baltic plankton (Erdtman, 1950) and in modern sediments from Pacific lagoons (McKee et al., 1959) and Mediterranean coastal areas (Rossignol, 1961, 1964). Intensive study of modern cysts, including identification of the thecate stages of some of the same species, began with the investigation by Wall (1966).

VARIABILITY

Variability in dinoflagellates is often great and occasionally spectacular. Its morphological expression in the theca involves chiefly differences in size, shape, and tabulation. Representative examples are shown in Figure 1.6. Discrimination of phenotypic and genotypic differences is difficult and the taxonomic value placed on differences thought to be genotypic has itself varied greatly. Consequently some taxa encompass a range of characters not tolerated in others of the same rank.

Species of *Ceratium* have received particular attention in studies of variation. Strikingly large ranges in cell size, general proportions, and, especially, the length and orientation of horns are known from both marine and freshwater species of the genus (Figure 1.6 A-F). These variations appear to reflect the combined effects of allometric growth, the particular cell shapes that dominate samples at any one time, and seasonal changes in water temperature. López (1955, 1966) found that, although Mediterranean samples of numerous *Ceratium* species contained a full range of cell forms and sizes throughout the year, a limited range (upper two outlines of Figures 1.6 A-C) dominated winter samples, and a contrasting range (lower outlines of same figures) dominated summer samples. Posterior horns in freshwater *Ceratium hirundinella* diverge increasingly as summer progresses (Hüber and Nipkow, 1923), but grossly dissimilar individuals may coexist at every season (although in differing proportions) in this and a similar species (Figure 1.6 D). Even within a single chain of vegetative cells—really a fraction of a clone in which all individuals, barring mutation, have identical nuclear genomes—contrasts may be so great that the terminal cells, if isolated, would be placed in different subgenera (Kofoid, 1909b; Figure 1.5 L). Cultures have produced many shape variations similar to those observed in nature, as well as apparent abnormalities presumed to be responses to genetic isolation or peculiarities of the medium (Hasle and Nordli, 1951; Figure 1.6 E-F).

Species of other genera may also show substantial variability in cell shape (Figure 1.6 G) although this may not be so conspicuous as when long horns are present. Whatever their cause, such obvious differences in size and shape among living cells having similar genotypes argue for caution in using these differences as principal criteria for distinguishing taxa—whether of thecae or cysts, modern or fossil.

Differences in tabulation—intraspecific, interspecific, and intergeneric—may involve fusion of plates normally separate (Figures 1.6 N-O; 7.4 G-H), appearance of "extra" plates produced by "spurious" sutures (Figure 1.6 Q-R), and topological differences in the tabulation mosaic without associated differences in number of plates and sutures (Figure 1.6 J-M). The contrast in taxonomic values accorded certain features in different taxa is well illustrated by the case of cingular plates in peridinioid and ceratioid dinoflagellates. Among peridinioid species, the number of cingular plates is a generic determinant (Balech, 1974, 1980), but this is not true among ceratioids (Bourelly, 1968), although the varied

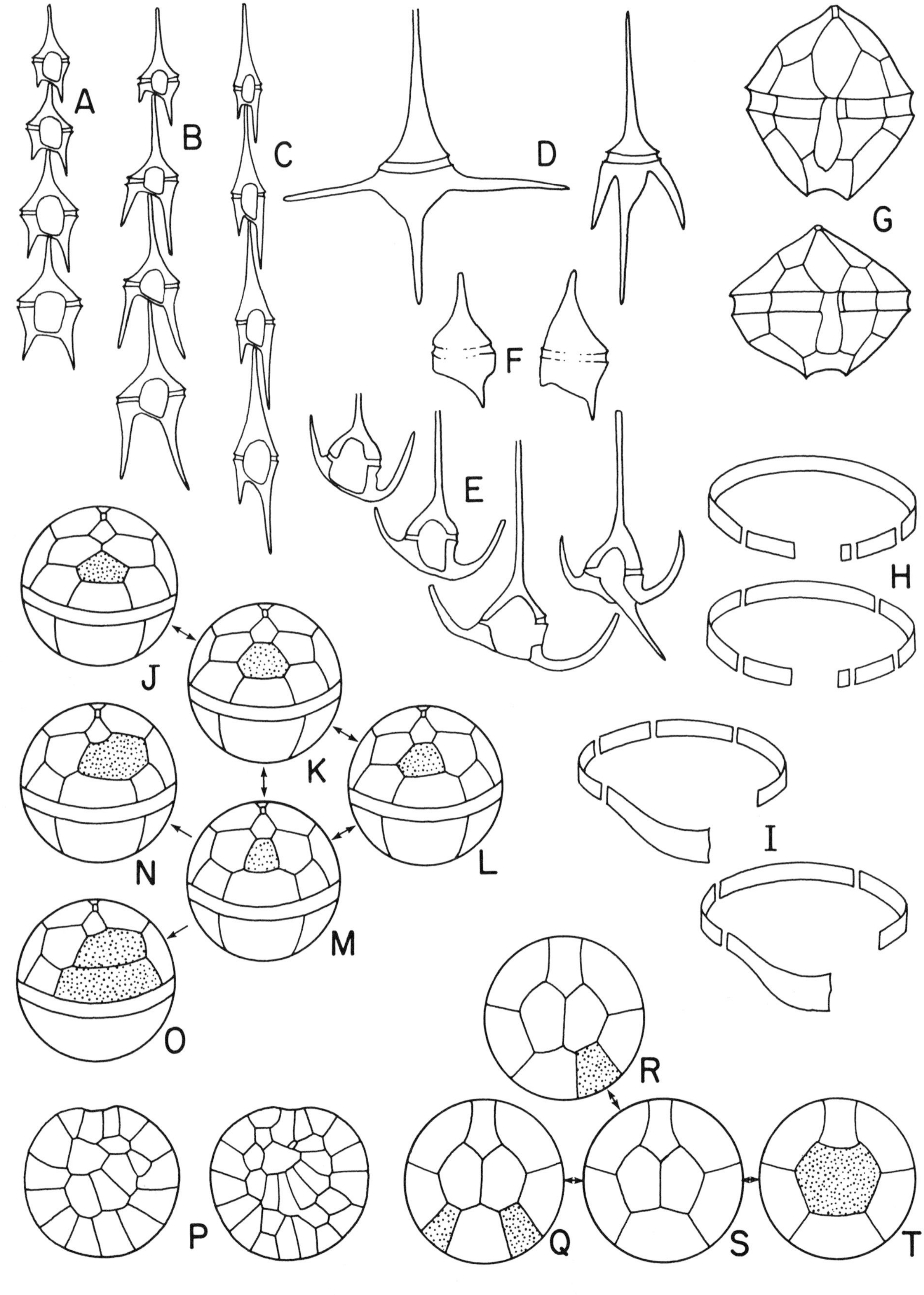

Figure 1.6

counts in both groups are explainable as the result of fusion—or lack thereof—along particular sutures (Figure 1.6 H-I). The manner in which such differences are accommodated in taxonomy is, of course, an entirely subjective matter and is often influenced significantly by individual monographs. Further comments germane to this point and an illustration of unusual tabulation variability in a living species will be found in Chapter 7 under Variable Variables.

While differences in a single suture may affect the determination of the genus in some instances and the species in other instances, it is clear that in still other instances similar variations may be only intraspecific and may occur with different frequency according to the species. The distinctive marine species, *Pyrophacus vancampoae*, provides an example of extreme intraspecific variability in tabulation (Figures 1.6 P); in contrast, the patterns of many species are virtually constant.

Given such variability in dinoflagellate thecae as has been cited in the foregoing paragraphs, it is not surprising that comparable differences occur among cysts. Although morphologic variability of modern cysts has not been studied extensively, it is altogether obvious that differences among fossil cysts present the paleontologist with problems quite as diverse as those facing the student of modern thecae. The important contrast with neontological studies is that paleontological taxonomy can never be responsive to properties of the protoplast or to changes that can be observed during the life cycle of a living species.

An intriguing question for students of dinoflagellate cysts—modern or fossil—is: Does the range of phenetic variation among vegetative thecae of a given species equal the range of phenetic variation among zygotic thecae of that species? It is variability among zygotic thecae that is most relevant to the morphology of preservable cysts. The fossil record cannot be expected to provide evidence of variability among long-vanished thecae of vegetative cells, however great that variability may have been. Extensive, but still only casual, observation suggests that tabulation variability among vegetative cells in living species exceeds, on the whole, variability in paratabulation to be found among fossil cysts.

Figure 1.6 - VARIABILITY IN SHAPE, SIZE, AND TABULATION

(All figures represent modern thecae except J)

A-C - Allometric growth in modern marine *Ceratium* species from the Mediterranean. Groups of thecae (vv) are figured as if in "chains" to show progressive increase in length of posterior horns and transdiameter. A -*C. pentagonum*, B - *C. candelabrum*, C -*C. furca*. Redrawn from Lópex, 1966.

D - Two examples of freshwater *Ceratium furcoides* showing extremes of shape present in a single plankton sample; dv. Felt Lake, Stanford, California, October, 1978.

E - Four examples of marine *Ceratium tripos* from plankton samples in Oslofjord (July, 1948); vv. Rightmost theca has supernumerary horn, a type of variant observed at a frequency of 0.29 per mille. Redrawn from Hasle and Nordli, 1951.

F - Two thecae of *C. tripos* observed in culture derived from same material as E; vv. Redrawn from Hasle and Nordli, 1951.

G - Shape contrasts in two thecae of antarctic marine *Protoperidinium parvicollum*, vv. Redrawn from Balech, 1958.

H - Differences in cingular series of plates accepted as distinguishing characters for *Protoperidinium* (above) and *Scrippsiella* (below). Based on Balech, 1974.

I - Cingular series of plates as represented in marine (upper) and freshwater (lower) species of *Ceratium*. Here a tabulation difference comparable to that in H has been accorded only interspecific value.

J-T - Schematic diagrams to illustrate differences in tabulation which have been variously evaluated as having intraspecific, interspecific, and intergeneric significance. The diagrams emphasize topologic aspects of the patterns; plate proportions do not faithfully represent the species mentioned. Stippling draws attention to areas of difference.

J-O - Apicodorsal views of epithecal patterns in *Protoperidinium* and similar genera. Patterns J-M have the same number of plates but plate contacts are locally different. Patterns K-M may occur as intraspecific variants, but are more commonly species-specific and distinguish three large groups of species within *Protoperidinium*. Although pattern J is extremely rare, a Cretaceous cyst (*Trithyrodinium striatum*) is characterized by intraspecific variation between the extremes represented by J and K. Patterns N-O involve plate "fusion" (i.e., absence of a suture normally present) and have fewer than normal number of plates; such variants occur in *P. saltans*, a species whose pattern varies notably from the mode represented by M.

P - Intraspecific variations in the hypothecal tabulation of marine *Pyrophacus vancampoae* from Phosphorescent Bay, Bahamas. After Wall and Dale, 1971.

Q-T - Antapical patterns in *Protoperidinium* and similar genera. The basic pattern is shown in S. Q and R are intraspecific variants with supernumerary sutures (and, therefore, plates) observed in cultures of *Scrippsiella faeroensis*; based on Dale, 1977. T shows the unique absence of the suture between two antapical plates that is a constant distinguishing feature of *P. unipes*; based on Balech, 1962.

SEVEN EXAMPLES OF CYST MORPHOTYPES

Seven different cysts—all but one from living species—are described briefly below and illustrated in Figures 1.5 B-E, 1.7 to give an impression of the range of morphological features that may be encountered. Cysts for five of the six modern species were identified either from thecae produced by cells excysted in culture or from thecae found still enclosing freshly formed cysts. Only those features that relate to the preservable sporopollenin wall (i.e., excluding the thick endospore wall and the protoplast) are described and illustrated. Descriptive terms used for the first time here and shown in boldface are fully defined in the chapters of Part II; others have been illustrated already.

1. *Protoperidinium obtusum* (Figure 1.7 A). The cyst wall consists of a single layer, the autophragm, formed in close proximity to the theca (**proximate cyst**) and therefore similar to it in size and shape. Although the zygotic theca displays prominent intercalary growth bands, the surface of the cyst is either unmarked or shows only narrow lines (**parasutural lines**) in the approximate position of the wide bands. The archeopyle is on the anterior slope of the dorsal surface and its operculum is completely surrounded by the archeopyle suture.

2. *Peridinium limbatum* (Figure 1.5 B-E). The cyst wall has two layers which are partly separated (especially at the horns) by an intervening space. The outer layer, or periphragm, is proximate, like the single layer in *Protoperidinium obtusum*, but in this case counterparts of intercalary growth bands are recognizable on it as unornamented strips between polygonal fields of coni that correspond to the plates of the juvenile theca. The inner wall, or endophragm, forms an ellipsoidal body which has a smooth external surface, but its internal surface shows a pattern of grooves defining a set of polygonal areas equivalent to the tabulation of thecal plates (i. e., a **paratabulation**). The archeopyle suture in each wall follows an identical course relative to the **parasutures** of the surface pattern, beginning at the **paracingulum** on one side, passing through the apex, and extending back to the paracingulum on the other side.

3. *Impagidinium patulum* (Figure 1.7 C). The exterior of the ellipsoidal cyst bears thin fencelike membranes (**septa**) arranged like thecal sutures and bounding unornamented polygonal areas, the **paraplates**, which correspond to thecal plates. Although wall layering is inconspicuous, locally one can distinguish two layers, the closely appressed periphragm and endophragm. The archeopyle corresponds to the middorsal plate just anterior to the paracingulum, and the operculum is fully isolated by a suture that runs within the outline made by the surrounding septa. Note: Although cysts of this type occur in modern deepwater sediments, the thecate stage has not been identified. However, the pattern of surficial septa implies a species closely similar to *Gonyaulax digitalis*.

4. *Gonyaulax digitalis* (Figure 1.7 D). The main cyst body is separated from the theca by a substantial distance spanned by spinelike projections (**chorate** cyst). The projections (**processes**) rise from ridges that interconnect their bases and form a pattern that defines the paratabulation. Periphragm and endophragm are present but closely appressed except under the processes and ridges, where the outer layer can be seen to form the projections. The archeopyle (not shown) corresponds to the middorsal precingular plate and its operculum is fully isolated.

5. *Gonyaulax polyedra* (Figure 1.7 B). The cyst is similar to that of *G. digitalis* in being chorate and two-layered (periphragm and endophragm), but the processes are not distributed in a way that suggests the tabulation (**nontabular** distribution) and their bases are not interconnected by ridges. The archeopyle represents three consecutive dorsal precingular plates and the operculum is correspondingly divided into three pieces, each of which is fully surrounded by archeopyle sutures. The shapes of the archeopyle and the opercular pieces are the only cyst features that suggest the shapes of thecal plates.

6. *Oligosphaeridium abaculum* (Figure 1.7 F, a fossil cyst of Cretaceous age). The cyst is chorate and two-layered, with processes formed by a periphragm which is closely appressed to the endophragm between process bases. The processes rise from the central portions of paraplates (**intratabular** distribution), which are outlined by parasutural lines. The parasutures define a complete paratabulation but certain paraplates consistently lack processes. The archeopyle is an apical opening whose undivided operculum (missing from top of figured specimen) has four processes corresponding to four apical plates. Note: Most fossil cysts of similar design lack all trace of parasutural markings, leaving the isolated processes as the only indication of paratabulation aside from the archeopyle suture. Neither of these types of cysts (i.e., having intratabular processes

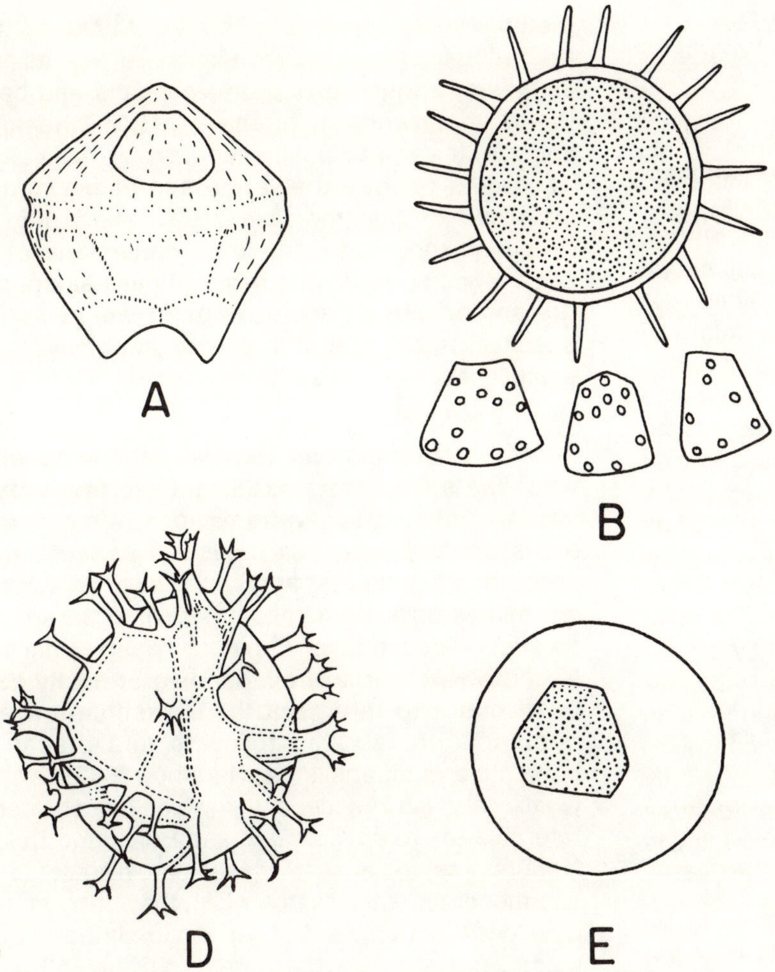

Figure 1.7 - REPRESENTATIVE CYST MORPHOTYPES

A - *Protoperidinium obtusum* (dv) from California coastal plankton, showing outline of archeopyle and traces of paratabulation on hypocyst.

B - *Gonyaulax polyedra* from Massachusetts coastal plankton; sectional view showing thin cyst wall underlain by thick (unpreservable) endospore (clear zone) surrounding the central protoplast (stippled). The three opercular pieces of the large archeopyle are shown below the main figure; circles mark bases of spinelike processes.

C - *Impagidinium patulum* (dv). Open archeopyle (stippled) occupies middorsal precingular area; opening is slightly smaller than the corresponding paraplate outlined by the parasutural septa. Specimen from Quaternary sediment, Yucatan Basin; species also occurs in modern open sea plankton.

D - *Gonyaulax digitalis* (slightly oblique vv), from Massachusetts coastal plankton.

E - *Protoperidinium conicoides* (slightly oblique dv), from Massachusetts coastal plankton, showing open archeopyle which corresponds to the middorsal intercalary paraplate.

F - Early Cretaceous *Oligosphaeridium abaculum* (left latv). Specimen lacks the operculum to the archeopyle which corresponds to all four apical paraplates.

C based on Wall (1967); B, D-E, based on Wall and Dale (1968b); F drawn from specimen from North Sea core material, courtesy R. J. Davey.

with or without parasutural lines) is known among living dinoflagellates.

7. *Protoperidinium conicoides* (Figure 1.7 E). The subspherical cyst consists of an autophragm with a nearly featureless surface. The only suggestion of paratabulation is the polygonal archeopyle, which has the shape characteristic of the central plate of the dorsal epitheca.

NOTES ON TAXONOMIC CRITERIA, NOMENCLATURE, AND CLASSIFICATION

Modern thecate dinoflagellates have been classified almost exclusively on the basis of thecal architecture. At the generic level the chief criterion has been the number and arrangement of thecal plates, supplemented by prominent features of cell shape (e.g., presence of prominent hornlike projections in particular locations). At the specific level, details of the shape, distribution, mutual contact, and sculpture of plates, as well as the overall size and shape of the cell, have been utilized. Until very recently no consideration whatsoever has been accorded cyst morphology in the definition of living taxa. The taxonomy of fossil dinoflagellates, in contrast, has been based almost exclusively on cyst morphology, and this frequently involves features quite dissimilar in detail to those of the theca. Also, the role of tabulation has been limited by its frequently incomplete or cryptic expression on the cyst. Consequently, two essentially disparate taxonomies and nomenclatures have grown up, and their coordination has received little serious thought. Not only does this situation have intrinsically provocative theoretical aspects, but recent developments are rapidly making it intolerable on practical grounds. Paramount among these are the following:

1. The identities of several dozen kinds of living cysts in terms of the names long applied to their thecate stages have now been established by culture experiments.

2. To a large extent, both the morphological and temporal gaps that had existed between the separate bulks of information on fossil and living dinoflagellates are being spanned by cysts of Late Tertiary and Quaternary age, which are being described in increasing numbers.

3. Comparison of living and fossil species, as well as communication about dinoflagellate evolution, is becoming ever more difficult because, when cysts and thecae are considered as separate entities, the criteria for their classification are described in incompatible terms.

Two contrasting solutions to the problems of cyst nomenclature have been proposed. According to one (illustrated by the proposals of Reid, 1974; Bradford, 1975, 1978a; and Harland, 1977), separate nomenclatures should be used for cysts and thecae when these objects are considered separately. However, in a single work dealing conjointly with both motile and encysted stages of the life cycle, the senior name would clearly take precedence. A procedure similar to this already applies to another group of organisms and bears the formal sanction of the I.C.B.N. (the fungi imperfecti as specifically covered by Article 59). According to the other approach (illustrated by proposals of Wall and Dale, 1971; Dale, 1978), no exception should be made to the general rule of the Code—one species, one name! By this token, only the senior name should be used whenever the specific identity of both cyst and theca have been established, regardless of the context in which the objects are cited.

A difficult but important step will have been taken when the many facets of this situation have been argued out and a generally acceptable *modus operandi* established. At present a *de facto* state approximating the first approach exists and, in this volume, established names based on cysts will not be suppressed in favor of names based on thecae. Recently, two partially overlapping attempts have been made to integrate the classification and nomenclature applied to cysts and thecae of peridinioid dinoflagellates. Harland (1982) revised Quaternary and Holocene taxa which he considered referable to *Protoperidinium*, and Bujak and Davies (1983) presented a more ambitious revision of fossil and modern dinoflagellates which they attribute to a new peridinialean suborder Peridiniineae. Both these items are discussed briefly in Chapter 13.

In the following paragraphs, the four classes of Pyrrhophyta and five orders of dinoflagellates listed in the Introduction (Table In.I) are characterized briefly. Thereafter (except in Chapter 12, where the orders Dinophysiales, Gymnodiniales, and Nannoceratopsiales are discussed), our attention through this volume will be confined essentially to the order Peridiniales.

Class DINOPHYCEAE - pyrrhophytes in which one flagellum is whiplike and extends longitudinally, while the second is ribbon-like and follows a circular path in a plane about at right angles to the first.

Order PROROCENTRALES - dinoflagellates in which the flagella are inserted terminally (desmokont condition), the longitudinal one extending in advance of the cell, and the transverse one encircling the other anterior to the cell body. Some forms have a cellulosic theca of distinctive structure. Preservable resting cysts are unknown and there is no certain fossil record, although the order is conceivably represented by some of the organic-walled fossils currently regarded as acritarchs. Representative genera: *Exuviaella* (nonthecate), *Prorocentrum* (thecate).

Note: In all three of the following orders for which the living cell is known, the flagella are inserted laterally (dinokont condition), the longitudinal one extends posteriorly, and both normally lie, at least in part, within channels (the so-called flagellar furrows) defined by various features on the cell surface. Fossil cysts of the extinct fourth order appear to record a similar organization.

Order DINOPHYSIALES - dinoflagellates having the transverse flagellar furrow near the anterior limit of the cell; cell normally shows moderate to strong lateral compression; two lateral plates of the cellulosic theca are much larger than any others. Preservable resting cysts are unknown and there is no unequivocal fossil record. Representative genera: *Dinophysis*, *Ornithocercus*.

Order PERIDINIALES - dinoflagellates having the transverse flagellar furrow normally located within the medial third of cell length; theca composed of several tens of cellulosic plates organized in several series paralleling the transverse furrow. Preservable resting cysts are found in some living species and there is an extensive fossil record. Representative living genera: *Peridinium*, *Gonyaulax*, *Ceratium*. Representative fossil genera: *Deflandrea*, *Gonyaulacysta*, *Odontochitina*. Silurian, Triassic-Holocene.

Order GYMNODINIALES - dinoflagellates having the transverse flagellar furrow usually located within the medial third of cell length; cellulosic thecal plates lacking or (rarely) thin, but corresponding vesicles more numerous than typical for Peridiniales, small, and all of about similar size. Preservable resting cysts are known in a few living species. Moderate fossil record of cysts and distinctive sporopollenin coverings of possibly motile cells. Representative living genera: *Gymnodinium*, *Polykrikos*. Representative fossil genera: *Dinogymnium*, ?*Distatodinium*, ?*Suessia*. Triassic-Holocene.

Order NANNOCERATOPSIALES - dinoflagellates having the transverse flagellar furrow near anterior extremity of cell; cell compressed laterally as in Dinophysiales; tabulation of inferred theca similar to Peridiniales in anterior part, similar to Dinophysiales in posterior part. Fossil; sole genus: *Nannoceratopsis*. Jurassic.

Class EBRIOPHYCEAE - nonphotosynthetic, biflagellate, free-living pyrrhophytes, lacking a resistant external covering but having a fossilizable internal siliceous skeleton. Representative genus: *Ebria*. Geologic range: Cretaceous to Holocene.

Class ELLOBIOPHYCEAE - attached parasitic pyrrhophytes without known fossil record.

Class SYNDINIOPHYCEAE - intracellular parasites without known fossil record.

Although it is a worthy objective, a widely accepted classification of fossil dinoflagellates at the family level has yet to be devised. Currently, divergent views on principles and criteria are more evident than is any general agreement on results. A comprehensive classification of fossil cysts that originated conceptually with Eisenack (1961) and was elaborated by Sarjeant and Downie (1966) has now been modified by them (Sarjeant and Downie, 1974; Sarjeant, 1974) and by others (Norris, 1978; Tappan, 1980) into several similar arrangements by which fossil cysts are distributed among about 40 families. While based mostly on cyst morphology, these families are regarded by Norris, at least, as approaching phylogenetically defensible entities. In contrast, Evitt (1975b) contended that a few modern genera collectively encompass the affinities of a majority of fossil cysts. In line with that view, but with modifications reflecting more recent interpretations, Table 1.1 lists 13 families, including nine from the hierarchy of modern taxa, that would appear to accommodate the great majority of fossil cysts (admitting that considerable uncertainty must attach to many fossils with highly "generalized" morphology). However, it is not the intent in this volume to pursue the problems of a phylogenetic classification. Instead, with obvious philosophical allegiance to the second approach mentioned above, we will focus attention in Chapter 8 on 17 morphological categories. While they will be defined without strict regard to family boundaries and will include cysts with "generalized" as well as "distinctive" morphology, their approximate correspondence to the families listed at the left in Table 1.1 is shown at the right.

HISTORICAL NOTE

The first comprehensive synthetic work on living dinoflagellates and the source of this familiar term by which we know the group today was Bütschli's review (1885) of the Dinoflagellata in Bronn's multivolume series: Klassen und Ordnungen des Thierreichs, wissenschaftlich dargestellt in Wort und Bild. Even in the hindsight of a hundred years, this stands as an impressive testimony to the collective observations that had been made on these organisms and to the level of understanding that had been reached regarding them. Some early steps in this progress are cited by Bütschli in an opening historical summary (which is followed by sections on morphology, biology, and systematics). The following are excerpts, with updated information in a few cases:

- –First description of dinoflagellate species (freshwater) and oldest specific names still in use today: O. F. Müller, 1773: *Bursaria hirundinella* (now *Ceratium hirundinella*, and probably the most-studied of all dinoflagellate species) and *Vorticella cincta* (now *Peridinium cinctum*, type of its genus).
- –First description of a marine dinoflagellate: O. F. Müller, 1777: *Cercaria tripos* (now *Ceratium tripos*).
- –First published illustrations of dinoflagellates (of the three preceding species): O. F. Müller, 1786.
- –Oldest generic name still in use today: *Ceratium* Schrank, 1793 (type: *C. tetraceras*, a marine species).
- –First recognition of dinoflagellates as a source of marine bioluminescence: Michaelis, 1830, in the species now known as *Peridinium divergens*.
- –Description of the living genus *Peridinium*: Ehrenberg, 1832.
- –First descriptions of fossil (Cretaceous) dinoflagellates and oldest names for fossil cysts (although not then recognized as cysts) still in use today: Ehrenberg, 1836: *Peridinium delitiense* (now *Spongodinium*) and *P. pyrophorum* (now *Palaeoperidinium*); also two others not then recognized as dinoflagellates but important in the spectrum of fossil cysts, as we will see in the next chapter): *Xanthidium ramosum* (now *Spiniferites)* and *X. pilosum* (now *Sentusidinium* and, along the way, attributed also to six other genera!).
- –Oldest generic name based on fossils (Cretaceous) still in use today: *Spiniferites* Mantell 1850, named by the versatile physician who, a quarter century earlier, had been the author of *Iguanodon*, the first-discovered representative of those other "dinos", the dinosaurs.
- –First observation of the distinctive structure of the dinoflagellate nucleus, of the insertion of the longitudinal flagellum in the sulcus, and of the common occurrence of resting stages: Allman, 1855.
- –First recognition of dinoflagellate-caused "red water" and of the resting stage as a regular part of the dinoflagellate life cycle: Carter, 1858, in a work describing *Peridinium sanguineum* with red bodies in its resting stage.

Table 1.1 - CORRESPONDENCE BETWEEN RECENT AND FOSSIL DINOFLAGELLATE FAMILIES AND MAJOR CYST MORPHOLOGICAL CATEGORIES

Division	Pyrrhophyta	
Class	Dinophyceae	
Order	Peridiniales	
Family	Peridiniaceae and Protoperidiniaceae	P-cysts and Rp-cysts (also calcareous and siliceous cysts); ?Rr-cysts (in part)
	Triadiniaceae	Gq-cysts (in part)
	Gonyaulacaceae	Gq-cysts (in part), Gs-cysts, Gi-cysts (in part), Gv-cysts (in part), Gn-cysts (in part)
	Ceratiaceae	Gc-cysts, ?Gv-cysts (in part),
	Microdiniaceae* (poss. = Cladopyxidaceae)	Gp-cysts (in part), ?Gi-cysts (in part), Gn-cysts (in part)
	Broomeaceae*	Rr-cysts (in part), ?Gn-cysts (in part), ?Gp-cysts (in part), ?Rp-cysts (in part)
	Pyrophacaceae	?Gi-cysts (in part), M-cysts (in part)
Order	Gymnodiniales	
Families	Gymnodiniaceae	D-cysts, ?Gn-cysts (in part), ?S-cysts (in part)
	n. fam. (*Suessia*)*	S-cysts, ?Gn-cysts (in part)
	Polykrikaceae	?Gn-cysts (in part)
Order	Nannoceratopsiales*	
Family	Nannoceratopsiaceae*	N-cysts

*No living representatives have been recognized.

- -Description of the living genus *Gonyaulax*: Diesing, 1866 (type: *G. spinifera*, a species whose cysts are now known to be morphological equivalents of *Spiniferites* Mantell 1850, thus leaving *Gonyaulax* as an unconserved but, nevertheless, widely used junior synonym.
- -First recognition that the transverse furrow houses a distinctively structured flagellum, instead of a band of cilia as had been thought previously: Klebs, 1883.

Thus, 1983 might be regarded as a sort of centennial of the full realization that dinoflagellates are a group of unicellular, biflagellate micro-organisms, unique in their nuclear structure and in the form of one of their flagella; that some of these organisms have cellulosic motile thecae alternating with nonmotile resting stages; that they occur in freshwater and marine environments today, occasionally in concentrations sufficient to color the sea by day and to brighten it by night; and that they also have a fossil record.

SELECTED REFERENCES

The following references are of a general or especially comprehensive nature. Full citations for the works identified here in abbreviated form will be found in the general list of references.

Abé, T. H., 1981, Studies on the family Peridinidae.

Balech, Enrique, 1974, El genero "Protoperidinium" Bergh 1881 ("Peridinium" Ehrenberg, 1831, partim).

Bold, H. C., and Wynne, M. J., 1977, Introduction to the algae. Structure and reproduction.

Bourrelly, Pierre, 1970, Les algues bleues et rouges, Péridiniens.

Brooks, James, and Shaw, G., 1972, Geochemistry of sporopollenin.

Brunel, Jules, 1970, Le phytoplancton de la Baie des Chaleurs.

Bütschli, Otto, 1885, Unterabteilung (Ordnung) Dinoflagellata.

Chatton, Edouard, 1952, Classe des Dinoflagellés ou Péridiniens.

Cox, E. R., 1978, Dinoflagellates.

Dale, Barrie, 1983, Dinoflagellate resting cysts: "benthic plankton."

Dodge, J. D., 1971, Fine structure of the Pyrrhophyta.

Dodge, J. D., 1982, Marine dinoflagellates of the British Isles.

Huber-Pestalozzi, G., 1950, Das Phytoplankton des Süsswassers, Systematik und Biologie, Teil 3, Cryptophyceen, Chloromonadinen, Peridineen.

Kofoid, C. A., and Swezy, Olive, 1921, The free-living unarmored Dinoflagellata.

Lebour, Marie, 1925, The dinoflagellates of northern seas.

Loeblich, A. R., III, 1970, The amphiesma or dinoflagellate cell covering.

Sarjeant, W. A. S., 1974, Fossil and living dinoflagellates.

Schiller, J., 1931-1937, Dinoflagellatae (Peridineae), in monographischer Behandlung.

Sournia, Alain, 1973, Catalogue des éspeces et taxons infraspécifiques de dinoflagellés marins actuels publiés depuis la revision de J. Schiller, I. Dinoflagellés libres (1973); III (1978); IV (1982).

Steidinger, K. A., & Williams, Jean, 1970, Dinoflagellates.

Tappan, Helen, 1980, The paleobiology of plant protists.

Taylor, F. J. R., 1976a, Dinoflagellates from the International Indian Ocean Expedition.

Taylor, F. J. R., 1980, On dinoflagellate evolution.

Taylor, F. J. R. (ed.), in press, The biology of dinoflagellates.

Wall, David, & Dale, Barrie, 1968b, Modern dinoflagellate cysts and the evolution of the Peridiniales.

Wall, David, Dale, Barrie, Lohman, G. P., and Smith, W. K., 1977, The environmental and climatic distribution of dinoflagellate cysts in modern marine sediments from regions in the North and South Atlantic Ocean and adjacent seas.

Chapter 2. Aspects of the Dinoflagellate Fossil Record

An historical review of studies of fossil dinoflagellates opens this chapter, which continues with a discussion of the special selectivity that strongly biases the fossil record in favor of certain taxa. Criteria for distinguishing dinoflagellate cysts from acritarchs are presented next and a summary of the stratigraphic distribution of fossil dinoflagellates follows. The chapter concludes with brief comments on taxonomy, nomenclature, and stratigraphic distribution. Terms for principal chronostratigraphic intervals are given in Table 2.1.

HISTORICAL REVIEW

Sarjeant (1974) has given a particularly comprehensive and clear account of the history of the study of fossil dinoflagellates. Accordingly, the discussion offered here will have a different focus, selectively emphasizing the contributions of a few individuals and recounting the development of ideas about the relationship between dinoflagellate cysts and thecae.

Early Period, Prior to 1930

Until recently, fossil dinoflagellates had two aliases, derived from what we now recognize as extremes of a single morphological continuum. Actually, fossils of both types were first reported nearly 150 years ago by Christian Gottfried Ehrenberg, the German microscopist commonly credited with fathering the discipline of micropaleontology, who found the fossils in thin flakes of Late Cretaceous chert from Silesia. In a succession of works, beginning with two notes (1836, 1837) and ending with his "Mikrogeologie" (1854), Ehrenberg named and figured morphologically varied species of fossil dinoflagellates, although he thought only some of them were dinoflagellates and did not realize any of them were cysts. Four of these species are familiar to most modern students of fossil dinoflagellates under their current binomials: *Palaeoperidinium pyrophorum, Spongodinium delitiense, Spiniferites ramosus*, and *Hystrichosphaeridium tubiferum*. Even with our current impressions of the broad spectrum of morphologies presented by fossil dinoflagellates, one could hardly select from among them any four that would better represent morphological diversity, stratigraphic significance, and interpretive challenge. The first and last of these four species, as well as the genus *Spiniferites*, will receive considerable attention in later sections of this volume.

Some of these fossils had external shapes or polygonal surface patterns so similar to features Ehrenberg had observed in living dinoflagellates that he recognized their affinities easily. Others, which were essentially spinose spheres, were not morphologically similar to familiar dinoflagellates. Ehrenberg attributed them to *Xanthidium*, a modern genus of desmid algae, and this determination, although incorrect, was applied to them as frequently as any other during subsequent decades of rather casual and infrequent study by a handful of natural scientists. During this time, similar objects recorded from Recent plankton were called "ova hispida" by Lohmann (1904) in the belief that they were eggs of planktonic crustaceans, but they were also reported under such noncommital labels as "unidentified marine cysts." Only occasional authors (e.g., Reinsch, 1905) correctly identified them as dinoflagellate cysts.

For virtually a century following Ehrenberg's announcement (1836) to the Royal Academy of Sciences in Berlin that microfossils occur extensively in sedimentary rocks, fossil dinoflagellates and the so-called xanthidians received little attention aside from the occasional description of a new species and periodic mention of their possible affinities.

Mostly Middle Period, 1930 to mid-1950s

A new period of study was inaugurated by Otto Wetzel (1933) with a major contribution that was

Table 2.1 - PRINCIPAL DIVISIONS OF GEOLOGICAL TIME SCALE, LATE TRIASSIC TO RECENT

Era/Period/Subperiod	Epoch	Age		Estimated Age (Ma) for beginning of selected interval*
CENOZOIC ERA				
Quaternary Period	Holocene or Recent			
	Pleistocene			2.0
Tertiary Period: Neogene Subperiod or Late Ter.	Pliocene			5.1
	Miocene			24.6
Tertiary Period: Paleogene Subperiod or Early Ter.	Oligocene			38.0
	Eocene			54.9
	Paleocene			65
MESOZOIC ERA				
Cretaceous Period	Late Cretaceous	Maastrichtian		
		Campanian	Senonian	
		Santonian	Senonian	
		Coniacian	Senonian	
		Turonian		
		Cenomanian		97.5
	Early Cretaceous	Albian		
		Aptian		
		Barremian**	Neocomian	
		Hauterivian	Neocomian	
		Valanginian	Neocomian	
		Berriasian	Neocomian	144
Jurassic Period	Late Jurassic	Tithonian (Portlandian)		
		Kimmeridgian		
		Oxfordian		163
	Middle Jurassic	Callovian		
		Bathonian		
		Bajocian		
		Aalenian		188
	Early Jurassic	Toarcian		
		Pliensbachian		
		Sinemurian		
		Hettangian		213
Triassic Period	Late Triassic	Rhaetian		
		Norian		
		Carnian		231

*ages based on W. B. Harland et al., 1982
**excluded from Neocomian by some authors

also his doctoral dissertation. From Late Cretaceous cherts collected in north Germany and of approximately the same age as the Polish cherts studied by Ehrenberg a century before, Wetzel described numerous examples of organic-walled microfossils, some of them clearly dinoflagellates and others having the morphology of Ehrenberg's fossil "xanthidians". Realizing that the latter were not desmids, Wetzel (1933) placed them in a new genus, *Hystrichosphaera*. From this name was derived "hystrichosphere," the informal term by which fossils of this type became widely known in subsequent years. Like Ehrenberg, Wetzel examined these dinoflagellates and hystrichospheres directly in thin sections and small flakes of chert mechanically chipped from nodules whose ultimate source was bedded Cretaceous chalks; he did not use the chemical extraction techniques that are employed almost universally today.

About simultaneously with Wetzel, two other European scientists also began studying fossil microplankton: Alfred Eisenack, a German, teaching secondary-school science in East Prussia, and Georges Deflandre, a French protozoologist at the University of Paris. To them must go the chief credit for dramatically expanding the horizons of this field during the next 20 years and for demonstrating the broad geographic distribution and potential stratigraphic importance of fossil dinoflagellates.

Eisenack had been forced from pursuing a career in geology by World War I and its aftermath. In his spare time as a science teacher, he began to explore the microfossil content of erratic limestone boulders of Ordovician, Silurian, and Jurassic age that had been quarried by Pleistocene ice sheets from the depths of the Baltic depression and left stranded on the north German plain (Eisenack, 1931, 1935). His technique, though simple, had not been applied extensively before. First, pieces of limestone were etched in hydrochloric acid, and the insoluble residue remaining was washed by allowing it to settle through several rinses of water in a tall cylindrical container. When the ion concentration fell to the point where most of the clay particles remained in suspension after the organic remains had settled, a great concentration of the fossils could be achieved simply by decanting the supernatant fluid. Thereafter, the fossiliferous residue was poured into a Petri dish and inspected under a low-power microscope. Desirable specimens were withdrawn individually in a pipette and transferred to a glass microscope slide, where they were allowed to dry, then mounted permanently by addition of a drop of canada balsam and a coverslip.

In a series of publications during the 1930s, Eisenack described many taxa of Early Paleozoic and Jurassic microfossils recovered from erratic boulders. They included spinose bodies which he later compared with Wetzel's *Hystrichosphaera*, eventually elevating that group to ordinal level as the Hystrichosphaeridea and firmly establishing the informal appellation of hystrichospheres. Also present in the Jurassic materials were fossils of a different kind, readily recognizable as dinoflagellates.

Just before World War II, Eisenack applied the same technique to phosphorite nodules from Early Tertiary amber-bearing strata of the Baltic coast near Königsberg. The promising results were heralded in a brief note (1938) which included the proposal of two new—and now familiar—genera, *Deflandrea* and *Wetzeliella*, named for his colleagues. However, only after the war was over and he had returned from imprisonment in Russia to eventual security in West Germany was Eisenack able to publish (1954) the complete work on this important material. This paper included a section stressing his conviction at that time that the Hystrichosphaeridea constituted a unified group of extinct organisms independent of the dinoflagellates, both groups being well represented by the fossils in the east Baltic materials.

Shortly thereafter began a collaborative association between Eisenack and the Australian palynologist, Isabel C. Cookson, which led to a remarkable series of publications on Cretaceous and Tertiary dinoflagellates and hystrichospheres from Australia and New Guinea. These were truly joint studies—sample preparation and preliminary microscopy in Australia; detailed analysis, description, and joint consultation in Tübingen; photography in part by each author; publication in Australian journals except for two which were completed and published in Germany after Cookson's death (the second also after Eisenack's). Spanning nearly 25 years (Cookson and Eisenack, 1958-1982), these works constitute an exceptional documentation of the richness of sporopollenin microplankton preserved in late Mesozoic and Tertiary deposits (Helby and Partridge, 1980). They also document gradual recognition of the dinoflagellate affinity of many "hystrichospheres."

Eisenack's considerable contributions also include his own studies on German Cretaceous assemblages (e.g., 1958), review articles on dinoflagellates (1959, in which confirmed fossil dinoflagellates were attributed to a total of 26 genera) and hystrichospheres (1963), an important catalog of genera and species (1964 and later), and investigations of Paleozoic samples (which continued beyond his 90th year). His influence was further extended through the work of students whom he encouraged (e.g., Klement, 1960; Alberti, 1961; Gerlach, 1961; Gocht, 1970).

Though Eisenack's methods remained simple, his observational power was acute and his standard of quality exemplary. Even the finest of modern photomicroscopes has seldom achieved the level of photographic excellence evidenced in his publications, yet he accomplished this with a camera of exceptionally simple design: a 35mm camera-back mounted on a tin can (for a bellows) which could replace another can of identical size provided with a ground glass focusing window, each resting on a can lid fixed on the tube of a vintage-model monocular microscope. His experience enabled him to judge by eye the desirable length of exposure, which he then controlled by switching on and off the microscope lamp.

Deflandre's early background (Evitt, 1975c) was somewhat less remote from micropaleontology than Eisenack's, although he had no formal training in geology. When his first paper on dinoflagellates appeared (1933, on siliceous forms from New Zealand Eocene diatomites), he was already an experienced systematist and had to his credit an impressive bibliography of publications on protozoology and phycology which included descriptions of many new taxa. While his work after that date was concentrated on microfossils, he directed his attention not only to organic-walled microplankton but also to calcareous nannofossils and siliceous radiolarians, ebridians, silicoflagellates, and archaeomonads. Although his most persistent interest lay with fossil radiolarians, his contributions to the other groups mentioned were as fundamental as to the one which primarily concerns us here.

His curiosity having been aroused by the remarkable fossils from north German cherts, Deflandre began to investigate cherts of similar character and age derived from the Late Cretaceous chalks that ring the Paris Basin. His technique, like Wetzel's, was simple and mechanical, not dependent on the hydrofluoric acid which most investigators use today for liberating sporopollenin fossils from a siliceous matrix. By deft blows of a hammer Deflandre flaked bits of chert from a nodule, then examined them under xylene to check for microfossils where the chips were sufficiently thin to be essentially transparent. Any flake that proved to be fossiliferous was then trimmed further, if necessary, and mounted without coverslip in a drop of canada balsam on a microscope slide. At this point specimens were ready for microscopic examination under a high-power dry objective (but with the chert surface oiled to reduce interference from the unpolished surface) or, in later years, under a X40-power oil-immersion objective. As might be imagined, it was only rarely that fossils so prepared were either ideally oriented or lay close enough to the surface of the chert chip to be accessible with a high-power immersion objective.

Given the impediment to high-magnification photomicrography which these conditions presented, Deflandre's considerable artistic ability played a large role in the success of his study. His illustrations are a monument to great skill in observing and faithfully recording the features of strange new objects, especially in view of the quality of the optical equipment with which he worked and the relatively meager knowledge of dinoflagellate morphology available at the time when he wrote his early papers. Only in one regard must some of his drawings be accepted cautiously: not knowing that dinoflagellate cysts usually have archeopyles, he often failed to show the subtle traces of archeopyle sutures and opercula, and, on occasion, he even "reconstructed" the missing parts of specimens which he thought had been damaged but which actually had open archeopyles.

Deflandre's major work on the Cretaceous cherts appeared in three papers; one on general biological considerations relating to the preservation of microfossils in cherts (1935), the second on dinoflagellates (1936), and the third mainly on hystrichospheres (1937). Thus, he, too, perceived a dichotomy between these two sets of fossils. Later, however, in a paper specifically addressing the "problem" of the hystrichospheres (1947), he pointed out the dinoflagellate-like aspects of some of them and concluded that the group was probably a heterogeneous mixture of the remains of various kinds of organisms in various life stages.

Although he occasionally returned to the subject of dinoflagellates from Cretaceous cherts (Deflandre and Courteville, 1939), Deflandre's subsequent systematic works touched on a variety of topics related to fossil dinoflagellates and other microplankton: Jurassic dinoflagellates and hystrichospheres (e.g., 1938, 1941), Paleozoic hystrichospheres (1945), and calcareous dinoflagellates (1948). They also include important summary articles, such as contributions to major synthetic treatises (1952a, 1952b). Like Eisenack, he contributed to the organization of knowledge in his field, publishing, over the years and with major participation by his wife, many series of fiches (1943-1966) abstracting the published systematic literature, not only on dinoflagellates, but on other groups of microfossils as well. His last extensive work on dinoflagellates, under joint authorship with Isabel C. Cookson (Deflandre and Cookson, 1955), was the paper that alerted the micropaleontological world to the wealth of organic-walled microplankton preserved in Australasian Mesozoic and Tertiary sediments (Helby and Partridge, 1980).

The "story of the cherts" would not be complete without mention of contributions by two other persons, although one of them is a member of the contemporary scene and, therefore, anachronous to this section. These are Maria Lejeune-Carpentier and Jean-Claude Foucher, respectively Belgian and French micropaleontologists.

Over a period of 15 years, Lejeune-Carpentier (1936a, as Lejeune, to 1951) published a series of "notes" in which she reported, in meticulous detail quite extraordinary for the time, her observations on selected species of dinoflagellates and hystrichospheres. The specimens she studied came from Belgian Late Cretaceous cherts similar to those that were the raw materials for both Wetzel and Deflandre, and she, too, examined chert flakes. As a supplement to the descriptive taxonomy of Wetzel and Deflandre, her careful and perceptive interpretations have been an important aid and inspiration to more recent investigators of dinoflagellate cyst morphology.

Foucher is one of a large number of more recent workers, far too numerous for individual mention, who are continuing and expanding the study of fossil dinoflagellate cysts pioneered before 1950. Under Deflandre's direction he completed a thesis which, to a large extent, made up for a major deficiency in Deflandre's own earlier work on assemblages from the cherts. At the time of his original studies, Deflandre, the biologist who had turned to paleontology, had no appreciation for the value of accurate information on the stratigraphic position of the fossils he described. His chert nodules had not come from areas in which stratigraphic relationships were known, or even from outcropping sequences, but from such convenient collecting sites as walks and driveways in and around Paris, which had been paved with this readily available and resistant material. Thus, the "type localities" for many of the species he described include the gravel-paved paths of the Tuilleries and the driveway to Wanda Landowska's villa traveled en route to a musical soiree. In most cases it was reasonable to date these specimens as Senonian, but to call them, say, Coniacian would have been purely speculative, and to cite an ultimate outcropping source was quite impossible. Through meticulous studies of microfossils in thin sections of chert nodules from well-dated strata, Foucher largely filled the critical gap in stratigraphic information present in Deflandre's earlier work (see Foucher, 1979, for summary of stratigraphic ranges and references to documenting articles). By this means, he provided an even more basic solution to the problem than the partial answer offered somewhat earlier by studies of assemblages from equivalent strata in nearby areas (Clarke and Verdier, 1967).

Ideas and Observations—late 1950s to the Present

After the pioneering works of O. Wetzel, Eisenack, and Deflandre, and the hiatus that resulted from World War II, the serious study of fossil dinoflagellates began again in the early 1950s (W. Wetzel, 1952; Klumpp, 1953; Eisenack, 1954). But soon a new influence became important. The study of sporopollenin microfossils had become a competitively important technique in the exploration for petroleum, and dinoflagellates were soon included among the "tools of the trade". Either directly or indirectly, the application of fossil dinoflagellates to petroleum exploration has provided the major impetus for their study over the last 25 years.

The consequence of this industrial influence has been manifold. There has been a great increase in the number of researchers involved with fossil dinoflagellates and other organic-walled microplankton. This has led, in turn, to vast amounts of new information on them, embodied in hundreds of publications devoted to the description of new taxa and determination of their stratigraphic and geographic distribution. Preparation procedures have been greatly improved and study techniques revised. Paleontologists have even realized the possible relevance of studies in the biology and ecology of living dinoflagellates to their own work!

This great burst of activity in the field makes it quite impractical to review here in any comprehensive way the contributions made by individual authors. But a further consequence of this activity that we should examine is the revolution in ideas about the nature of fossil dinoflagellates that has also taken place.

As the pace of studies on fossil sporopollenin microplankton quickened in the late 1950s, new observations began to provide evidence for two ideas, neither totally novel, but neither previously susceptible to such extensive testing as was now possible. One was that the fossil record of dinoflagellates included fossil cysts, not just fossil thecae. The other was that many of the so-called hystrichospheres were, in fact, also dinoflagellate cysts. Critical to both these ideas were two discoveries. First, many fossils—both typical hystrichospheres and "obvious" dinoflagellates—were found to possess an opening, the archeopyle, which we now know to be an excystment aperture. This opening was observed to be normally operculate, to have a constant shape and location in any one species, and to exhibit a moderate variability from species to species. Second, the distal extremities of spines, processes, or other projections on the exterior surface of many of the fossils—again, in hystrichospheres and dinoflagellates alike—had shapes suggesting that they formed in contact with some surrounding surface. Further important observations were that the walls of the fossils, unlike modern dinoflagellate thecae, are not divisible into discrete plates and that the features long identified as cingulum and sulcus often are not furrows in which flagella could have functioned effectively.

These observations led to publication (Evitt, 1961) of the outline of a theory that many hystrichospheres are fossil dinoflagellates and that many fossil dinoflagellates represent the encysted rather than the motile stage in the life cycle. Following initial presentation earlier (Evitt, 1958, orally to field trip participants, Paleobotanical Section, American Institute of Biological Sciences annual meeting), these ideas were developed further during a period of study and consultation with Eisenack and Deflandre in 1959. Since the genus *Hystrichosphaera* (from which "hystrichosphere" had been derived) was among those fossils now interpreted as dinoflagellate cysts, a new name—"acritarchs"—was proposed (Evitt, 1963) for those hystrichospheres, mostly from the Paleozoic, that do not possess a minimum of features considered diagnostic of dinoflagellates.

The most significant modification of these ideas since 1961 has involved the recognition that most, if not all, fossil dinoflagellates are cysts, including even those with the most theca-like aspect. As will be seen in Chapter 9 (under *Palaeoperidinium pyrophorum*) and in Chapter 12 (under *Dinogymnium*), some further qualifications on this point now seem appropriate.

The theory presented in 1961 received vigorous support from Deflandre and others, but it appeared to have a major shortcoming. While it was still fair at that date to include fossil dinoflagellates among the "little-known" microfossils, modern dinoflagellates were a group long familiar to neontologists, and yet, observations that could be cited in support of the theory were not a part of the "working knowledge" of recently or currently active specialists. In fact, the archeopyle had been so named because it was thought to be restricted to fossil dinoflagellates. Furthermore, hystrichospheres were widely held to be extinct, despite reports of similar bodies from Recent sediments by McKee et al. (1959), observations of spiny cysts in Scandinavian waters (Braarud, 1945; Nordli, 1951; Erdtman, 1950, 1954), and significant, but forgotten, much earlier observations on spiny cysts in modern dinoflagellate species (e.g., Schütt, 1887, 1895).

This perceived deficiency of evidence from modern dinoflagellates was shortly eliminated, not by rediscovering reports of earlier observations, but by new observations made nearly simultaneously on samples of modern marine microplankton in two widely separated laboratories. Evitt and Davidson (1964), working at Stanford, found cysts that would qualify as "obvious dinoflagellates" (i.e., with peridinioid outline and traces of cingulum, sulcus, and plate patterns) both within thecae of a species of *Protoperidinium* and as isolated specimens. They also extracted from thecae of *Gonyaulax digitalis* and *G. polyedra* cysts which had the morphological properties of typical hystrichospheres. The cysts of all three species were equipped with archeopyles similar to those known from the fossil record. Furthermore, the cyst walls withstood severe acetolysis treatment, presumably indicating potential preservability as fossils. At almost the same time, Wall (1965) at Woods Hole, in the course of an approach that closely approximated "experimental paleontology," watched a living specimen of *Hystrichosphera* as it excysted. He observed and photographed the cyst as the archeopyle "popped open" and the protoplast then squeezed through it, thereafter developing into a typical motile thecate cell of *Gonyaulax spinifera*.

These observations made by paleontologists on living dinoflagellates effectively confirmed the explanation of the nature of the dinoflagellate fossil record that had been theorized a few years earlier. Although reported at the time as new discoveries, it is now apparent that some of them were renewals of observations made originally many years before (summarized in Tappan, 1980). Nevertheless, it now became widely recognized for the first time that objects morphologically and compositionally like fossil dinoflagellates are produced in the modern biota and that these objects are resting cysts. The era during which studies of fossil and living dinoflagellates had followed almost entirely independent paths was near its end, and many investigators during the next decade and a half contributed to knowledge on both fronts. For example, Wall and Dale (1967) established that many kinds of dinoflagellate cysts are produced in the sea today, that they are basically comparable to fossil dinoflagellates, and that they can be identified by inducing them to excyst and to develop into the thecate stages on which the taxonomy of living species has been based. These investigators also accumulated extensive information on ecological aspects of cyst distribution (Wall et al., 1977), greatly extending earlier observations on the subject (e.g., D. B. Williams and Sarjeant, 1967; D. B. Williams, 1971a, 1971b; Davey, 1971).

Von Stosch, after conclusively demonstrating syngamy in marine and freshwater *Ceratium* (1964, 1965), proceeded to further illuminate the role of sexual reproduction in the life cycle (1972, 1973). Pfiester (1975 and later) showed the interrelationship between zygotic thecae and resting cysts in several species whose preservable cysts are morphologically similar to many fossils. Peculiarities of the pattern of thecal growth visible directly in living species (Dürr and Netzel, 1974) were shown to have counterparts in previously misinterpreted markings on fossil cysts (Gocht, 1979).

In morphological analyses of thecae and cysts, Evitt and Wall (1968) threw new light on the relationship between tabulation and paratabulation in *Peridinium limbatum* and confirmed (Wall and Evitt, 1975) the suspected general affinity of a group of Cretaceous cysts to *Ceratium*.

In 1978 a Penrose Conference sponsored by the Geological Society of America (Evitt et al., 1979) brought together for the first time substantial numbers of both neontologists and paleontologists concerned with dinoflagellates. Marked by reviews of separate problems and discovery of common opportunities, this conference may come to be regarded in retrospect as a rather significant event in the study of dinoflagellate cysts. It also defines a convenient point at which to close this review.

SELECTIVITY OF THE DINOFLAGELLATE FOSSIL RECORD

The fossil record of any group of organisms presents the paleontologist with highly selected data that have been processed by a complicated interplay of many factors. Quantitatively, the record is only an infinitesimal sampling of the individuals that once lived. Both quantitatively and qualitatively, it is strongly biased with respect to certain organisms by circumstances of habit and habitat during life; by the size, shape, and composition of their preservable parts; by the nature of the enclosing sediment; by the postdepositional history of the site of deposition; even by the matter of chance that determines the discovery and study of particular fossils. The constraints on paleontological interpretations imposed by these factors apply to dinoflagellates no less than to other types of fossils, and their significance is widely appreciated. But the dinoflagellate fossil record embodies a further and as yet little appreciated selectivity that has fundamentally important implications for the perceived geologic history and evolution of the group. This selectivity is a consequence of the fact that the sporopollenin structure on which the fossil record of dinoflagellates largely depends is an adaptive feature peculiar to a certain stage in the life cycle of only certain species. That not all species produced cysts has profound consequences for palynologists and for the conclusions they and others draw from their observations.

Four generalizations derived from collective observations on living and fossil dinoflagellates have bearing on this selectivity:

1. Biological considerations imply that dinoflagellates were present throughout the Phanerozoic and probably much earlier, but from the rock record of much of this time we have no recognized dinoflagellate fossils. —-The primitiveness of dinoflagellates, as reviewed in Chapter 1, makes it highly probable that dinoflagellates evolved long before the metazoans appeared about 700 million years ago. That we see no record of their early existence obviously has two possible explanations: either no traces of them have been preserved, or we have not recognized them as dinoflagellates. The second possibility is dealt with later in this chapter under the discussion of dinoflagellates *versus* acritarchs.

2. Production of preservable resting cysts is a limited phenomenon among dinoflagellates. —-Despite recognition that sexual reproduction is probably a normal part of the life cycle of dinoflagellates, only in a minority of species living today is the cyst enclosed in a wall of geologically resistant material. In the majority, either the sexual phase of the cycle is completed without formation of a resting cyst or the encysted cell lacks a geologically preservable container. In other words, only a relatively few living species produce fossilizable remains. Since there is no reason to believe that a fundamentally different situation existed in the geologic past, we must conclude that many dinoflagellate species lived without leaving fossilizable traces.

3. Production of preservable resting cysts is a quantitatively variable phenomenon. —-Although we do not know precisely what stimuli lead to encystment or what controls determine the number of cysts produced, numerous observations on modern plankton and bottom sediments show that, in any given area, no simple or constant relationship exists between the number of resting cysts, on the one hand, and either the number of actively metabolizing vegetative cells or the number of cyst-producing species, on the other. Extrapolated to the fossil record, this implies that, for those dinoflagellate species that did produce preservable resting cysts, the relationship between number of cysts, number of vegetative cells, and number of cyst-producing species varied from time to time and place to place. If that was the case, nothing in the preserved record

of dinoflagellates can now reveal to us the changing quantitative values of this relationship. We cannot determine numbers of motile cells from counts of cysts.

4. Even within a single reasonably low-level taxon—say, a genus or family—cyst production may be characteristic during one interval of geologic time, but not another. —-This important conclusion cannot be based on observation of the living biota alone, which amounts to drawing samples from a single time plane, but must also take into account evidence that comes from the fossil record. The following three examples germane to this point involve the genera *Ceratium*, *Arpylorus*, and *Protoperidinium*.

Ceratium, one of the major genera of modern peridinialean dinoflagellates, comprises a large number of species most easily recognized generically by their distinctive outline, with three or four prominent horns. Among the hundred or more living species of the genus, not one is known to form a preservable resting cyst. Presumably, therefore, remains of these species are not fossilizable. However, the fossil record from the latest Jurassic to the latest Cretaceous contains an assortment of cysts with strikingly *Ceratium*-like features (Wall and Evitt, 1975), of which about 35 species have been described. No fossil cysts similar to any of these Cretaceous examples have been recognized as indigenous elements in any Tertiary sediments. This means that about 70 million years of Cenozoic time separate the occurrence of the youngest fossil ceratioid cysts from the abundant thecate but nonfossilizable representatives of *Ceratium* today. In other words, the fossil record would have us believe that ceratioid dinoflagellates became extinct at the end of the Cretaceous, a conclusion obviously at odds with the facts.

The Late Silurian species *Arpylorus antiquus* provides further evidence of cyst formation during only a limited interval of geologic time. Alone in all the Paleozoic, *Arpylorus* appears to this author to be a very dinoflagellate-like dinoflagellate—so much so in fact, that, were it to be found in a Mesozoic assemblage, it might attract no more attention than any other distinctive species. Unlike other Paleozoic microfossils discussed later in this chapter that may also represent dinoflagellate cysts but lack a minimum of features that would establish their affinity with relative certainty, *A. antiquus* was described (Calandra, 1964) as a dinoflagellate because it looks like one. Restudy of the type material from subsurface Algeria by Evitt (1967) and Sarjeant (1978) led these authors to reaffirm the basic identification, although the original material is not ideal, and their interpretations of it are not identical. However, the dinoflagellate nature of this fossil is not unquestioned. Bujak and Williams (1981) and Bujak and Davies (1983) have urged an open mind on its identification and suggested it may not be a dinoflagellate. Beyond that, they discount its bearing on the matter of a selective dinoflagellate fossil record. It is clear that one occurrence of such a potentially important fossil, consisting of less than perfectly preserved specimens, is insufficient to resolve the matter satisfactorily. The need is for new material, including better, or at least differently, preserved specimens and coming from another area, which will enable the characters of this fossil species to be determined afresh.

But it is not the recovery of a dinoflagellate from Silurian strata that is surprising, for we have already considered the biological reasons to believe that dinoflagellates were probably present in the Precambrian. What is spectacular in this case is the absence of fossil dinoflagellates from younger Paleozoic strata. After the Silurian, there are no other fossils definitely identifiable as dinoflagellates for about 200 million years, through all the rest of the Paleozoic and part of the Triassic. This is a span of time approximately equal to the entire subsequent and essentially continuous fossil record of dinoflagellates from the Carnian to the present. In light of what we now know about the production of preservable cysts among modern dinoflagellates, we can probably best regard *A. antiquus* as an especially "precocious" species, which carried out a successful early experiment in sporopollenin production long before that technique really "caught on" as the "fashionable" dinoflagellate thing to do in the early Mesozoic.

The living genus *Protoperidinium* and its fossil relatives provide somewhat more complex evidence regarding the selectivity of the dinoflagellate fossil record. As we will have occasion to explore later, the many species included in *Protoperidinium* today show certain differences in the tabulation on the dorsal and ventral surfaces of the epitheca. In each area three distinct patterns are recognized; those on the ventral surface are termed ortho, meta, and para, and those on the dorsal surface quadra, penta, and hexa. The meanings of the terms need not concern us here (see Chapter 4). Suffice it to say that 189 species of *Protoperidinium* recently reviewed by Balech (1974) and found to have epithecal tabulations determinable in both the critical areas showed the distribution of patterns summarized in Table 2.2.

Table 2.2 - THECAL PATTERNS IN *PROTOPERIDINIUM* The matrix shows percentages of nine possible combinations of three dorsal and three ventral patterns among 189 species of living *Protoperidinium* (based on Balech, 1974). Single and double underscoring identify, respectively, the less and more important of the two combinations that dominate the fossil record; in addition, meta hexa and para hexa patterns are known among fossils from one species, and one group of allied species, respectively.

Ventral pattern	Dorsal Pattern: quadra	penta	hexa	Total
ortho	9.5	4.8	12.2	26.5
meta	23.3	18.0	12.7	54.0
para	4.2	1.1	14.3	19.6
Total	37.0	23.9	39.2	100.0

Now, as already pointed out, cysts which appear to have formed within *Peridinium*- or *Protoperidinium*-like thecae are abundant in the fossil record. For many of them, but by no means all, the dorsal and ventral epithecal tabulation can be inferred with reasonable confidence from the paratabulation. A comparison of patterns among these fossil and Recent dinoflagellates shows that:

a. Among the living species, 74% have a meta or para ventral arrangement, as opposed to 26% ortho. However, the ortho style is the overwhelmingly dominant ventral pattern in the fossil record where the meta style is exceedingly rare (Edwards and Bebout, 1981), and the para style is represented by one closely knit group of Paleogene species, the *Wetzeliella* complex (Goodman, 1983).

b. Of living species with an ortho ventral tabulation, 61% show a quadra or penta dorsal arrangement. However, the ortho quadra combination is unknown among fossils (although, until recently, it was thought to characterize the *Wetzeliella* complex), and the ortho penta combination is apparently limited to a few Quaternary to Holocene sporopollenin cysts (Harland, 1982) and to some siliceous cysts of Eocene to Miocene age (Lefèvre, 1933; Dale, 1978).

c. The ortho hexa combination is by far the dominant pattern in the fossil record, but only 12% of the living species in Balech's list show this combination.

Certainly, these relationships could be explained by normal evolutionary change during the long history of this group. We might easily imagine, for example, that living meta and para types evolved relatively recently from an older ortho hexa ancestral stock. But the distributions of the ortho quadra and ortho penta groups are not so easily explained. Although both are represented in the early *and* late Cenozoic, similar fossils are conspicuously missing in between. Only an iterative evolutionary process that gave rise to these morphologies once in the Early Tertiary and a second time later would explain the observed distribution. While such an explanation is conceivable, variation in cyst-producing capabilities within the peridinioid group seems a far simpler explanation and one quite consistent with the examples of *Ceratium* and *Arpylorus*, and with observations on cyst production by living species today.

The impression that emerges from all these considerations is that formation of geologically preservable cysts has been and continues to be an inconstant feature of dinoflagellate life cycles. Two generalizations seem justified. First, some dinoflagellates have produced fossilizable remains while many have not. Second, those groups that have produced fossilizable cysts have done so through intervals of geologic time which vary in length and contrast in continuity, according to the taxa.

Although these conclusions in themselves mark some advance in our knowledge of dinoflagellates, their implications, which are less than immediately obvious, are of greatest importance. They can be summarized as follows: When preservable dinoflagellate cysts are recovered from a sediment, we can conclude that (barring redeposition, stratigraphic leakage, or contamination) motile cells of at least the species represented by those cysts were around when the sediment was deposited (although we cannot guess what other species may have been represented by motile cells or nonpreservable cysts). But, what does the *absence* of preservable cysts mean? This is less easily interpreted than the absence of clams, or brachiopods, or foraminifera. In those cases, providing subsequent removal by solution or other processes can be ruled out, we can often be confident that no such organisms lived in the depositional area of the sediment under study. However, when no fossil dinoflagellates turn up, we have an important alternative to the possibility that no living dinoflagellates were there; namely, that only dinoflagellates that failed to produce preservable remains were there. If extrapolation from observations on modern species is correct in this regard, this was the prevalent situation. That conclusion has some interesting ramifications which we will pursue in the following paragraphs. They pertain to five general topics:

1. Determining population size and assemblage composition.

2. Determining the evolutionary succession of characters or taxa.

3. Judging the role of dinoflagellates in their ecological setting.

4. Estimating rates of speciation and extinction.

5. Qualifying simple statements about the dinoflagellate fossil record.

First. Paleontologists are accustomed to estimating assemblage size and population density by counting numbers of species and individuals. But, from counts of the number of fossil dinoflagellate cysts, one can never arrive at the composition of an assemblage in terms of dinoflagellate species and one cannot estimate the population density in terms of numbers of living individuals per unit of sediment. All that one can determine is the number of species or individuals of dinoflagellate cysts. This is a very different matter, with peculiarities that require special understanding to be appreciated. Lists of dinoflagellate species look like any other species lists. That they tell a different story from the usual one is in no way apparent by inspection. Conceivably 10 species of fossil cysts could represent 20, 50, or 100 dinoflagellate species, and 10 individual cysts could represent 100, 1,000, or 10,000 motile cells—and each at different times or in different samples For example, Dale (1976) found only 26 cyst types in sediments of Trondheimfjord where 57 motile dinoflagellate species had been reported. He also reported that for another locality (Woods Hole) the ratio of cysts to thecae in the plankton at the time of encystment ranged, according to the species, from 1:2 to 1:500 Similarly, Pollingher (1981) found that only one percent of individuals in the population of *Peridinium cinctum* f. *westii* in Lake Kinneret encyst.

Dale (1976) also pointed out the varied susceptibility of cysts from different species to damage or destruction during the chemical and mechanical processing used to recover them from sediments.

Although comparisons based on counts of individuals and species of dinoflagellate cysts may be useful in studies of local sections over short intervals of time, the limitations must be borne carefully in mind. A higher number of species of dinoflagellate cysts may actually be associated with a lower number of total dinoflagellate species. At the present time, the tropical oceanic environment contains relatively many species but almost none of them are cyst-producers; whereas, relatively shallow waters at higher latitude contain fewer species but many more cyst-producers (Dale, 1976; Wall et al., 1977).

Second. It is fundamental to most kinds of paleontology that the progressive appearance of distinctive fossils at successively higher levels in the stratigraphic column records the sequence in which they evolved. This relationship serves as the chief guide to theories of phylogeny within particular groups of organisms, and imposes a considerable constraint on the fertile imaginations of would-be phylogenists. But, the temporal sequence in which distinctive morphological features appear among fossil dinoflagellates—and this includes the diversity of highly useful and taxonomically important paratabulation patterns—cannot be accepted as an accurate record of the evolutionary sequence in which these features developed. Thus, it cannot serve either as a reliable guide to phylogenetic ruminations or as an effective constraint upon them. The earliest appearance of a dinoflagellate character in the fossil record establishes only the latest date by which it had been developed. The much earlier presence of the same features may be unrecorded only because they happened to occur in a non-cyst-producing species that left no detectable remains. Similarly, the seeming "disappearance" of a particular feature from dinoflagellates in stratigraphically younger samples does not necessarily mean that the feature was absent from contemporary or younger species. It may only be that such species, however numerous they may have been, did not produce preservable cysts.

Of course, it is always possible that the first (or last) occurrence of a feature in the dinoflagellate fossil record may accurately record its first (or last) occurrence in dinoflagellates at large, but this will never be certain. Therefore, any theory based on the mistaken assumption that the record can be taken at face value has a built-in uncertainty factor that should not be overlooked.

The modern biologist seeking evidence of phylogenetic relations among organisms living today is handicapped by having to rely on data collected from a single time plane. He must estimate primitive and advanced states inductively through comparative observations and theory. Usually the paleontologist is considerably more fortunate. Knowing the ages of his specimens, he can construct phylogenies with some confidence that observed traits appeared, changed, and disappeared in a particular succession. But, in regard to dinoflagellate evolution, the paleontologist's advantage over the neobiologist all but disappears. He *does* work with fossils from different time planes, but his ability to determine the temporal sequence in which the features *displayed* by those fossils really appeared is confounded by the missing record in a way that affects few other groups of fossils.

Third. Conclusions about the role played by fossil dinoflagellates in their environments must be drawn cautiously. Selectivity of the record need not bring into question ecological interpretations covering short intervals of space and time, for these will be based on individual species and assemblages. However, "megathinking" with dinoflagellates is another matter. For example, fossil cysts cannot give reliable evidence on the role of dinoflagellates as primary producers in the geologic past (compare Tappan and Loeblich, 1973). Nor do changes in observed numbers of fossil cysts have any validity as evidence for amounts of photosynthetic activity and consequent changes in levels of atmospheric oxygen or carbon dioxide, which have been speculated to play a role in major events of speciation and extinction. Care should also be exercised in using relative abundances of fossil dinoflagellates for interpreting oxygen minimum (depletion) zones, areas of upwelling, or other characteristics of ancient water masses.

Fourth. As a few authors like Bujak and Williams (1979) have noted, rates of speciation and extinction based on fossil dinoflagellate cysts pertain to cyst morphotypes only (compare Tappan, 1971, 1980; Tappan and Loeblich, 1971, 1972). We have no basis for supposing that these rates have a fixed relation to the rate of change in dinoflagellate genotypes. One cannot reliably estimate either rates of evolution in dinoflagellates at large, or the "standing crop" of dinoflagellate taxa at any one time, from counts of cyst taxa as they appear and disappear through the stratigraphic column. Since such counts have no decipherable relation to the numbers of dinoflagellate genera or species that actually lived, they are also useless for comparison with evolutionary patterns in other groups of organisms. Since the ratio of preserved cysts to actively metabolizing motile cells cannot be determined from the fossil record, counts of fossil specimens provide no basis for judging either the direct contribution of dinoflagellates to the energy cycle of the evolving ecosystem or their indirect contribution to the production of free atmospheric oxygen. Qualitative judgments about the kinds of evolutionary change in dinoflagellate cysts may be realistic, especially when they involve short lineages at the species level (e.g. Eaton, 1971; Habib, 1973; Fensome, 1981), but quantitative estimates in answer to such questions as "How many taxa were part of the lineage and at what times?" are fatuous.

Fifth and finally. Many simple and direct statements about fossil dinoflagellates—even when made by knowledgeable authors—require insertion of qualifying phrases if they are to avoid grossly misleading readers who are not specialists in the field. For example, none of the following statements can be justified by evidence from the fossil record, although one might expect to read any of them in current literature:

-Dinoflagellates were rare in the Paleozoic.

-Dinoflagellates of *Ceratium* type were not present during the Tertiary.

-The dominant Tertiary peridinioid dinoflagellates were ones with ortho hexa tabulation.

-The fossil record shows a great increase in diversity of dinoflagellates during the Mesozoic and early Tertiary, followed by a general decrease during the Neogene.

Even if each statement were edited into technical accuracy by specifying "dinoflagellate cyst" in place of "dinoflagellate," the message perceived by the unwary reader is likely to be grossly incorrect. The high potential for misunderstanding indicates the need for meticulous care in phraseology and emphasis.

In sum: the success of dinoflagellates and their role in the world around them cannot be assessed from the record of fossil cysts. At one theoretical extreme, all cyst-producing species could be extinguished and dinoflagellates could still be a thriving group, but without a fossil record. At the other extreme, one or more cyst morphotypes could exist for every dinoflagellate genotype. Nothing in the fossil record can tell us what position between these extremes was represented in the dinoflagellate biota at any one time. Only at the moment of today is it *theoretically* possible to make such a determination, through a census of all living species. But, even were we able to complete that census, we could not then judge whether it is typical of times in the past or quite unique.

This recital of many things that cannot be told about or done with fossil dinoflagellates is not meant to be discouraging. Knowing what we now know does not diminish their value and still unrealized potential in palynology. Nor does it necessarily bring into question the reliability of small vistas on evolution at the level of species and genera that will appear as fossil assemblages are followed carefully through local stratigraphic successions. Already exemplifying such studies are those on *Areosphaeridium* (Eaton, 1971), *Druggidium* (Habib, 1973), *Energlynia/ Wanaea* (Fensome, 1981), and several on taxa in the Protoperidiniaceae as that family was defined by Bujak and Davies (1983). The essential point is to realize that, when interpreting a group as unusual as the dinoflagellates, we need to exercise unusual caution. Otherwise, one would conclude from the disappearance of fossil ceratioid dinoflagellates at the close of the Cretaceous that *Ceratium* is today extinct!

CYSTS FROM FRESHWATER DEPOSITS

Reliably identified freshwater deposits are relatively uncommon in the Mesozoic-Cenozoic stratigraphic record, and rather few fossil dinoflagellates have been described from them. Churchill and Sarjeant (1963) reviewed the history of their study, and Bradford (1978a) gave extensive references to Pleistocene and Holocene records from sediments. Pre-Pleistocene freshwater cysts were described from the Eocene by Krutzsch (1962) and Harland and Sharp (1980), from the Oligocene by Evitt (1974), and from the Miocene by Engelhardt (1976). None are known with certainty from Mesozoic deposits.

Sporopollenin cysts are common in many lake sediments today. Chiefly, they represent species of *Peridinium* (e.g., Evitt and Wall, 1968; Boltovskoy, 1979; and other works cited by those authors), but cysts presumed to represent *Gonyaulax apiculata* are also known (WRE, pers. obs.). Cysts of freshwater *Ceratium*, while widely reported (e.g., Entz, 1925; Chapman, et al., 1982; Happach-Kasan, 1982), are insufficiently resistant to enter the geologic record.

FOSSIL DINOFLAGELLATES VERSUS *ACRITARCHS*

When the morphology of a living specimen suggests that it is a dinoflagellate cyst, at least the possibility exists to test the suspicion by culturing the specimen or examining the cell contents for critical properties. However, identification of empty cysts, modern or fossil, must rest on morphological criteria alone. Comparison of modern cysts with one another quickly teaches us an important lesson in this regard: some have more dinoflagellate-like properties than others, and some have no morphological properties diagnostic of dinoflagellates at all. That is to say, there is potential overlap in size, composition, and form between nondescript dinoflagellate cysts and the fossils that have been called acritarchs (Evitt, 1963; Downie et al., 1964). While the boundary between the two groups is a matter of some practical importance, the criteria for drawing it are not objectively determinable and must be set by agreement. Since calling a specimen an acritarch is to point out that its biological affinity is unknown, the critical question is: What minimum set of particular characteristics will be required of a specimen to justify calling it a dinoflagellate? Of the many possible answers, the one that will guide us here is the following: To be considered a dinoflagellate, a fossil specimen having appropriate size, shape, and composition should also show either traces of features having the position, extent, and mutual relationships of a cingulum and sulcus, or traces of a pattern of polygons consistent with a thecal tabulation, or both. The same requirement should be made of modern specimens that cannot confidently be linked to identifiable motile or thecate cells. As will become clear in Part II, the traces referred to may be quite incomplete and they may take various forms, some of them quite unlike the thecal features referred to. Therefore, it is pattern, position, extent, and mutual relationship—not form—that are important.

These criteria, even if widely accepted, will not distinguish between dinoflagellates and nondinoflagellates, but only between objects that can be identified as dinoflagellates with a certain degree of confidence and those that cannot. Even cysts of some living dinoflagellates "fail" the test! For example, the cyst of the toxic species *Protogonyaulax excavata* is a smooth ellipsoidal body (Anderson and Wall, 1979) and that of *Scrippsiella faeroensis* is a spinose sphere (Dale, 1978). In each case, morphological features distinctive of dinoflagellates are lacking. While these objects are demonstrably dinoflagellate cysts on biological grounds, they can be considered only speculatively so on morphological grounds. We must assume that some dinoflagellate cysts in the past were equally featureless or noncommittal about their identity, leaving us to speculate about their affinities but quite unable to demonstrate them on the basis of their morphology.

Almost certainly, cysts that might have been produced by dinoflagellates with terminally inserted flagella, like the modern Prorocentrales, would not be recognizable as dinoflagellates. The dinoflagellate affinity claimed for *Zosterosphaera tripunctata* from the Precambrian (Schopf, 1968) is a speculative possibility but is completely unsupported by morphological evidence.

It is quite possible—perhaps, quite probable—that some of the early Paleozoic fossils that have been called acritarchs are, in fact, dinoflagellate cysts (Lister, 1970; Bujak and Williams, 1981). This seems especially likely in the cases of many Tremadocian species with neatly formed operculate openings, distinctive and consistent polygonal patterns, and regularly arranged processes (Deunff, 1961; Górka, 1969). All that these fossils lack is an organization of the visible elements into an arrangement that is recognizably dinoflagellate-like. Conceivably the "missing" dinoflagellate-like arrangement was a feature acquired only at an evolutionary level more advanced than that represented by these fossils. Bujak and Williams (1981) suggest this may have happened in the Late Triassic, thus accounting for the beginning then of the extensive record of acknowledged dinoflagellates. However, these possibilities do not render the earlier fossils the least bit more demonstrably dinoflagellates on a morphological basis. Those that are not identifiable with some other group can be considered acritarchs. Although this term is misleading if it is understood to imply knowledgeable identification in the usual sense, it can be convenient as a general label for a large and morphologically varied complex of fossils whose affinities cannot be established from the available morphological evidence.

TAXONOMY AND NOMENCLATURE

Fossil dinoflagellates have been treated by different authors under both the zoological and botanical codes of nomenclature. Generally speaking, works before 1963 utilized the Zoological Code and after that date the Botanical Code. Although few authors have been explicit about their intent, it can be inferred at times from the family endings they have used. Currently the Botanical Code is applied by almost all paleontologists, and this is also the dominant practice among students of living dinoflagellates. Although the basic goals of the two codes are identical and most of their provisions are closely similar, some differences are significant and have been a source of continued uncertainty or debate regarding the standing of certain names. A concise and useful comparison of the codes as they apply to the algae is given by Sournia, Cachon, and Cachon (1975).

Quite understandably, fossils originally recognized as dinoflagellates were initially attributed to those modern genera which they most closely resemble. These modern genera (including *Peridinium*, *Gonyaulax*, *Glenodinium*, *Ceratocorys*, *Podolampas*, and *Ceratium*) were all based on the motile stage of the life cycle. As appreciation of the morphological differences in detail between fossil and modern dinoflagellates grew, new generic names were proposed for the fossils. This practice was favored by realization that the fossils were cysts and that virtually all of them differed from the small number of cysts known in Recent assemblages. A certain "inertia" created by the large number of genera (over 500) that are now based on fossil cysts has also favored application of distinct sets of generic names for fossil and modern forms, even in the face of a growing appreciation that some fossil dinoflagellates are congeneric or conspecific with living ones. In addition, very real uncertainties about detailed biological relationships have generally inhibited gross mergers of modern genera based on thecae with genera of fossils based on cysts. Contrasting philosophies on the subject were mentioned in Chapter 1. Pertinent considerations include the following:

a. The greater morphological complexity of the encysted stage, combined with the stratigraphic application of fossil cysts, has favored much narrower circumscription of genera based upon cysts than is typical for genera based on modern thecae.

b. Although culture experiments have yielded virtually identical thecate cells from markedly different cysts (Wall and Dale, 1968b), it is not yet clear whether this means that a single species may have more than one cyst morphotype, or that the cysts of two different species may be much more distinct than their thecae.

c. If cyst morphology as a generic criterion were applied to modern dinoflagellates as it has been to fossil ones, such modern genera as *Gonyaulax* and *Protoperidinium* would be subdivided into several components. Conversely, if observed or inferred thecal morphology were to govern generic distinctions, many genera from the fossil record would be merged. To devise a compromise taxonomy that utilizes both sets of characters is a major undertaking for which paleontologists and neontologists have not yet united, although a certain "splitting" of modern genera and "lumping" of fossil ones can already be recognized.

d. The fact that many living species do not produce cysts threatens to leave ambiguous the place of those species in a joint cyst-theca taxonomy.

e. The demise of some familiar names can be delayed by inaction. The case of *Spiniferites* and *Gonyaulax*, which has already been mentioned, is an example. If the I.C.B.N. were rigidly applied in the framework of current taxonomy, modern species of *Gonyaulax* would become species of *Spiniferites*. If, to avoid this, a successful appeal were made to conserve *Gonyaulax* against *Spiniferites*, then the many fossils now known as *Spiniferites* would become *Gonyaulax*. Ultimate nomenclatural conflict between these two names appears inevitable, but the resulting disturbance would be less severe (and the always controversial call for conservation, perhaps, would be less tempting) if the scope of *Gonyaulax* were first reduced. This might be either on the basis of thecal morphology, as has already transpired with the separation of *Protogonyaulax*, or on the basis of cyst morphology, which has not yet been attempted.

In this volume, names of genera based on cysts will be used without deference to possibly senior names based on thecae, and vice versa. However, equivalence of genera based on corresponding cysts and thecae will be pointed out where appropriate.

STRATIGRAPHIC DISTRIBUTION

In assessing reports of the stratigraphic distribution of described dinoflagellate species in the fossil record, ranges established by nonselective compilation of species occurrences cited in the literature must be regarded skeptically. Many early reports reflect much broader interpretations of species limits than is current practice and include misidentifications stemming from a less critical analysis of morphology.

The oldest fossil known to date that has been accepted widely as a dinoflagellate (i.e., excluding speculative dinoflagellates discussed earlier in this chapter) is *Arpylorus antiquus* from the Late Silurian of North Africa (Evitt, 1967; Sarjeant, 1978; but note uncertainty of Bujak and Davies, 1982). Taxa originally reported from the Permian are either not dinoflagellates (Tasch, 1963; Stover and Evitt, 1978) or probably represent contamination from younger sediments (Jansonius, 1962; Jansonius, pers. comm.). No other dinoflagellates have been reported from the Paleozoic or Early Triassic.

Triassic dinoflagellates, first reported from the Rhaetian by Sarjeant (1963), are now known to be abundant and varied beginning with strata of Carnian age and possibly older (Anisian; L. E. Stover, pers. comm.). Many show no paratabulation, although shapes and wall structure (proximate and cavate) are more or less "conventional". Some associated spinose bodies may also be dinoflagellate cysts, but so far they have not been shown to include diagnostic dinoflagellate characters. The earliest fully paratabulate cysts include *Suessia* from the Carnian and *Rhaetogonyaulax* and *Dapcodinium* from the Rhaetian.

Until recently, only a relatively few dinoflagellates had been described from the Early Jurassic but their numbers are now growing rapidly and they show considerable morphologic diversity. From the Middle Jurassic onward, abundant dinoflagellates are common features of marine sedimentary strata. Individual works on distribution are too numerous to mention; useful syntheses include those by Evitt (ed., 1975b), Riley and Sarjeant (1972), Sarjeant (1967b), Williams (1977), and Woollam and Riding (1983). The references cited give data on ranges and further references beyond the scope of this discussion; however, users of the earlier works should be aware of the varied accuracy of specific identifications on which the reported ranges depend.

A decrease in overall cyst diversity characterizes the Late Tertiary but precise timing of the decrease is not yet established. Pliocene assemblages are not yet well known (e.g, Baltes, 1967) and Quaternary assemblages, although rapidly becoming so (see Bradford, 1978a, for comprehensive bibliography), contain neither the large numbers of species nor the range in morphology typical of earlier times as recent as the Miocene. The sparse record of fossil dinoflagellates from freshwater deposits, which is confined to Cenozoic sediments, was mentioned earlier.

How far back in the fossil record do dinoflagellate cysts occur whose detailed morphology is closely similar to that known in cysts of living species? Table 2.3 gives some examples.

Table 2.3 - CLOSELY SIMILAR FOSSIL AND RECENT CYSTS

Fossil cyst (oldest known)	Modern cyst and theca
Spiniferites spp. (Early Cretaceous)	*Gonyaulax spinifera* group
Nematosphaeropsis labyrinthea (Early Cretaceous)	*Gonyaulax spinifera* group
Lingulodinium machaerophorum (Paleocene)	*Gonyaulax polyedra*
Operculodinium sp. (Early Cretaceous)	*Gonyaulax grindleyi*
Tuberculodinium sp. (Early Miocene)	*Pyrophacus horologium*
Polysphaeridium spp. (Eocene)	*Pyrodinium bahamense*

SELECTED REFERENCES

The following references are of a general or especially comprehensive nature. Full citations for the works identified here in abbreviated form will be found in the general list of references.

Artzner, Darrah, et al., 1979, A systematic illustrated guide to fossil organic-walled dinoflagellate genera.

Bradford, Martin, 1978a, An annotated bibliography and geographic review of Pleistocene and Quaternary dinoflagellate cysts and acritarchs.

Bujak, J. P., and Davies, E. H., 1983, Modern and fossil Peridiniineae.

Davey, R. J., et al., 1966, Studies on Mesozoic and Cainozoic dinoflagellate cysts.

Eisenack, Alfred, 1964 and 1967, Katalog der fossilen Dinoflagellaten, Hystrichosphären und verwandten Mikrofossilien.

Eisenack, Alfred, and Kjellström, Göran, 1971-1981, (same title as preceding entry).

Evitt, W. R., 1975b (ed.), Proceedings of a forum on dinoflagellates held at Anaheim, California, 1973.

Lentin, J. K., and Williams, G. L., 1973-1981, Fossil dinoflagellates: index to genera and species.

Lentin, J. K., and Williams, G. L., 1976, A monograph of fossil peridinioid dinoflagellate cysts.

Lister, T. R., 1970, A monograph of the acritarchs and Chitinozoa from the Wenlock and Ludlow Series of the Ludlow and Millichope areas, Shropshire.

Loeblich, A. R., Jr., and Loeblich, A. R., III, 1966-1974, Index to the genera, subgenera and sections of the Pyrrhophyta.

Stover, L. E., and Evitt, W. R., 1978, Analyses of pre-Pleistocene organic-walled dinoflagellates.

Tappan, H. 1980, Paleobiology of plant protists.

Wall, David, and Dale, Barrie, 1968b, Modern dinoflagellate cysts and the evolution of the Peridiniales.

Williams, G. L., 1977, Dinocysts. Their classification, biostratigraphy and paleoecology.

Wilson, G. J., and Clowes, C. D., 1980, A concise catalogue of organic-walled fossil dinoflagellate genera.

Part II - MORPHOLOGY OF THE CYST

To the untrained eye many dinoflagellate cysts look similar. Accurate discrimination of species depends upon an eye sensitized to detail and trained to recognize, among an assortment of features, those that are likely to be of value in distinguishing one species from another, those that are intraspecifically variable, and those that are taxonomically useless artifacts of preservation or preparation.

The features to be considered when identifying dinoflagellate cysts include the following:

General features (e.g., size, shape, criteria for orientation)
Wall structure and surface features
Indications of paratabulation
The archeopyle

While the mind of the microscopist enables him simultaneously to assess the bearings of these features on one another as he examines a specimen, written communication is less versatile. Unfortunately, the printed word requires a sequential presentation that leaves integration of ideas mostly up to the reader. We are doomed, therefore, to consider the features separately, in a staid and systematic way that inevitably lacks the excitement and challenge of seeing a specimen under the microscope.

In addition to the morphological features visible in an individual specimen, variability of features among many specimens is also of major importance. Variability is unpredictable and differs in degree from one taxon to another. Intraspecific variation in important features must be evaluated before a firm concept of a species can be formulated from either inspection of specimens under the microscope or analysis of a published description and its accompanying illustrations.

In Part II we will be concerned almost exclusively with the order Peridiniales; exceptions will be noted as encountered. We will explore the morphology of the peridinialean cyst, and the relation of the cyst to the theca insofar as we know this from observations on living dinoflagellates or can infer it from observations on fossil cysts. Descriptive terminology will be presented, equivalence of cyst and thecal structures pointed out, and suggestions offered to assist interpretation under the microscope.

Chapter 3 will introduce the basic morphology of the theca and the widely used method for designating particular thecal plates and patterns of plate arrangement. The point will be made that a satisfactory descriptive terminology is independent of both a knowledge of precise plate homologies and an understanding of the evolution of plate patterns—matters to be discussed in greater detail later.

The subject of Chapter 4 is the cyst wall, its structure and surface ornament. Here we will first explore the complex matter of wall layering, consider the varied elements of low and high surface relief that constitute what is usually referred to as the ornament, and mention several additional wall features of different form. Then we will discuss the still unsettled matter of wall development.

In Chapter 5 we will analyze several patterns of thecal tabulation which are widely represented by patterns of paratabulation on cysts. Beginning with the relatively simple and symmetrical pattern of *Peridinium*, we will proceed to the asymmetrical patterns of *Gonyaulax* and other genera (*Triadinium*, *Pyrodinium*, and *Protogonyaulax*) whose patterns are very *Gonyaulax*-like, closing with the patterns of *Ceratium* and fossil *Microdinium* which are, at once, *Gonyaulax*-like and distinctive in their own rights.

Chapter 6 will deal with openings in the cyst wall. Most important among these is the archeopyle, the exit for the living dinoflagellate at the end of its period of encystment. We will review some of the seemingly endless variety of archeopyle types, consider some aspects of recognizing them under the microscope, and discuss the taxonomic significance of the archeopyle.

Note on Descriptive Terminology

The list of descriptive terms for dinoflagellate cysts is already a long one (Williams et al., 1978, for terms introduced to that date) and continues to grow (e.g., Stover and Evitt, 1978; Sarjeant, 1982a; Bujak and Davies, 1983). The intent in the following chapters is to keep the special vocabulary relatively small but sufficiently inclusive to cover common and readily distinguished features. However, the participation of many individuals in the recent rapid expansion of our knowledge of dinoflagellate cyst morphology has led, understandably, to a diversity of terms and usages which as yet lack the sanction and stability acquired with time. In fact, relatively strong and divergent opinions on the matter have been recorded (Norris, 1978a; Sarjeant, 1982a). Therefore, the selective exclusion from this work of many available terms gives the vocabulary used here a certain subjective flavor which will not appeal to some whose viewpoint differs markedly from my own. Not included here are many terms which, I believe, are of doubtful utility, having been suggested for narrowly delimited states within intergrading continua, for distinctive but rarely encountered features, or for features which are likely to require extensive further description despite special names. Newcomers to the field should be aware of the unsettled aspects of this situation and should consult the sources cited for more comprehensive lists of terms.

Chapter 3. Basic Organization of Theca and Cyst

MAJOR FEATURES AND SPECIMEN ORIENTATION

To the principal morphological features of the peridinialean theca summarized in Chapter 1 (Figure 1.2) we can now add those of the cyst (Figure 3.1 A). Of course, the diagrams in these figures are highly schematic and many dinoflagellates depart markedly (even wildly!) from these relatively prosaic designs. The terminology for cyst features which we will begin to develop at this point will be a selective one (compare Williams et al., 1978). Besides many terms long in use and certainly "secure" in the descriptive literature, others whose "durability" has not yet been established (e.g., terms introduced by Norris, 1978a; Stover and Evitt, 1978; Sarjeant, 1982a; Bujak and Davies, 1983) will be included, and a few will be newly introduced here.

The main regions of the cyst that correspond to the epitheca, cingulum, hypotheca, and sulcus of the theca are, respectively, the **epicyst**, **paracingulum**, **hypocyst**, and **parasulcus**. Whereas the theca proper never has more than one visible layer, any one cyst may be constructed of from one to three layers out of a total of at least four different layers. Also unlike the theca, none of these cyst walls is separable into plates along sutures (except in the special case of the archeopyle, to be mentioned shortly). However, the surface of the cyst may bear **parasutures** in the form of ridges or other linear elements, which, to a sometimes confusing degree, simulate thecal sutures and divide the surface into more or less polygonal areas that resemble thecal plates. Such polygons on the cyst are especially clear and simple examples of **paraplates**, but parasutures and paraplates are often expressed by a variety of other features (notably fencelike septa and spine- or pillarlike processes) not at all suggestive of thecal sutures and plates (see Chapter 4). Any features on the cyst whose distribution approximates that of the thecal sutures are said to be **parasutural**.

Just as the tabulation comprises the number and arrangement of plates on the theca, the **paratabulation** encompasses the number and arrangement of paraplates on the cyst. Thecal tabulation has been the single most important criterion for the recognition of genera and species of living thecate dinoflagellates. Cyst paratabulation has also been important in the taxonomy of fossil dinoflagellates, but, as we shall see, it retains much unexploited potential, both for taxonomy in the formal sense and for day-to-day microscopy in applied palynology.

Directional terms that apply to cyst and theca are shown in Figure 3.1 B, D. The **apical**, or anterior end, and the **antapical**, or posterior end, are defined with respect to the principal direction of motion of the swimming cell. The sulcus and parasulcus are taken arbitrarily to mark the **ventral** surface. **Right**, **left**, and **dorsal** then follow automatically from the other directions. **Adcingular** is usable anywhere on the surface for expressing direction or location toward the cingulum.

Dimensional terms **length**, **width**, and **thickness** apply, respectively, to the anterior-posterior, right-left, and dorsal-ventral measurements.

Terms such as "polar," "longitudinal," "latitudinal," and "meridional," which relate to geometric figures of revolution, are frequently useful and their meaning is normally obvious; however, "equatorial" should be used with caution. The prominence of the encircling cingulum makes it a convenient reference for locating many other features, but careless equation of "cingular" and "equatorial" can be misleading, for the cingulum is often closer to one pole than to the other and its plane may be oblique to the polar axis.

Telling right from left might seem such an elementary matter as to require no particular attention here, but this is not the case. It is true that in many fossil peridinioid cysts the distinction makes little difference, for bilateral symmetry often dominates the shapes and surface patterns encountered. But,

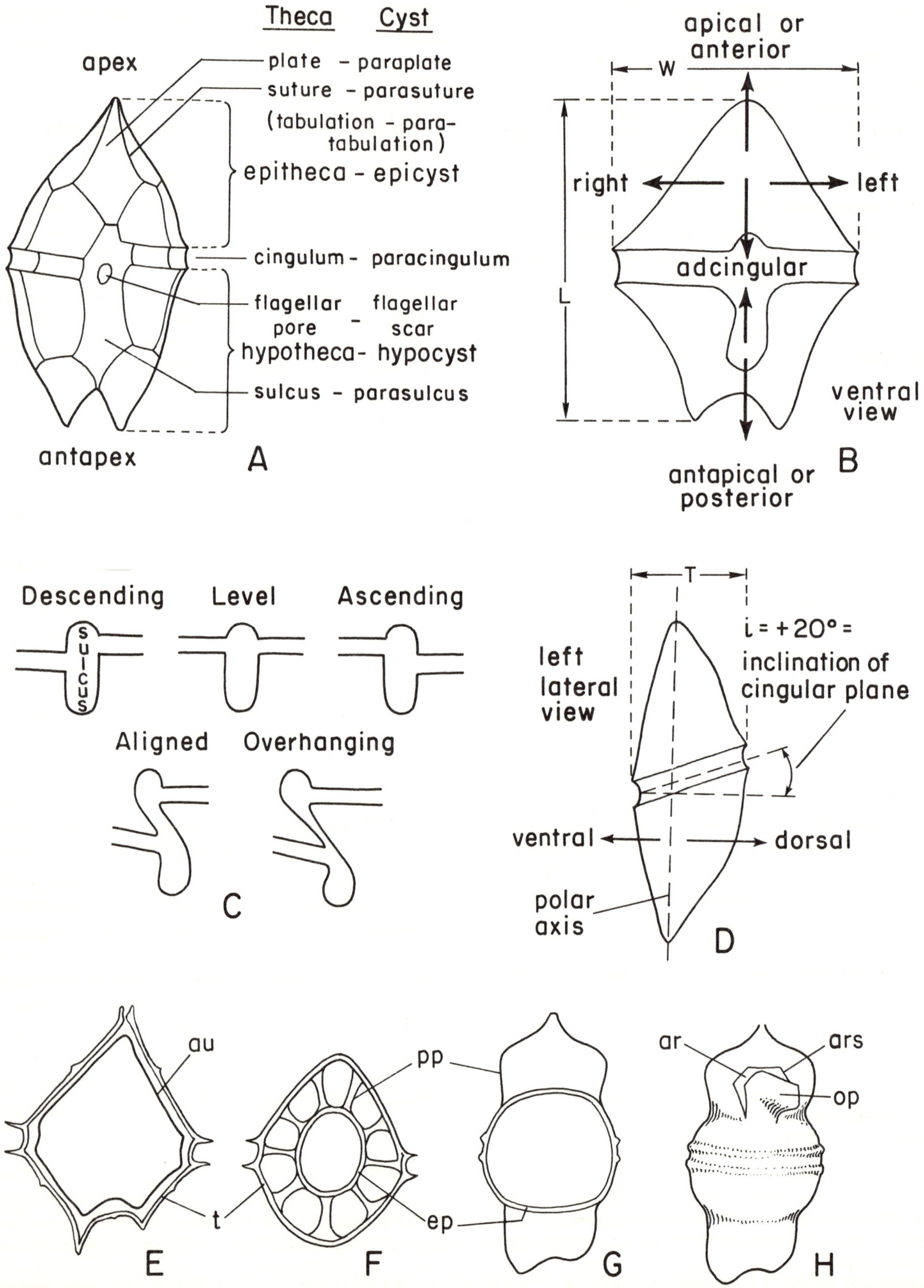
Theca
Cyst
apex
plate - paraplate
suture - parasuture
(tabulation - paratabulation)
epitheca - epicyst
cingulum - paracingulum
flagellar pore - flagellar scar
hypotheca - hypocyst
sulcus - parasulcus
antapex
A
apical or anterior
W
right
left
adcingular
L
ventral view
antapical or posterior
B
Descending
Level
Ascending
sulcus
Aligned
Overhanging
C
T
left lateral view
ι = +20° = inclination of cingular plane
ventral
dorsal
polar axis
D
au
t
pp
ep
ar
ars
op
E
F
G
H

Figure 3.1

when outlines and patterns are asymmetrical, the taxonomist/microscopist who cannot tell right from left is certain to misidentify critical features and misconstrue their spatial relationships. The difficulty occurs both when viewing a specimen under the microscope and when viewing a photomicrograph or published figure. In each instance, a correct interpretation depends upon two pieces of information about the specimen: the identity of the surface under inspection, and the direction from which it is being seen. The microscopist can check the requisite factors as he works and, with moderate experience, he normally acquires a reliable subconscious sensitivity to specimen orientation. The problem is much greater for the author or reader who is considering illustrations and who must avoid giving or receiving ambiguous information about the specimen being portrayed but not actually present for inspection. It is not difficult to select words that will convey the necessary information correctly, but the continuing appearance of published examples deficient in this respect demonstrates the need to draw specific attention to the matter.

Published figure captions contain many phrases in which words such as "view," "surface," and "focus" are ambiguously combined; for example, where "ventral view" is applied to a figure for which any one of the following would be a more accurate alternative: "dorsal view focused on the ventral surface," "dorsal view in low focus," "ventral surface in low focus," or "ventral surface in dorsal view." The requisite information is virtually certain to be clear if "view" is taken always to mean "direction of view" and the identity of the "surface" or "focus" is always added to specify the location of the detail shown. Other phraseology can be equally clear, but the choice must be deliberate and careful.

For the illustrations in this book, an exterior view at high focus is to be assumed unless indicated otherwise, and the following abbreviations will be used in the figure captions, appearing in lower case and run together as the meaning may make appropriate:

a - apical (or apico-)	p - polar
aa - antapical (or - antapico-)	r - right
d - dorsal	v, in terminal position - view; in nonterminal position - ventral (e.g., dv = dorsal view; dvv = dorso-ventral view)
eq - equatorial	
l - left	
lat - lateral	
o - oblique	

Figure 3.1 C shows schematic configurations of the ventral surface where the ends of the cingulum meet the sulcus. The cingulum may lie entirely in a plane or its course may be somewhat spiral and its ends consequently displaced or offset. The degree of spirality is customarily specified by recording the number of cingulum widths by which the ends are displaced. In most living and fossil species the cingulum is about **level**, or the right end is posterior to the left as the consequence of a **descending** spiral (also called left-handed, sinistral, levorotatory, or levogyre). The opposite condition of an **ascending** cingulum is uncommon among living species and virtually unknown in fossil ones. Therefore, in dealing with fossil specimens it is safe to assume a descending cingulum unless otherwise specified. Depending upon the width and inclination of the sulcus, one end of a spiral cingulum may or may not **overhang** the other.

Most commonly the cingular plane (or its approximation if the cingulum spirals) is about normal to the polar axis. Appreciable departures from this condition merit emphasis. If the cingulum lies closer to the apex on the dorsal surface than it does on the ventral, the cingular plane is said to have a **positive inclination**, the opposite inclination being **negative**. While the sense and severity of inclination can be detected in a dorsoventral view, a lateral view is required for measuring the angle relative to the position of a plane normal to the polar axis (i = +20° in

Figure 3.1 - THECA AND CYST - BASIC TERMINOLOGY AND ORGANIZATION

A - Terms for theca and cyst.
B - Directional terms, ventral view (L - length, W - width).
C - Displacement and overhang at ends of cingulum.
D - Directional terms, left lateral view (T - thickness).
E - Longitudinal section of unornamented proximate cyst which fills the theca (t) and consists of a single resistant wall (a - autophragm).
F - Longitudinal section of chorate cyst whose relatively contracted main body is separated from the surrounding theca (t) by hollow processes. These are constructed of an outer wall layer (pp - periphragm) which closely surrounds the thicker inner layer (ep - endophragm).
G-H - A cavate cyst without its theca, seen in longitudinal section (G) and dorsal exterior view (H) with archeopyle. Periphragm (pp) and endophragm (ep) are separated over polar areas but in contact equatorially. A polygonal archeopyle (ar) opens through the periphragm; its operculum (op) is flaplike because the archeopyle suture (ars) is not developed along its adcingular edge (adnate operculum). Paratabulation is expressed only by the archeopyle and faint paracingulum.

Figure 3.1 D). Positive inclination of the cingular plane is more common than negative.

If the cyst wall forms as a layer of material just inside the theca (Figure 3.1 E), the cyst necessarily resembles the theca more or less closely in size and shape. Such cysts are termed **proximate** and their external surfaces may be smooth or bear only relatively short projections (up to about 10 percent of the diameter of the main body). In contrast, **chorate** cysts (Figure 3.1 F) have a spherical to ellipsoidal or lenticular central cavity of relatively small volume compared to that of the theca, and bear projections whose length may nearly equal the diameter of the central capsule. Most fossils identified in the literature before 1961 as dinoflagellates were proximate cysts, while chorate cysts were called hystrichospheres. We now know that a complete integradation unites these extremes of cyst structure. Furthermore, either proximate or chorate cysts may also be **cavate** (Figure 3.1 G-H); that is, having a wall that is separated, at least locally, into an inner and outer layer by an enclosed cavity. Important aspects of all these properties will be explored in Chapter 4.

The **archeopyle** is the opening in the cyst wall through which excystment takes place. We will give special attention to this opening in Chapter 6, but it should be mentioned here as one of the especially important morphological features of the cyst (Figure 3.1 H). It is normally outlined by the **archeopyle suture**, a distinct line of separation in the cyst wall. Rupture of the wall along this suture commonly isolates a portion of the wall, the **archeopyle operculum**, which is dislodged or lost when the archeopyle opens. The sutures that define the archeopyle are normally the only true sutures (i.e., cuts through the cyst wall) that occur on the cyst. They are usually approximately parasutural in position and in some cysts they are the best, or even the only, evidence of paratabulation.

KOFOID'S SYSTEM FOR ANALYZING THECAL TABULATION

Even casual observation of peridinialean thecae and of many cysts thought to represent this order shows that the subdivision of the theca into plates and of the cyst into paraplates follows an orderly plan with differences from one taxon to another. It is natural that taxonomic descriptions of dinoflagellates should focus on this feature when it is apparent. Kofoid (1907, 1909a) summarized earlier attempts by others to analyze tabulation and, drawing especially on the ideas of Bütschli (1885), devised the special procedure most widely used today for naming and symbolizing the plates of the theca and analyzing their distribution over its surface. We will call this the Kofoidian system. It is only one of many ways in which plate distributions might be analyzed, and some subsequent authors have used other methods (e.g., Fauré-Fremiet, 1908; Lefèvre, 1933; Graham, 1942), but no other system has been adopted so widely. Nevertheless, Kofoid's scheme has not persisted without some change, first by Kofoid himself and later by others. Until recently, the changes proposed (and widely accepted) have been relatively minor. Now, however, more fundamental changes are being urged on the grounds that they are necessary to make the Kofoidian system more expressive of homologies between plates in different taxa, but the wisdom of this approach is arguable.

Certainly, the recognition of plate homologies is an essential precursor to understanding the evolution of dinoflagellate thecal patterns, a topic with current appeal and future promise justified by recent advances in studying both fossil and modern dinoflagellates. Further, it has become apparent that Kofoid's system tends to obscure some relatively obvious homologies and to make for confusion when plate patterns of different taxa are compared or their tabulations discussed in the context of possible evolution. However, the great advantage of the Kofoidian system as an aid in describing dinoflagellates lies in its relative objectivity and stability, in contrast to the subjectivity and variability of a system based on such an interpretative matter as plate homology. Recent proposals (e.g., Loeblich III and L. R. Loeblich, 1979; Balech, 1980; Eaton, 1980; Bujak et al., 1981) would alter the system precisely in order to make it conform to perceived homologies—each proposal reflecting the different perceptions of its particular author! Such proposals threaten to destroy the considerable utility of this widely understood and long-used system by making it the subject of continual debate and amendment.

The following discussion of tabulation adheres to Kofoid's principles except in minor details. However, in Chapter 5, we will explore the very different approach to tabulation suggested by Taylor (1979, 1980) and, in Chapter 12, we will touch on the still different hypotheses of Eaton (1980).

Figure 3.2 summarizes the names and symbols employed in the Kofoidian system. Kofoidian symbols will be shown in italics in the text. In a

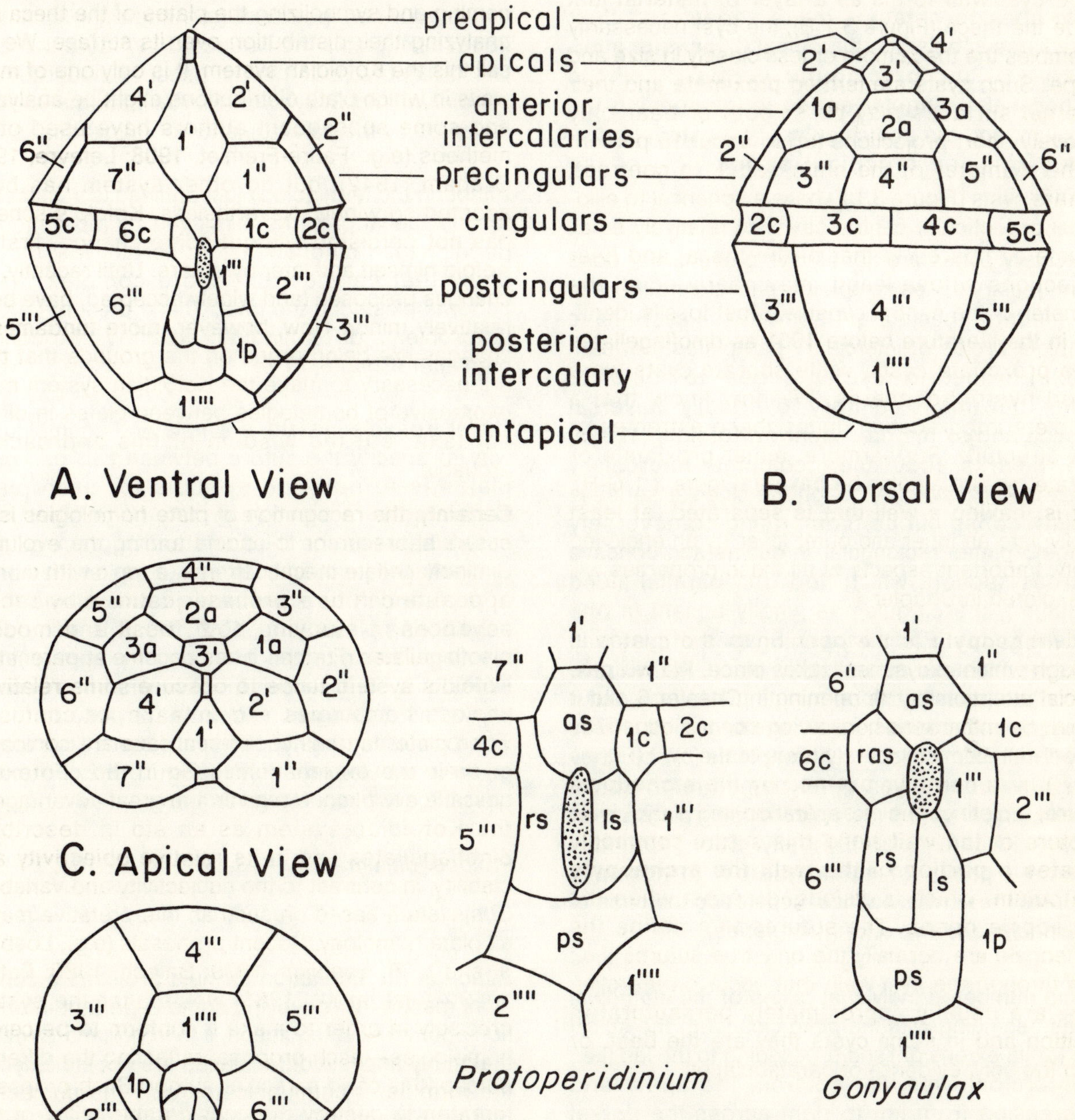

Figure 3.2 - KOFOIDIAN SYSTEM OF TABULATION

A-D - Tabulation of a hypothetical peridinialean theca. This pattern would be represented by the formula: *1pr, 4′, 3a, 7″, 6c, 6‴, 1p, 1⁗, 5s*.

E - Principal patterns of sulcal plates typical of *Protoperidinium* and *Gonyaulax*. Additional smaller plates may also be present. Flagellar pore region stippled.

typical peridinialean theca, a majority of the plates occur in seven series arranged approximately latitudinally with respect to the apical-antapical axis. From apex to antapex, these series (and their Kofoidian symbols) are the **apical** (′), **anterior intercalary** (*a*), and **precingular series** (″) of the epitheca, the **cingular series** (*c*) composing the cingulum, and the **postcingular** (‴), **posterior intercalary** (*p*), and **antapical series** (⁗) of the hypotheca.

Intercalary plates are not always present, nor does an intercalary series completely encircle the theca in any modern species, but some fossil cysts have complete cycles of intercalary paraplates. Sarjeant (1982a) argued to restrict "series" to complete cycles, but this is contrary to virtually universal practice and to the precedent of Kofoid's (1909a) own remarks about the incomplete intercalary series in *Peridinium*. Although the cingular series is usually conspicuous (comprising transversely elongate plates rectangular in outline and concave in cross-section, which floor the parallel-sided cingulum), the cingulum is wholly absent in one modern genus (*Podolampas*). Shapes of plates in the other major series are highly varied. At the apex, one or two (possibly more) minute plates may occur within the apical series; following Loeblich III (1970), we will call them collectively **preapicals** (*pr*), but they have also been given other names (e.g., apical platelet, apical pore plate, apical closing plate, cover plate, and canal plate). Finally, a longitudinally extended group of **sulcal plates** (*s*) normally occupies the ventral surface, separating the ends of the cingular and two adcingular series.

Kofoid numbered individual plates of the latitudinal series beginning with the plate closest to the sulcal axis (or its extension) and proceeding to the left (i.e., counterclockwise viewed from the apex, and progressing from left to right across the dorsal surface). This "Kofoidian direction" of numbering corresponds with the extension of the transverse flagellum from its proximal point of emergence in the sulcus, around the cell to its free distal extremity in the cingulum to the right of the sulcus. Note that one must determine from the context whether a particular combination of number and symbol (e.g., *6″*) means the collective plates of a series (i.e., "six precingulars") or an individual plate (i.e., "the sixth precingular"). Contrary to the plates of the latitudinal series, the plates of the sulcus are designated by their position, employing terms and abbreviations that vary somewhat depending upon the number and position of plates present, as well as upon the author. We will call them the right (*rs*), left (*ls*),anterior (*as*), posterior (*ps*), and right accessory (*ras*) sulcal plates, as illustrated in Figure 3.2 E for two distinctive patterns that occur in modern *Protoperidinium* and *Gonyaulax*. Similar patterns are also indicated by the paratabulation of many fossil cysts.

Although the terminology just outlined is satisfactory for the thecal patterns displayed by most modern peridinialean dinoflagellates, certain fossil cysts show that considerably more complex thecal patterns existed in the past. These present particular problems of terminology which will be considered later (Chapter 5).

Under Kofoid's system, a simple and unambiguous way to specify the suture between two particular plates is to note the symbols for each plate, separated by a slash (e.g., *3″/4″* to signify the suture between the third and fourth precingulars). Similarly, the junction of three sutures at the corner of a plate can be specified in terms of the three neighboring plates (e.g., *1′/2′/1″*). It is also possible to summarize the number of plates in each series of a peridinialean theca in what has become known as a **tabulation formula** (or plate formula). The formula for the hypothetical pattern illustrated is given in the explanation to Figure 3.2. Formulae typical for two modern genera are:

Protoperidinium - 2pr, 4′, 3a, 7″, 4c, 5‴, 0p, 2⁗, 4s

Gonyaulax - 2pr, 4′, 0a, 6″, 6c, 6‴, 1p, 1⁗, 5s

Although the tabulation formula provides a convenient means for conveying some important information about thecal tabulation, it is important to realize that many aspects of tabulation are not indicated by the formula. It communicates only the number of plates in each series, saying nothing about the shapes and relative sizes of individual plates, about the mutual contacts between them, or about their position relative to a reference feature on the surface of the theca. As will become apparent in Chapter 5, a formula must be supplemented by additional information on these items if the tabulation is to be characterized completely. Nevertheless, tabulation formulae have great utility for certain purposes, and many genera of modern Peridiniales are distinguished by differences in those aspects of tabulation that are conveyed adequately and succinctly by such formulae.

PARATABULATION

So far, we have discussed the Kofoidian system with respect to the theca. Now we can ask: How does this system apply to the cyst—modern or fossil? In principle the application is the same, but, in practice, several factors influence it strongly.

First, the paratabulation of the cyst is commonly expressed incompletely. Whereas the theca is composed of plates whose limits are open to inspection, given the appropriate techniques and equipment, small to large portions of the cyst—occasionally virtually all of it—may simply lack any evidence of paratabulation.

Second, while thecal tabulation is indicated almost exclusively by the sutures between plates (or by surficial ridges or lists that lie close to the sites of these sutures), paratabulation of cysts may be indicated by a much greater variety of features. On some cysts, surficial ridges or grooves on some cysts closely simulate thecal sutures or sutural lists; but on other cysts paraplates may be indicated by any of a great variety of features, including rows of short projections, fencelike membranes (septa), isolated larger projections (processes), clusters of such projections, local enlargement or reduction in general surface ornament, structures located between two wall layers rather than on the exterior surface, or the archeopyle sutures alone.

Third, the paratabulation tends to be "abbreviated" by comparison with the tabulation. Generally speaking, the smaller a thecal plate is, the less likely it is to be recorded clearly as a paraplate on the cyst (Minority Suppression, Chapter 7). From observations on the cyst, it may be impossible to resolve features that would be closely spaced but distinct on the theca. This is true even when the elements that express the paratabulation are relatively clearly developed. Examples to be found in Part III will provide ample support for this generalization; however, exceptions do occur.

Fourth, even when paraplates are outlined on the cyst by distinct parasutural lines, the shapes of individual paraplates may differ more or less appreciably from their thecal counterparts. Generally speaking the differences will be greater where the overall shape of the cyst contrasts with that of the theca. For example, the shapes and sizes of paraplates on a spherical cyst formed within a biconical theca tend to differ from those of the corresponding thecal plates as if they had been produced by projecting the thecal outlines inward onto the cyst surface (Centripetal Projection, Chapter 7).

Finally—and important, although of a somewhat different nature—there is the considerable influence of postmortem degradation, especially in fossil cysts where this may result from conditions of temperature, pressure, and pore-fluid composition not to be found in recent sediments.

Within limits that take these factors into account, the Kofoidian system can be applied to cyst paratabulation, using most of the terms and symbols we have already reviewed, and the appropriate information can be stated in a **paratabulation formula** (or paraplate formula) in similar manner. However, because the cyst features may indicate paratabulation incompletely, or the evidence for it may be ambiguous or modified by postmortem degradation, means are needed for expressing degrees of uncertainty regarding the existence, or number, of paraplates. An "X" in the place of a numeral in a paraplate formula means that there is evidence for the presence of the paraplate series specified by the following symbol but no evidence for its subdivision into several elements. For example, the formula *4″, 3a, 7″, Xc, 5‴, 2⁗* would apply to some peridinioid cysts in which the paracingulum as a whole is indicated but its subdivision is not clear. On the other hand a query in the place of a numeral indicates uncertainty as to whether the specified series is represented at all. A query in conjunction with a numeral questions the identification of the series or the count, usually the latter.

If processes are very numerous (substantially over 30) and are uniformly distributed over the surface, their relation to the paratabulation is likely to be quite obscure. However, when the number is about 30 or less, it is usually worthwhile to report the distribution of processes in detail. For this, a **process formula**, following the same Kofoidian principles as applied when paraplates are indicated by parasutural features, can be useful. In *Hystrichosphaeridium tubiferum*, an abundant and widespread chorate cyst in Upper Cretaceous and Lower Tertiary strata, a one-to-one equivalence of cyst processes and thecal plates appears to have existed, judging by the observed similarity between the distribution and relative sizes of processes, on the one hand, and the well-known gonyaulacoid pattern of tabulation, on the other. The origin of this inferred relationship between processes and plates is depicted in the diagrams of Figure 3.3 A-C. The process formula for this species—*1pr, 4′, 0a, 6″, 6c, 6‴,1p, 1⁗, 5s*—is identical to the gonyaulacoid tabulation formula. However, it is important to realize that a process formula, while useful, may contrast with a paratabulation formula for the same species based on other evidence. For instance, the process formula for *Areosphaeridium multicornutum* (Figure 3.3 D)— *0pr, 4′, 0a, 5″, 4-5c, 5‴, 1p, 1″ ″, 1s* —records only five precingulars and one sulcal (*ps*), whereas the angulations in the margin of the archeopyle clearly indicate a sixth precingular paraplate and an anterior sulcal. Once alerted to the fact that certain paraplates may not be represented by processes, we can account for other seeming "deficiencies" in the representation of processes, although the collective evidence may still justify inferring a full gonyaulacoid tabulation.

An even more striking example of contrasting formulae based on different evidence from the same specimen is provided by *Oligosphaeridium abaculum* (Figure 1.7 F), a Lower Cretaceous species mentioned already in Chapter 1 because of its virtually unique features. The parasutural markings on the surface of this cyst support a paratabulation formula identical to that of *Hystrichosphaeridium tubiferum* and an inferred thecal tabulation of gonyaulacoid type (although topological differences evidently existed in the the tabulations; see Chapter 11). However, 12 particular paraplates out of the total of 30 that are defined by parasutural lines—the preapical, six cingulars, one postcingular, and four sulcals—lack processes. The resulting process formula— *0pr, 4′, 0a, 6″, 0c, 5‴, 1p, 1⁗, 1s*—thus differs sharply from both the paratabulation formula based on parasutures and the formula for inferred thecal tabulation.

In view of the potential for a single specimen to be represented by contrasting formulae (even when strict Kofoidian principles have been applied in a uniform way—a condition that does not always exist, as we shall see in Chapter 5), authors should take care to make clear what kind of a formula they are presenting. Readers, also, must be alert to the distinctions, but, in addition, they must be prepared to discern what an author intends by a formula even if he has stated it ambiguously. In much of the existing descriptive literature on fossil dinoflagellate cysts, it is up to the reader to decide whether the terms "tabulation," "tabulation formula," or simply "formula" actually mean "paratabulation formula," "process formula," or "inferred tabulation formula."

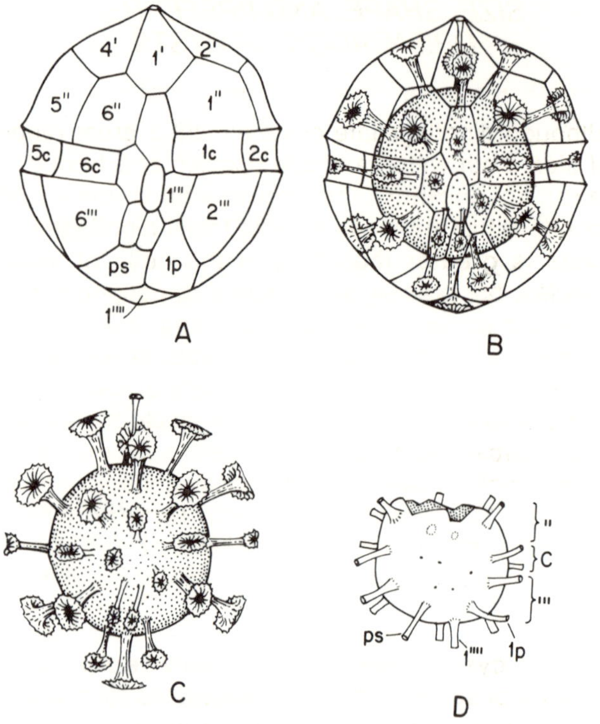

Figure 3.3 - COMPLETE AND INCOMPLETE PARATABULATION ON CHORATE CYSTS

A-C - Schematic, to show derivation of a chorate cyst—in this example, Upper Cretaceous *Hystrichosphaeridium tubiferum* —from a theca with gonyaulacoid tabulation. A - Theca in ventral view. B - Cyst inside the theca, one trumpetlike process associated with each thecal plate. C - Cyst in same orientation after theca has disintegrated. Identical Kofoidian formulae would describe the number and distribution of processes and thecal plates: *1pr, 4′, 6″, 6c, 6‴, 1p, 1⁗, 5s*.

D - *Areosphaeridium multicornutum*, a cyst from the British Eocene, showing archeopyle suture and processes (tips omitted) in ventral view. The process formula (including processes on operculum not shown here) is: *4′, 5″, 4c, 5‴,1p, 1⁗, 1s*; but two further paraplates (*6″* and *as*) are clearly indicated by the archeopyle outline although they are without processes. In addition, comparison with the gonyaulacoid pattern shown by similar cysts of other species suggests that six more paraplates are unrepresented by processes (or by any other features that permit their delineation). These include one in the middorsal region (*3c*) not visible here, and five on the ventral surface (approximate centers marked by single dots: *6c*, *ras*, *rs*, *ls*, and the usual *1‴*).

In this book "tabulation formula" pertains only to the theca and "paratabulation formula" only to the cyst; a "process formula" is based solely on process count, whatever other evidence for paratabulation may or may not be present; and an **inferred tabulation formula** is a formula of thecal tabulation not observed directly, but estimated from observations on a cyst.

SIZE, SHAPE, AND PREFERRED ORIENTATION OF CYSTS

Although dinoflagellate cysts exhibit a great range of size, both within a single species and from one taxon to another, the impression seems well founded that the mean size of fossil dinoflagellate cysts is greater than that of the cells of living species, but precise data needed to document this assertion are not available. The reason for this size difference is not clear. Fossil cysts in excess of 100μm are not uncommon but, with a few exceptions, cells with such a dimension would be considered relative giants among modern species. Large fossil dinoflagellates are especially conspicuous in the Late Cretaceous and Early Tertiary, where examples of the peridinioid *Deflandrea* and *Wetzeliella* groups are especially noteworthy for their large size.

Zygotic cysts observed in sexually reproducing cultures of some modern dinoflagellates form within thecae that grow considerably larger than vegetative cells of the same species. Therefore, cyst size may not be a safe guide to the dominant size of motile cells in a living population. Also, cyst size may be influenced by environmental conditions. In any event, distinctions between species based solely on differences in cyst size seem likely to be ill founded. Convenient general categories for overall size are: small - under 50μm, intermediate - 50-100μm, large - over 100μm.

Subspherical to ellipsoidal cysts are common, just as are similarly shaped motile cells among living dinoflagellates. However, moderate to extreme departures from these "generalized" shapes also occur widely. The possible adaptive significance of the more elaborate shapes is unknown.

Cyst shape plays an important role in paleopalynology, quite apart from its biological significance, by influencing specimen orientation at two stages: first, as the cyst settles on the original depositional interface, and again later, as it settles on the microscope slide (if prepared as a strew mount). Any postdepositional compaction that may follow original sedimentation will normally serve only to accentuate those primary departures from equidimensional shapes that determined an original selective orientation. Orientations dependent upon shape are so strongly favored in some species that specimens are virtually never seen in other positions. This has considerable practical significance when taxonomically critical morphological features are thereby concealed. If specimens have suffered post-depositional compression, the microscopist often can do little to improve the situation, but when such secondary compression has been minimal, manipulation of individual specimens in uncovered microscope mounts provides a means to overcome the unfavorable effects of preferred orientation determined by original cyst shape.

Motile individuals of some peridinialean dinoflagellates are virtually spherical and some spherical fossil cysts probably also represent spherical motile cells. However, the virtual sphericity of some cysts has little to do with the shape of the original theca; it may only mean that the protoplasmic mass contracted into a body with minimal surface area. Nearly spherical cysts are common and, of course, can be expected to present random views to the microscopist, whether or not there has been post-depositional compression. In fact, the degree of randomness of orientation is a useful guide to original shape and to primary (i.e., biological) as opposed to secondary (i.e., mechanical = post-depositional) compression.

Cysts commonly deviate from sphericity (at least in amounts sufficient to affect orientation on a depositional interface) in one or more of three ways (Figure 3.4): by elongation in the apical-antapical direction (= axial or longitudinal elongation); by primary compression or flattening, most frequently in a dorsoventral direction; and by the presence of one to a few major projections, or horns. Obviously, axial elongation alone will favor essentially equatorial views, while dorsoventral views are favored by both the dominance of dorsoventral primary compression and the frequent occurrence of horns on the hypocyst approximately in a plane extending from right to left through the antapical pole. Lateral compression, unlike dorsoventral compression, is extremely rare in dinoflagellate cysts, notable examples being *Dingodinium* and the non-peridinialean *Nannoceratopsis*. Consequently, apical, antapical, and lateral views are relatively uncommon and, for many species, truly exceptional. While lateral views are seldom critical for species identification, something approaching a polar view may be essential to reveal details of paratabulation critical for a reliable identification. Here is another argument in favor of manipulating individual specimens.

Although length:width ratios may reach 5 or more in such cysts as *Palaeocystodinium* and *Odontochitina*, axial elongation usually has a value of less than 2.

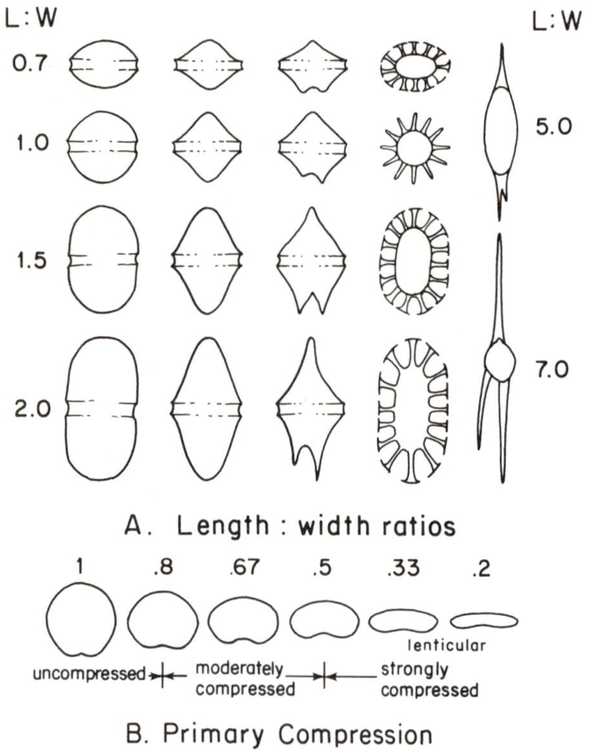

Figure 3.4 - DIMENSION RATIOS AND COMPRESSION

A - Outlines showing selected length:width ratios combined with varied expressions of horns and projections. Extreme elongation shown at right by two Cretaceous forms: *Palaeocystodinium* (upper, eqv), and *Odontochitina* (lower, vv).

B - Outlines of cysts in polar view showing representative amounts of primary compression or flattening (i.e., deviation from a circular cross-section not related to postdepositional processes such as sediment compaction).

Axial shortening—in other words, polar compression—which naturally favors polar views, occurs only rarely.

The transverse cross-section of the cyst may depart from circular in varying amounts. Commonly the parasulcus lies in a local excavation near the ventral midline, and much of the ventral surface may be a broad concavity. Alternatively, dorsal and ventral surfaces may be nearly planar and parallel, and the lateral surfaces reduced to narrow strips. The term **lenticular** has been applied to cysts whose main body shows strong dorsoventral compression without pronounced elongation.

One to several major projections, or **horns**, rise at specific locations from the thecae of many species of dinoflagellates. Although they are particularly prominent features of *Ceratium* (Figure 1.6), they are also present, usually in less elongate form, in many other genera, including *Protoperidinium*. Corresponding projections on cysts are also called horns. While located similarly to thecal horns, these cyst horns appear as more or less "abbreviated" versions of their thecal counterparts in those species for which both cyst and theca are known. In both cases, horns vary greatly in size and in the distinctness with which they are set off from the rest of the body. When least distinct, they are mere angulations in the outline, and the term "horn" is hardly fitting, although correspondence to the longer structures may be apparent. Horns on the cyst may also include short, solid projections, in addition to hollow extensions of the main body. Probably all horns on cysts represent what were slender continuations of the thecal cavity, rather than the solid projections (sometimes referred to as horns) that rise from the apical and antapical extremities of the theca in some modern species of *Protoperidinium* and *Gonyaulax*.

The "cardinal" positions that are the normal location of horns in dinoflagellate cysts are shown in Figure 3.5. Apical and antapical horns favor preferred orientation of settled specimens by increasing the length:width ratio. Lateral, postcingular, and paired antapical horns favor a dorsoventral view since they are normally about coplanar among themselves and with the apical horn.

The presence of an apical horn (or pronounced angulation in the outline), combined with two antapical horns in *Peridinium* and its allies, or one antapical and one postcingular horn in *Ceratium*, is so typical for these genera that the terms **peridinioid** and **ceratioid** are commonly used to describe these distinctive outline shapes. In cysts with a peridinioid outline (Figure 3.5 E-K), the two antapical horns may be of equal size or one may be smaller, even to near disappearance. Curiously, among modern *Protoperidinium*, when there is a clear difference in the sizes of the antapical horns, the righthand horn is consistently the larger; but among fossil cysts having evident affinity to *Protoperidinium* and *Peridinium*, exactly the reverse is true, virtually without exception. Thus, inequality of horns in fossil cysts may provide a means for immediately and reliably identifying the direction of view. This criterion is useless if relative horn size cannot be judged accurately, as when specimens are viewed obliquely.

Among fossil ceratioid cysts (Figure 3.5 L, N-Q) a single postcingular horn usually lies on the right side, corresponding to the thecal horn constructed

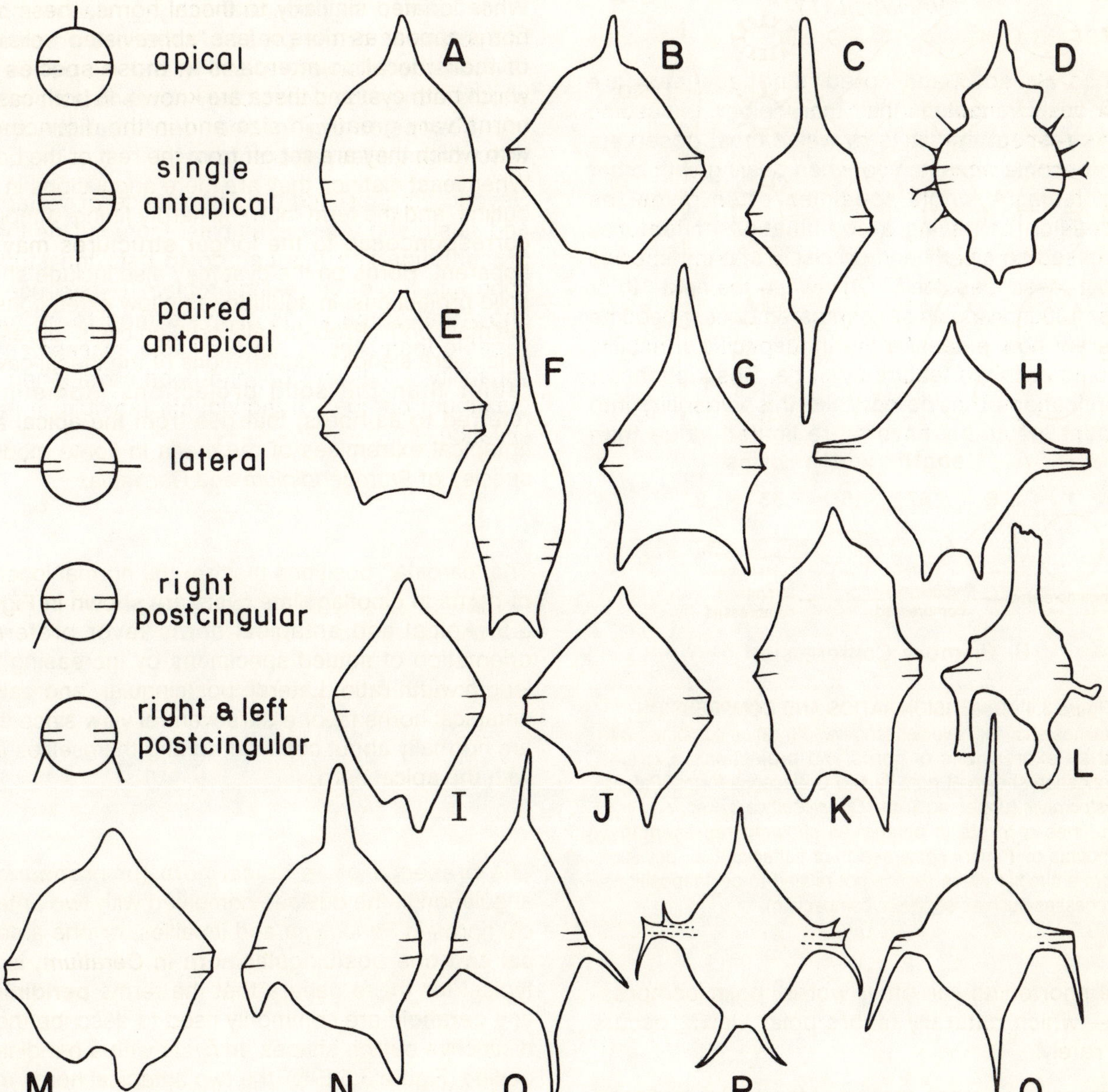

Figure 3.5 - HORNS: CARDINAL LOCATIONS AND COMBINATIONS

At left - diagrams showing "cardinal" locations for horns, applicable to both thecae and cysts.

A-Q - Outlines of an assortment of fossil cysts (all in vv), illustrating a variety of horns; accessory ornament and projections omitted except for a few in D and P. Short parallel lines at margins show position of paracingulum. A -*Pareodinia*, B - *Gonyaulacysta*, C- *Biconidinium*, D - *Triblastula*, E - *Alterbia*, F - *Batioladinium*, G -*Deflandrea*, H - *Rhombodinium*, I - *Isabelidinium*, J - *Ginginodinium*, K - *Chatangiella*, L - *Xenascus*, M - *Canningia*, N - *Endoceratium*, O - *Pseudoceratium*, P - *Phoberocysta*, Q -*Muderongia*. (A-B - Late Jurassic; F, N-Q - Early Cretaceous; D, I-M - Late Cretaceous, E - Paleocene; C, G-H - Eocene)

by the fourth and fifth postcingular plates of modern *Ceratium*. In some species a smaller left postcingular horn is present, or this may be a variable character within a single species. In fossil *Muderongia*, whose paratabulation clearly indicates affinity with *Ceratium* (Wall and Evitt, 1975), and in *Phoberocysta*, two postcingular horns (or cingular projections with postcingular extensions) are combined with one or two antapical horns in a distinctive combination without modern counterpart.

VARIABILITY

As has already been noted, dinoflagellates are notoriously variable in their morphology, exceeding in this respect the limits to which most observers have become accustomed when dealing with other organisms. A single specimen often gives the impression of having a combination of features expressed so clearly and precisely that the species almost "describes itself." Only when the next 10, or 20, or 100 individuals are compared does it become apparent how extensive the intraspecific variability of some of these features can be. Descriptions or identifications that do not take this variability into account inevitably have more limited value than others which do. Understandably, limited numbers of specimens or the press of time may preclude a full consideration of variability, but the deficiency should always be consciously appreciated and the "confidence rating" of the identification or the description assessed accordingly.

Beyond emphasizing the importance of variability and illustrating a few examples, there is little that a presentation like this can do to help the student cope with this often vexing problem. Experience will show that some kinds of cysts tend to be more variable than others, but a reliable assessment of variability will always depend upon considering the maximum number of individuals feasible in each situation.

Chapter 4. The Cyst Wall: Stratification and Structure, Surface Relief, and Formative Processes

INTRODUCTION

In many respects we know least about these most obvious features of dinoflagellate cysts. The development of the cyst wall has never been studied in living dinoflagellates. The relation of the wall to the biological processes that go on within the cell during encystment is unknown, as is also the functional significance of cyst shape and surface relief in the hydrodynamic environment of the sea or lake. Study of fossil cysts especially has revealed great diversity in wall layering, internal structure of the layers, and relief on their surfaces, but the complex details of these features lie at the limit of resolution with light optical systems. Electron microscopic examination of the wall has been tremendously informative. Many features previously unobservable have been reported (e.g., Jux 1968 and later; Gocht, 1975a; Piel and Evitt, 1980), and scanning electron microscopy is becoming routine in conjunction with taxonomic studies. Nevertheless, up to this time only a small fraction of known species have been studied at the high magnification and with the special preparation (e.g., investigation of broken or cut sections) that will be required before the wall gives up even those secrets that relate to its physical structure.

Gocht (1983), whose remarks are paraphrased with slight modification in this and the following paragraph, pointed out that features in and on the cyst wall, which can be distinguished as thecamorphic and cystomorphic, combine to produce a mixture of properties that is distinctive for each species. **Thecamorphic features** are always homologs of thecal features, as is evident from an outward similarity that may depend upon similar topography and relief (e.g., elevated margins of the paracingulum, depression of the parasulcus) or upon a similar position or distribution with altered relief (e.g., "negative" parasutures as grooves incised on the inner surface of the endocyst, or processes arranged along parasutural lines). Unlike the thecal features to which they correspond, thecamorphic cyst features appear to be functionless. Examples include: paracingulum and parasulcus, paraplates and paratabulation, parasutures, the flagellar scar, features recording the overlap patterns of thecal plates, and traces of such thecal details as the fine surface reticulum of plates, trichocyst pores, the "ventral pore" (or adelopore, as defined later in this chapter), and bands of incremental growth. Thecamorphic features are frequently (but not exclusively) expressed on proximate cysts.

In contrast to thecamorphic features, **cystomorphic features** (of which processes are a prime example) have no equivalents in outwardly similar features on the theca. They may be homologs of thecal features, either obviously or only possibly but unprovably, or not. They are the real form-giving elements of the cyst and range from prominent projections to much less striking features. They tend to be distributed in concert with thecamorphic elements to whatever extent these are present or recognizable (e.g., intratabular processes, parasutural denticulate septa), and they may be function-related, even if the function in specific cases is unknown. Examples include: processes (either theca homologs or not), septa and flanges, trabeculate outer walls (ectophragm), inner bodies (endocysts), all kinds of structural elements in the cyst wall(s), protrusions of all sorts (blisters, funnel-like features), and openings in the periphragm (claustra). Cystomorphic features are especially (but not exclusively) characteristic of chorate cysts, and their importance is particularly clear when they give the cyst its distinctive form (e.g., as in the intratabular processes of *Hystrichosphaeridium*).

Three closely interrelated matters are involved in a consideration of the morphology of the cyst wall: the gross organization of the wall into one or more layers, the internal microstructure of the wall material that makes up these layers, and the relief features (most commonly various kinds of projecting structures) developed upon the external and, more rarely, the internal surfaces of these layers. The reasonable ease with which at least the coarser aspects of the first and last items can be seen in the light microscope has made them important considerations in descriptive taxonomy. They will receive most of our attention here. The internal microstructure of the wall remains at present a largely unexplored but potentially important matter. It will be mentioned here mainly in relation to the other two topics, and the reader will be left to pursue it further in the references already cited.

Except where the context indicates otherwise, comments about the appearance of wall features refer to their appearance in the light microscope.

WALL LAYERS

The terminology adopted here is a modification (Evitt et al., 1977) of an earlier one (based largely on Davey et al., 1966). It employs three word stems to distinguish between a wall layer (-**phragm**), the cavity which it encloses (-**coel**), and the three-dimensional body which it forms (-**cyst**). To these stems may be added a variety of prefixes which identify particular wall layers: **auto**-, **endo**-, **peri**-, **meso**-, and **ecto**-. The additional prefixes **epi**- and **hypo**- may also be used to specify the particular portion (i.e., "hemisphere") of the layer, cavity, or body under consideration. Examples: autocyst, endophragm, epipericoel. Various types of wall stratification and terms applied to them are shown in Figure 4.1.

In cysts with the simplest wall construction, a single wall, the **autophragm**, surrounds an internal cavity, the **autocoel**; and the three-dimensional body so formed is an **autocyst** (Frontispiece D). Many cysts have a wall composed of only one layer, but, unfortunately, the existence of one, and only one, layer is not always easy to confirm in the absence of sectional views either prepared artificially or exposed by accidental breakage. The junction between closely appressed layers may be nearly impossible to see clearly, whereas phenomena of light refraction (i.e., Becke lines) may create the appearance of layering where none exists. The approach favored here is to consider the wall single layered (i.e., an autophragm) unless division into more than one layer can be seen clearly, not just suspected or assumed. At least this leads to a certain "honesty" in descriptions, although there can be little doubt that future studies with new techniques and better materials will show that some walls early taken to be simple are, in fact, complex.

Endophragm and **periphragm** are, respectively, the inner and outer layers in many cysts with definitely two-layered walls. These layers may be in close contact or separated by a **pericoel** (Frontispiece B); locally they may even be fused. More than one of these conditions may exist in different parts of a single cyst. Where a pericoel is present, the layers are usually not connected by structures spanning the gap, or if such connections appear, as they do in

Figure 4.1 - SCHEMATIC EXAMPLES OF WALL STRUCTURE

(Dorsoventral views in optical section except as noted)

A - Acavate cyst with single wall layer (as in Recent *Trinovantedinium*).

B - Holocavate cyst (polar view below) with autophragm and ectophragm (e.g., Late Jurassic-Early Cretaceous *Gardodinium*).

C-E - Cyst with endophragm, periphragm and ectophragm; cornucavate with respect to pericoel; irregularly cavate with respect to ectocoel. D - lateral view in optical section. E - Detail section of wall showing continuity of periphragm and ectophragm and variable extent of pericoel into process shafts. (e.g., Eocene *Kisselovia*).

F - Circumcavate cyst (polar view below; e.g., Late Jurassic-Early Cretaceous *Sirmiodinium*).

G - Bicavate cyst (e.g., Late Jurassic *Gonyaulacysta jurassica*).

H - Cornucavate cyst (e.g., Late Cretaceous-Early Tertiary *Palaeocystodinium*).

I - Bicavate cyst with mesophragm (e.g., occasional species of Late Cretaceous *Isabelidinium*).

J - Circumcavate cyst (e.g., Early Cretaceous *Endoceratium ludbrookiae*).

K - Acavate autocyst (outline as in Early Cretaceous *Aptea*), suggesting that this is really an endocyst, as in J, without pericyst.

L - Epicavate cyst (e.g., Late Cretaceous *Nelsoniella*).

M - Camocavate cyst (left lateral view on left; e.g., Early Tertiary *Thalassiphora pelagica*).

N - Autocyst with calyptra (e.g., some Late Jurassic *Pareodinia*).

O - Oblique detail and section of apiculocavate wall, showing pericoel completely restricted to interior of processes.

P-R - Suturocavate wall. P - Oblique detail and section, showing pericoel restricted to parasutural ridges and interior of open-tipped processes (as in some *Spiniferites*). Q-R - Pericoel confined to base of parasutural septa. Q - Oblique view and section; R- plan view, showing extent of pericoel (stippled), septa (paired parallel lines), and line of contact between periphragm and endophragm encircling central portion of paraplate.

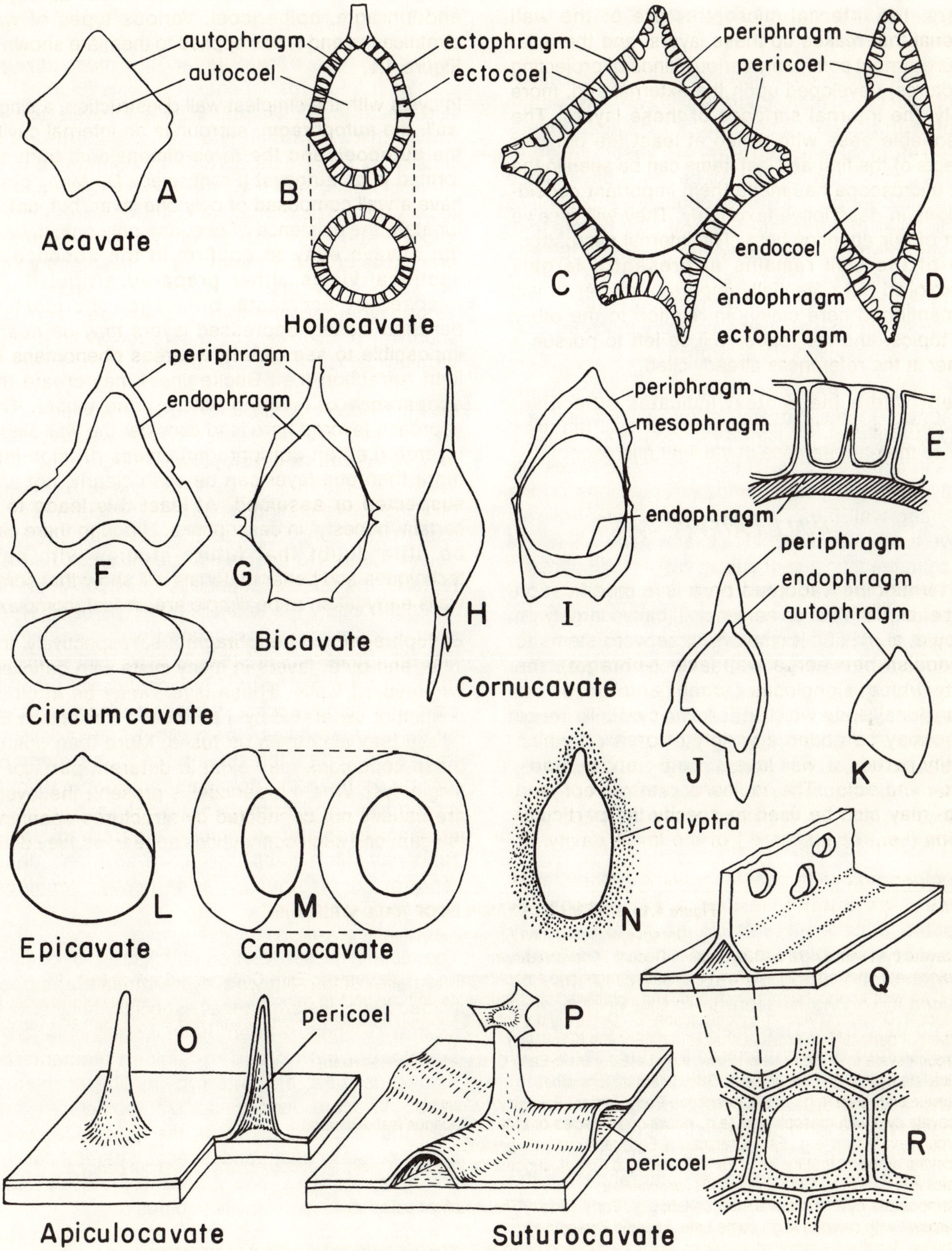
autophragm
autocoel
A
Acavate
ectophragm
ectocoel
B
Holocavate
periphragm
pericoel
endocoel
endophragm
C
D
ectophragm
periphragm
mesophragm
E
endophragm
periphragm
endophragm
F
G
H
Bicavate
I
Cornucavate
Circumcavate
periphragm
endophragm
autophragm
J
K
L
Epicavate
M
Camocavate
calyptra
N
Q
O
pericoel
P
pericoel
R
Apiculocavate
Suturocavate

Figure 4.1

occasional species, their sporadic occurrence and irregular distribution usually mark them as quite incidental features.

Under the light microscope the septa or processes that project from many chorate cysts are easily seen to be composed of periphragm rising above a substratum of endophragm. However, in other species, either the layers may be too subtly defined for their boundary to be recognizable or only a single layer (differentiated or not) may be present, as confirmed by TEM studies (e.g., Jux, 1968 and later). Therefore, the presence of processes or septa, alone, should not be taken as prima facie evidence for two wall layers, contrary to the implication of some published descriptions.

Diagnostic but easily overlooked indicators of the presence, at least locally, of two wall layers are narrow strips of pericoel that run along the parasutures in some species. These elongate cavities are easily masked by thickenings or elevations on the overlying wall (i.e., ridges or septa) immediately above them. However, close examination will reveal the characteristic pair of subparallel lines (Figure 4.1 P-R) where the wall layers come into contact at the limits of the narrow pericoel. If only one line is noticed, it may be mistaken for a parasuture, but close inspection will show that it does not reach the triple junction at the corner of the paraplate, as true parasutures would do. Lines having a similar origin may also be visible encircling the bases of projecting structures that rise from a point (i.e., processes), demonstrating both the hollow construction of these projections and the presence of two wall layers (= apiculocavate cyst).

An **ectophragm**, a layer of highly varied character, is present external to either a periphragm or autophragm of some cysts (Figure 4.1 B-E). It may consist of a membrane that is continuous, perforate, or interrupted along parasutures, or it may be reduced in varying measure to rodlike or ribbonlike trabeculae between the tips of supporting processes, or even to remnants at the margins of individual process tips. An ectophragm is almost always supported above the underlying layer by processes or other projecting structures, and these are normally numerous, distributed extensively over the surface, and clearly form an integral part of the wall design. Although the ectophragm, its supporting structures, and the underlying layer usually are more conveniently considered individually (owing to their physical separation and contrasting morphology) the continuity of their substance indicates that these three features really represent a single complex layer. Projections from the ectophragm are unknown.

A further type of wall layer, the **mesophragm**, occurs in a few fossil peridinioid cysts (Figure 4.1 I, and discussed in Chapter 9). It is a thin layer lying between two other layers easily identified as periphragm and endophragm by comparison with other species. Although the mesophragm may be locally in close contact with the periphragm or endophragm, it is not joined to either of them by processes or similar structures. A mesophragm has not yet been reported in fourfold combination with ectophragm, periphragm, and endophragm, but such wall structures (and others for which present terminology is inadequate) are likely to be found.

One fairly common but previously unlabeled type of wall complexity is expressed in what we will refer to as a **differentiated autophragm** (Figure 4.2 C-F). This is a wall in which an obvious difference in microstructure distinguishes two (or more) regions essentially concentric to the cavity of the cyst; however, the wall substance is continuous between regions, the textural change is transitional, and no internal boundary is defined. One type of differentiated autophragm is associated with cysts whose wall and processes are usually described as fibrous (e.g., *Cordosphaeridium, Turbiosphaera*). Fiberlike elements from the process bases merge with others on the outer surface of the wall between processes, and, as the packing density of these elements increases inwardly, the wall may become quite continuous and homogeneous toward its inner surface. Other types include a differentiation of relatively thick walls into more and less dense "spongy" or "frothy" zones (*Aldorfia, Canningia, Spongodinium*).

The distinction between one thick layer of complex internal structure and two layers intricately interconnected becomes a matter of especially subjective or arbitrary judgment when the inner and outer parts are close together and the linking elements are numerous, fine, and intergrown. The problem is especially acute when the scale of the details is finer than the resolving power of the light microscope. Such complex constructions require careful description on an individual basis; reliance on terms coded to precise definitions invites confusion.

Although the obvious diversity of wall structures has evoked an abundance of descriptive terms, our knowledge is by no means proportionate to the apparent sophistication of the terminology. As yet we can really only speculate about the true homologies among the features we describe and name, and different authors will apply some of the names differently. For example, one of the combinations that is here termed ectophragm and autophragm (Figure 4.1 B) will also be found described as periphragm and endophragm.

While admittedly imperfect, the following sequence of questions and answers provides a "rule-of-thumb" procedure for identifying the most frequently encountered types of wall layers in a reasonably consistent manner: To decide what wall layers are present, first ask: How many wall layers are visible?

If the answer is "One", then it is an autophragm.

If the answer is "Two", then ask the further question: Are the two layers linked by numerous structures that are obviously an essential part of the wall design, being distributed extensively over the surface?

If the answer is "Yes", then the layers are ectophragm and autophragm.

If the answer is "No", then the layers are periphragm and endophragm.

If the answer is "Three", then ask the same question as when two layers are visible, but apply the question to the two outermost layers now being considered.

If the answer is "Yes", then the layers are ecto-, peri-, and endophragm.

If the answer is "No", then the layers are peri-, meso-, and endophragm.

A different and historically interesting approach to descriptive terminology for wall layering was suggested by R. L. Cox (1971). Although having the merit of studied objectivity, his proposal was eclipsed by the earlier appearance of the comprehensive work by Davey et al. (1966) that did so much to set the style for contemporary analysis of dinoflagellate cyst morphology and to establish the terminology for describing it.

Almost certainly, the complexity of wall composition and structure is far greater than we yet appreciate. For example, scattered evidence suggests differential distribution of sporopollenin in the cyst wall. Absence of a sporopollenin periphragm would explain autocysts that are morphologically extremely similar to endocysts of other species (Figure 4.1 J-K), whereas other autocysts resemble pericysts but cannot be so called for want of an endocyst (Figure 4.1 A). Differentially preservable wall-building substances in a single species may account for walls with localized thin spots or perforations (e.g., *Odontochitina* spp., *Thalassiphora reticulata*), walls consisting of reticulate frameworks only (e.g., *Evittosphaerula, Eatonicysta*), and thick walls with internal "struts" or "spongy" filling already mentioned. Features that apparently reflect inhomogeneities in original wall composition may be relatively random and irregular in areal distribution, or highly systematic and closely coordinated with the paratabulation.

Both thickness and continuity of wall layers vary, but certain tendencies are apparent. The endophragm or autophragm is usually thicker than more external layers, and this innermost wall is virtually never penetrated by an opening other than the archeopyle (but note *Dinogymnium*, Chapter 12). The periphragm also is continuous in many cysts, but in many others it is penetrated by large or small openings of variable number and arrangement (discussed later in this chapter). Attention already has been drawn to the extreme range in character of the ectophragm. The mesophragm is a thin but continuous layer in the few species where it has been observed.

The foregoing discussion has emphasized the wall layers, but at times it may be just as useful to focus attention on the **cavation** of the wall (Stover and Evitt, 1978); that is, the form and distribution of the major cavities between layers, especially the pericoel (or its analog, the ectocoel, in those cysts with a prominent and continuous ectophragm). Thus, the terms **epicavate**, **hypocavate**, and **bicavate** may be used to describe cysts in which one cavity is confined to either the epicyst (Figure 4.1 L) or the hypocyst, or two cavities occur, one at each pole (Figure 4.1 G, I). In **holocavate** cysts an encircling cavity is evident in all directions of view, and supporting structures of some kind extend between the wall layers (Figure 4.1 B). By contrast, in **circumcavate** cysts, the cavity is continuous around the margin of the cyst as seen in dorsoventral view, but the two layers are in contact middorsally and midventrally (Figure 4.1 F). **Camocavate** cysts have the wall layers in contact or close together on one surface, but more widely separated elsewhere (Figure 4.1 M). Cysts in which cavities occur only at or near the bases of horns or equivalent structures are **cornucavate** (Frontispiece J, Figure 4.1 H); they are **suturocavate** (Benedek et al., 1982) if the cavities are restricted to the lines of parasutures (Figure 4.1 P-R) and **apiculocavate** (new) if restricted to the bases of spines or processes (Figure 4.1 O). **Acavate** cysts are without wall cavities (Figure 4.1 A). Sarjeant (1982a) proposed new special terms for particular varieties and combinations of pericoel distribution, but they appear to this author to have limited utility.

A final kind of enveloping material, although hardly to be considered a part of the wall, is a **calyptra** (Figure 4.1 N). Looking like a gelatinous mass, irregular in thickness and closely surrounding the cyst, a calyptra is often made visible by a "cloud" of minute foreign particles enmeshed in the "gel". Calyptras are encountered infrequently and are associated especially with certain pyriform to fusiform cysts of Late Jurassic and Early Cretaceous age.

SURFACE RELIEF

Form of Relief

The inner surfaces of autophragm, endophragm, and periphragm are usually smooth, and the outer surface of the endophragm is either smooth or has features of only low to moderate relief. In contrast, the outer surface of an autophragm or periphragm may be extended into a great variety of projecting structures (Figure 4.2, 4.3). As already noted, an ectophragm can be considered to result when trabeculae or membranes interconnecting these projections distally are sufficiently continuous to be regarded as a discrete layer.

Most projections from the autophragm or periphragm fall into two general categories: those which rise essentially from a point or very limited basal contact with the underlying surface, and those which rise from a linear base. The former, depending upon their height, include granules, verrucae, spinules, spines, and a host of structures collectively termed **processes**. Similarly, the linear elements grade from ridges or low folds into higher, fencelike **septa**, which have also been termed ledges, crests, or lists.

In chorate cysts (Frontispiece E), the projecting structures are especially prominent, exceeding about a tenth of the diameter of the main cyst body at the lower limit and even surpassing that diameter at the other extreme. For cysts having projections in the lower part of the range (between about one to three-tenths of the main body diameter), the term **proximochorate** has been used. Although precise quantitative limits for these terms have been suggested, variability and intergradation seem too great to justify this, and we will use the terms only in a general descriptive sense. Depending on the kind of projections that dominate, chorate cysts may be **skolochorate**, with processes, and **murochorate**, with septa.

Tips of adjacent projecting structures are normally of closely similar height. The resulting general conformity of process tips or septal crests over a

Figure 4.2 - SCHEMATIC DETAILS OF WALL FEATURES

A-B - Grana, coni, and processes of several types, standing singly and interconnected. Conformity of tips of neighboring processes suggests limiting effect of once-surrounding theca.

C-F - Differentiated autophragm. C-E - With "fibrous" structure and different internal density. F - With internal vesicles of different relative volume (as in Late Cretaceous *Spongodinium delitiense*).

G - Projecting structure with distal branches which expand into irregularly perforate ectophragm (as in some Late Cretaceous-Early Tertiary *Glaphyrocysta*).

H - Perspective detail and section of wall with local separation of periphragm and endophragm, layers closely appressed elsewhere (e.g., Early Tertiary *Pentadinium*).

I - Detail longitudinal section of apical horn and adjacent wall in an autocyst (as in many Jurassic-Tertiary *Millioudodinium* and Jurassic-Cretaceous *Cribroperidinium*). Stippling shows sporopollenin cyst wall; unstippled areas show hypothesized thecal plates (ap - apical plate, pp -preapical pore-bearing plate). Diagram suggests origin of distinctively shaped solid apical horn of cyst as a "plug" filling the apical pore and subjacent extremity of the thecal cavity. Neighboring short processes derive their expanded extremities from contact with inner surface of the theca.

J - Perspective sections through theca (top), periphragm (middle), and endophragm (bottom) of encysted specimen of modern *Peridinium limbatum*. Theca shows striate growth band between areas with trichocyst pores. Exterior of periphragm shows corresponding smooth pandasutural band and adjacent fields of coni. The endophragm is externally smooth, but internally a narrow groove corresponds to the suture and the rest of the surface is finely rugulate. (See also Figure 1.5 A-E and discussion in Chapter 1).

K-N - Perspective diagrams showing several expressions of ectophragm. K - A continuous layer supported by numerous processes (e.g., Late Jurassic-Early Cretaceous *Gardodinium*). L - Ectophragm interrupted along parasutures (e.g., some Eocene *Kisselovia*). M - Highly perforate ectophragm (e.g., Early Tertiary *Eatonicysta*). N - Ectophragm reduced to pairs of trabeculae extending along parasutures between trifurcate processes at paraplate corners (e.g., Tertiary-Holocene *Nematosphaeropsis labyrinthea*). Small circles in K-N indicate positions of processes beneath the ectophragm.

O - Isolated process with expanded polygonal tip (of ectophragm) suggesting outline of thecal plate (e.g., Eocene *Areosphaeridium diktyoplokus*).

P - Several types of interconnecting or supporting structures such as occur between ectophragm and underlying periphragm or autophragm.

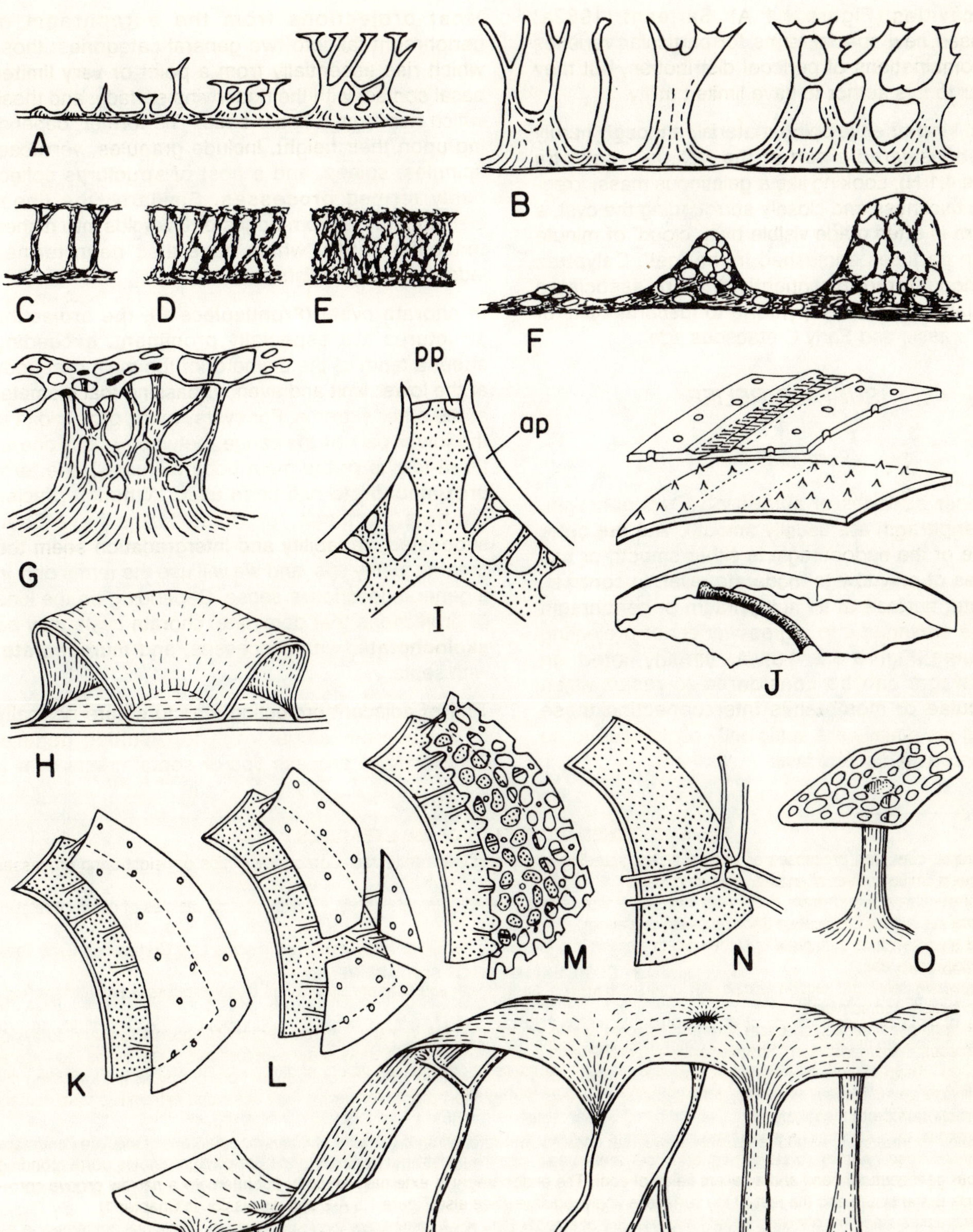

Figure 4.2

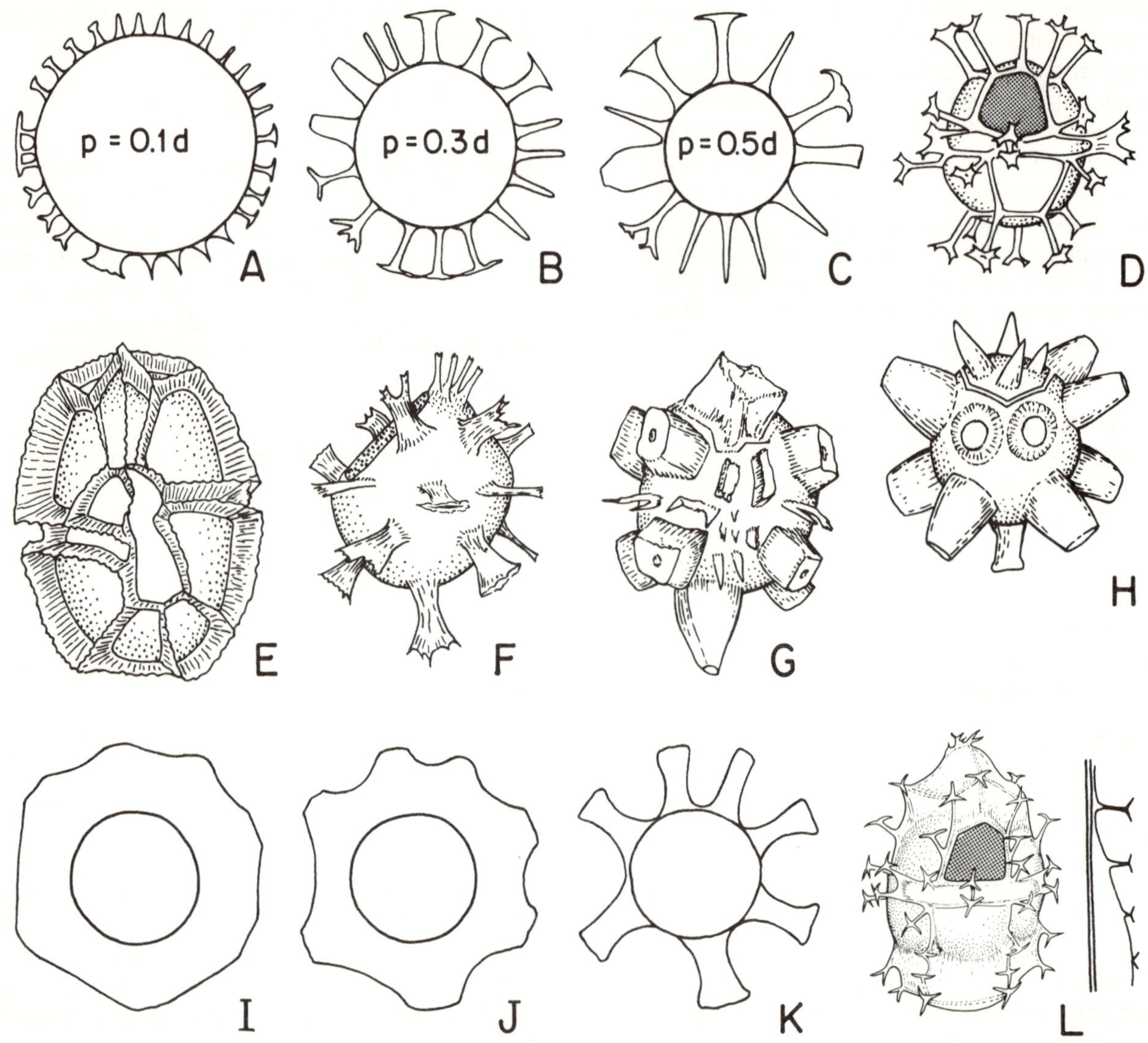

Figure 4.3 - SOME EXAMPLES OF SURFACE RELIEF

A-C - Diagrams to show surface relief of different height relative to the diameter of the main body. A - Maximum height of relief (less than 10% of main body, or endocyst, diameter) present on proximate cysts. B - Maximum height of relief (about 30% endocyst diameter) on proximochorate cysts. C - Chorate cyst with relief height equal to 50% endocyst diameter.

D - Skolochorate cyst with parasutural ridges giving rise at their intersections (i.e., at corners of paraplates) to trifurcate processes (Early Tertiary *Spiniferites pseudofurcatus*); dv, showing archeopyle (patterned).

E - Murochorate cyst having paratabulation indicated by parasutural septa (Early Cretaceous *Prionodinium*), vv.

F-H - Skolochorate cysts without clear parasutural markings but with processes situated in the central portions of paraplates (intratabular). F - Solid, fibroid processes, differing in size but of similar design (Early Tertiary *Cordosphaeridium inodes*), rlv (archeopyle stippled). G - Boxlike processes represent pre- and postcingular paraplates, slender conical processes are single in parasulcus and paired in paracingulum; apical processes are merged into a large irregular unit, and antapical process is uniquely long (Early Tertiary *Hystrichokolpoma*), vv. H - Conical apical processes (on archeopyle operculum surrounded by archeopyle suture) and large, hollow, approximately cylindrical processes elsewhere, but no cingular processes (mid-Cretaceous *Litosphaeridium siphoniphorum*), dv.

I-K - Optical sections through precingular region of cysts seen in apical view, showing extremes of morphology in a single species. Endophragm has constant size and form, but processes are variably defined by periphragm that "sags" toward the endophragm from peripheral loci that are the tips of the fully formed processes in K (undescribed Miocene species).

L - Bicavate cyst with surface relief, in principle, like that in D, but with the length of process shafts proportioned inversely to the height of the pericoel (clarifying diagram at right); processes, in effect are formed from the tips down (Early Tertiary *Rottnestia*), dv (archeopyle patterned).

specimen is usually conspicuous, even when gradation toward longer or shorter projections in particular regions (e.g., polar area, paracingulum, parasulcus) is taken into account. The shafts of processes may be hollow or solid, or hollow proximally and solid distally. When hollow throughout their length, the ends may be open or closed. It is not unusual for a moderate to strong difference in form (especially in the direction of smaller and simpler processes than occur elsewhere) to distinguish the processes of the parasulcus and paracingulum. Also, a uniquely constructed process may stand at either pole.

Process tips and bases are often expanded. In occasional species with relatively large and few processes (under 30), the outline shapes of the distal expansions are clearly polygonal, matching the shapes appropriate to the thecal plates represented by the process-bearing paraplates. When processes are more numerous, they may be about uniformly spaced or associated in groups. Assemblages of processes representing a single paraplate have been called **process groups**.

It is not always easy to decide what to accept as the limits of a single process, for intricate gradations in form may occur. Imagine, for example, a gradation from a simple process with a single shaft to a large process whose shaft includes several structural elements closely associated, then to a cluster of more widely spaced elements more loosely united, and finally to a group of discrete projections joined at their bases by ridges or at higher levels by trabeculae or membranes. In some species such gradations may be observed in a single individual, although more frequently the processes on one specimen are of the same or closely similar construction.

In a number of cysts that are otherwise morphologically different (e.g., some species of *Chlamydophorella, Dinopterygium, Gardodinium, Heteraulacacysta, Membranophoridium, Pseudoceratium)*, the multiple supporting structures of an ectophragm, or the sporadic connectors between periphragm and endophragm, may have a similar and especially distinctive form. Giving the impression of long or short processes, they are partly or entirely hollow, closed at the base, and expand to a funnelform distal opening, the "process" wall merging smoothly with the external layer. Some examples appear almost as if they had been formed by pressing a slender object through one elastic layer until it touched an underlying layer, the layers adhering at the point of contact after the object was withdrawn. Examples vary in dimensions, proportions, and spacing (Figure 4.2 P).

Distribution of Relief

The local elements of wall structure, whatever may be their exact form and whether or not they involve surficial or internal detail, may be distributed in different ways (Figure 4.4). Combinations of distribution patterns may occur on a single cyst, and the clarity and completeness with which the relief features indicate the paratabulation may range between great extremes.

A distribution which gives no indication of paratabulation is termed **nontabular** and its elements may be uniformly or irregularly spaced. In contrast, a **tabular** distribution does indicate paratabulation and is said to be **parasutural** when it emphasizes the boundaries between paraplates or **intratabular** when it accentuates the central portions of paraplates more or less remote from these boundaries. Intermediate states have received a variety of special terms.

If the basal diameter of an intratabular feature is imagined to increase until it encloses most of the area of the paraplate, then the special case of a **penitabular** feature results. Penitabular features on adjacent paraplates may approximately parallel the paraplate boundaries. In principle, they form closed lines around each paraplate but penitabular elements associated with large pre- and postcingular paraplates of some species tend to be interrupted on their adcingular sides. Strictly speaking, penitabular features are contained within the limits of single paraplates. However, when they run very close to the parasuture, the pairs of lines from adjacent paraplates may give the appearance of a single parasutural feature.

The simplest and most apparent parasutural features are ridges or septa which mark off paraplates much as thecal sutures separate thecal plates. The dinoflagellate-like appearance imparted to a fossil by well-developed parasutural features is obvious (Figure 4.3 E). It was responsible not only for the first identification of dinoflagellates in the fossil record, but also for the early acceptance of their remains as fossilized thecae. Parasutural features may also comprise various kinds of discrete elements arranged in rows (e.g., granules, spinules, or processes), narrow strips of pericoel, and local modifications of internal wall structure; and these may occur alone or in combination with ridges or septa. Parasutural projections on a single

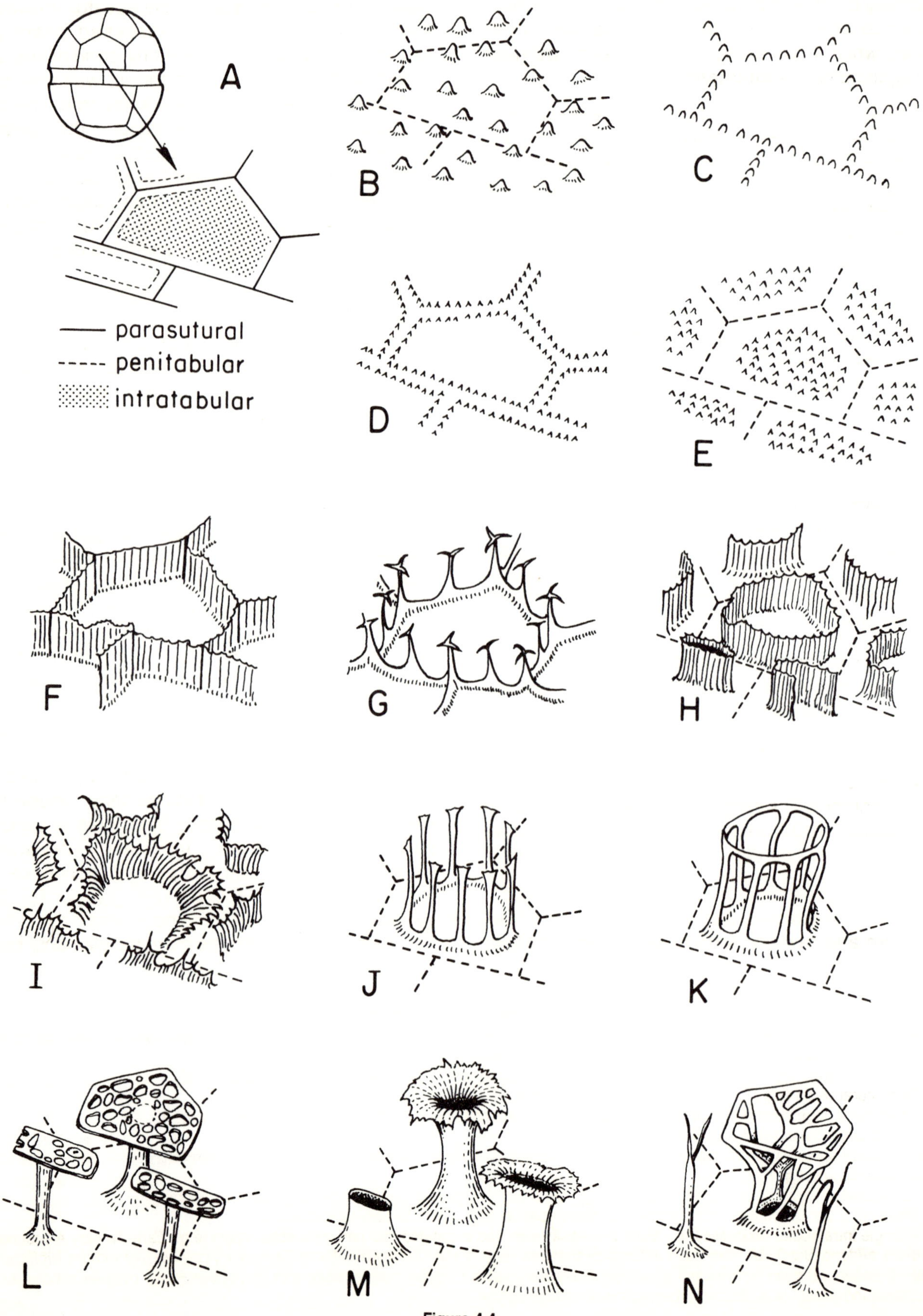

Figure 4.4

cyst may be of uniform or varied design. Projections localized at the angles of paraplates are termed **gonal**, while those along paraplate sides between triple junctions are **intergonal**.

Since features in intratabular arrangement lie at, or roughly encircle, the centers of paraplates, rather than being aligned along the bordering parasutures, they do not bring out the paratabulation pattern as obviously as parasutural features. Nevertheless, it still may be possible to interpret the paratabulation with confidence on the basis of an intratabular arrangement (recall *Hystrichosphaeridium tubiferum*, Figure 3.3 A-C, whose processes, one for each paraplate, are graduated in size to match the relative sizes of the thecal plates they represent; and *Oligosphaeridium*, in which some paraplates are simply not represented by processes, Figure 1.7 F). Process form may also differ substantially according to location on the cyst. As noted in Chapter 7 (Minority Suppression), processes associated with the smaller paraplates tend to be selectively reduced in size or modified in structure. In some instances, such differences, which must be distinguished from intraspecific variations, have been used as grounds for generic distinctions.

The zones of incremental plate growth intercalated along the thecal sutures in many modern species are clearly represented in some fossils by what have been called **pandasutural bands**. Like their thecal counterparts, these may even be transversely striate and vary in width from one part of the cyst to another. The strips of wall between adjacent penitabular structures may also correspond to these intercalary bands.

In cysts that are nearly circular in equatorial section, external relief features and internal features of wall structure and cavation tend to be developed about equally in all directions about the equator. However, among dorsoventrally compressed cysts (an assemblage which includes many peridinioids and certain groups in other categories) wall structure and surface relief may vary markedly from one part of the cyst to another. In principle, the variably expressed features are least apparent or of simplest structure in the central portion of dorsal and ventral surfaces, and most prominent or complex in a marginal band around the cyst (over the poles and along the right and left sides).

In cavate peridinioid cysts, which generally lack strong surface relief, a frequent expression of this regionally differentiated wall development is a central suppression of the pericoel, but this may not be very obvious in the dorsoventral view favored by the compression. By contrast, the effect is striking in many dorsoventrally compressed gonyaulacoid cysts with strong surface relief (e.g., in such species-rich genera as *Cyclonephelium*, *Glaphyrocysta*, and others, Figure 4.5). Both external relief features and features within the wall (e.g., the height of the pericoel or of elements connecting inner and outer wall surfaces) may be affected. In the extreme case, the central portions of dorsal and ventral regions may be essentially featureless, regardless of the form and prominence of the features marginal to these areas. In less extreme cases, the central area merely bears features that are somewhat "subdued" relative to those present elsewhere. External features may include elaborate processes or septa, often intricately expanded or interconnected. Internal features may include the empty pericoel or connecting structures of various types extending between inner and outer walls. Central reduction and marginal prominence of features is usually appreciably greater for the ventral surface than for the dorsal one. The dominating effect which the marginal elements have

Figure 4.4 - DISTRIBUTION OF FEATURES

A - Schematic dorsal view of a theca (above) and portion of dorsal surface of its cyst (below), to identify parasutures shown by dashed lines in other figures and to show location of parasutural, penitabular, and intratabular ornament.

B-E - Four principal types of distribution: nontabular (B), parasutural (C), penitabular (D), and intratabular (E).

F - Parasutural septa (e.g., Cretaceous-Holocene *Impagidinium*).

G - Parasutural ridges combined with gonal and intergonal processes (e.g., Cretaceous-Holocene *Spiniferites*).

H - Closed penitabular septa (e.g., Tertiary *Alisocysta*).

I - Penitabular septa with adcingular interruptions on larger paraplates; paracingular projections of different design (e.g., some Late Cretaceous *Areoligera*).

J - Penitabular to intratabular process group rising from a basal ridge (e.g., Jurassic-Tertiary *Systematophora*).

K - Similar to J, but processes united distally by ring-trabecula (e.g., Early Tertiary *Emmetrocysta*).

L-N - Single intratabular processes. L - Solid processes with tips expanded into perforate ectophragm (e.g., Early Tertiary *Areosphaeridium diktyoplokus*). M - Hollow tubiform processes (e.g., Late Cretaceous-Early Tertiary *Hystrichosphaeridium tubiferum*). N - Complex process (or process group) on major paraplates (condition intergrades with that shown in K), cingular processes of contrasting design (e.g., Early Cretaceous *Hystrichosphaerina*).

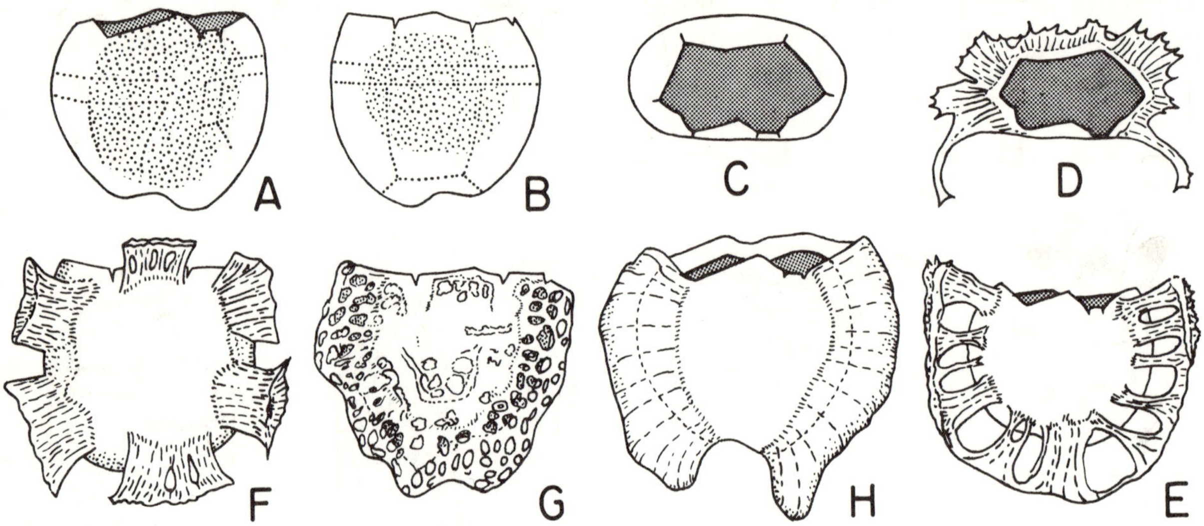

Figure 4.5 - RELIEF DISTRIBUTION IN LENTICULAR GONYAULACOID CYSTS

A-B - Diagrammatic ventral and dorsal views of cyst with open apical archeopyle, showing parasutures (dotted), area in which surface relief (or internal wall structure) is generally reduced or absent (stippled), and marginal area in which features are especially prominent.

C - Diagrammatic apical view of same, showing dorsoventral compression and transverse elongation of apical archeopyle associated with marginal accentuation of wall structures (marginate cyst).

D-H - Schematic simplifications based on Cretaceous and Early Tertiary species. D-E - Apical and ventral views of cyst with modified septa distally interconnected around margin (e.g., *Glaphyrocysta*). Ventral view shows only those elements immediately surrounding ventral surface. F - Boxlike marginal processes and short lines of septa toward poles on middorsal pre- and postcingular paraplates (undescribed form), dv. G - Highly perforate marginal periphragm represented centrally only by low and incomplete reticulum (undescribed form), dv. H - Marginal pericoel absent in central region of ventral surface (e.g., *Membranophoridium*), vv.

on the appearance of the cyst (which has suggested the term **marginate** applied by some authors to certain non-peridinioid cysts of this type) may obscure the basic plan of relief distribution, but, with close study, this may still be distinguishable as nontabular, parasutural, penitabular, or intratabular.

MISCELLANEOUS WALL FEATURES

Openings in the Wall

In addition to the archeopyle (Frontispiece B, D-E; Chapter 6), a great variety of other openings have been observed in the walls of dinoflagellate cysts. Some exhibit a constancy and distinctiveness that suggest potential taxonomic significance. Others are so variable in size, shape, position, and occurrence as to seem relatively unimportant. Until recently, openings apart from the archeopyle have been widely ignored or mentioned only in passing; rarely have they received special attention or descriptive terminology. Probably a more critical concern for them will be rewarding. Some of the relatively obvious features are pointed out in the following survey.

The term **claustra** (sing. claustrum) was introduced (Norris, 1978a) as a collective appellation for openings other than the archeopyle that occur in the outer wall (usually periphragm, sometimes ectophragm). It is adopted here to include openings varied in size and distribution, and in part of uncertain origin. Claustra do not penetrate the innermost wall layer (i.e., endophragm or autophragm) and have not yet been reported in cysts of living dinoflagellates.

As sometimes prominent and consistently localized features, these openings merit attention and careful description, but we will not consider here the specific terms suggested for some of them (see Sarjeant, 1982a) according to their precise form and location. A variety of claustra are illustrated in Figure 4.6. Some have the appearance of holes formed by excessive stretching of a thin membrane; others may result from selective degradation of an inhomogeneous wall (in vivo or postmortem?); still others may mark small areas of contact between cyst and theca; and a few appear to have been

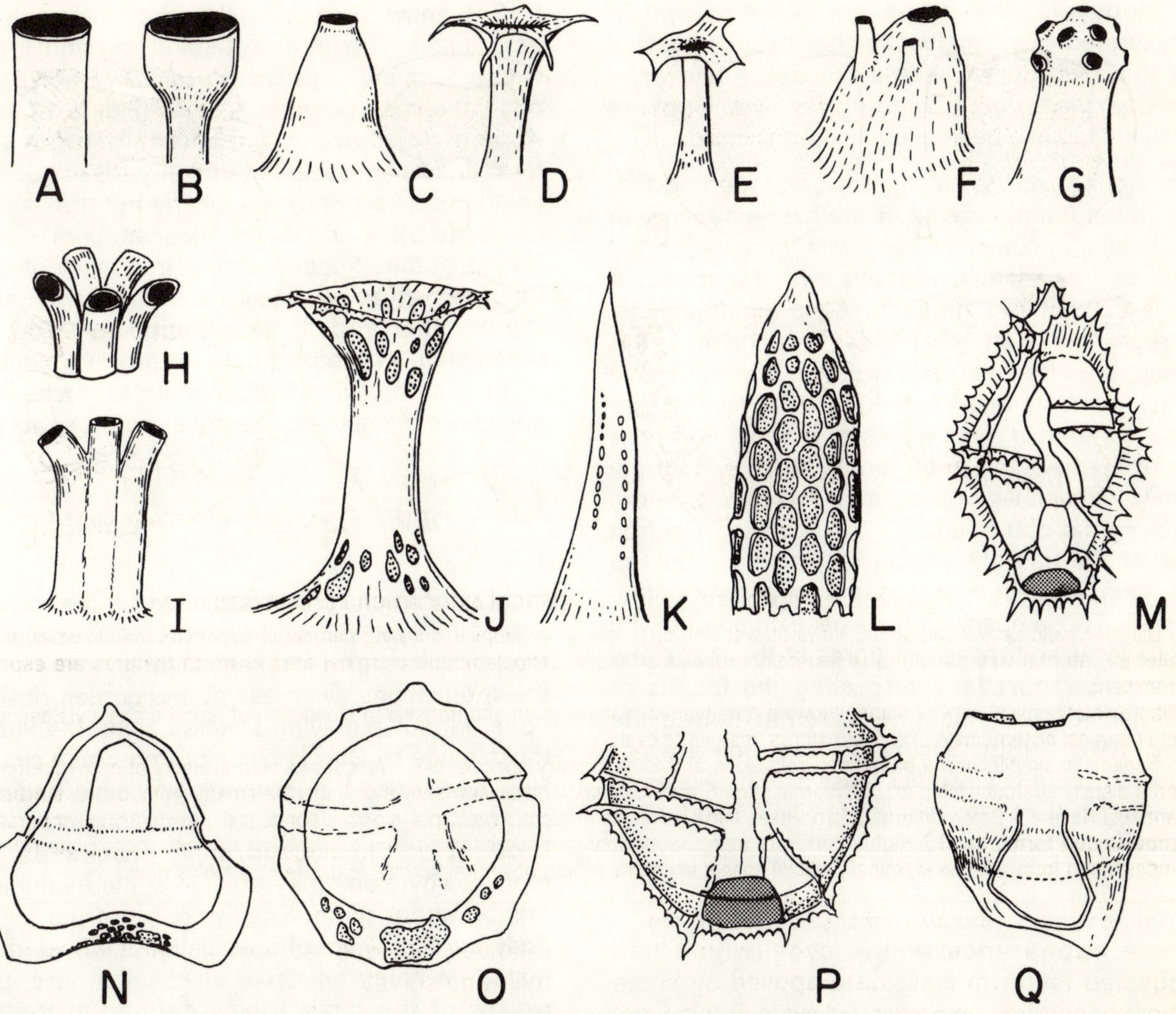

Figure 4.6 - CLAUSTRA - OPENINGS THROUGH THE OUTER CYST WALL

A-I - Openings at tips of processes into the pericoel of the process interior. A-C - Mid-Cretaceous *Litosphaeridium*, D - Cretaceous *Oligosphaeridium*, E - Early Tertiary *Spiniferites pseudofurcatus*, F - Early Tertiary *Hystrichokolpoma*, G -antapical process in Early Tertiary *Diphyes*, H-I - Early Tertiary *Homotryblium floripes*.

J-L - Perforations through outer portions of the periphragm. J - Process of Jurassic-Tertiary *Hystrichosphaeridium*, K-L - horns of Cretaceous *Odontochitina* species.

M - Opening in antapical paraplate in Late Jurassic-Early Cretaceous *Tubotuberella eisenackii*, vv.

N-O - Openings in antapical paraplate. N - Cribrate and irregular openings in Late Jurassic-Early Cretaceous *Sirmiodinium grossii*, dv. O - large opening in antapical paraplate, and small openings into other parts of pericoel in Late Jurassic *Sirmiodinium* sp., vv.

P - Opening in posterior sulcal paraplate in Late Jurassic *Gonyaulacysta jurassica*; portion of endocyst visible through opening, vv.

Q - Distinctive operculate opening in antapical pericoel in mid-Cretaceous *Ascodinium verrucosum* (apical end truncated by archeopyle), vv.

functional, although it is not easy to identify the purposes they may have served. Some are widely distributed over the surface, while others are limited to species-specific locations, and variability in the degree of their development in different individuals may be great.

An especially distinctive opening occurs in *Ascodinium verrucosum*, placing the hypopericoel in communication with the exterior (Evitt, 1967). The opening is located on the ventral surface, is defined by a suture in the periphragm, and is covered by a flap-like operculum attached along its adcingular end. The shape of the operculum strongly suggests general equivalence to the posterior sulcal paraplate in this peridinioid form, but lack of parasutural markings on the surface leaves the interpretation speculative. The opening resembles an archeopyle, but its location is unique. It leads only into the hypopericoel, and a well-developed archeopyle is already present in the epicyst. A pressure-equilibrating function has been suggested but this hypothesis seems fanciful; however, a better explanation is wanting for what appears definitely to have been a functional structure.

The archeopyle is usually the only opening that brings the interior cavity of the cyst (endocoel or autocoel) into communication with the exterior, but there are interesting exceptions, two of which occur in the distinctive Late Cretaceous dinoflagellate, *Dinogymnium* (Evitt et al., 1968; Evitt, 1968). In this genus, the single-layered wall is penetrated by a lobate ventral opening in the sulcal region and by a large number of widely distributed minute pores (under 1µm). Some of the pores clearly penetrate the wall completely; others may end blindly before reaching the outer surface. The scanning electron microscope has documented their details but, so far, has been used for examining only a few species. These openings have been compared with the flagellar and trichocyst pores of the theca and cited as support for interpreting the fossils of *Dinogymnium* as motile cells, rather than cysts (May, 1976, 1980). Similar perforations are not known to occur in any other fossil dinoflagellates or in resting cysts of any modern species. See further discussion in Chapter 12.

The Porichnion and Its Thecal Counterpart, the Adelopore

In species of living *Triadinium* and many similar gonyaulacoids, a distinctive porelike opening about the size of an apical pore penetrates the ventral thecal surface in a right anterolateral location. The function of the opening is unknown, and it has been mentioned infrequently in publications, usually under the general and somewhat ambiguous designation of "ventral pore" (e.g., Taylor, 1976, 1980; Dodge, 1982). It seems always to be located in approximately the same position, either within the limits of a single plate (e.g., in *Triadinium*, Figure 5.7 E) or along the junction between two plates (e.g., *Gonyaulax spinifera*, Figure 5.12 E). The term **adelopore** (Gr. adelos = obscure) is suggested here for this opening in the theca.

A variety of thecamorphic structures in fossil cysts evidently represent the adelopore, but little attention has been paid to them. Probably they have usually been overlooked entirely, even where they are prominent enough to be detected with the light microscope. Scanning electron micrographs of *Cribroperidinium ventriosum* (Gocht, 1979a; Figure 5.12 D) show the feature clearly on the apical horn as a small pit with a craterlike rim, and it has a similar form in *Gonyaulacysta jurassica* (WRE, pers. obs.) and in some other species (Figure 10.9 B-C). Apparently equivalent structures of more varied form are being observed in many fossil gonyaulacoid species but they are clearly not present in all (J. Lucas-Clark and Javier Helenes, pers. comm.). Details of the structure and its frequency of occurrence remain to be established, but it is certainly a feature that should be watched for. The term **porichnion** is proposed here for any such localized feature on the cyst which appears to represent the adelopore of the theca, regardless of its exact form.

DEVELOPMENT OF THE WALL

What phenomena are associated with the construction of the cyst wall, and what processes determine the particular form it takes? Definitive answers to these questions await as yet unrecorded observations on multiple living species along the lines of those made by Dürr (1979c) for *Peridinium cinctum*. However, since the morphology visible under the microscope—even for fossil cysts long dead—is a product of the cyst-forming processes, this morphology alone may provide some useful information about them. Space does not permit (nor does the high level of speculation it would involve justify!) a comprehensive discussion here, but a review of the three topics defined in the next paragraphs will serve to introduce some of the issues. The summarily stated "answers" are tentative in varying degrees and are intended as challenges to thinking in an area where more soundly based answers lie in the future.

Situation and question: Since the cyst wall forms between the protoplast and the theca, both the DNA within and the enclosing theca without are in positions possibly to influence the form of the final cyst. While recognizing that controls for all aspects of cyst formation reside ultimately in the genetic material, one may ask: Does the theca play a direct role of any consequence in determining the character of features on the cyst? Answer: Yes, and in some species the effects of endogenous (genetic) and exogenous (thecal) influence can be discriminated.

Situation and question: The exterior surface of many cysts bears prominent projecting structures in the form of spines, processes, septa, etc. As the cyst forms, do these projections develop in a centrifugal sense (i.e., outward from the encysting mass of the cell), or in a centripetal sense (i.e., inward from the level of the theca)? Answer: Some evidence positively supports development in a centripetal sense; some evidence that has been argued to support centrifugal growth is equally compatible with centripetal development; some evidence is ambiguous or noncommittal, but strong support for centrifugal growth seems to be lacking.

Situation and question: Morphological variability of cysts is conspicuously great in many species. Are all the differences observed among conspecific individuals "frozen" into the substance of the wall as it forms, or is the wall of a single cyst subject to progressive reorganization in form as the cyst matures? Answer: A majority of examples seem adequately explained without calling upon what appears to be a complex process of reorganizing wall material once it has formed. However, the possibility of such reorganization is difficult to exclude, and a few fossil cysts appear to have involved a major amount of reorganization during cyst development.

Influence of Theca and Protoplast

It is obvious that many dinoflagellate cysts, both modern and fossil, resemble thecae in a variety of ways. In Chapter 1 it was suggested that this similarity stems from two sources. Some cyst features are thecalike simply because both containers of the cell are constructional responses to a single genotypic control. Others owe their thecalike aspect to a direct, essentially mechanical influence, by which the theca serves as a sort of template for the cyst forming within it, thus extending or mediating the still fundamental genetic control.

The evidence for thecal influence on cyst morphology lies in the form and, in part, the distribution of features on the external cyst surface (often at the extremities of projections), which point to their having been molded against an external surface, presumably the surrounding theca. These features include the following:

1. Abrupt termination and distal expansion of small-scale projections (Figure 4.2 A-B, I).
2. Polygonal outlines of the expanded tips of much larger, individual intratabular processes which are appropriate for outline shapes of thecal plates in the same positions (Figure 4.2 O; 4.4 L, N; 11.3 I).
3. Cavities or interruptions in extremities of processes or septa which apparently represent the loci of globules (fluid or gaseous?) caught within the sporopollenin of the cyst but contacting the surrounding surface (Figure 4.2 I).
4. Concordant elevations of the tips of neighboring processes, often combined with adjustment of process length so that the tips collectively imply a body of more thecalike shape than that from which the processes rise (Figure 4.1 C).
5. Ectophragmal continuations from process tips, which consistently tend to define a three-dimensional surface of the form that would be expected if it had originated as a layer applied against the inner surface of the theca (Figure 4.1 C).
6. Details of the apical horn in cysts of some gonyaulacoid species, which suggest that it formed as a plug of sporopollenin inside the apical horn of the theca (Figure 4.2 I).
7. A great variety of features expressing intercalary growth bands (= pandasutural bands on cysts); in other words, characters that develop during cell ontogeny and that, therefore, have phenotypic aspects peculiar to the individual mature theca, as well as genotypic aspects relating to fundamental cell geometry (Figure 4.2 J, L). See further discussion of these features later in this chapter.

Although these apparently contact phenomena are widespread, they certainly are not found at distal extremities of all surface features or in all cysts. For example, acuminate spines without trace of distal expansions also occur, and variations in length among closely spaced processes obviously preclude equal contact with a surrounding surface of more or less even convexity. Also, it should be quite obvious that mere compatibility between the form of a feature and formation in contact with the theca is not proof that such contact existed.

Unfortunately, cysts of most species living today lack surface features ideally suited for recording this template effect in an unambiguous way. Nevertheless, some critical observations on modern specimens have been recorded. For example, the contact between cyst and theca was observed in *Gonyaulax digitalis* (Evitt and Davidson, 1964), a species whose cyst is morphologically equivalent to widespread fossils referred to *Spiniferites*. In this case, modern cyst and theca were seen to have the close contact at process tips that had been hypothesized earlier (Evitt, 1961) on the basis of the fossils.

As mentioned in Chapter 1, the cyst and theca of the living freshwater species, *Peridinium limbatum* (Figure 1.5 A, C; 4.2 J) provide compelling evidence that both the template effect of the theca and genetic control from the protoplast play important roles in determining certain features of the cyst. The patterns on the external surface of this cavate cyst faithfully match those peculiar to the mature zygotic theca, including the intercalary growth bands acquired during its enlargement. In contrast, a paratabulation on the inner surface of the endophragm, while essentially comparable to that on the periphragm, is defined by single, linear parasutures only, instead of broad pandasutural bands. In other words, the innermost surface that was adjacent to the protoplast appears to show the geometrical pattern of tabulation/paratabulation typical for the species, unmodified by partly phenetic features (i.e., intercalary growth bands of differing width) which are recorded only on the outer cyst surface that formed immediately within the theca.

A slightly different kind of template effect is suggested in the case of the fossil species described by Ehrenberg as *Peridinium pyrophorum* and now referred to *Palaeoperidinium* (Frontispiece C; Figure 9.6 I-J). This fossil was recently interpreted (Gocht and Netzel, 1978) as an especially impressive example of how genetic control from the protoplast may lead to a remarkably thecalike pattern on the cyst. However, a different interpretation (already considered and rejected by Gocht and Netzel) is supported in Chapter 9; namely, that *P. pyrophorum* represents a unique case of a sporopollenin wall formed external to the theca, and displays on its inner surface relief that originated through direct contact with the outer surface of the theca.

Centripetal or Centrifugal Progression in Cyst Formation?

If the question relates only to whether the outer wall or the inner wall of an obviously multilayered cavate cyst forms first, the answer is simple: centripetal development has been observed in modern species. Pfiester (1975 and later) reported periphragm forming before endophragm in species of living *Peridinium*, including *P. limbatum* (pers. comm.) in which this sequence was inferred earlier from empty cysts that lacked the endocyst present in most specimens (Evitt and Wall, 1968). Fossil cysts of similar appearance probably formed similarly, but this cannot be established beyond doubt from observations on the cysts alone.

If, however, the subject of attention is not the wall layers but the prominent processes and other major elements of surface relief, and the question is whether these form from the base outward or from the tip inward, then we find no recorded observations on formative processes in living cells, and contradictory opinions expressed by investigators who have studied only completed cysts of different kinds. The uncertainty might be reduced if more were known about the distribution of cytoplasm in living cysts during the period of wall development. Although the bulk of the cytoplasm appears to occupy only the innermost cavity of the cyst, its restriction to that position would preclude centrifugal growth of features on the exterior of a continuous periphragm or following the appearance of an endophragm. A jellylike substance has been observed around cysts of some living species (e.g., *Scrippsiella faeroensis*, = *Peridinium faeroense* as reported by Dale, 1977), but it is not clear that this was living protoplasmic material.

With respect to development of processes (or other projecting structures of substantial length) at least three possibilities come to mind. First, processes may develop in a purely centripetal sense, forming progressively from their tips downward as the protoplast contracts, and coming to stand eventually as supports between the condensed protoplast and the theca. This manner of forming process tips in contact with the interiors of thecal plates would readily explain the concordant lengths of neighboring processes, as well as the thecalike details often displayed when these tips are large and expanded. Necessarily, the overall dimensions of cysts formed in this way would be limited by the maximum dimensions of the theca.

In contrast, a second possibility is that processes elongated in a purely centrifugal sense, growing outward from a previously contracted protoplasmic mass by addition of wall material at distal extremities. In this case, neighboring processes could easily be of unequal length and indications of cyst-theca contact would be limited to those processes whose developing tips reached the level of the theca before its plates dissociated. After this dissociation, the theca would impose no limit on the size which the cyst might attain.

In a third possibility of somewhat "hybrid" character, the process tips might form first in contact with the theca, and subsequently be displaced outward by redistribution of cytoplasm at process bases. As a result, process tips might show indications of original contact with the theca, while lengths of neighboring processes could vary, and overall dimensions of the completed cyst might exceed those of the theca.

Choice among these alternatives is hindered by lack of evidence from living specimens and by the multiple interpretations consistent with the circumstantial evidence from empty cysts. Features observed on many cysts give conflicting or inconclusive evidence about direction of cyst growth, and even when their implication has seemed clear, addition to our limited knowledge of cysts and life cycles in living species has shown some conclusions to be unjustified. A few examples will illustrate the problem.

Features that suggest centripetal process development especially strongly are well illustrated by the cavate species of *Spiniferites* and a fossil species probably related to *Hystrichokolpoma* (discussed more thoroughly in Chapters 10 and 11, respectively). In cavate cysts of some species of *Spiniferites* and similar genera, an inverse relationship between the lengths of process shafts and pericoel height is clearly evident (Figure 4.3 L). The process shafts extend from the furcate tips to the "local base level" determined by the periphragm. Where the periphragm contacts the endophragm (i.e., in an acavate area), the processes are fully formed, but where a pericoel is present, the shafts exist nowhere below the periphragm; consequently the lengths of the upper portions vary inversely with the height of the pericoel.

In the case of the somewhat *Hystrichokolpoma*-like cyst, intraspecific variants range from individuals with a nearly spherical pericyst to ones with well-developed processes. Gradations between the extremes (Figure 4.3 I-K) appear to record progressive degrees of periphragm collapse between small areas (intratabular patches) where theca and periphragm remained in contact.

Many chorate and proximate cysts are characterized by a conspicuous concordance in the lengths of neighboring processes, combined with expanded tips that appear to have contacted a surrounding surface. These relations might result from either centripetal or centrifugal process growth. Processes that differ markedly in length from specimen to specimen but occur on central bodies of similar diameter within a single species (a condition observed in many fossil cysts and some living ones), may seem to suggest stages in centrifugal process development, but they could as well result from encystment of thecate cells of greatly different diameter. Since we do not know how nearly constant the ratio of theca volume to cyst volume is for a species, it is difficult to choose between these alternatives. Failure to find mature cysts (i.e., cysts with the longest processes typical for the species) in unruptured thecae of living *Scrippsiella faeroensis* suggests that process elongation continued centrifugally after thecae broke apart (Dale, 1977).

A scenario that would result in acuminate processes which lack contact-modified tips and differ markedly in length over short distances might be imagined under either growth regime.

The occurrence, in modern samples, of cysts having diameters greater than those of thecae for the same species has been cited as evidence for centrifugal growth (Wall and Dale, 1968b). However, such evidence is not sufficient in itself for that conclusion, since it is now known that (for some species, at least) the zygotic thecae within which cysts form are consistently larger than the much more numerous and more frequently observed vegetative thecae. Dale (1982) pointed out the need in such cases to be sure that cyst and theca of the zygotic stage are being compared.

Chorate cysts whose processes have expansive polygonal tips, such as *Hystrichosphaeridium* sp. cf. *H. tubiferum* (see discussion in Chapter 11, and Figure 11.4 I) and *Areosphaeridium diktyoplokus* (Figures 4.4 L, 11.1 A), may seem also to offer support for an hypothesis of centrifugal process development. Well-preserved multiple specimens of these cysts easily can be seen to show different degrees of separation between the edges of adjacent polygonal process tips. If cyst development was indeed centrifugal, this separation and its variable expression would be reasonable consequences of increasing overall cyst diameter. However, an alternative explanation is consistent with the notion of exclusively centripetal process development and with the observed enlargement of zygotic thecae by accretionary growth of plates before encystment. According to this view, the material that forms the distal polygons of process tips underlay the original thecal plates but not the accretionary bands around them. The different distances between process tips, then, reflect different amounts of thecal enlargement before encystment, comparable to the pandasutural bands of varied width in some species of proximate cysts.

Other Cyst Features Related to Thecal Growth

Thecal growth in peridinioid species is interestingly different from that in gonyaulacoid and ceratioid species. As a typical peridinioid zygotic cell grows, each of the major thecal plates acquires a strip of incremental material along its sutural edges except adjacent to the cingulum. Although additions of new plate material are similarly marginal in gonyaulacoid and ceratioid species, in these cases the additions are limited to those edges of a plate along which it overlaps its neighbors. Whereas intercalary bands in peridinioid thecae lack trichocyst pores and are often transversely striate, the strips of new material in gonyaulacoids and ceratioids lack striae but may have trichocyst pores and the surface pattern of the strips may be either quite similar to or quite different from that of the primary plate area.

On fossil peridinioid cysts, pandasutural bands occur commonly, although not in all species by any means. Frequently they resemble the bands already described on the cysts of modern *Peridinium limbatum* (Figure 1.5 C). In none of these cysts is the direction of plate overlap suggested by the character of the pandasutural bands. Fossil cysts with apparently gonyaulacoid and ceratioid affinities also show a variety of features which are confined to roughly parasutural strips and are possibly related to intercalary bands on the theca. Examples (Figure 4.4) include the gaps between penitabular lines of ornament, bands defined by contrasting surface texture or by elements within the wall, and (as suggested in the preceding section) spaces between the opposed and approximately parallel edges of the expanded polygonal tips on intratabular processes of some chorate cysts. As in peridinioid cysts, these features usually provide no basis for judging which plate may have received a larger growth increment or which plate was the overlapping one. However, there are interesting exceptions to this generalization, some of which are discussed in a section on plate overlap and growth in Chapter 10. One includes a group of fossil cysts currently referred to the genus *Cribroperidinium*. Gocht (1979a) showed that surficial ridges which occur in addition to the strictly parasutural ones in these cysts reflect the growth of thecal plates and, indirectly, the direction of plate overlap. More recently, Gocht (1981) showed in other groups of fossil cysts that the direction of plate overlap can sometimes also be sensed by careful attention to even more subtle properties of the surface or the wall along parasutures.

All these observations suggest that in the future additional features of fossil cysts will be recognized as sources of information about the character of thecal plates or the relationship between them. By such indirect means, the difficulty of comparing fossil cysts with modern thecae may gradually be somewhat reduced. The opportunity remains for experiments with living cells to demonstrate how characters acquired during the growth of a theca are recorded on the cyst that forms within it.

Reorganization of the Wall

No observations yet reported on cyst formation in modern dinoflagellates suggest that the cyst wall, once it has been formed, is restructured to a significant degree during its later development. Although some of the differences in detail of form within species of fossil cysts might be perceived to represent different stages in cyst maturity, a simple and more satisfactory explanation in most cases is that they are expressions of intraspecific variability among individuals that were, essentially, fully formed. Occasionally, however, successive developmental stages may be indicated.

Mesocysts of some peridinioid species, which seem to record the changing and decreasing volume of the cell during encystment, are discussed in Chapter 9. Possibly analogous stages in centripetal contraction of the protoplast may be represented by the two examples of chorate cysts cited in the preceding section (Figure 4.3 I-L). While these examples appear to document developmental changes of a sort, they involve what was probably an easily deformed layer of a plastic nature that had not yet attained its final degree of hardness, rather than removal and reformation of more solid materials.

Ontogenetic changes of a somewhat different and much more impressive kind have been suggested (Gocht, 1968; Benedek and Gocht, 1981a) to explain the peculiar contrasts observed among fossil cysts referred to the genus *Thalassiphora*, especially *T. pelagica*. In this remarkable case, development appears to progress from a stage characterized by an endocyst compactly enclosed within a robustly fibrous pericyst that bears prominent parasutural ridges to a more mature and much more expansive structure consisting of a thin, nonfibrous pericyst without paratabulation (except for the archeopyle), surrounding a relatively small endocyst to which it may be joined by a few to numerous pillarlike processes. In the course of this development, a slit opens in the ventral surface of the pericyst and enlarges to form an opening whose diameter eventually exceeds that of the endocyst.

Morphologically varied cysts attributed to a second species of this genus, *T. reticulata*, appear to offer another example of extensive cytological activity involving wall modification during cyst development (Benedek and Gocht, 1981b). In this case, it seems that the outer wall developed a perforation which became increasingly extensive and elaborate in progressively more mature cysts.

Although the mechanism for such apparently ontogenetic changes is difficult to imagine, and no comparable phenomena are known from modern dinoflagellates, it is difficult to explain the observed morphologic diversity in this species except by processes that involve a significant redistribution and restructuring of wall material.

Summary Remarks on Wall Development

In concluding this discussion, the inadequacy of our knowledge of life processes in modern dinoflagellates must, again, be emphasized. A small number of new observations could easily reduce the choices of explanations now before us, or show that what seem to be reasonable possibilities are only erroneous speculations. Nevertheless, the morphological evidence from the cysts themselves suggests that the theca does have a significant template-effect on cysts in some species, that processes and other large projections develop frequently in a centripetal sense, and that wall material is not normally subject to major restructuring. These conclusions are consistent with the formation of the cyst wall around a protoplasmic mass that is losing water—hence, also volume—and is withdrawing from the thecal surface over large areas. Centrifugal growth of processes at their tips and major restructuring of the cyst wall during its ontogeny may occur, but suitable mechanisms to account for these changes are not obvious; therefore, it is especially important to seek and scrutinize evidence that might establish their frequency and importance.

Chapter 5. Tabulation and Paratabulation

INTRODUCTION

In this chapter we will explore in some detail the distinctive features of several patterns of tabulation and paratabulation. Tabulation is far and away the most significant morphological character in the systematics of living peridinialean dinoflagellates. Genera are distinguished primarily by differences in the number and arrangement of plates, and minor differences in plate shape and arrangement are often significant at the species level. Aside from tabulation, the taxonomic criteria based on thecal morphology which are available to the systematist are limited to such features as thecal shape and surficial relief of the plates including the distribution of trichocyst pores. Normally, the tabulation of many modern peridinialean thecae is at least partially apparent in an untreated specimen, although full exploration of the details of plate arrangement usually requires "dissection" of the theca with the aid of sodium hypochlorite solution. This procedure makes essentially the complete tabulation "accessible" to the interested investigator.

In the study of dinoflagellate cysts the conditions are rather different. Here two factors assure that paratabulation will never be so uniquely important a taxonomic criterion as is tabulation for dinoflagellate thecae. First, paratabulation is frequently expressed incompletely or imperfectly and, when this is so, no technique equivalent to dissection can make obscure paraplate boundaries more readily visible. Second, the cyst commonly shows a greater diversity of morphological detail than the theca (in, for example, wall structure, surface ornament, and the archeopyle) and this detail is frequently more obvious and easier to analyze than paratabulation.

Since paratabulation is only imperfectly and variably expressed on the cyst and so many other features are available for study, why should a paleontologist worry at all about these surficial patterns? The answer that justifies the great emphasis placed on tabulation and paratabulation throughout the rest of this volume is threefold: In the first place, the importance of paratabulation will inevitably increase as descriptions of new taxa reveal a diversity of patterns heretofore unsuspected and as the eyes of microscopists become more sensitive to small differences among patterns.

In the second place, paratabulation, which is an expression of fundamental cyst geometry, has prime utility as a guide to specimen orientation and, therefore, to the identification of the particular portion of the cyst surface that may be under inspection. It is a key to locating and identifying features through a simple reasoning sequence: if the spatial relations of features A, B, and C on the surface of a specimen are known from previous observations, then, in a similar specimen newly under study, the location of C can be anticipated assuredly if features A and B can be recognized. In other words, paratabulation provides a vitally important "conceptual" framework for organizing into a consistent whole the often incomplete observations assembled from multiple specimens.

In the third place, details of paratabulation patterns are important taxonomic criteria in those cases where the patterns are clearly and more or less completely displayed, just as tabulation patterns are for living thecate forms.

Actually, the manner in which paleontologists have dealt with paratabulation in fossil dinoflagellates, especially the validity they have accredited to paraplate shape for identifying particular paraplates, has been criticized by specialists on modern dinoflagellates (e.g., Balech, 1967b; Taylor, 1976a). Given the broad spectrum of variation in plate shapes and tabulation patterns among modern thecate dinoflagellates, this criticism is understandable. However, as was stressed in Chapter 2, the high selectivity of the fossil record has effectively narrowed the concern of paleontologists to only a fraction of the many patterns which confront the specialist on modern dinoflagellates. Within this severely more limited scope, the usefulness and reliability of shape as a clue to recognition of individual paraplates is greatly enhanced. Of course, it is not infallible; care must be exercised, and other criteria used when available.

In rare cases in which an uncompressed cyst has ideally prominent parasutures, it may be relatively easy to analyze a paratabulation fully. However, the task is more likely to be slow and frustrating, even if traces of parasutures are reasonably clear, demanding an amount of time that is exorbitant for anyone facing an imminent deadline. Fortunately, full analysis is not necessary for paratabulation to be a highly useful interpretive tool. The objective of the following pages is to make paratabulation a more familiar feature, so that the practicing microscopist can exploit it to the fullest extent that he may judge consistent with his particular interest, the time available, and the quality of the specimens before him.

Differences in paratabulation depend mainly on one or a combination of the following four variables: 1) the presence or absence of paraplates, 2) the failure of certain parasutures to be recognizable, resulting in a lowered paraplate count, 3) differences in the particular contacts made by a given paraplate with those near it, and 4) differences in paraplate shapes resulting from different relative lengths of their sides. Comparison of paratabulation formulae will give some information about differences involving the first two variables, but not necessarily about the third and certainly not about the fourth. Most published descriptions of paratabulate cysts accord neither the third variable nor, especially, the fourth the attention they merit. Many refinements in species identifications can be expected as these details are observed and described more carefully. It is a primary objective of examples cited in the illustrations throughout this volume to point out some of the features that might be considered.

A fifth variable, of a somewhat different kind, also plays a role at times in the communication of ideas about tabulation. This involves the use of contrasting designations for the same plate or paraplate, reflecting inadvertent author error, different opinions as to the best terminology to use, or a change in an author's own best judgment. Adjustment to these irregularities requires that the reader have an intelligently critical eye, a certain familiarity with current jargon, and a sensitivity to the particular vocabulary and style of individual authors.

Before considering particular patterns important in the study of dinoflagellate cysts, the nature of these patterns—i.e., the concept of tabulation or paratabulation—should be examined somewhat more closely. Tabulation may be thought of as having four distinct but interrelated aspects. Three of them involve properties inherent in the patterns themselves. They can be determined objectively and are basic to the analysis or description of individual specimens or species. The fourth, which involves a considerable degree of subjective judgment, is significant when comparing patterns from different taxa and contemplating their possible evolutionary relationships.

The first, or what may be called the distributional aspect of tabulation, pertains to the number of plates and their general placement in the theca. The second aspect is essentially topologic and concerns the mutual relationship of plates in the tabulation mosaic. Of course, the distributional aspect plays a role in this relationship, but it is the pattern of plate contacts and sutural intersections which is the critical factor in the topologic aspect. Here the question is not Where is a plate located? but What other plates does it contact? The third aspect is strictly geometric, involving the sizes and shapes of plates. Although these factors are obviously related to both plate distribution and topology, the concern here is specifically the number and dimensions of the sides and angles on the individual plates. Finally, the matter of plate homology is fundamental to the analysis and comparison of tabulation in different taxa, although it is a less important consideration at the descriptive stage. Here the question is What plate in species B is the homolog of a particular plate in species A? Clearly, this question will be difficult to answer, and the subjective nature of the considerations involved assures a variety of responses.

For the tabulation of a theca or the paratabulation of a cyst to be fully characterized, the first three of these aspects must all be considered, either explicitly or implicitly. Only the first, or distributional, aspect can be indicated by a tabulation formula in Kofoidian style. Unfortunately, the connotation given to "tabulation" is commonly limited to just this distributional aspect, and failure to give adequate attention to the topologic and geometric aspects of tabulation as well is responsible forthe serious deficiency of many taxonomic descriptions.

The distributional aspect of tabulation has already been covered in Chapter 1, and the geometric aspect involves quite simple and familiar matters of size and shape. However, the topologic aspect concerns properties that are, probably, less familiar or less obvious. Topology, as a field of mathematical inquiry, is concerned with those properties of geometric configurations which are unaltered by elastic deformations. Two dinoflagellate tabulation patterns may be regarded as topologically identical if they can be transformed into one another by

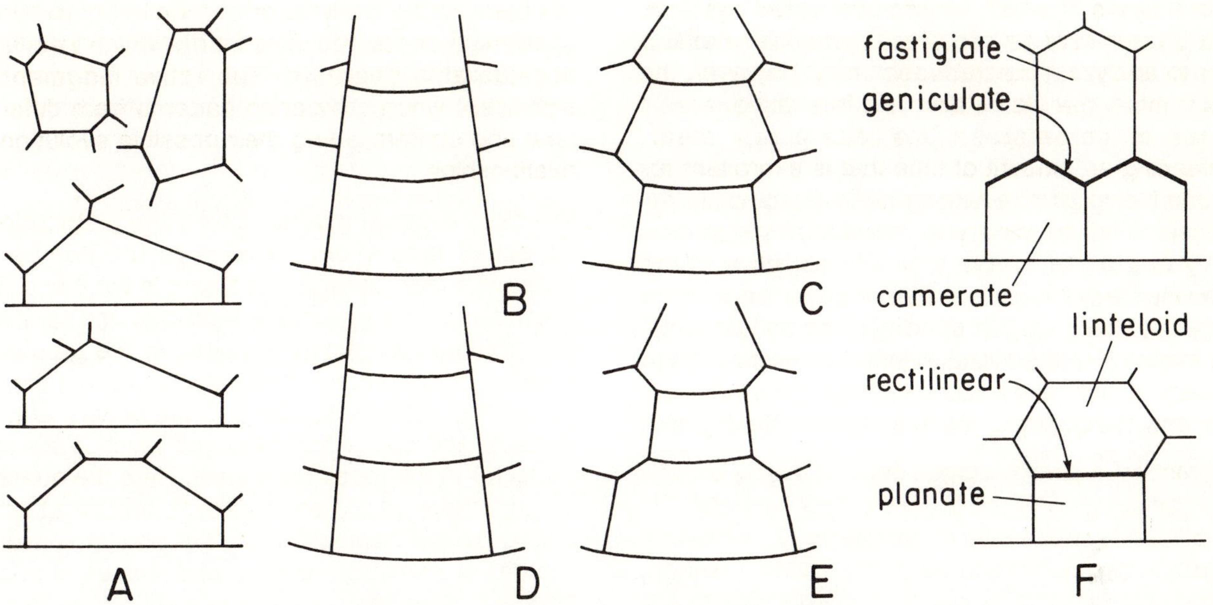

Figure 5.1 - TOPOLOGY AND SHAPE OF PLATES

A-E - Topologic identity and contrast. A- Fragments of five hypothetical plate patterns, four of which (upper two and lower two) are topologically identical, whereas no amount of elastic deformation would convert the middle figure into one of the others. B-E - Schematic portions of middorsal epithecal patterns like those in living *Protoperidinium conicum* (B), *P. obtusum* (C), a hypothetical pattern (D), and *P. depressum* (E). The patterns in B and D are geometrically similar but topologically different; those in B and C and in D and E are topologically identical but geometrically different; whereas C and E are different both topologically and geometrically.
F - Diagrams to illustrate differences between camerate, planate, fastigiate, and linteloid plates, and between geniculate and rectilinear plate boundaries (heavy lines). Basal line in each diagram represents edge of cingulum. See text for definitions and discussion.

imagined elastic deformation of suture lines and plate areas—in other words, if the same plates are present in both patterns and every plate in one pattern contacts the same adjacent plates as its counterpart does in the other, regardless of the actual size and shape of the individual plates or of the theca as a whole (Figure 5.1). Topologically identical tabulation patterns will always be represented by identical tabulation formulae (providing they are based on the same methods of analysis and symbolism), but, inasmuch as no information about contacts among individual plates is contained in formulae, identical formulae may represent patterns that are topologically distinct. The importance of distinguishing the geometric and topologic aspects of tabulation patterns is illustrated in Figure 5.1 A-E.

If one wishes simply to describe a tabulation pattern, it is relatively easy to avoid the difficult and provocative question of plate homologies. However, if one wishes to compare patterns exhibited by different taxa, then the issue of plate homologies is difficult to escape. The Kofoidian system of plate designations has proved to be a highly effective descriptive device, and, within the major groups of peridinioid dinoflagellates, Kofoidian labels apply with reasonable consistency to what also appear to be homologous plates in different taxa. In some other groups the efficacy of the Kofoidian "language" for this purpose is not an important issue, because either species are too few or tabulations too poorly known for pursuit of the subject yet to be productive. However, in the major group of dinoflagellates that includes *Gonyaulax*, the discussion of homologies and evolution justified by our present state of knowledge is severely handicapped by limitations of Kofoid's system. Therefore, in treating this group of dinoflagellates, we will consider an alternate system.

Although the shapes of plates and paraplates are almost endlessly varied in detail, an effort to distinguish two general shapes that have proved descriptively useful has led to several alternative terms among which the "fittest" have yet to be selected. The somewhat overlapping scopes of these terms should be clear if the following definitions are considered along with the diagrams in Figure 5.1 F: **Planate** (Bujak, 1980) describes a plate that appears to contact the next more poleward series along a line essentially parallel to the cingulum (synonym: flat-topped). The companion term, **camerate** (Bujak, 1980), distinguishes a plate that appears to contact two plates of the next more poleward series along two lines meeting at an angle (synonym: **gabled**). Corresponding approximately to these terms, but describing the boundary between plate series, rather than the shapes of individual plates, are two further terms. **Geniculate** (Evitt, 1978) refers to the angular boundary between two latitudinal series where a single plate of one series meets two plates of the other. In contrast, **rectilinear** (Evitt, 1978) describes the boundary where the separation between adjoining plates in two series is a single line segment about parallel to the cingulum. Finally, **fastigiate** describes an anterior intercalary whose adcingular boundary is geniculate, while **linteloid** applies to an intercalary in the opposite case where that boundary is rectilinear (Norris, 1978; Bujak and Davies, 1983).

The recurrence of particular shapes, especially in plates occupying critical positions in the tabulation, invites an explicit terminology, but confusion is inevitable unless special terms are used—and read!—with caution and with an acute awareness that intermediate or "non-conforming" cases are bound to occur.

In our systematic analysis of tabulation, we will first consider the peridinioid patterns, emphasizing those few of the many variations known among thecate species today that are important in the fossil record. This will limit our concern almost entirely to patterns that are strongly bilaterally symmetrical. Then, following a few general comments on the major categories of non-peridinioid dinoflagellates in the fossil record, we will turn to several groups of cysts, all of which display what we will refer to as gonyaulacoid patterns.

The diagrams that illustrate various views and details of each pattern show many relationships which it is impractical to discuss fully in the text. Besides simply looking at these patterns or labeling the plates in the unmarked figures, the imaginative reader seriously interested in learning the patterns may wish to proceed more "aggressively", using them as guides for preparing various learning aids, some of which can be immeasurably helpful when interpreting specimens under the microscope. Originally many of the figures were drawn from projected images of gesso-coated styrofoam balls marked with india ink; ping pong or polo balls offer more sporting possibilities. Copies of the unlabeled circular diagrams, cut out and mounted on cards in random orientation, can be the basis for a very instructive "solitaire" which will develop skill and confidence in recognizing randomly oriented patterns. If the patterns are drawn on transparent glass spheres (such as round-bottom distillation flasks, from which markings can be removed easily with a solvent), they can be compared easily with either exterior or interior views of actual specimens. Alternatively, the glass surface can be the base on which to gradually construct a complete paratabulation from many isolated fragments observed in either exterior or interior view. The usefulness of such devices should not be dismissed as trivial without serious trial. Of course, many fossil specimens were not originally spherical and many others have been distorted—matters that must be considered when constructing an idealized model or using it as an interpretive aid.

Although the "real world" presents a virtually boundless diversity of variations on the themes illustrated by these diagrammatic patterns or models based upon them, it is surprising how many specimens can be compared with them reasonably closely. A comfortable familiarity with these patterns, however it is gained, will be a great aid to the paleontologist, whether he may be trying to interpret a specimen under the microscope or a published figure that is incompletely or confusingly annotated.

THE PERIDINIOID PATTERNS

Representative Paratabulation Formula:

2pr, 4′, 3a, 7″, Xc, 5‴, 2⁗, Xs

Outstanding Features: Bilateral symmetry, three large anterior intercalaries, seven precingulars, two antapicals; outline commonly peridinioid. See Figure 5.2.

Figure 5.2 - PERIDINIOID PATTERNS

A-B - Schematic theca (ortho hexa) with typical peridinioid outline, in ventral and dorsal views. Conventional labeling for sulcal plates: *as* -anterior, *rs* - right, *ls* - left, *ps* - posterior; *1c* - first cingular plate (="transitional" of some authors).

C - Alternative styles of ventral epithecal tabulation: ortho (upper left), meta (upper right), reversed meta (lower left), and para (lower right).

D-I - Schematic tabulation in views of hypothetically spherical theca (ortho hexa). D - Apical view, sulcus down. E - Ventral view. F - Antapical view, sulcus up. G - Left lateral view. H - Dorsal view. I - Right lateral view.

J-L - Oblique views of anterior intercalary region, showing plates in quadra (J), penta (K), and hexa (L) arrangement.

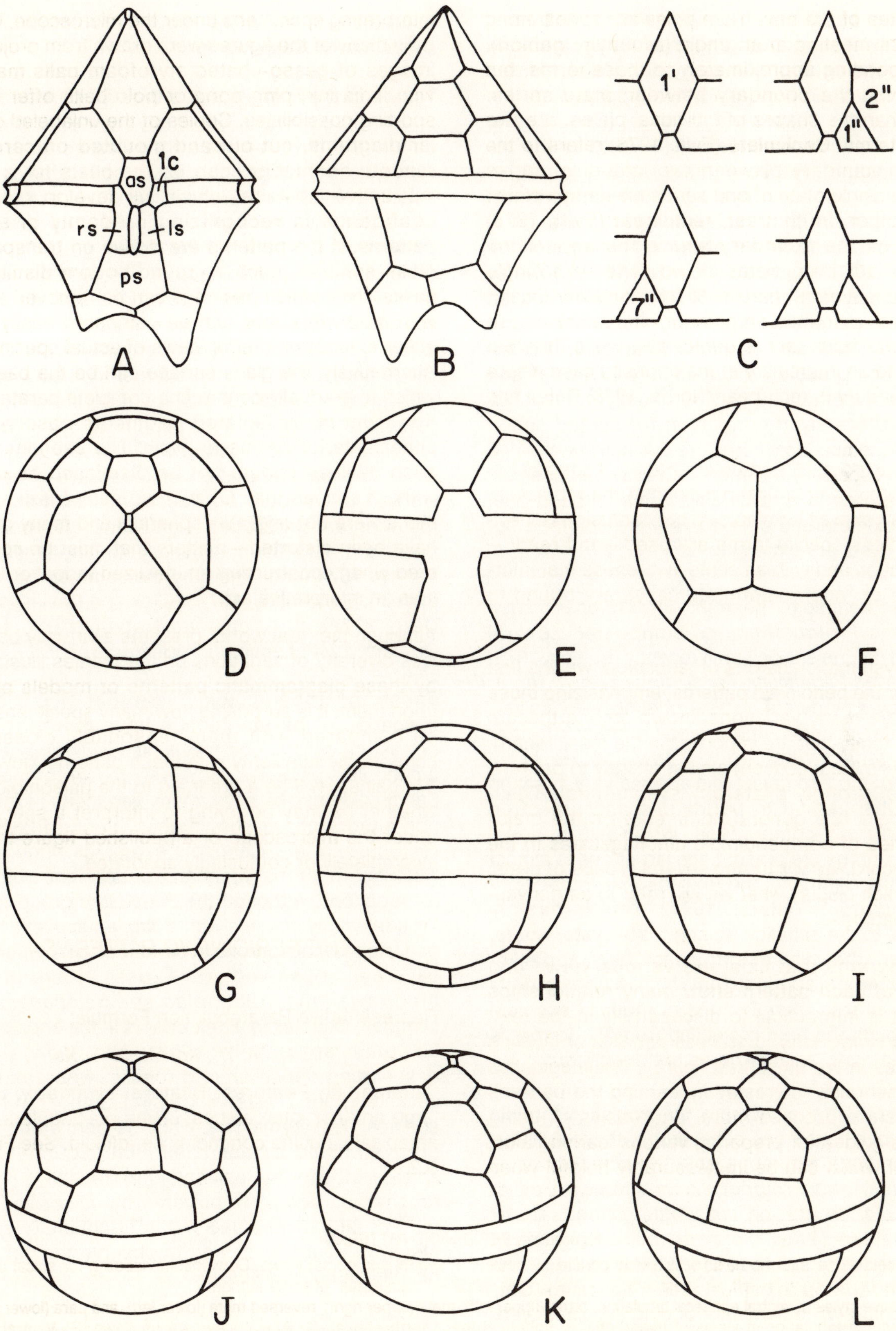

Figure 5.2

Fossil Representation. This is one of the two most commonly encountered groups of patterns among fossil cysts, occurring in many distinctive species with diverse morphologies. Most important are the genera of the *Deflandrea* complex (Early Cretaceous-Recent) and the *Wetzeliella* complex (Paleocene-Miocene). Representatives are rare in pre-Albian sporopollenin cysts but the existence of dinoflagellates with typical peridinioid tabulation at least by the Hauterivian is shown by patterns on the interior of calcareous cysts of *Pithonella patriciagreelyae* (Keupp, 1980). No fully expressed peridinioid pattern is known from the Jurassic (but note Early Jurassic *Liasidium*, Chapter 9). Dorsoventral compression (occasionally extreme) and customary peridinioid outline usually establish a decidedly preferred orientation for settled specimens. If the antapical horns are distinctly unequal, a situation especially common in Cretaceous species, the larger one is virtually always the lefthand one. Both proximate and cavate cysts are common but chorate cysts are extremely rare.

Modern Examples. These include *Protoperidinium* subgenus *Protoperidinium*, comprising several hundred marine species until recently (Balech, 1974) attributed to *Peridinium*; *Peridinium s.s.*, now represented by a moderate number of almost exclusively freshwater species; and a small number of other marine genera with one to a few species each. Thecae of modern peridinioid species vary in shape from subspherical to peridinioid and are often dorsoventrally compressed. When two unequal antapical horns are present, the larger one is almost always the righthand one (exception: *Protoperidinium adulterum*; Balech, 1974). This is exactly opposite to the situation among fossil cysts, where, however, the rare examples of fully apparent paratabulation suggest that *1''''* is consistently smaller in area than *2''''* (as in modern species) even though the horn projecting from the former is longer. In many species horns are about equal.

Analysis of Patterns. Actually, the tabulation formula given above is only one of several, with small differences among them, applicable to living species of *Peridinium* and *Protoperidinium*. Indications of paratabulation are, on the whole, rather rare or incomplete in peridinioid cysts, being confined in many species to sutures associated with the archeopyle on the dorsal epicyst (see Chapter 6). Only exceptionally will the student of dinoflagellate cysts encounter a pattern represented by another formula; for example, with two instead of three intercalaries, as in the modern *Protoperidinium* subgenus *Archaeperidinium**.

The *X*s in the formula mean that the paracingulum and parasulcus are generally indicated but the paraplates that compose these areas cannot be distinguished. Although details in both cingulum and sulcus are now considered sensitive taxonomic characters at the generic level in the study of modern thecae from the *Peridinium* group, comparable details of paracingulum and parasulcus are rarely visible on cysts. Exceptions include a few species of *Kisselovia*, *Palaeoperidinium*, *Phthanoperidinium*, and *Trithyrodinium* from the Upper Cretaceous and Lower Tertiary, but their numbers do not yet warrant generalizations about paracingular and parasulcal details in fossil peridinioid cysts. Therefore, it is especially important that any fossils in which the paratabulation of these areas can be worked out should be described carefully. The following are particularly significant references on aspects of tabulation in peridinioid dinoflagellates: Balech, 1974 (and others by the same author cited therein); Boltovskoy, 1979; Bujak and Davies, 1983; Dürr, 1979a-c; Gocht and Netzel, 1974, 1976; Graham, 1942; Lentin and Williams, 1976.

Those fossil peridinioid cysts that give a clear indication of paratabulation have a surprisingly consistent design, one that is also well represented among living species. However, thecae of modern peridinioid species show many variations in tabulation either not yet known at all among fossils or restricted to late Cenozoic examples. Figure 5.2 illustrates both the "standard" pattern and some of its variations. Although the *Peridinium* group today includes many species with asymmetrical tabulation patterns, few distortions in paratabulation symmetry have been observed in the fossil record of this group (Harland, 1982; also see *Selenopemphix* complex, Chapter 9). Even offsets in the ends of the cingulum are usually slight, and always in a descending sense among fossils, whereas both ascending and descending spirals appear in living species. This symmetry, which is persistent—but always should be assumed with caution— is a convenience for the paleontologist, enabling him to reconstruct complete patterns by mirror-imaging right or left halves observed in fragmentary specimens, without first having to observe the patterns directly on both halves.

---

* Unless contrary intent is explicitly clear, the generic name *Protoperidinium* in this work refers to *Protoperidinium* subgenus *Protoperidinium*.

In modern peridinioid thecae, preapicals may be quite obvious, but equivalent paraplates in cysts are only rarely apparent, for they are small and occur in an area where cyst details are commonly difficult to observe. Thecal plate *l′* is usually large and often conspicuously diamond-shaped, meeting the sulcus close to the cingulum; *3′*, on the dorsal surface opposite *1′*, is much smaller. Species with two, three, and four anterior intercalaries occur in the *Peridinium* group of genera today, but three is the nearly universal number in fossils (for exception, see Chapter 9). Thus, the dorsal midline normally passes through paraplates *3′*, *2a*, *4″*, and *3‴*. Paraplate *3‴* is large and pentagonal, and parasuture *1⁗/2⁗* extends from its antapical extremity to the posterior extremity of the sulcus. The first and last postcingulars form most of the ventral surface of the hypocyst flanking the parasulcus.

Basic variations in tabulation in modern species affect especially the midventral and middorsal areas of the epitheca. On the ventral surface (Figure 5.2 C) the variations involve the shape of the first apical plate (depending upon whether it contacts two, three, or four precingulars) and these three styles are known, respectively, as **ortho**, **meta**, and **para**. On the dorsal surface (Figure 5.2 J-L) the second intercalary contacts one, two, or three precingulars, and these arrangements long have been referred to as **quadra**, **penta**, and **hexa** (Evitt, 1978), according to the number of sides on *2a*. On each surface, intermediate styles marked by quadruple junctions (termed **neutra** on the ventral surface) are theoretically possible, but they rarely occur. These dorsal and ventral patterns have proved convenient criteria for segregating the many modern species of *Protoperidinium*, but their variability and intergradation (see example of *P. depressum* under Principle of Variable Variables, Chapter 7) suggest a negligible biological significance, although some selective representation in the fossil record is evident.

The dominant ventral arrangement among living species is the meta style, but to date this has been reported (in the meta hexa combination) in only a single fossil from pre-Neogene strata (Eocene *Phthanoperidinium brooksii*, Edwards and Bebout, 1981). This species has the further peculiarity that the critical "extra" contact between *1′* and a third precingular is between *1′* and *6″* (i.e., **reversed meta**) rather than between *1′* and *2″* as is the case in virtually all modern species (rare exception today: some individuals of *Protoperidinium latissimum*; Balech, 1974). In contrast, the ortho hexa combination clearly is most important among fossil cysts, although the para quadra combination characterizes one important group of Neogene species (see *Wetzeliella* complex, Chapter 9). The ortho penta organization, with its inherent dorsal asymmetry (e.g., five-sided *1a* and *2a*, six-sided *3a*), is known (but as yet poorly) from a few modern and Quaternary organic-walled cysts in which the *2a* paraplate is represented by the archeopyle in otherwise relatively featureless spheroidal autocysts (Harland, 1982). It also occurs in a few species of Tertiary siliceous cysts (all referred to *Peridinites*) which fall outside the scope of this work. Consequently, in the great majority of fossil peridinioid organic-walled cysts, three middorsal paraplates will be found in the epicyst, one above the other, and, whenever it is visible, the medial portion of the intercalary-precingular boundary (i.e., *2a/4″*) will be essentially rectilinear, between a linteloid *2a* and a planate *4″*. These are important contrasts with the pattern in another major group of cysts (R-cysts; see Chapter 8) having prominent anterior intercalaries and including *Pareodinia*, *Rhaetogonyaulax*, and others. As will be described later, intercalaries and precingulars in these forms tend to alternate in position, the medial boundary between them is geniculate, the intercalaries are fastigiate, and the precingulars are camerate.

Although the nominal difference between the quadra and hexa arrangements lies in paraplate *2a*, the shapes and relative sizes of the adjoining paraplates are also usually distinctive. For example, Figures 5.2 J and L show the contrasting lateral extent of *2a* relative to *4″*, a relationship that can facilitate distinguishing ortho and hexa arrangements even if the shape of *2a* itself is uncertain.

Although the configuration of plates and sutures along the intercalary-adcingular boundary has been the subject of published comment, little notice has been given to the intercalary-apical boundary. This is, perhaps, an even more critical area for distinguishing the typical peridinioid dorsal patterns from the rhaetogonyaulacoid ones. In peridinioid patterns with three intercalaries and seven precingulars, the middle intercalary is nearly always planate, regardless of whether its adcingular boundary is rectilinear, as in many examples, or geniculate, as in those penta examples in which *2a* has "shifted" to a position between adjacent precingulars (e.g., living freshwater species of the *Peridinium cinctum* group whose pattern is termed "cinctioid" by Bujak and

Davies, 1983, in contrast to the hexa patterns exemplified by the *Peridinium bipes* group, which they term "bipesioid"). In contrast, geniculate boundaries appear to characterize intercalary-apical contacts in cysts of rhaetogonyaulacoid type. A special problem arises in making the distinction in those rare cases where only two intercalaries are present—e.g., fossil *Umbodinium* (Figure 6.5 T) which suggests *Protoperidinium* subgenus *Archaeperidinium*—but this does not destroy the validity of the generalization.

Detailed comparisons of paratabulation are usually not possible within this group of cysts owing to the incompleteness or faintness of parasutures. It is often chiefly the archeopyle that, by documenting the presence and relationship of anterior intercalaries, provides the best evidence that the paratabulation of a particular fossil cyst is peridinioid. For comparing fossil peridinioid cysts, we will assume that the Kofoidian system consistently yields identical labels for homologous paraplates in different taxa. The few examples with contrary indications on this point can be handled as special cases on an individual basis (see remarks under *Angustidinium acribes*, Chapter 9).

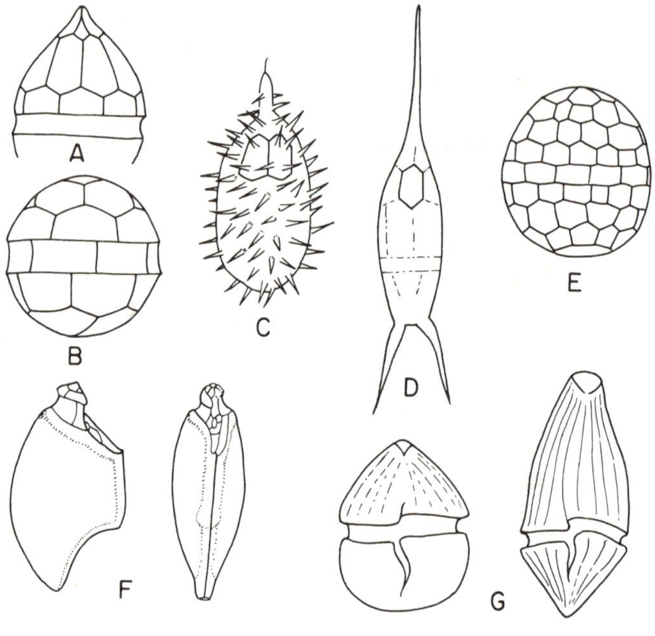

Figure 5.3 - OTHER MAJOR GROUPS OF CYST-PRODUCING DINOFLAGELLATES

A-D - Dorsal views, showing large anterior intercalaries with medially geniculate adcingular limit. A - *Rhaetogonyaulax* (Late Triassic). B - *Dapcodinium* (Late Triassic-Early Jurassic). C -*Gochteodinia* (Late Jurassic-Early Cretaceous). D - *Broomea* (Late Jurassic).

E - *Suessia* (Late Triassic), highly schematic and generalized dorsal view, showing numerous paraplates arranged in several series in each hemisphere.

F - *Nannoceratopsis* (Jurassic) in right lateral and ventral views. Paratabulation is peridinialean in the epicyst and dinophysialean (just two large and two small paraplates) in the hypocyst.

G - *Dinogymnium* (Late Cretaceous). Ventral views of two examples of the sporopollenin test, which lacks paraplates but is perforate (trichocyst pores?) and has well-formed flagellar furrows and possible flagellar pores, and may represent covering of cell in motile stage.

OVERVIEW OF OTHER MAJOR PATTERNS IN FOSSIL CYSTS

Although the paratabulation patterns of many fossil cysts are incompletely known, several distinct types of patterns besides the peridinioid ones are relatively easy to categorize (Figure 5.3). These will be mentioned only briefly here and discussed in greater detail in Chapter 12. One of them, exemplified by *Rhaetogonyaulax*, includes an obviously varied but as yet poorly understood group of cysts from Triassic-Albian strata. Their paratabulation resembles the peridinioid design in having large and symmetrically placed dorsal anterior intercalaries. Despite this similarity, it seems unlikely that, in the long run, these cysts will prove to be more than superficially similar to the *Peridinium* group. At any rate, they are distinguished morphologically by having fastigiate intercalaries alternating with camerate subjacent precingulars, rather than having a linteloid median intercalary positioned directly above the planate precingular; thus, the precingular-intercalary boundary is middorsally geniculate, rather than rectilinear.

Although many modern peridinioid species depart from the symmetrical quadra and hexa dorsal designs presented here as "standards" for fossil peridinioid cysts, examples of such departures are exceedingly rare among middle Tertiary and older cysts (e.g., *Selenopemphix*). Therefore, the contrast between peridinioid species with linteloid middorsal anterior intercalaries and other species with fastigiate anterior intercalaries (e.g., *Rhaetogonyaulax*, etc.), especially from Upper Triassic to Lower Cretaceous strata, is conspicuous and highly useful.

The genus *Suessia* represents another major group of patterns from the fossil record. The distinctive feature of this group, from the Triassic, is a striking multiplicity of more or less similar paraplates, arranged in a greater number of latitudinal series than normal for most modern Peridiniales, but suggesting the arrangement of amphiesmal vesicles in certain Recent freshwater species (e.g., *Amphidinium carterae*; Dodge and Crawford, 1968). Although *Suessia*-like patterns are highly distinctive, their restricted stratigraphic occurrence and relative unfamiliarity (despite pioneering work by Morbey, 1975, and attention in conjunction with revisions in descriptive terminology suggested by Sarjeant, 1982) remove them from our present focus on more frequently encountered and better known patterns.

Two further groups of fossil dinoflagellates, both with limited stratigraphic ranges, are also special. One is the Jurassic genus *Nannoceratopsis*. The paratabulation of these highly distinctive cysts is a curious combination of pattern components known otherwise only in living dinoflagellates referred to two different orders (Peridiniales and Dinophysiales). Unfortunately, the paratabulation can be seen clearly only with the scanning electron microscope (Frontispiece H). In the second case, which involves the Late Cretaceous genus *Dinogymnium* (Frontispiece F), no paratabulation is visible, a situation which certainly places the group outside the scope of this chapter, while other characteristics discussed in Chapter 12 suggest that these highly distinctive fossils may represent motile cells rather than cysts (May, 1976). The genus is mentioned here only to include it in this enumeration of major fossil groups.

Having thus, at least temporarily, accounted for four major groups of cysts in addition to those with peridinioid paratabulation, we find that a large majority of the remaining paratabulation patterns known from the fossil record can be viewed as constituting a single "family" of patterns. We will call these the gonyaulacoid patterns, and they are the subject of the remainder of this chapter.

THE GONYAULACOID PATTERNS

General

Since the patterns we will include under this heading are at once too diverse to be treated as a single group and too numerous to be treated individually, we will discuss them in four groups, which we will define by utilizing a different system for analyzing tabulation, replacing Kofoid's approach with that recently suggested by F. J. R. Taylor (1979, 1980). This Taylorian system is a new departure, and those readers already familiar with the application of Kofoid's concepts to gonyaulacoid patterns will need to readjust their thinking. Those who are encountering these patterns for the first time will not face that problem at the start but, in view of the extensive existing literature which treats gonyaulacoid patterns in Kofoidian terms, they must also be able to deal easily with both approaches. Therefore, we will return to Kofoid later in this chapter.

Before making the initial break, however, we can take advantage of Kofoidian tabulation formulae and symbols to make a gross comparison between the peridinioid patterns we have just finished considering and the gonyaulacoid patterns which lie ahead. The two following formulae represent a particular tabulation pattern in each group which is extensively indicated by paratabulation patterns among fossil cysts:

peridinioid: *2pr, 4′, 3a, 7″, Xc, 5‴, 0p, 2⁗, Xs*
gonyaulacoid: *0pr, 4′, 0a, 6″ 6c, 6‴, 1p, 1⁗, 5s*

Remember, however, that some aspects of tabulation are not at all revealed by such formulae. When representative specimens, rather than formulae, are compared, some of the differences seem greater, some smaller than the formulae suggest. All-in-all the two types of pattern are quite distinct. This is demonstrated by the following enumerative comparison and by the diagrams in Figure 5.4, which contrast peridinioid and gonyaulacoid patterns as expressed in some species of the modern genera, *Peridinium* and *Gonyaulax*:

1. Anterior intercalaries in *Peridinium* are large, dorsal, and symmetrically situated, but in *Gonyaulax* they are absent (alternatively, small and asymmetrically situated).
2. Seven precingulars are a characteristic of *Peridinium*, whereas the typical number for *Gonyaulax* is six.
3. Five postcingulars in *Peridinium* contrast with six in *Gonyaulax*, but the effect of this difference is diminished if, as is true for many species of this genus, the first postcingular is small and, essentially, "lost" through incorporation into the sulcus.
4. No posterior intercalaries are present in *Peridinium*, whereas a single such plate constitutes a more or less conspicuously asymmetric element on the left side of the hypotheca in *Gonyaulax*.
5. Two nearly symmetrical antapicals in *Peridinium* contrast with one in *Gonyaulax*.
6. The Kofoidian designations of plates in "key" positions (especially along the dorsal midline,

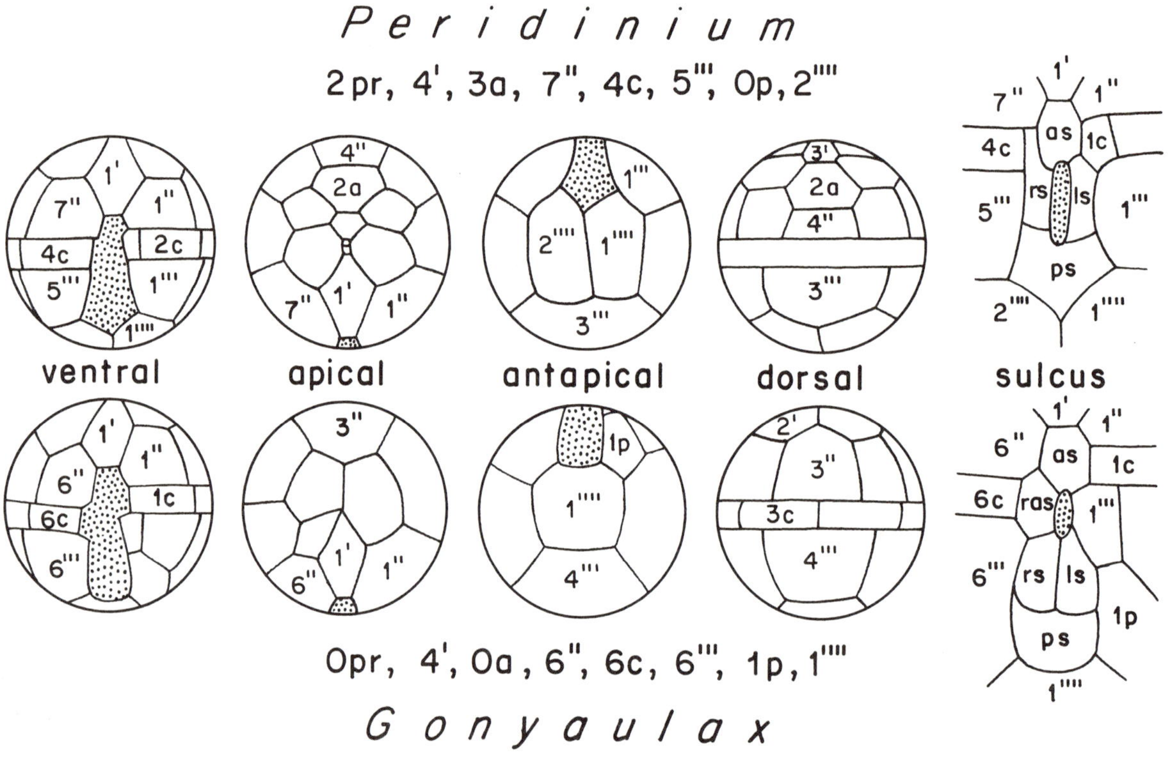

Figure 5.4 - *PERIDINIUM* AND *GONYAULAX*: COMPARISON OF TABULATIONS

Schematic representation of one pattern for each genus. Diagrams show different views of entire pattern and detail of the sulcus. Kofoid tabulation formulae are shown, and key plates on the diagrams are labeled with the Kofoid designations conventionally used. Designations for sulcal plates are: *as* - anterior, *rs* - right, *ls* - left, *ras* - right accessory, *ps* - posterior. Sulcus patterned in principal views; flagellar pore regions stippled in details of sulcus.

although the symmetry in individual species may deviate from that diagramed) are characteristically different in the two genera.

The preceding statements must be qualified by reservations that apply whenever generalizations are made about large and varied groups of natural objects. Some of the gonyaulacoids we will soon consider are exceptions to these statements, just as are a significant number of the peridinioids which are important today but inconsequential in the fossil record. Also, recent recommendations to amend the way Kofoid's method is applied to gonyaulacoid patterns have been ignored for purposes of this discussion.

The taxa characterized by the gonyaulacoid patterns we are about to review appear to be rather closely linked genetically. However, the Kofoidian system for designating individual plates is not satisfactory for comparing them, because plates that are almost certainly homologs receive confusingly different Kofoidian labels in different genera, or even among species that have been referred to a single genus. This is the principal justification for using the alternative system suggested by Taylor.

For convenience, we will distinguish four groups of gonyaulacoid patterns (and genera), primarily on the basis of differences in the arrangement of plates posterior to the cingulum (i.e., the hypotheca and posterior portion of the sulcus). One of these, typified by the modern genus *Triadinium*, (known as *Heteraulacus* and *Goniodoma* before the recent work of Dodge, 1981), we will refer to as **quinqueform**, in reference to the hypothecal configuration which shows a major plate at the antapex surrounded by five others. A second we will call **sexiform**, referring to the six plates around the antapical in genera like *Gonyaulax*. Those having the antapical area divided between two paraplates, as in the fossil genus *Microdinium*, we will label **partiform**. And, finally, for patterns such as characterize living *Ceratium*, we will use **corniform**, in reference to the horns formed by pairs of hypothecal plates. Genera based on thecae and cysts which are representative of these four groups are listed in Table 5.1.

Table 5.1. REPRESENTATIVE GENERA WITH GONYAULACOID PATTERNS

Group term	Genera based on thecae	Genera based on cysts*
quinqueform	*Triadinium*, *Pyrodinium*, *Protogonyaulax*	*Dinopterygium*, *Eisenackia*, *Eocladopyxis*, *Heteraulacacysta*, *Polysphaeridium* (=*Hemicystodinium*)
sexiform	*Gonyaulax*	*Canninginopsis*, *Cribroperidinium*, *Ctenidodinium*, *Gonyaulacysta*, *Hystrichokolpoma*, *Leptodinium*, *Lithodinia*, *Millioudodinium*, *Oligosphaeridium*, *Rhynchodiniopsis*, *Spiniferites*,
corniform	*Ceratium*	*Endoceratium, Muderongia, Odontochitina*
partiform	possibly: *Cladopyxis*, *Palaeophalacroma* (including *Sinodinium*)	*Carpathodinium*, *Cladopyxidium*, *Druggidium*, *Eyachia*, *Fibrocysta*, *Gillinia*, *Histiocysta*, *Luehndea*, *Microdinium*, *?Paragonyaulacysta*, *Phanerodinium*

*All the genera listed include at least some species in which paratabulation is shown by parasutural lines. Many other genera not listed here can be referred confidently to these groups, although paratabulation patterns in them may be indicated less obviously (e.g., by intratabular features) or only incompletely.

Within each of these groups there is some variation, and the generalized remarks and schematic diagrams on the following pages will not fit all examples neatly. Possibly all the patterns will eventually be found to intergrade, but that prospect does not detract from the present utility of the distinctions. Many different epithecal designs occur in combination with the four basic hypothecal patterns. This is consistent with the commonly held impression that the hypotheca has been the relatively more conservative region in an evolutionary sense (Graham, 1942). Summary comparisons of the gonyaulacoid patterns with each other and with the peridinioid hypothecal design will be found in Table 5.4 and Figure 5.19 at the end of this chapter. A key to the identification of the gonyaulacoid patterns is also presented there.

With regard to its fossil representatives, the corniform group stands somewhat apart from the others. Fossil cysts apparently allied with modern *Ceratium* are usually recognizable by their distinctive shape, but the corniform paratabulation pattern is often indistinct or incomplete. It is more commonly the case among fossil cysts from the other gonyaulacoid groups that paratabulation is sufficiently visible to serve as a useful basis for comparison.

Although the modern genera included in these groups and listed in Table 5.1 usually have been placed in four different families—Heteraulacaceae (for *Triadinium* under the name of *Heteraulacus*), Gonyaulacaceae, Ceratiaceae, and Cladopyxidaceae)—other interpretations are possible. Thus, Taylor (1976, 1980) has drawn attention to a morphological intergradation which he believes to represent an evolutionary progression from *Triadinium* to *Gonyaulax*; Wall and Evitt (1975) have presented an analysis of *Ceratium* that emphasizes its similarities to *Gonyaulax*; and we will explore here the possible derivation of the partiform patterns from a *Gonyaulax*-like base. Paratabulation patterns from the fossil record can be construed as strengthening the arguments for regarding all these taxa as closely related, and we will utilize this concept as the basis for a unified consideration of the gonyaulacoid genera. However, interpretations that contrast strongly with this one have been proposed by others (Norris, 1978; Eaton, 1980; Dörhöfer and Davies, 1980). These competing hypotheses will be discussed in Chapter 12, but, until that point, we will follow Taylor's ideas without complicating the discussion by interjecting alternative explanations at every step. A major opportunity in dinoflagellate research in the next decade will be the amplification, testing, and refinement of these diverse new ideas pertaining to major evolutionary patterns in the Peridiniales.

According to Taylor's hypothesis, the morphological gradients that can be resolved from a phenetic comparison of those living dinoflagellates we are referring to as quinqueform and sexiform gonyaulacoids are strongly suggestive of evolutionary sequences. Specifically, he envisages thecal evolution among these dinoflagellates as having begun with a primitive design characterized by a relatively small number of symmetrically arranged plates (vesicles). This design is not matched precisely by the pattern of any known living or fossil species, but the principal geometrical features attributed to it are approximated by the tabulation of living *Triadinium* and similar paratabulations known from fossil cysts. From these relatively primitive designs, which incorporate quinqueform hypothecal patterns, progression was through forms having thecae organized as in living *Pyrodinium* and *Protogonyaulax*, to ones with sexiform hypothecal patterns like those typical of living *Gonyaulax*. Consistent with Taylor's ideas, we will extend this hypothetical progression to two further but separate specializations beyond *Gonyaulax*; namely, to *Ceratium* and *Microdinium/Cladopyxidium*, whose hypothecal organizations are, respectively, corniform and partiform.

It cannot be overemphasized that the advantage of the Taylor approach to analysis and description of gonyaulacoid patterns is wholly independent of any similarity that may exist between the evolutionary progression he hypothesizes and the historical course of events. It is the principal purpose of the following pages to demonstrate the superiority of Taylor's method for comparing one gonyaulacoid pattern with another, regardless of what may have been their evolutionary connections. For simplicity and consistency, however, we will assume that he has correctly hypothesized the actual sense of evolutionary change, returning (Chapter 13) to touch briefly on other possibilities that have been put forward.

In the following discussion, we will first explore the features of Taylor's hypothetical starting point and note the changes by which the *Triadinium*-to-*Gonyaulax* sequence of patterns might be derived from it. Then we will discuss the quinqueform and sexiform patterns themselves. Thereafter, we will analyze corniform and partiform patterns, while pointing out the distinctions of *Ceratium* and *Microdinium/Cladopyxidium* from *Gonyaulax*.

Figure 5.5 shows the tabulation of the initial stage essentially as hypothesized by Taylor. The terms and symbols adopted for the plates and the series in which they are arranged have been selected to contrast with the familiar terms of Kofoid; identical designations do not occur in the two systems. The presentation here incorporates minor deviations from, or extensions to, the terminology originally suggested by Taylor. In the text, Taylor and Kofoid symbols are shown in **boldface** and *italics*, respectively.

The **primary tabulation** of the primitive model comprises five latitudinal series of individually symmetrical pentagonal and hexagonal plates, symmetrically arranged. Around a small **prepolar** (**P**) at the anterior extremity, or **pole**, the **polar** series includes three plates (**A-C**). Next come three series of six plates each—the **preequatorial** (**1-6**), **equatorial** (**a-f**)*, and **postequatorial** (**I-VI**) series**. The arrangement is completed by a final series of three plates, the **antipolars** (**X-Z**), around the posterior extremity, or **antipole**. The flagella are assumed to have emerged through an opening (conceivably covered by one or more separate plates) at the **f/a/I** triple junction (i.e., the common point of plates **f**, **a**, and **I**), and this point is considered ventral.

In oral communication, confusion between the two series of numerals that appear as plate symbols can be avoided easily by calling the arabic numerals of the preequatorial series by their vernacular names ("one," "two," "three," etc.), while referring to the roman numerals of the postequatorial series by their Latin names ("unus," "duo," "tres," "quattuor," "quinque," and "sex").

Subdivision of a primary plate into two or more **secondary** plates plays a critical role in Taylor's ideas, for it is by this means that additional plates appear. Secondaries resulting from divisions along essentially meridional lines are labeled "**i**," "**m,**" and "**u**" (for "initialis," "medialis," and "ulter,") in reference to their position relatively toward the beginning or end of the cingulum***. Symbols for secondaries

*Instead of the Greek letters used by Taylor for the equatorials, lower case Roman letters have been used here in the interest of ease of typing. The resultant slight overlap with other symbols is not likely to prove disturbing, especially since the initial equatorial is almost always subdivided and then denoted by **ai** or **au**, as explained later.

**Originally, Taylor hypothesized a meridional alignment of the intraseries sutures between preequatorial, equatorial and postequatorial plates. Consequently, the plates of these three series were in longitudinal rows, had quadrangular outlines, and were separated by sutures that met at quadruple junctions. In the arrangement adopted here, the plates in adjacent series alternate in position, none is quadrangular, and all suture intersections are triple junctions. Support for the greater plausibility of this arrangement lies in the widespread occurrence of similar patterns of polygonal units in nature, in contrast to the rarity of the gridlike plan suggested by Taylor. Furthermore, many dinoflagellate cysts and thecae have a pattern of sutures on the dorsal surface which compares closely to the alternating arrangement. This area, which is most remote from the evolutionarily labile ventral region, may give a reasonably reliable indication of the primary plate arrangement.

***The letters "**i**" and "**u**" replace, respectively, "**d**" and "**s**" (for "dextral" and "sinistral") used originally by Taylor, notations which become confusingly contradictory when applied to subdivisions of primaries on the dorsal surface. Similarly, "**v**" and "**h**" replace, respectively, "**a**" and "**p**" (for "anterior" and "posterior") of Taylor, in order to avoid confusion with "*a*" and "*p*" of Kofoid which have very different meanings but may appear in similar combinations of symbols.

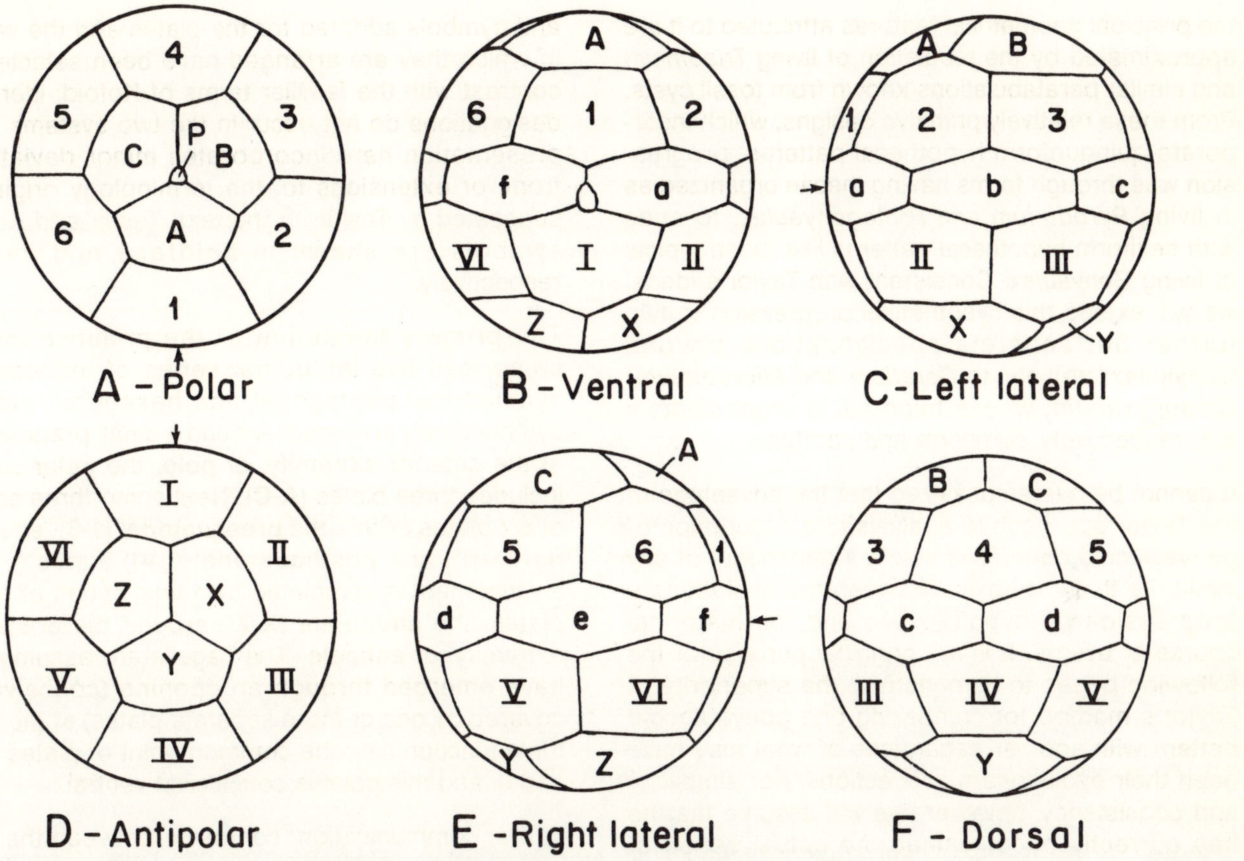

Figure 5.5 - - TAYLOR HYPOTHESIS: PRIMARY TABULATION

P = prepolar, **A-C** = polars, **1 -6** = preequatorials, **a-f** = equatorials, **I-VI** = postequatorials, **X-Z** = antipolars. Arrows in lateral and polar views mark area of flagellar insertion shown by open ellipse in ventral view. If diagrams B, C, F, and E are viewed in that sequence, the effect is as if the specimen were being observed at successive positions 90 degrees apart while being rotated about the polar axis.

that result from divisions along essentially latitudinal lines follow an analogous convention: "**v**" and "**h**" (for German "vorne" and "hinten" in the absence of classical roots with initials not used elsewhere among the Kofoid or Taylor symbols), and, again, an intermediate "**m**", relative to their positions along a meridian. Secondaries requiring "**v**" and "**h**" play no role in the derivation of patterns that will occupy our immediate attention, but they are important in a few examples to be discussed later.

Figure 5.6 shows thecal diagrams of modern taxa arranged according to the evolutionary progression from the primary tabulation to *Gonyaulax* as visualized by Taylor. This arrangement implies the following evolutionary changes:

1. Transformation of the series of hexagonal equatorials into the parallel-sided cingulum of rectangular plates typical of modern thecae and most cysts (*Spiniferites* being a striking exception; see Chapter 10).

2. Early subdivision of certain primaries into two or three secondaries, and associated changes in size and shape (as illustrated in Figure 5.6). The particular developments hypothesized are:

   a. Division of **1** into **1i** and **1u** (Figures 5.6 A-B, 5.7 A-C). Thereafter (Figures 5.6 C-E, 5.7 D) secondary **1i** maintains its preequatorial position although evidencing a variety of sizes and shapes. In contrast, the polar extremity of **1u** (stippled in Figure 5.6) appears to "migrate" along **A/2** and then **A/B** until it contacts the prepolar, thereby fully separating **A** from **B**. This change in **1u** is accompanied by a significant change in the size and position of **A**, but by little change in **B**. The relationship at an early stage in the progression, when **A** contacts **2**, is termed **exsert** (Figure 5.6 B). The advanced relationship, when contact between **1u** and **P** precludes contact of **A** with either **B** or **2** is termed **insert** (Figure 5.6 D-E).

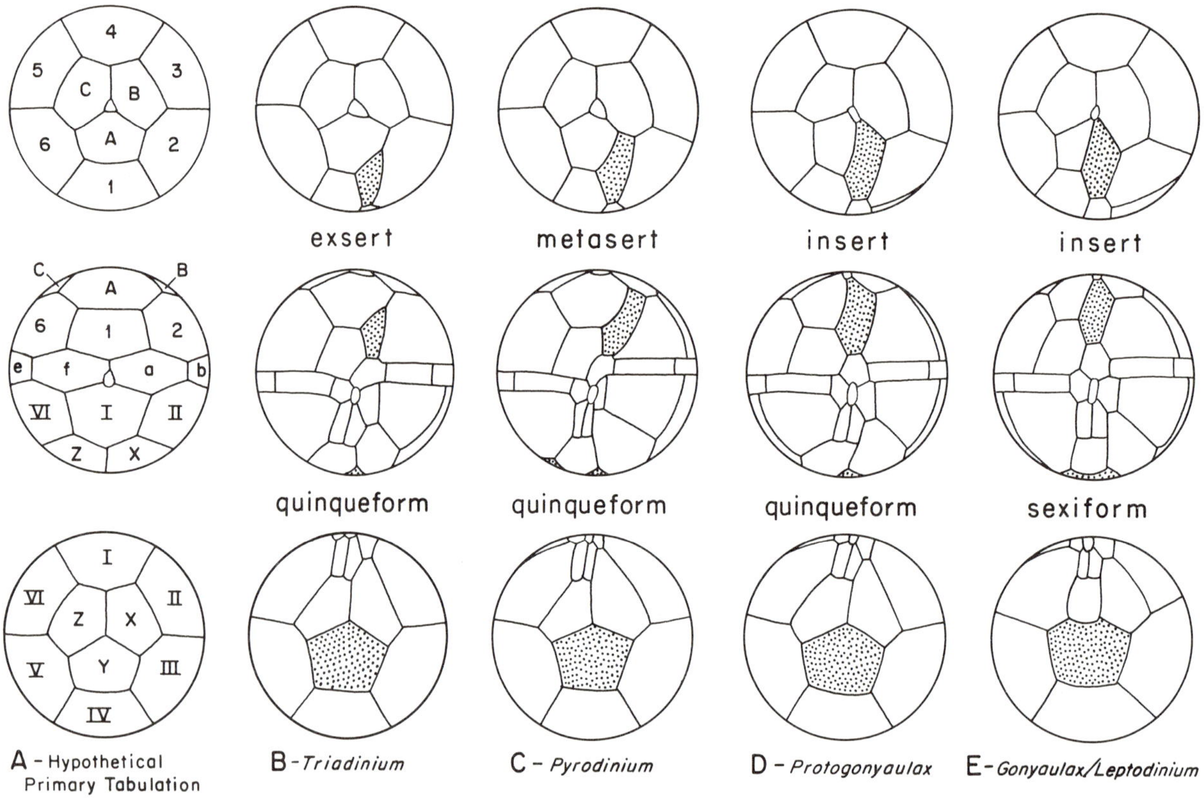

Figure 5.6 - HYPOTHESIZED DERIVATION OF QUINQUEFORM AND SEXIFORM PATTERNS FROM TAYLOR'S PRIMARY MODEL
Top row - polar views; middle row - ventral views; bottom row - antipolar views. Plate **1u** is stippled in polar and ventral views to emphasize exsert (B), metasert (C), and insert (D-E) conditions. Plate **Y** patterned in ventral and antipolar views to emphasize quinqueform (B-D) and sexiform (E) conditions.

The intermediate relationship, when **A** contacts **B** but not **2** is termed **metasert** (Figure 5.6 C).

b. Division of **a** and **f** each into two secondaries (Figure 5.7 A-D), of which **au** and **fi** remain in line with the other equatorials to form the cingulum, whereas **ai** and **fu** change somewhat in shape and become constituents of the sulcal region. The changes in **ai** and **fu** result in development of a ventral sinuousity in the equatorials and, eventually, a distinct offset in the ends of the cingulum proper as marked by **au** and **fi**. *Triadinium* (Figure 5.7 C, E-F) records an early stage in the relationship of plates in this area.

c. Division of **I** into three secondaries (Figure 5.7), **Ii** and **Im**, which regularly form part of the sulcal region, and **Iu**, which may also be part of that area, but which commonly lies on the flank of the hypotheca distinctly outside the limits of the sulcus.

3. Reorganization of relationships between primaries in the antipolar and ventral regions. **Z** decreases in relative size and shifts from a truly antipolar to a more ventral position, losing its contact with **V** and becoming integrated into the sulcus. The same change is reflected by the shift in the antipolar end of **V/VI** from its original termination against **Z** to a termination against **Y**, as **Z** "withdraws" and **VI** gains contact with **Y**.

Quinqueform Hypothecal Patterns

Quinqueform patterns occur in some significant modern and fossil species, but they are generally less familiar than the sexiform patterns which we will consider in the subsequent section, and the paratabulation of the fossil species seldom has been described in detail.

Characteristics: See Figures 5.6 B-D Primary **Y** is five-sided, contacting primaries **III**, **IV**, **V**, **X**, and **Z**; **V** has a definite contact with **Z**, precluding contact between **VI** and **Y**. **Z** is variably distinct from, or

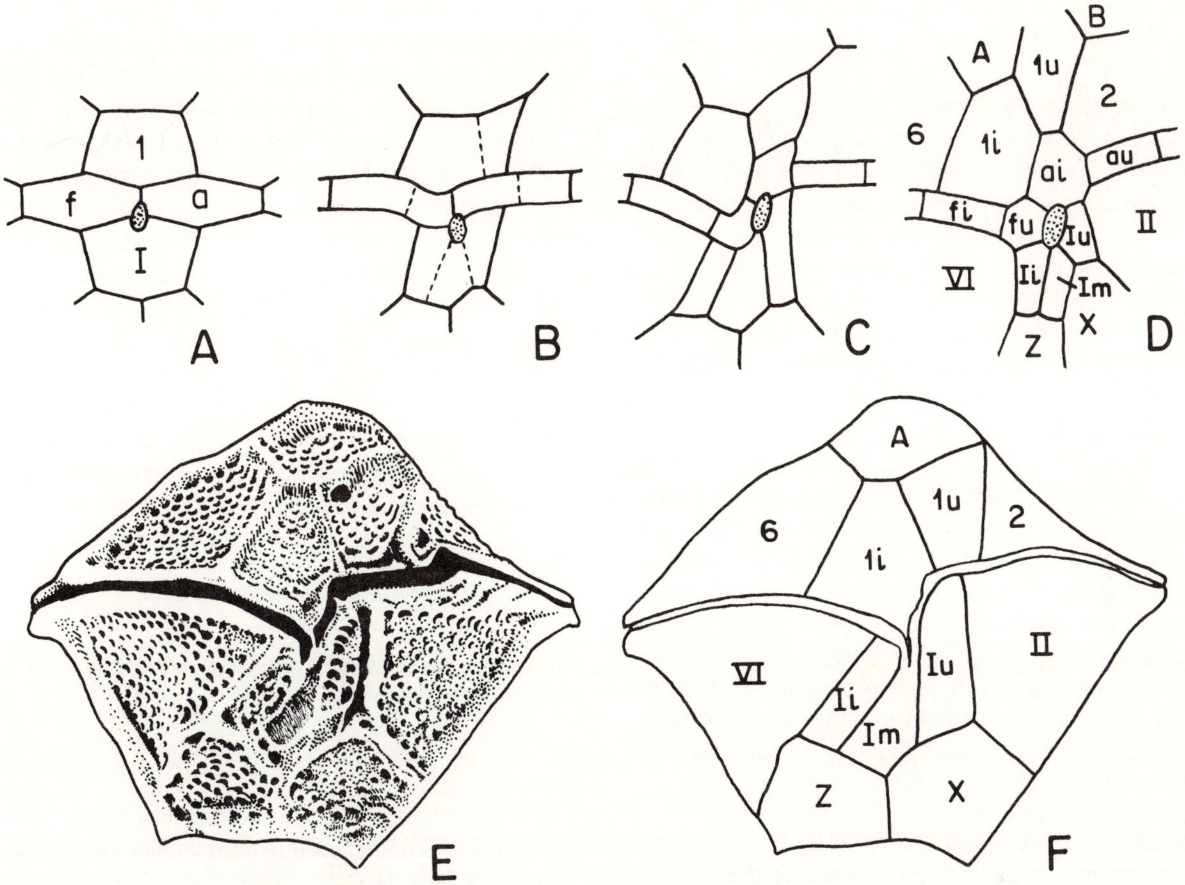

Figure 5.7 - HYPOTHESIZED ORIGIN OF SECONDARY VENTRAL PLATES

A-B - Primary model and stage showing incipient subdivision (dashed lines) of primaries **1**, **a**, **f**, and **I**.

C, E-F - Theca of modern *Triadinium polyedricum*. C -Tabulation of sulcus and adjacent area; sulcal plates based on dissection by Graham (1942). E - Drawing of ventral view; note conspicuous adelopore (based on SEM in Taylor, 1977, fig. 513). F - Diagram of same with plates labeled to accord with the origins hypothesized. Sutures separating the cingular plates in E and F are obscured by prominent sutural lists and compression along cingulum.

D - Diagram of corresponding ventral pattern in modern *Gonyaulax* or fossil *Leptodinium*.

Patterned = flagellar pore.

integrated into, the sulcal region. Secondary **Iu** is decidedly smaller than **X** and lies within the general depression of the sulcal region, rather than outside this region on the flank of the hypotheca.

Modern Examples: In *Triadinium* the hypothecal pattern is almost radially symmetrical and differs from the hypothetical model chiefly in regard to **I**, which has an asymmetrical outline and is subdivided into three secondaries. In *Pyrodinium* and *Protogonyaulax*, the symmetry of the antipolar area is reduced; **Z** is integrated into the posterior part of the sulcal region and **X** is more or less ventral. Although the central part of the sulcal region includes **Ii**, **Im**, and **Iu**, it is more sharply defined and markedly narrower than in *Triadinium*.

All the modern species of these genera are marine, and most occur in relatively shallow water. A tendency toward toxicity is conspicuous. Although *Triadinium* species are nontoxic, toxicity has been recorded at times in *Pyrodinium*, and *Protogonyaulax* includes the principal toxic species of thecate dinoflagellates. Thecae of all species are of relatively compact shape, without horns or conspicuous spines, but prominent lists are developed on some of them.

Associated Epithecal Features: In *Triadinium*, three similar polars surround a central prepolar which is slightly elongated along **A/B**; secondaries **1i** and **1u** are distinct, the latter with a large pore close to **1i/1u/A**, and a wedgelike extremity along **A/2**. The arrangement is exsert, since **A** definitely contacts **2** beyond the tip of **1u**.

In *Protogonyaulax* and *Gonyaulax*, the epithecal arrangement is normally insert, although the range of variation in some species of *Protogonyaulax* includes metasert individuals (e.g., *P. kutnerae*, Balech, 1979). It appears that **1u** has "migrated" poleward along **A/2** and **A/B**, and now contacts **P** in the manner of a typical Kofoidian apical. Primary polars **A** and **B** are completely separated by **1u**. In *Pyrodinium*, both exsert and metasert arrangements have been observed (Wall and Dale, 1969; Steidinger et al., 1980).

Cysts and Fossil Examples: Living *Triadinium* and *Protogonyaulax* are not known to produce preservable cysts like any fossils currently recognized as dinoflagellates, but cysts of some species of *Protogonyaulax* are relatively featureless ellipsoidal bodies (Dale, 1977; Wall and Anderson, 1978; in both cases reference is to species at that time referred to *Gonyaulax*), without wall morphology that would permit their identification as dinoflagellates. In paleontological terms they would have to be regarded as acritarchs.

Nevertheless, substantial documentation of quinqueform patterns exists in the fossil record, extending back to the Early Cretaceous and most varied in the Early Tertiary. Examples include cysts of both proximate and chorate design. *Heteraulacacysta* and *Dinopterygium* are fossil proximate cysts whose parasutures delineate a hypocystal paratabulation similar to the tabulation of *Triadinium*, while *Homotryblium* and *Polysphaeridium* (= *Hemicystodinium* of authors) are chorate cysts closely resembling those of living *Pyrodinium*. The general affinities of both *Homotryblium* and *Polysphaeridium* with *Pyrodinium* have been accepted for some time, but little notice has been given to the occurrence of other representatives of this group. Although *Polysphaeridium* (as *Hemicystodinium*) has been equated to *Pyrodinium* (Wall & Dale, 1969), it, like other cysts of this group from the fossil record, has an insert epicyst paratabulation. Thus, they resemble *Protogonyaulax* more closely than they do either *Pyrodinium* or *Triadinium*. Information on sulcal details is not available from any of these cysts, except for consistent indications (just as in the modern thecae) that secondary **Iu** was not a part of the hypocystal flank outside the sulcal region.

Hystrichosphaeridium tubiferum (Figure 11.2 D-I, 11.3 A-B) is another example of a chorate cyst in which a quinqueform paratabulation in the hypocyst is combined with an insert **1u** in the epicyst. However, it is currently uncertain how many taxa out of the many having a general resemblance to this species—that is, a chorate cyst with isolated intratabular processes and an apical or precingular archeopyle—also show a quinqueform arrangement of elements in the hypocyst. In at least one species (*Oligosphaeridium abaculum*, Figure 1.7 F, 11.3 C-I) ascribed to a genus that was previously thought to differ from *Hystrichosphaeridium* only by lacking cingular processes, the paratabulation in the hypocyst is unquestionably sexiform. See Chapter 11 for further discussion of this topic.

Sexiform Hypothecal Patterns

These include the best known *Gonyaulax*-like patterns from the fossil record.

Characteristics: See Figure 5.6 E. Primary **Y** contacts six other primaries: **X**, **III**, **IV**, **V**, **VI**, and **Z**; **V** has no contact with **Z**, the posterior plate of the sulcal region. Secondary **Iu** is decidedly smaller than **X** and may lie either within the general depression of the sulcal region or on the flank of the hypotheca definitely outside this depression. Plates **Iu**, **II**, and **III** are of progressively increasing size and **III** is the only one of these to contact **Y**. The eccentric, left ventral position of **X** makes the ventral hypothecal tabulation definitely asymmetrical.

Modern Examples: This pattern is characteristic of thecae in the major modern genus *Gonyaulax* (as emended by Taylor, 1979, by removal of species now referred to *Protogonyaulax*). It is the only modern genus with a sexiform hypotheca that need concern us and includes many living marine species, as well as a single freshwater one (*G. apiculata*). Thecae of most living species are compact, without conspicuous hollow horns, and are seldom strongly compressed dorsoventrally; however, a number of species have spinelike horns and a few are strikingly fusiform.

Cysts and Fossil Examples: Sexiform hypocystal patterns defined by parasutural relief occur in a broad morphological spectrum of examples distributed in sediments ranging in age from Middle Jurassic to Holocene. Although the cysts are variously proximate, chorate, and cavate, most are of

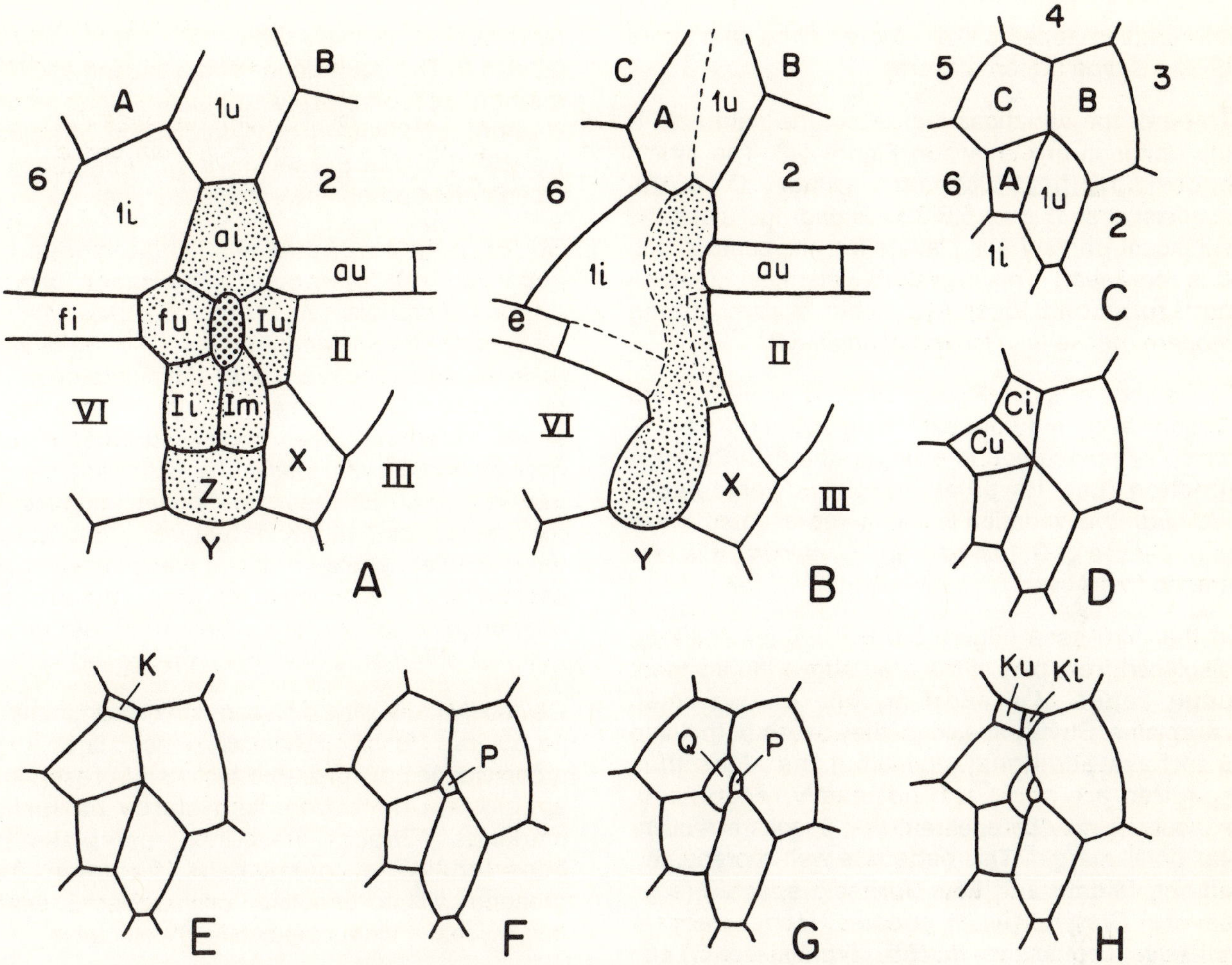

Figure 5.8 - GONYAULACOID PATTERNS: VENTRAL AND APICAL VARIATIONS

A-B - L-type and S-type ventral organization. Dashed lines mark parasutures commonly suppressed, as discussed in the text.

C-H - Variations in the apical region involving subdivision of **C** (Figure D), appearance of one or two incidentals (Figures E, H), and appearance of one or two prepolars (Figures F-H).

compact shape. While long horns are exceptional, a distinct, short apical horn is a common feature. Dorsoventral compression, which is generally slight to moderate, is pronounced in one varied and important fossil group (*Cyclonephelium* complex, Chapter 10).

As in the case of species having quinqueform hypothecal patterns, those with sexiform patterns show varied details of tabulation in other areas. However, contrary to the former situation, a single evolutionary line does not seem to be represented here. Instead, the combinations appear nearly random, but the differences tend to be concentrated in three areas; namely, the sulcus, the right half of the epitheca (in curious contrast to the left, where the pattern is nearly constant), and the extreme apex. We will consider these areas in turn.

Two contrasting types of ventral organization affecting the sulcus can be recognized if attention is directed to the overall shape of the parasulcal area, to the amount of separation between the ends of the paracingulum in both longitudinal and transverse directions, and to the immediately adjacent paraplates. These are diagramed in Figure 5.8 A-B. We will refer to them as **L-type** and **S-type** patterns, in allusion both to the taxa in which they are typically developed—*Leptodinium*, a genus of fossil cysts; and *Spiniferites*, a genus of modern and fossil cysts, as well as the thecae of the modern spinifera group of *Gonyaulax*—and to the longitudinal *versus* sigmoidal aspect of the parasulcus. Both ventral arrangements are widely represented in the fossil record, have counterparts in thecae of living species, and are usually quite readily distinguishable, although eventual documentation of complete intergrading appears likely. More will be said about L- and S-type patterns shortly.

Three of the variations that affect the right half of the epitheca are shown in Figure 5.8. The differences specifically concern primary **C**. In the simplest case (Figure 5.8 C), it is undivided, and the subjacent primary **5** is planate (i.e., its contact with **C** is rectilinear). This is probably the most common pattern among fossil cysts, but is rare among modern thecae (cf. *Gonyaulax grindleyi*).

In the second case (Figure 5.8 D), **C** is divided into **Ci** and **Cu** by a suture extending poleward from **5**, which is now camerate owing to the **5/Ci/Cu** triple junction (i.e., its polar margin is geniculate). Although this variation is widely represented today (e.g., thecae of *G. polyedra* and *G. milneri*), it is rare among fossil cysts.

In the third case (Figure 5.8 E, H) **C** is, in effect, displaced from part of the area above the adjacent polar edges of **4** and **5** by one or two small paraplates. Strictly speaking, they cause both **4** and **5** to have additional sides, but the sides thus acquired are short and not easily recognized, although it may be apparent that **5** lacks a rectilinear polar margin. This pattern is well represented among Middle and Late Jurassic species (e.g., several *Ctenidodinium* species, *Gonyaulacysta jurassica, Leptodinium mirabile*; Frontispiece D) and rare in the Early Cretaceous (e.g., *Rhynchodiniopsis aptiana*). It does not appear to be known thereafter and comparable plates have not been reported in thecae of species living today. It is not clear that these small paraplates represent subdivision of a primary, or, if so, of which one(s), and their limited occurrence makes it unlikely that evidence on this matter will soon be discovered. It is proposed here to call them **incidentals** and to denote them by **K**, using **Ki** and **Ku** when two adjacent paraplates possibly result from subdivision of a single area. While obviously intercalaries in a Kofoidian sense, they are quite different from the intercalaries of peridinioid patterns.

At the extreme apex two arrangements are commonly encountered. In one of them (Figure 5.8 C-E), the four parasutures meet at a point or nearly so. In the other (Frontispiece D, Figure 5.8 F-H), one or two small paraplates occur here. In the latter situation, the larger paraplate lies characteristically to the right of the smaller, and **C/A** (or **Cu/A**) terminates against it. The single paraplate, or the smaller and more strictly apical one when two are present, is taken to represent the single prepolar of the hypothesized primary tabulation. Accordingly, it is labeled **P**. The equivalence of the second and larger element is completely uncertain (independent element, subdivision of **P**, sudivision of adjacent polar?). For the present, we will consider it an independent prepolar element and label it **Q**.

So far, authors of published descriptions have accorded little attention to the exact form and position of prepolars and to their contacts with each other and with subjacent polars. The areas where these features occur are often difficult to observe in direct view and the dimensions of the features are small. However, preliminary studies of fossil gonyaulacoid cysts using the scanning electron microscope (J. Helenes, pers. comm.) suggest that the intricacy and variety of features which occur in the immediate vicinity of the apical pole will reward careful study with new insights into probable relationships among taxa. Only the most general features of the area have been mentioned here.

Obviously the ability to discern small incidentals and prepolars in fossil material will depend largely upon a coincidence of favorable factors of preservation, specimen orientation, and clarity of surficial markings. Although these small paraplates may occasionally be conspicuous, frequently their presence will go unnoticed or imperfect preservation will leave them completely undetectable.

The gonyaulacoid patterns most abundantly represented in the fossil record are those that combine the L-type or S-type ventral organizations with one of the polar designs in which **C** is undivided. The effort necessary to become thoroughly familiar with the principal features of these patterns, as reviewed in the following subsections of this discussion, will be rewarded by greatly increased facility in interpreting a great many fossil cysts under the microscope.

Cysts with L-type Ventral Organization. (Figures 5.8 A, 5.9-5.12). These are common in the fossil record and include proximate cysts with clear parasutural patterns such as are assigned to *Cribroperidinium, Ctenidodinium, Hystrichogonyaulax, Leptodinium* (Frontispiece G), *Lithodinia, Meiourogonyaulax, Millioudodinium, Occisucysta, Rhynchodiniopsis*, and *Sirmiodinium*. Many other proximate and chorate fossil cysts also give more or less strong indications of the L-type paratabulation, although incomplete expression of the paratabulation leaves various degrees of uncertainty in regard to many individual species; examples include *Areosphaeridium, Cordosphaeridium, Hystrichodinium,*

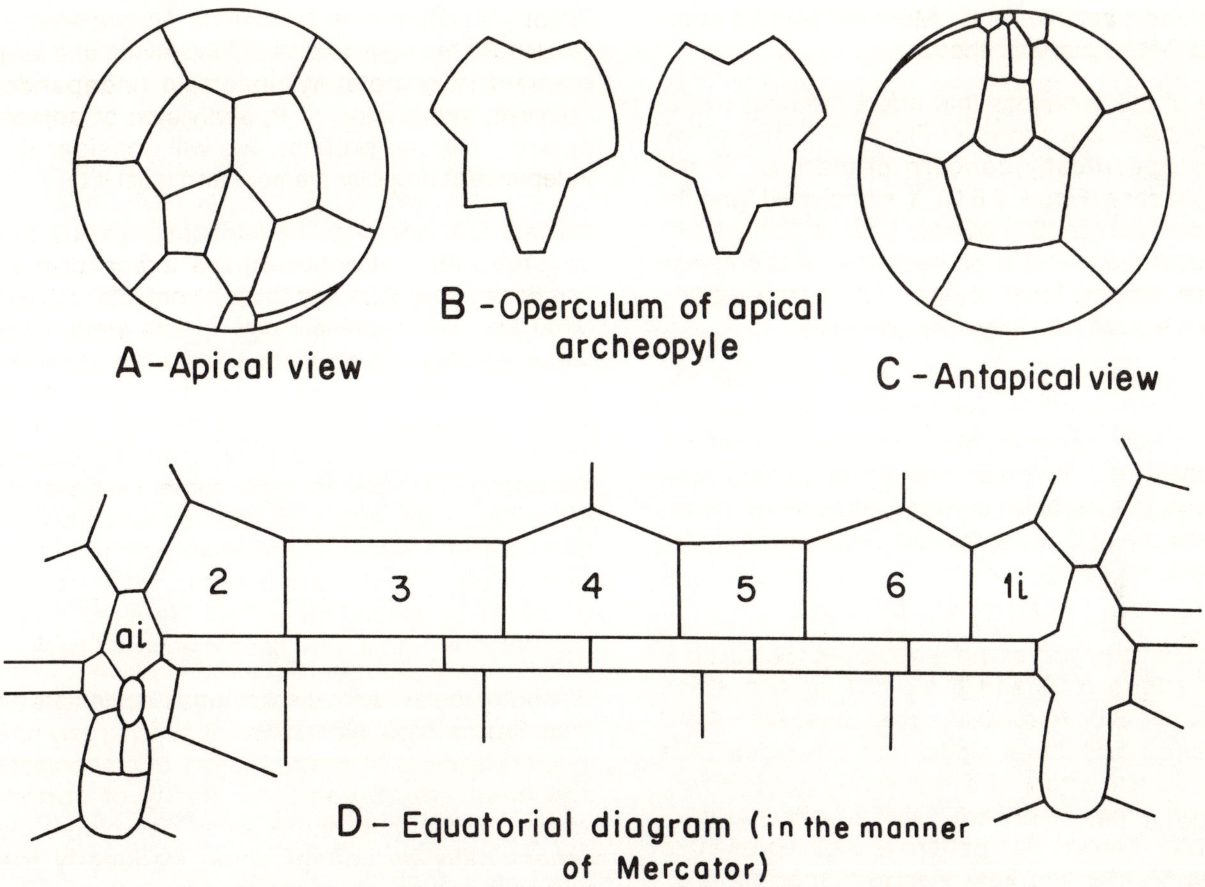

Figure 5.9 - IMPORTANT ASPECTS OF L-TYPE SEXIFORM PATTERNS

A-C - Diagrams based on an ideally spherical individual. A, C - Apical and antapical views, showing general tabulation. B - Differences in outline of the operculum of an apical archeopyle as it would appear in exterior (left) and interior (right) views.

D - Tabulation as it might appear in a Mercator-like projection; ventral region repeated at ends, polar areas not shown. Tabulation of sulcal region detailed at left. Note sequence of camerate and planate precingular plates; and incorporation of **Iu** into the sulcal region at the right.

Hystrichokolpoma, *Scriniodinium*, and *Turbiosphaera*. Living species with comparable L-type ventral organization of the theca include *Gonyaulax grindleyi*, *G. milneri*, and *G. polyedra*, but these species do not produce cysts with readily decipherable paratabulation. Except for the first named, they also differ from typical fossil cysts in the further respect that **C** is divided.

In cysts with L-type ventral organization, the paracingulum normally is distinctly descending and its ends are offset up to one paracingulum width or more but do not overhang (cf. Figure 3.1 C). The parasulcus is more or less straight and approximately meridional. Some of the variations in paraplate arrangement within and near the parasulcus are discussed in Chapter 10. Asymmetry of the paratabulation adjacent to the sulcus is normally pronounced, with a trio of paraplates (**1u**, **II**, and **X**) on the left flank of the hypocyst opposed to just **VI** on the right flank. On the other hand, bilateral symmetry is approximated in the dorsal paratabulation in many species.

Figure 5.9 illustrates a combination of features that is especially common among the large number of fossil cysts with L-type sexiform gonyaulacoid patterns. The sequence of parasutures along the apical boundary of the precingular series shows a particular and useful asymmetry: geniculate segments occur above **2**, **4**, **6**, and **1i**, whereas rectilinear segments occur above **3** and **5**. The **1u/ai** parasuture forms a further (and definitely the shortest) rectilinear line segment in this succession. Parasutural ornament on the cyst surface may make this sequence of parasutures visible, and it is also apparent in many cysts that lack such ornament but have an apical archeopyle, for this is the course of the archeopyle suture. In either case, the different sequences of precingular shapes encountered in progressing from the parasulcus toward the right (camerate-camerate-planate) and left (camerate-planate-camerate) are useful criteria for distinguishing interior and exterior views, especially with regard to isolated opercula (Figure 5.9 B).

The relatively large sizes of **B** and **3** are characteristic for a large number of fossil species, as also is the path of the dorsal midline approximately along **B/C** and **c/d** and across the central areas of paraplates **4** and **IV**.

Paraplates **Iu**, **II**, and **III** are successively larger. Not only is **III** much the largest of these three, but it is also the first paraplate of the postequatorial series that contacts **Y**. Paraplates **III-V** are of similar size (**IV** often slightly the largest); **VI** is, again, smaller. **Y** is six-sided, and its ventral contact with the posterior end of the parasulcus is usually conspicuously concave. **Y** is often, but by no means always, approximately symmetrical, and commonly **Y/X** and **Y/VI** are approximately symmetrical equivalents of one another.

Paratabulation is often faint or incomplete in the parasulcus. The anterior and posterior sulcal paraplates (**ai** and **Z**) are the most consistently and strongly delineated ones, and the intermediate small paraplates (including **Iu** at times) may merge into a single area. The parasutural line marking the adcingular boundary of **Z**, which is actually a combination of two parasutures (**Z/Ii** and **Z/Im**), virtually always meets **X** along its right margin, confirming that the term "posteroventral" (*pv*), which has been applied in some published works to an "extra" paraplate in this area, is, in fact, an alias for the posteriormost sulcal (i.e., **Z**). Even in those extreme cases in which **Iu** appears to be fully integrated into the parasulcus, without a trace of parasutures on its right side, a diagnostic offset corresponding to **Iu/X** is almost always visible in the outline of this enlarged "parasulcal" region (righthand side of Figure 5.9 D). In dorsoventrally compressed cysts of the *Cyclonephelium* complex (Chapter 10), the parasulcus is offset to the left side of the ventral surface.

The L-type sexiform pattern, combined with an apical arrangement lacking both prepolars and incidentals, is shown in a variety of orientations in Figures 5.10 (in exterior view) and 5.11 (in interior view). An ability to identify with relative ease the plates shown in these views will be a great asset when examining specimens under the microscope.

Cysts with S-Type Ventral Organization. (Figures 5.8 B, 5.12-13). The best known and most common examples include proximate to murochorate cysts referred to *Hystrichosphaeropsis* and *Impagidinium*, and skolochorate cysts belonging to *Spiniferites* and other genera of the *Spiniferites* complex (Chapter 10). Their combined fossil record is long, extending from the Middle Jurassic to the Holocene. Modern examples of *Impagidinium* and *Spiniferites* are cysts belonging to species of the spinifera group of *Gonyaulax*.

The S-type ventral organization presents a complex of contrasts with the L-type organization. Many of its typical features may be visible in a single species. Especially characteristic are a quasitriangular **1i** and suppression of the markings that identify certain parasutures (primarily on the ventral surface, but also at other specific locations in some species). Principal contrasts with L-type patterns include the following:

—Ends of paracingulum strongly offset, tending also toward overhanging.

—Parasulcus narrowest between ends of paracingulum and often gently sigmoid.

—Lateral boundaries of **1i** apically convergent, rendering the outline of this paraplate quasitriangular or even definitely three-sided in conjunction with reduced contact of **1i** with **A** and **1u**; contact between **1i** and **A** may even be lost entirely.

Figure 5.10 - L-TYPE SEXIFORM TABULATION: EXTERIOR VIEWS

Schematic tabulation on a hypothetically spherical cyst with moderately spiral paracingulum. The features are seen here as they would appear in a direct view (high focus) of the exterior.

A - Apical view, sulcus down
B - Ventral view
C - Antapical view, sulcus up
D - Left lateral view
E - Dorsal view
F - Right lateral view
G - View about normal to **2/3**
H - View about normal to **4**
I - View about normal to **6**
J - View about normal to **V**
K - View about normal to **Ii/Im**
L - View about normal to **II/III/X**

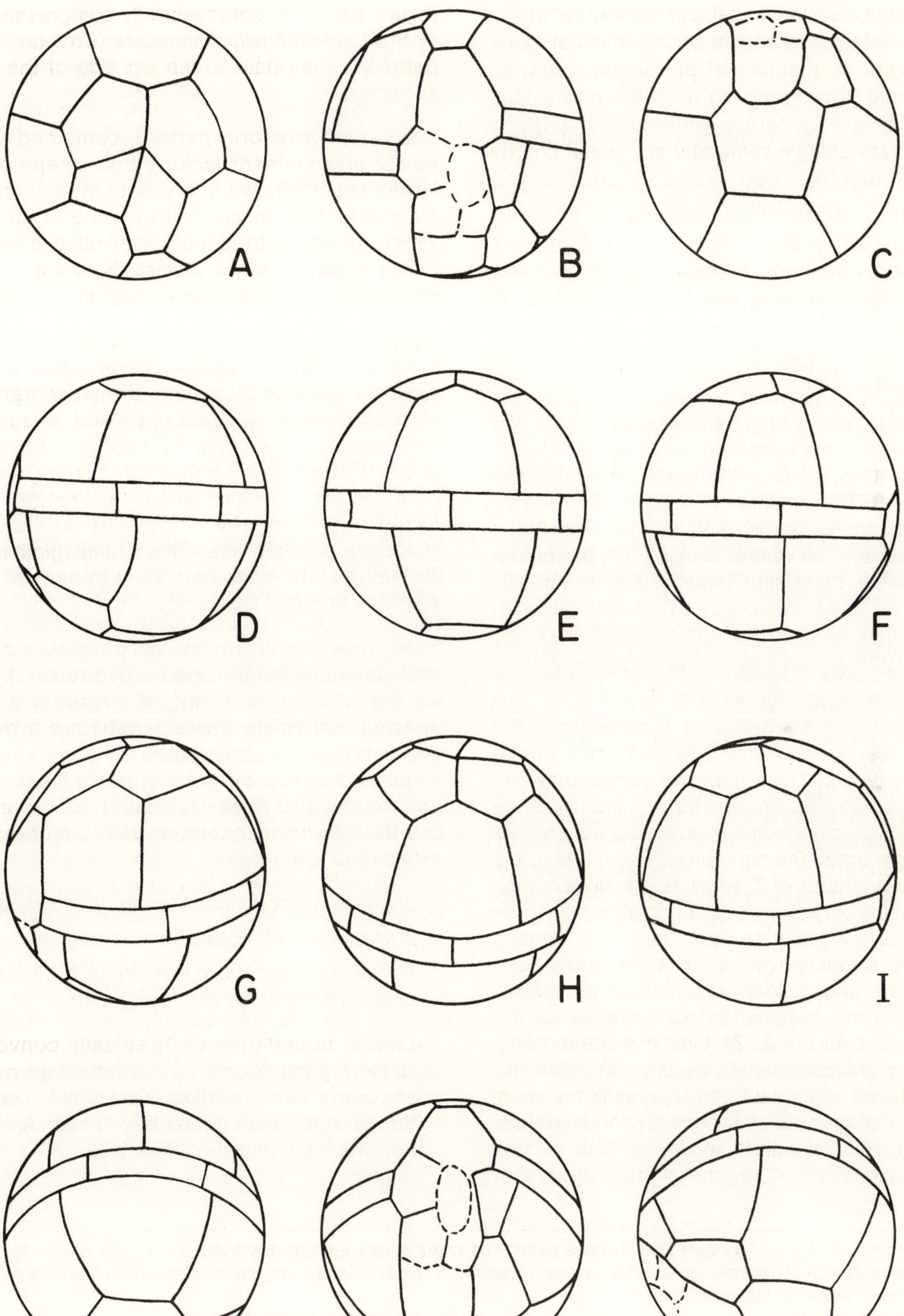

Figure 5.10

—**C** enlarged, commonly nearly symmetrical to **B**, the two forming the dorsal polar surface (Figure 5.12 L).

—**A** reduced to either 1) a long, narrow, fully ventral and approximately symmetrical equivalent of the adjacent **1u** (Figure 5.12 L),or 2) a small paraplate between **C** and **1u**, having minimal or no contact with **1i** (Figure 5.12 M). These cases are similar in that only three paraplates (one ventral, two dorsal) may be visible around the apex, but the explanations are different. In the first case, **A** may be overlooked because of its partial or complete merger with **1u** (i.e., **A/1u** is faint, incomplete, or absent). In the second, **A** may be overlooked because of its diminutive size and minor effect on adjoining paraplates.

—Parasutures within the parasulcus commonly reduced in strength; those that form the posterior boundary of **ai** (combination of **ai/fu** and **ai/lu**) and the anterior boundary of **Z** (combination of **Z/li** and **Z/lm**) being the most persistently present (Figure 5.12 G).

—Parasutures separating parasulcus from certain adjacent areas (especially **1i** and **lu**) may be severely reduced (Figure 5.8 A).

—Parasuture **1i/fi** may not be visible (Figure 5.8 A).

A further small but interesting feature is that the lower end of **A/1u**, whenever it is visible, tends to abut against **ai** rather than **1i**, making the margin of **1i** rectilinear and that of **ai** geniculate, just the opposite to their typical condition in the L-type pattern (compare Figures 5.12 K and L).

As a comparison of Figures 5.8 A and B will suggest, some of the distinctive features of the S-type patterns might be imagined to result from an elastic contraction of an L-type pattern that would bring into greater proximity the triple junctions: **A/6/1i**, **A/1i/1u**, **1i/1u/ai**, and **1u/ai/2**. While specific patterns are known that could be taken for intermediate stages in this "contraction" sequence, they are not so numerous as others close to the L-type and S-type extremes. Further comments on these ventral arrangements will be found in Chapter 10.

The S-type sexiform pattern, accompanied by strong selective suppression of parasutures, is shown in a variety of both exterior and interior views in Figure 5.13.

Corniform Hypothecal Patterns

Distinguishing the major modern genus *Ceratium*, these patterns are occasionally visible in fossil cysts of similar aspect, including *Muderongia, Endoceratium*, and others (Wall and Evitt, 1975).

Characteristics: See Figure 5.14. The sulcus lies to the left of the ventral surface, and, in some species, includes plates which are speculatively homologs of **fu** and **ai**, but no other well-developed plates are present in this area. Neither **Z** nor **I** is present, unless represented by membranous sulcals. **X** is chiefly ventral; **Y**, which is not clearly polygonal, makes contact with six and seven areas in modern freshwater and marine species, respectively. There are six postcingulars and, in the absence of **Iu**, these include **VIi** and **VIu** as subdivisions of **VI**. The two hypothecal horns almost always present are composed of **X** and **Y**, and **V** and **VIi**, and a third horn, present in some species, is composed of **II** and **III**. Modern marine species of *Ceratium* have only five cingular plates, apparently as a result of the fusion of **c** and **d** to make an unusually long dorsal cingular. Individual cingular paraplates are rarely discernible in fossil cysts of corniform design; when a complete count is possible, there are six, as in living freshwater *Ceratium*. The sulcus in the modern theca is a specialized area of mostly thin plates, complexly arranged and associated with a large excavation beneath (i.e., dorsal to) the major plates of the ventral surface (Dodge and Crawford, 1970). Paratabulation details have never been seen clearly in the parasulcus of fossils.

Figure 5.11 - L-TYPE SEXIFORM TABULATION: INTERIOR VIEWS

Schematic tabulation of a hypothetically spherical theca with moderately spiral cingulum. The features are seen here as they would appear in low-focus view of a three-dimensional, transparent specimen. Note the distinction between the stated direction of view and the surface or feature shown (i.e., in focus).

A - Antapical view, sulcus down
B - Dorsal view
C - Apical view, sulcus up
D - Right lateral view
E - Ventral view
F - Left lateral view
G - View about normal to **V**
H - View about normal to **li/lm**
I - View about normal to **II/III/X**
J - View about normal to **2/3**
K - View about normal to **4**
L - View about normal to **6**

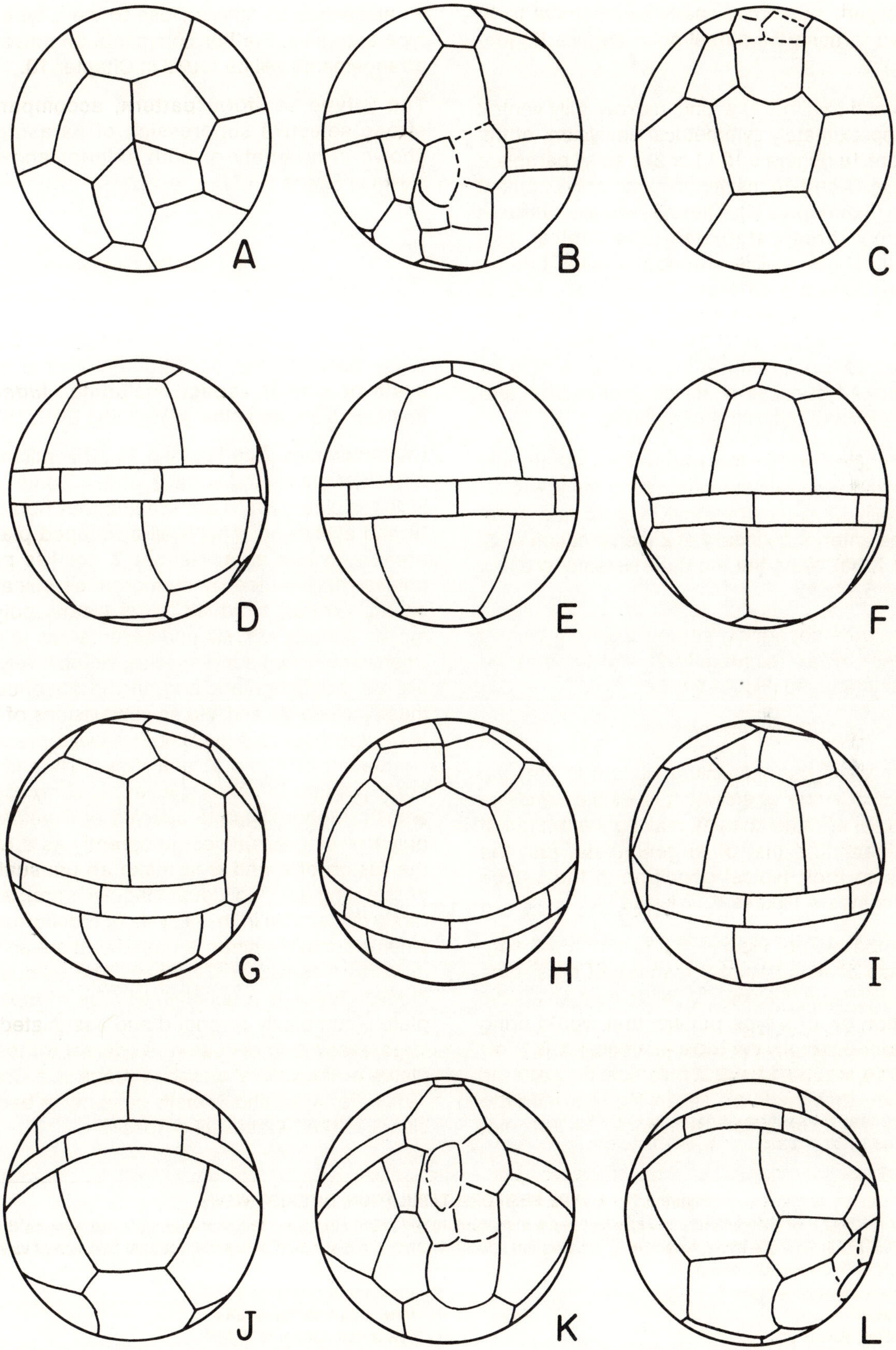

Figure 5.11

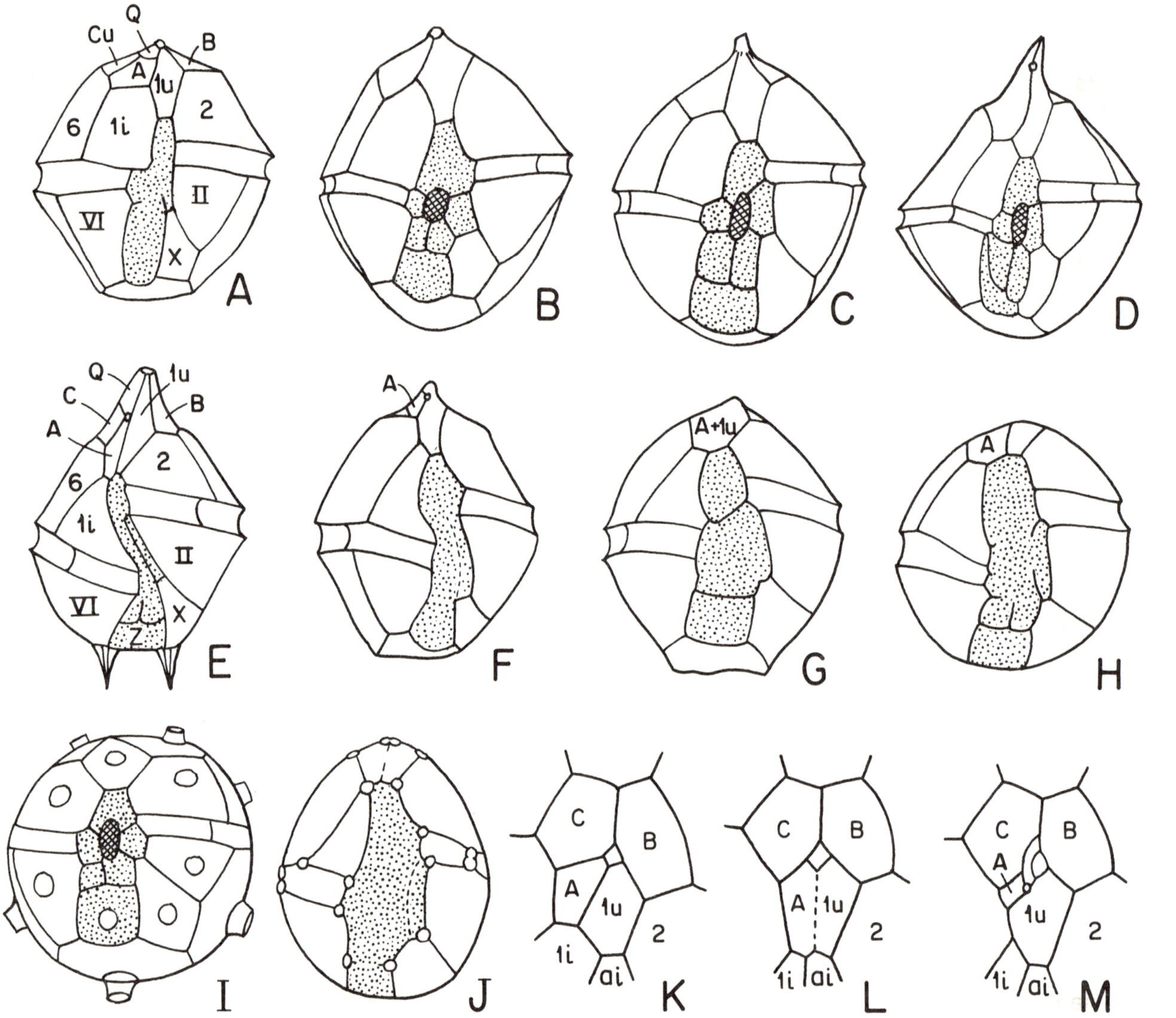

Figure 5.12 - L-TYPE AND S-TYPE VENTRAL PATTERNS

(light patterning - sulcal region including **lu**; heavier patterning - flagellar scar; porichnion as small circle near apex in D, E, F, and M)

A-D - L-type ventral organization in theca of modern *Gonyaulax polyedra* (A), and in fossil proximate cysts of Late Jurassic *Leptodinium mirabile* (B), Early Cretaceous *Occisucysta* sp. (C), and Early Cretaceous *Cribroperidinium ventriosum* (D, drawn from Gocht, 1979)

E-H - S-type ventral organization in theca of modern *Gonyaulax spinifera* (E, after Balech, 1977), and in fossil cysts with parasutural ridges or low septa (septa not shown). F - Early Cretaceous undescribed form, with minute **A** (for discussion of apical arrangement, see M) and suppression of parasutures in the parasulcus. G - Eocene *Impagidinium maculatum*, with parasuture **A/1µ** suppressed in addition to those in central part of parasulcus. H - Eocene *I. californiense*, with incomplete parasutures in parasulcus, and relatively large **A**.

I-J - L-type and S-type ventral patterns in Early Cretaceous skolochorate cysts (only process bases shown). I - L-type pattern in *Oligosphaeridium abaculum*. J - S-type pattern in *Spiniferites* sp.

K - One of the apical arrangements typically associated with L-type ventral patterns in the fossil record. Note size and distinctness of **A**.

L-M - Two apical arrangements present in fossil cysts with S-type ventral pattern. L - The most commonly encountered pattern, with strong tendency for **A/1µ** to be suppressed. M - A second pattern, in which the strong reduction of A results in an unusual topologic relationship of the paraplates (associated with the ventral pattern shown in F), typical of Early Cretaceous *Gonyaulacysta cretacea* group.

If the tabulation of *Ceratium* is examined with a fundamentally gonyaulacoid pattern in mind (following the views of Wall and Evitt, 1975), features homologous to those in the epitheca of *Gonyaulax* seem obvious. However, the sharply contrasting hypothecal tabulations in the two genera leave some uncertainty about the particular homologies in that part of the theca. The interpretation adopted here rests heavily on the premise that the first plate in the sequence of postcingulars which contacts the antapical, **Y**, is correctly identified as **III**. From this it follows that the initial plate of the postcingular series in *Ceratium* is **II**, and that **Iu** is not indicated on the hypotheca of that genus outside the sulcus. Nevertheless, despite this absence of a plate normally present in the gonyaulacoid pattern, there is a total of six postcingulars. This is taken to imply the presence of a plate not represented in the hypotheca of *Gonyaulax* and hypothesized to result from secondary division of **III**, **IV**, **V**, or **VI**. There are few guides or constraints to the speculative choice among these possibilities. Relative to plate proportions in *Gonyaulax*, the breadth of the last precingular (**1i**) in *Ceratium* is notable. Perhaps, associated with this, it was a comparable "extension" in the hypotheca through approximately the same longitudinal span that resulted in division of primary **VI**. On the basis of this hypothesis, the tabulation in Figure 5.14 has been labeled to show **VIi** and **VIu**. Although other relationships can be cited as consistent with this choice (e.g., plate overlaps in *Ceratium* determined by Happach-Kasan, 1982; the position of **6/1i** and **e/fi** relative to **VIi/VIu** in *Ceratium* as compared to their position relative to **V/VI** in *Gonyaulax*), the significance of that consistency is also only speculative.

Happach-Kasan (1982), in reporting a detailed study of living freshwater *Ceratium cornutum*, did not accept the gonyaulacoid interpretation of *Ceratium* tabulation suggested by Wall and Evitt (1975). She also argued against the notion advanced by Taylor (and strongly supported here; see Chapter 13 for extended discussion) that a fundamental organizational difference distinguishes the gonyaulacoid and peridinioid patterns.

Modern Examples: *Ceratium* is represented by a few freshwater and many estuarine to oceanic marine species. In the freshwater species, **VIu** contacts **X**, separating **Y** from the extremity of the sulcus; in the marine species, the sulcus terminates against **Y** in more conventional fashion. Almost all species have a prominent anterior horn, as well as two hypothecal horns (one from the right postcingular area and one antapical). Rarely the postcingular horn is reduced, and a third hypothecal (i.e., left postcingular) horn occurs in freshwater *C. hirundinella* and *C. furcoides*. Although thecal shape is highly varied, largely in conjunction with diverse development of horns, tabulation is virtually constant. Cysts produced by living freshwater species—whether they are vegetative or zygotic cysts is a matter of debate—lack sporopollenin and are not preservable in sediments beyond a few years (Wall and Evitt, 1975).

Associated Epithecal Features. Epithecal organization is normally like the insert gonyaulacoid pattern with undivided **C**: four plates (**A**, **1u**, **B**, and **C**) normally reach the end of the apical horn, but in *C. furcoides* **A** is restricted to the lower portion of the horn (becoming a Kofoidian intercalary).

A peculiarity of the corniform tabulation in general is the modification of **1i** and **fi**, which, together with **VIu**, occupy much of the midventral surface. Plate **1i** is very large and **fi** is sinuous, with an expanded sulcal end. In modern marine species these three plates of the ventral area are much thinner than the other plates of the theca.

Fossil Examples. Many species with a pronouncedly ceratioid appearance occur in uppermost Jurassic to Upper Cretaceous strata. However, paratabulation is too seldom visible in this group of cysts to be a generally useful character. The archeopyle is consistently apical. Its normally zigzag margin defines the opposed edges of six precingular and four apical paraplates, as well as the anterior sulcal which marks the strong offset of the parasulcus to the left side of the ventral surface. An essentially complete paratabulation like that in modern *Ceratium* has been observed in exceptional cases (*Muderongia* cf. *M. mcwhaei*, Wall and Evitt, 1975). Important fossil representatives are discussed under Gc-cysts in Chapter 10.

Variability. Modern species of *Ceratium* vary extensively in cell shape, especially with respect to the length, curvature, and orientation of the horns. In contrast, the tabulation is nearly constant, the principal differences being ones that distinguish marine from freshwater species (five versus six cingulars, zero versus two thick sulcal plates, and contact between the sulcus and **Y**). This conservatism in tabulation, which is also evident among fossil cysts insofar as the paratabulation is actually distinct, contrasts strongly with the highly labile designs among those gonyaulacoids with quinqueform, sexiform, and partiform patterns.

Kofoid and Taylor Systems for Gonyaulacoid Patterns

Both the extensive use of Kofoid's system for gonyaulacoid patterns in published literature and

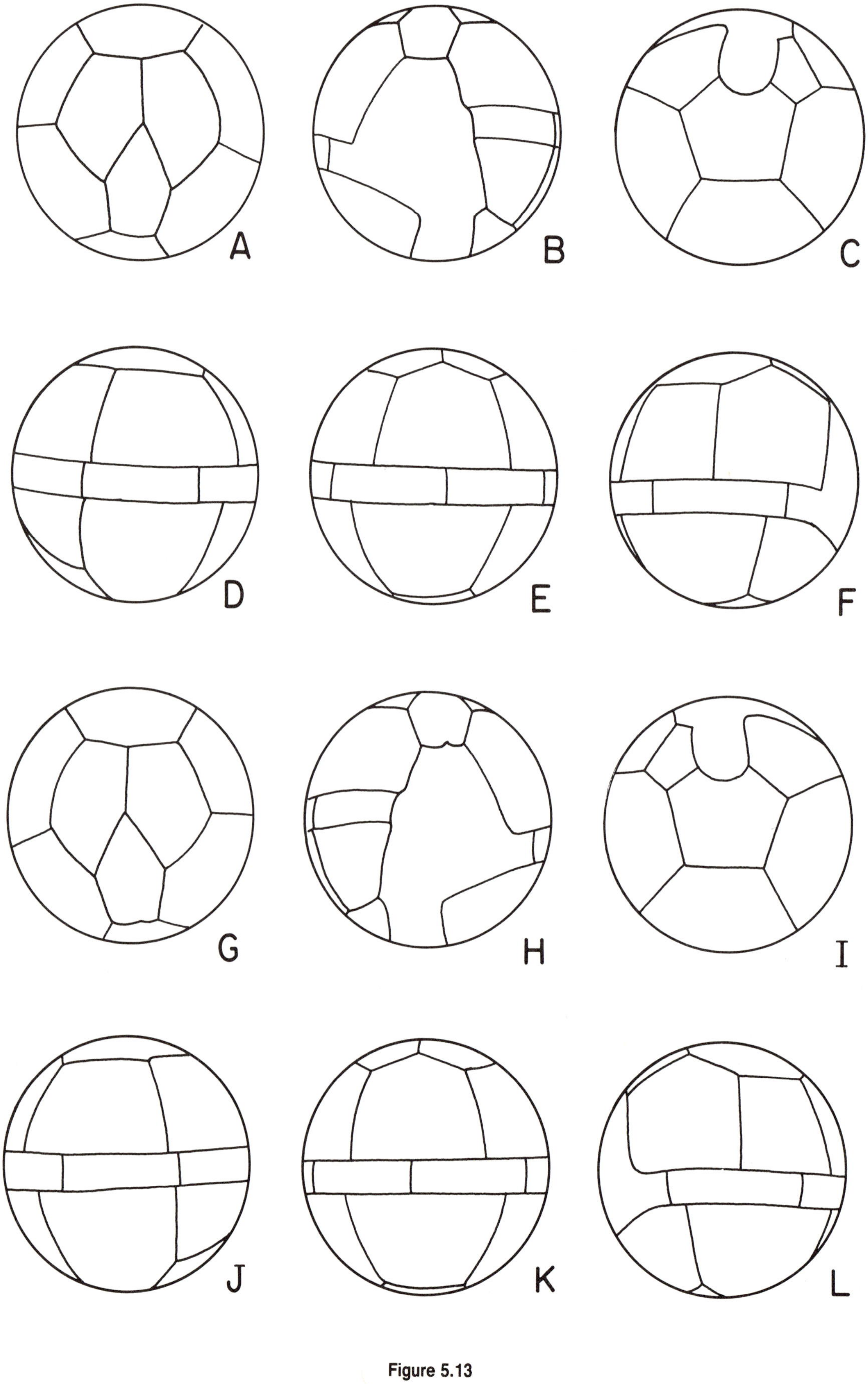

Figure 5.13

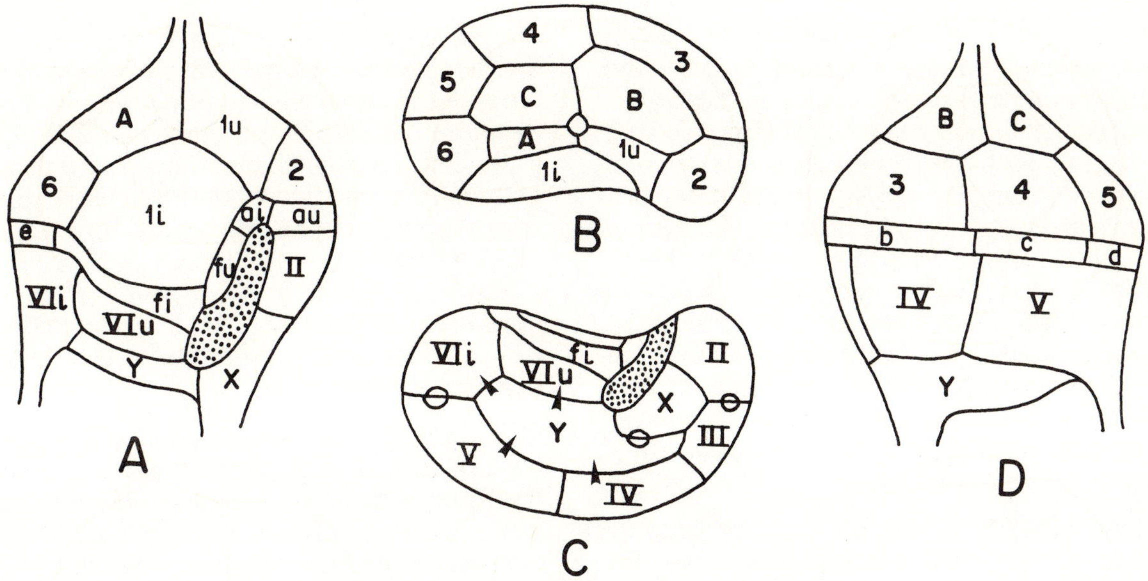

Figure 5.14 - CORNIFORM TABULATION PATTERN

A - Ventral view. B - Apical view. C - Antapical view. D - Dorsal view. Circles along sutures in antapical view show locations of hypothecal horns, only two of which are indicated in the dorsal and ventral views. See text for discussion of "extra" postcingular plate implied by **VIi** and **VIu** and plate overlap directions shown by solid angles on sutures in C.

In contrast to earlier figures in this chapter, the diagrams deviate from circular in order to suggest the three horns and dorsoventral compression so characteristic of the species whose tabulation/paratabulation is organized as in the modern genus *Ceratium*. In the diagrams, the two plates characteristic of the sulcal region in modern freshwater species are combined with contacts at the posterior end of the sulcus that occur in modern marine species. The cingular plates shown here separately as **c** and **d** are fused along **c/d** in modern marine species but are separate in freshwater species.

the continuing utility of the system assure that it will not soon be replaced by any other. By providing an essentially objective basis for designating plates according to their position in the theca, it facilitates description and communication when plate homologies are unsettled. Consequently, a thorough familiarity with how the Kofoid system applies to the patterns here termed gonyaulacoid is essential.

So long as there is no concern for recognizing equivalent plates in different taxa or for comparing similar but different patterns, each pattern can be described independently in terms applied arbitrarily according to Kofoid's rules. However, inconsistencies between Kofoidian labels soon appear when a judgment about plate homologies enters the picture, and confusion rapidly follows if descriptions and comparisons are phrased in terms of the conceived homologies. In descriptive works on modern dinoflagellates, few authors have followed Kofoid's procedure with absolute rigor. As the following examples will illustrate, most have tolerated some deviation or have suggested redefinition of Kofoidian terms in order to maintain the same labels for what they regard as "equivalent" plates: a) Kofoid himself applied "apical" to some plates that may not reach the apex (e.g., the first apical in many *Peridinium*). b) Balech (1967, 1980) suggested that the posterior intercalary in *Gonyaulax* should be considered an antapical, a suggestion taken up by Loeblich (1980). c) Although *Triadinium* was interpreted to have three apicals and seven precingulars (Graham, 1942) and all the latter are remote from the apex, the "obvious" equivalence of one of the

Figure 5.13 - S-TYPE SEXIFORM PARATABULATION

A hypothetically spherical cyst showing strong suppression of certain parasutures.

A-F - Exterior views:

A - Apical view, parasulcus down
B - Ventral view
C - Antapical view, parasulcus up
D - Left lateral view
E - Dorsal view
F - Right lateral view

G-L - Interior views, as if focused low on a transparent entire specimen:

G - Antapical view, parasulcus down
H - Dorsal view
I - Apical view, parasulcus up
J - Right lateral view
K - Ventral view
L - Left lateral view

"precingulars" to the first apical in *Gonyaulax* has led to restatement of the tabulation as including four apicals and six precingulars (e.g., Taylor, 1976). d) Others have carried the reinterpretation of *Triadinium* a step further, suggesting a revision of criteria for recognizing the first apical (Loeblich, 1980), or a special notation for denoting that it is not a "true" apical (Bujak, 1980). e) An epithecal plate in certain modern freshwater *Ceratium* (alternatively identified as *Ceratium furcoides* or a form of *C. hirundinella*) clearly fits the definition of a Kofoidian anterior intercalary, but it has been called 4′ (Wall and Evitt, 1975) because of its apparent equivalence to the fourth apical in virtually all other examples of modern *Ceratium*.

Such departures from Kofoid's system are arguably useful, and, in small numbers, they may not seriously impair the system's value. However, as concern grows for recognizing homologous plates among different taxa, serious complications in communication are likely if a clearly arbitrary, but notably logical, system is continually modified, especially if this is done in response to individual differences of opinion. Taylor's system, with its different approach, neither compromises nor replaces Kofoid's; but, within this gonyaulacoid group of dinoflagellates, it provides a complementary way of comparing patterns on the basis of apparent homologies. This complementary relationship between the two systems will be exploited by using them together in the following section on partiform patterns.

In many cases, use of the Taylor system avoids the almost impossibly intricate "conversion tables" that would be required for comparisons if only Kofoidian labels were applied. The complexities of the problem will be obvious if the reader will take the trouble to reread the foregoing discussion of gonyaulacoid patterns, substituting Kofoid plate designations throughout for the Taylor designations that now appear in the text. This will bring out that the postcingular region and the right side of the apical region are the two areas where confusion is most likely when species having quinqueform, sexiform, and corniform patterns are compared. The contrasting Kofoidian labels in these areas suggest differences in plate number, size, and mutual relationships much more profound than those that actually exist. In both areas the difficulty is obviated by using the Taylorian system. The diagrams in Figures 5.15 and 5.16 and the summaries of information in Tables 5.2 and 5.3 are designed to make this point.

In the postcingular region (Figure 5.15), **lu** as a thecal plate may appear to be a part of the sulcus or an initial postcingular; on cysts it may not be delineated at all. As a result, the total of postcingulars on both theca and cyst may be either five or six. The paleontological literature documents two different approaches to this situation. One has been to interpret **ll** objectively as either *1‴* or *2‴*, depending upon whether it is the first or second *delineated* postcingular area. One consequence of this approach is that the first postcingular in contact with the antapical will be numbered *2‴* in one species and *3‴* in another, even though all the principal topologic relations in the two species are identical, excepting the difference in the small area at the beginning of the series. The second approach has been to consider **ll** to be *2‴*, even when it is the first of the recognizable postcingulars, following a subjective judgment that the "true" first postcingular is "absent". In this case, the visible postcingulars are numbered *2‴-6‴*, and each bears the topologic relation to its neighbors which it does in a pattern with six postcingulars.

Proponents of each approach can advance good arguments in their support, but that does not reduce the consequent confusion. At the root of the problem is a failure to recognize that one set of symbols—Kofoid's—is being called upon to serve two purposes: to provide objective labels for

Table 5.2 - TAYLOR-KOFOID COMPARISONS: VENTRAL REGION

Listed in the columns below the Taylor symbols are the Kofoid symbols for plates in four modern gonyaulacoid genera: *Triadinium* with quinqueform exsert pattern, *Pyrodinium* with quinqueform metasert pattern, *Gonyaulax* with sexiform insert pattern, and *Ceratium* with corniform pattern. This table should be studied in conjunction with Figure 5.15.

	Kofoid symbols listed under Taylor symbols												
Item in Fig. 5.15	**A**	**1i**	**1u**	**2**	**ai**	**fu**	**li**	**lm**	**lu**	**ll**	**X**	**Y**	**Z**
A. *Triadinium*	*1′*	*7″*	*1″*	*2″*	*as*	*ras*	*rs*	*ps*	*ls*	*1‴*	*1⁗*	*2⁗*	*3⁗*
B. *Pyrodinium*	*4′*	*6″*	*1′*	*1″*	*as*	*ras*	*rs*	*ls*	*1‴*	*2‴*	*1p*	*1⁗*	*ps*
C. *Gonyaulax*	*4′*	*6″*	*1′*	*1″*	*as*	*ras*	*rs*	*ls*	*1‴*	*2‴*	*1p*	*1⁗*	*ps*
D. *Ceratium*	*4′*	*6″*	*1′*	*1″*	*o**	*m**	–	–	–	*1‴*	*1p*	*1⁗*	–

**These symbols have been used only for sulcal plates in Ceratium.*

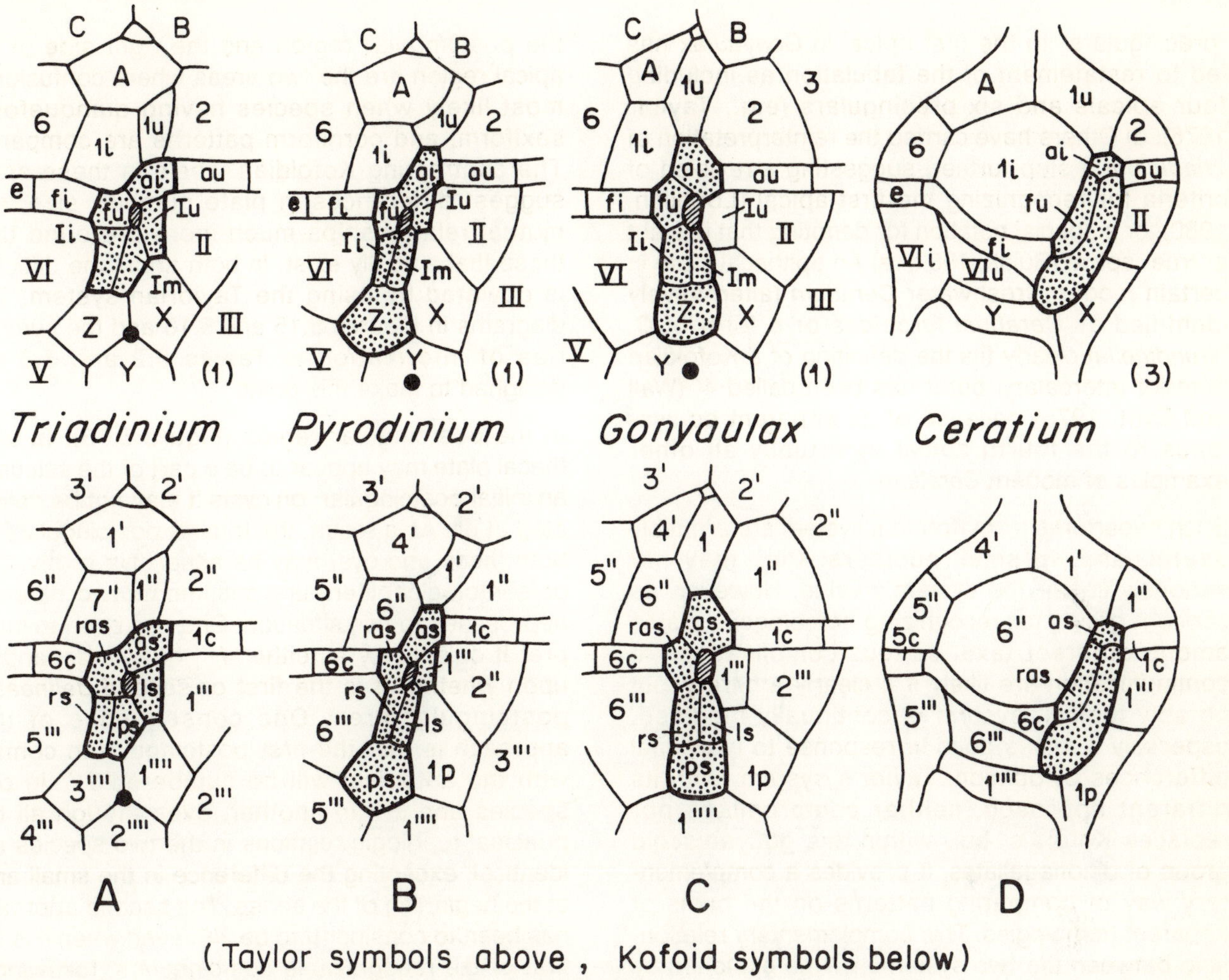

Figure 5.15 - TAYLOR-KOFOID COMPARISONS: VENTRAL REGION

Portions of the ventral tabulation diagramatically represent the following: a quinqueform exsert pattern in *Triadinium* (A), quinqueform metasert in *Pyrodinium* (B), sexiform insert in *Gonyaulax* (C), and corniform in *Ceratium* (D). Taylor symbols appear in the upper diagram of each pair and Kofoid symbols in the lower. The latter reflect the terminology of Graham (1942) for *Triadinium* (as *Goniodoma*), and Wall and Evitt (1975) for *Ceratium*.

The solid circle in the first three diagrams locates the approximate topographic antapex; the heavy line surrounds the area conventionally considered to constitute the sulcus, and the number in parentheses shows the number of postcingular paraplates no part of which is shown on the diagram. The contrast in symbols and the inconstancy of the Kofoidian symbols for apparently homologous plates is emphasized in Table 5.2.

geometric patterns and to represent quite particular concepts of homology among the pattern areas.

Similarly vexing conflicts between symbols differently applied arise on the right side of the epicyst (Figure 5.16). Here, relatively large and conspicuous areas (plates/paraplates) having obviously consistent topologic relationships to one another become either apicals or intercalaries (and may receive different numbers in each series) depending solely upon the presence or absence of one or more quite small areas in positions that happen to be critical for establishing the way the labels are applied. This situation is especially troublesome for paleontologists because the details necessary for determining the objectively correct Kofoidian labels are often indeterminable in fossils in which the apparent homologies of major plates are relatively clear.

In the early paragraphs of the discussion of gonyaulacoid patterns, some of the characteristics of the Taylor and Kofoid systems were contrasted. After the rather extensive discussion that has ensued, a retrospective comparison may be the best way to conclude this part of our discussion on tabulation.

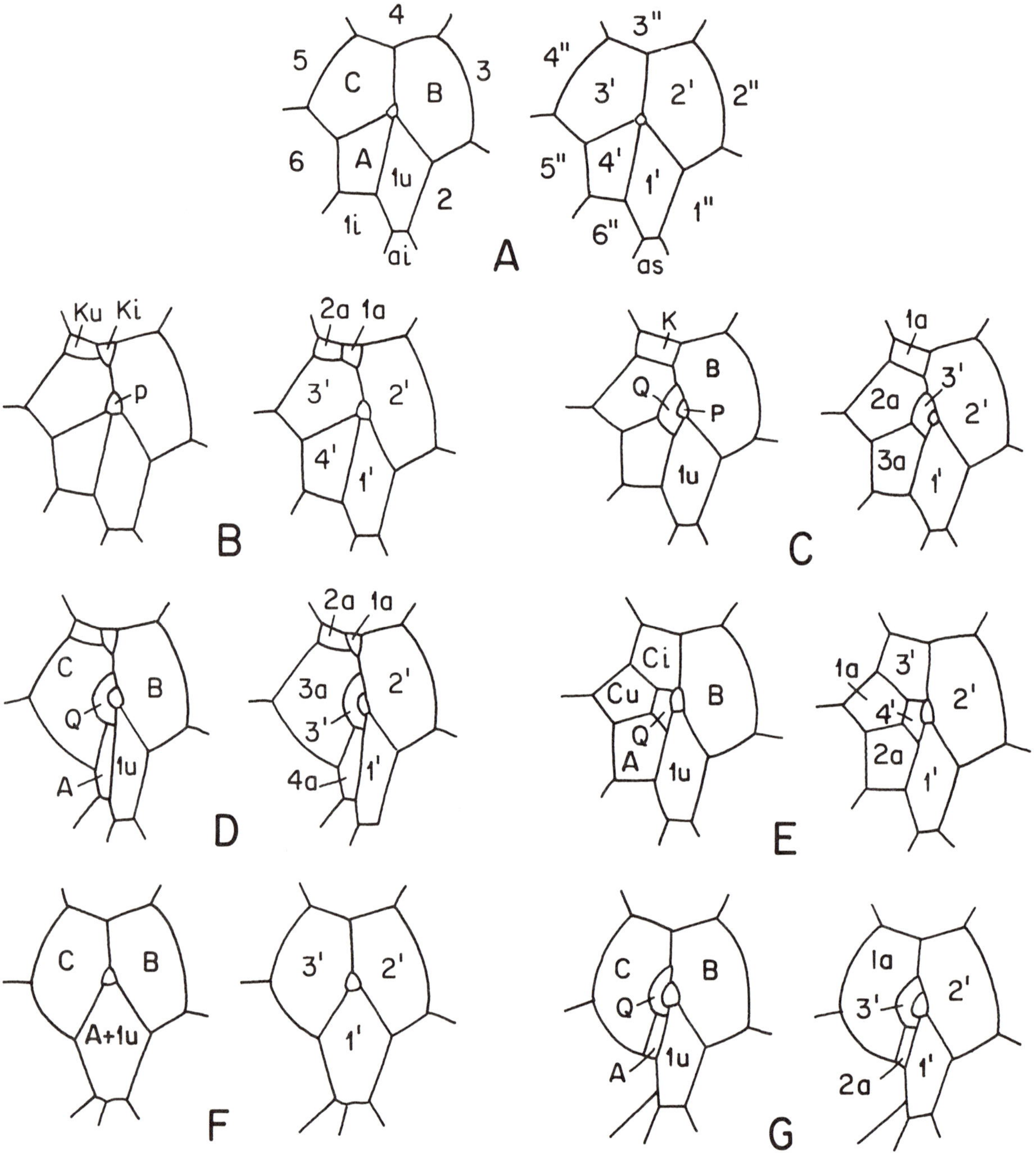

Figure 5.16 - TAYLOR-KOFOID COMPARISONS: APICAL REGION

Diagrams of partial patterns on epitheca/epicyst in apical view. Taylor symbols are shown on the lefthand diagram of each pair and Kofoid symbols on the right. In all the diagrams, the incompletely delineated peripheral areas represent the anterior sulcal (**ai** = *as*) and the six conventional gonyaulacoid precingulars.

A - The "basic" pattern approximated in many fossil cysts but rare in modern thecae.

B-G - Variations of the basic pattern, distinguished (in Taylor terms) by the differentiation of one or two incidentals (**Ki** and **Ku**), a secondary division of **C** into **Ci** and **Cu**, fusion of **A** and **1u**, and appearance of **Q**. B-C -Jurassic-Cretaceous *Ctenidodinium* spp. D - Late Jurassic *Gonyaulacysta jurassica*. E - Modern *Gonyaulax polyedra*. F - Cretaceous-Holocene *Spiniferites* spp. and *Impagidinium* spp. G - Early Cretaceous *Gonyaulacysta cretacea* (also very similar to Recent *Gonyaulax spinifera*).

The inconstancy of the Kofoid symbols for apparently homologous plates is summarized in Table 5.3.

In a Kofoid analysis, peridinialean tabulation is objectively resolved into a small number of plate series; that is, groups of plates arranged into latitudinal sequences (longitudinal, in the case of the sulcus). Prescribed symbols are used in descriptive text and illustrative figures to identify the plates and series, and to specify the numbers of each in a tabulation formula. The key operational question in a Kofoid analysis is: How many plates occur and in which series? Geometric properties and topologic relationships of plates are not critical to the answer (beyond the organization of plates into series)—they are essential additional attributes of the tabulation but not of the Kofoid analysis. The usual procedure in making a Kofoid analysis of a pattern is: 1) to distinguish the plate series, 2) to determine the number of plates in each series, and 3) to apply the appropriate numerals and symbols to describe or illustrate the results of the analysis.

In a Taylor analysis (limited in this discussion to gonyaulacoid patterns as defined here), tabulation is diagnosed by comparing it to a conceptual standard in the form of an hypothesized primary tabulation. The analysis is essentially subjective, being dependent upon acceptance of a certain set of homologies. Prescribed symbols are used in descriptive text and illustrative figures to identify plates, but a method for summarizing the tabulation in a formula has not been proposed. The key operational question in a Taylor analysis is: Which plate in the primary model is the homolog of each plate in the pattern being analyzed? Geometric properties and topologic relationships are critically important in arriving at the answer, but the number of plates in each series is not. A practical procedure in making a Taylor analysis of a pattern is: 1) to determine the points of identity or near identity with the primary model, 2) to determine the minimum steps (i.e., by division, fusion, and migration of primary plates) needed to transform the primary model into the observed pattern, and 3) to apply the appropriate symbols to describe or illustrate the results of the analysis.

The inadequacy of Kofoid analyses for comparing patterns is well illustrated by two distinctive modern species, both attributed by Loeblich (1980) to *Fragilidium* but first described under monospecific genera: *Helgolandinium subglobosum* and *Fragilidium heterolobum* (Figure 5.17). Neither species was originally considered to have gonyaulacoid properties. *Fragilidium* was simply said to be "quite different from all previously described dinoflagellates" (Balech, 1959). *Helgolandinium* was

Table 5.3 - TAYLOR-KOFOID COMPARISONS: APICAL REGION
At the left are listed the letters and names corresponding to the diagrams in Figure 5.16, which should be studied in conjunction with this table. In the first three columns headed by Taylor symbols are the Kofoid symbols for these plates in the diagrams. In the fourth column is the Kofoid symbol(s) for the plate(s) immediately subjacent to the plate which would be considered the Kofoidian third apical (*3′*) in the particular diagram to which reference is made.

Item in Figure 5.16	Kofoid symbols under Taylor symbols C or Ci	Cu	A	Kofoid symbols for plate(s) subjacent to Kofoid 3′
A. *Leptodinium/ Millioudodinium*	*3′*		*4′*	*3″-5″*
B. *Ctenidodinium*	*3′*		*4′*	*1a, 2a, 4″,5″*
C. *Ctenidodinium*	*2a*		*3a*	*2a, 3a*
D. *Gonyaula-cysta jurassica*	*3a*		*4a*	*3a, 4a*
E. *Gonyaulax polyedra*	*3′*	*1a*	*2a*	*3″, 4″*
F. *Spiniferites/ Impagidinium*	*3′*		*1′*(part)	*3″-5″*
G. *Gonyaula-cysta cretacea*	*1a*		*2a*	*1a, 2a*

placed next to *Pyrophacus* (Figure 1.6 P) in the Pyrophacaceae (von Stosch, 1969) on the basis of a plate formula showing an unusually large number of plates in certain areas; no comparison was made with *Fragilidium*. The Kofoid analyses which accompanied the original descriptions fail completely to bring out the many points of similarity and the limited but quite specific points of contrast between these two species, and also between them and a quinqueform gonyaulacoid pattern, such as in *Triadinium*.

The Kofoid tabulation formulae for the two taxa are:

Fragilidium: *1pr, 4′, 0a, 7″, 12c, 9‴, 1p, 2⁗*

Helgolandinium: *1pr, 4′, 0a, 9″, 10c, 7‴, 0p, 3⁗*

Although these formulae suggest that the greatest differences between these taxa lie in the hypothecae, this is not true. These areas are nearly identical, whereas the epithecae are distinguished from one another and from other species by features that are especially notable because they appear on the normally "conservative" left side. If one overlooks the symbols on the diagrams, the similarities among the patterns are easy to appreciate. However, to phrase an intelligible and incisive verbal comparison of *Fragilidium* with *Helgolandinium* using only the Kofoid designations of the original authors, and to include a further comparison with *Triadinium* (Kofoid symbols given in Figure 5.15 A), would be a difficult undertaking.

In contrast, a concise summary of the Taylor analyses is easy to generate. It might read as follows:

Both *Fragilidium* and *Helgolandinium* are gonyaulacoid dinoflagellates with quinqueform tabulations closely similar to that of *Triadinium*, but with differences that involve the following contrasts with the primary gonyaulacoid models. In *Fragilidium*, three primary epithecal plates are subdivided: **B** into **Bi** and **Bu**, **2** into **2i** and **2u**, and **3** into **3i** and **3u**; **Bi/Bu** lies toward the sulcus from **3i/3u**. In *Helgolandinium*, **1i** is metasert, and two primary epithecal plates are subdivided: **B** into **Bi** and **Bu**, and **2** into **2i** and **2u**. In both genera; 1) The tabulation of the hypotheca is as in *Triadinium*. **Ii**, **Im**, and **Iu** are distinct. 2) Several additional small sulcals (omitted from the diagrams) are present and possibly include the minute *9″* in *Helgolandinium*. 3) A prominent prepolar is like that in *Triadinium* if allowance is made for the additional angle in its outline at **2i/2u**. 4) The dorsal and right sides of the epitheca are topologically identical to *Triadinium*.

The small plates of the sulcus and the "supernumerary" cingulars (plausibly subdivisions of primary equatorials) in *Fragilidium* and *Helgolandinium* have not been accounted for in this discussion. These matters require further attention before a close comparison with the Taylor primary model or with *Triadinium* can be considered complete. However, this example will have served its purpose if it has made clearer how Taylor analysis can facilitate comparison of gonyaulacoid patterns and provide new insights into the nature and extent of similarities and differences among them.

Other systems of pattern analysis based on similar principles (i.e., homologies) have similar potential. For example, Eaton (1980; see Chapter 13) applied a system developed from a very different concept of homologies, to a broader spectrum of tabulation patterns than has been attempted here. The contrasting pictures of peridinialean evolution that emerge from Eaton's and Taylor's interpretations of tabulation cannot both be correct. However, we are greatly in debt to these authors for inviting and inspiring a fresh look at tabulation and paratabulation by proposing novel and provocative methods of pattern analysis.

Partiform Hypocyst Patterns

Fossil and Modern Representation: Whereas the three groups of gonyaulacoid patterns discussed so far are well known among both fossil and living dinoflagellates, this is not true of the partiform patterns we are about to consider. These patterns occur in a large number of fossil cysts, but they have received only slight attention to date (e.g, Morgenroth, 1968; McLean, 1972; Goodman and Stover, 1981). Although rather similar patterns occur in thecae of modern *Cladopyxis* and *Palaeophalacroma* (including *Sinodinium*, Loeblich III, 1980; Balech, 1964, 1967a), these are not familiar taxa and, apparently, they do not produce preservable cysts. The small size (mostly under 40 µm) of the modern and most fossil examples has inhibited observation of details, and study of the small fossils is often made difficult by a relatively thick and complexly structured wall having a high degree of sphericity. Cysts which are exceptions to these generalizations are likely to be especially informative.

Although having a basically gonyaulacoid design, the partiform pattern is distinct from the previously discussed quinqueform, sexiform, and corniform patterns. Since few species showing partiform patterns have been studied carefully and differences among them are obvious, precise comparisons are not yet possible. For the same reason, much of what can be said currently about the patterns must be regarded as tentative and, in part, speculative.

Partiform patterns have been observed to date in cysts ranging in age from Early Jurassic to Miocene, and, as already remarked, a few living species have similar tabulation patterns (Table 5.1). Some of the described fossil species have already proven useful in stratigraphic studies and others appear to have similar potential. Careful documentation of obscure details of wall structure and paratabulation can be expected to add rapidly to our understanding of this group and to enhance its stratigraphic usefulness.

For discussing partiform patterns we will draw upon both Kofoidian and Taylorian terminology to point out features of particular interest and to signal apparent homologies with paraplates in other gonyaulacoid designs. Selected fossil species that illustrate this interesting group of patterns are discussed in Chapter 10.

Characteristics: See Figure 5.18 and 5.19. As was true for quinqueform and sexiform patterns, the critical features of the partiform design lie in the hypocyst, and epicyst patterns of several kinds occur in combination with the distinctive hypocyst. The hypocyst, epicyst, midventral features, and overall features will be discussed in turn.

Hypocyst features. Two differences—one topologic and absolute, the other geometric and relative—

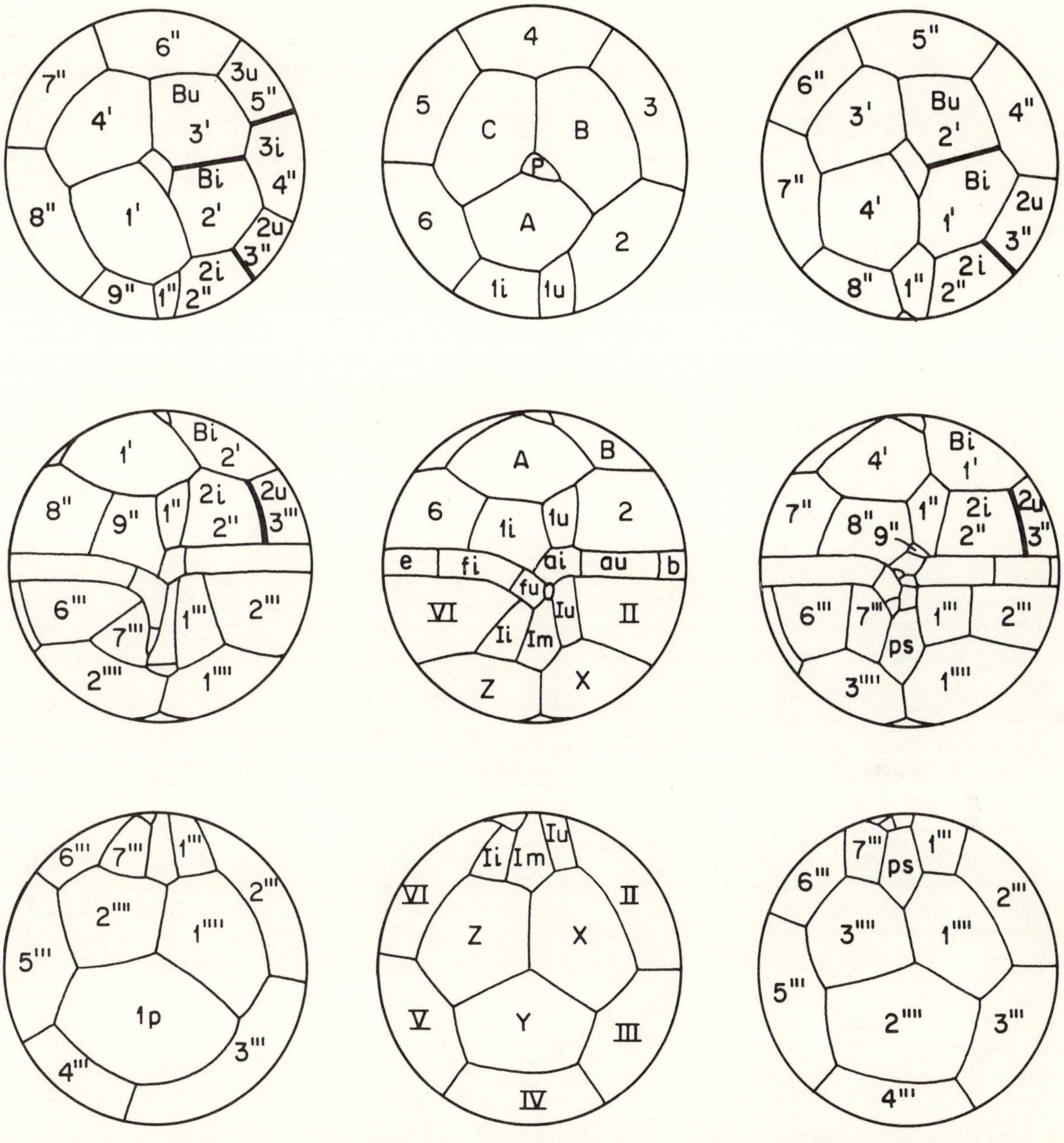

A. *Fragilidium* B. *Triadinium* C. *Helgolandinium*

Figure 5.17 - COMPARISON OF *FRAGILIDIUM*, *TRIADINIUM*, AND *HELGOLANDINIUM*

Apical (upper), ventral (middle), and antapical (lower) views. Taylor symbols appear on plates of *Triadinium*; on *Fragilidium* and *Helgolandinium* Kofoid symbols appear as used by original authors, but Taylor symbols have been added to those plates interpreted here to be subdivisions of Taylor primaries, as shown by heavy sutures. Note corresponding topology of all major plates except for these subdivisions. See text for discussion.

distinguish partiform hypocyst patterns from sexiform patterns, the group with which they have most in common: 1) Parasuture **Im/Z** within the parasulcus abuts against **Iu**, rather than against **X**. 2) **Z** is long and large, forming the major part of the parasulcal area, whose width is constricted toward the paracingulum where relatively small **Ii** and **Im** are located.

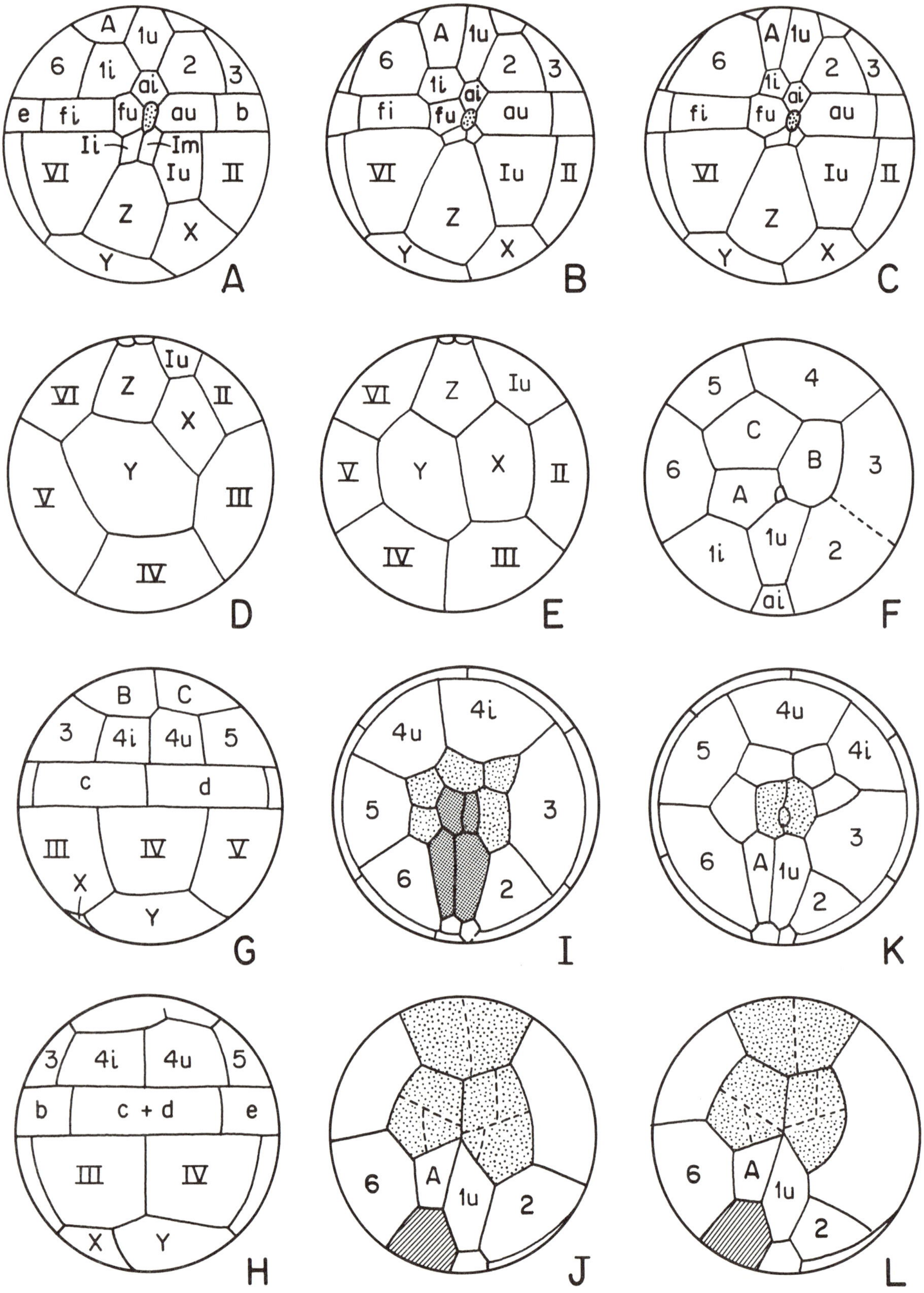
A
A 1u 6 1i 2 3 ai e fi fu au b Ii Im VI Iu II Z X Y
B
A 1u 6 1i 2 3 ai fi fu au VI Iu II Z Y X
C
A 1u 6 1i 2 3 ai fi fu au VI Iu II Z Y X
D
Iu Z VI II X Y V III IV
E
VI Z Iu V Y X II IV III
F
5 4 C 6 B 3 A 1u 1i 2 ai
G
B C 3 4i 4u 5 c d III IV V X Y
I
4i 4u 5 3 6 2
K
4u 5 4i 3 6 A 1u 2
H
3 4i 4u 5 b c + d e III IV X Y
J
6 A 1u 2
L
6 A 1u 2

Figure 5.18

In combination, the two diagnostic features are usually easy to recognize, providing the parasulcus is clearly outlined. Even when parasutures within the parasulcus are generally unclear, the locus of the **Im/Iu/Z** triple junction (partiform) may be indicated by an angulation along the left margin of the parasulcus and thus distinguished from the **Im/X/Z** junction in the sexiform pattern. In all observed cases, **VI** contacts **Y**, consistent with the notion that this arrangement is derived from a sexiform one. However, if transitions from the sexiform pattern occur (i.e., with **Im/Z** in line with **Iu/X**), they are surely rare and have not been reported.

A number of other features, varying somewhat from species to species, are related to these characteristics. One or more of them are usually distinct:

1) **Iu**, although of variable size, is normally larger than in a sexiform pattern and may approximately equal **VI** on the opposite side of the parasulcus from it. If **Iu** is decidely smaller than **VI** (Figure 5.18 A, D), overall similarity to the sexiform pattern is relatively great. If **Iu** and **VI** are nearly equal (Figure 5.18 B, E), a relatively high degree of symmetry exists on the ventral surface. In one sense this recalls the peridinioid pattern, but that has only five postcingulars. In another sense it suggests the symmetry of the quinqueform gonyaulacoid pattern, but in that case it is **II**, not **Iu**, which is the first postcingular, and, again, there is a total of five. Relative to the sexiform pattern, the large partiform **Iu** occupies an area analogous to the combination of **Iu**, **II**, and **X** (cf. Figure 5.19).

2) The role of **X** is also dependent upon the relative size of **Iu** (Figure 5.18 A-E). If **Iu** is small, **X** resembles its homolog in a sexiform pattern; that is, **X** lies mostly on the ventral surface, **X/Y** extends strongly to the left from the parasulcus, and **Y** is antapical. If, however, **Iu** is as large as **VI**, then **X** is displaced posteriorly, shares the antapical area with **Y**, and **X/Y** may be nearly sagittal, extending from the posterior extremity of **Z** to the broad posterior angulation of **III** on the posterodorsal surface.

3) Both the position of **III** and the identity of the feature that occupies the dorsal midline are affected by the size of **Iu** (Figure 5.18 D-E, G-H). When it is relatively small, **III** is lateral and **IV** is middorsal, but when it is large, **II** and **IV** are displaced and **III/IV** (or even a portion of **III**) may be shifted into a middorsal position. In all cases, however, **III** retains the important topologic characteristic it has in all the gonyaulacoid patterns: it is the first of the postcingulars to contact **Y**.

The partiform hypocyst is seen here as a derivative from the sexiform hypocyst, distinguished by the single but critical positional relationship of two parasutures: **Im/Z** and **Iu/X**. Variation of pattern within the hypocyst concerns primarily the relative size of **Iu** and associated effects on other paraplates.

Figure 5.18 - SCHEMATIC DIAGRAMS OF PARTIFORM PATTERNS

A-C - Ventral views, showing variants of the hypocyst pattern combined with different epicyst patterns. Note effects of different size of **Iu** on ventral symmetry, on the position of **Iu/X** and of **X**, and on the orientation of **X/Y**. A to C show three stages in the reduction of **1i**, from a major precingular element to a small sulcal paraplate no longer in contact with **fi**. A is representative of Early Tertiary *Glyphanodinium facetum* and Early Cretaceous *Microdinium opacum*. The combination of features shown in B occurs in undescribed Late Cretaceous species; the hypocyst pattern of B-C and E occurs in Early Jurassic *Eyachia prisca*, Late Cretaceous *Histiocysta palla*, and *Phanerodinium cayeuxii*. The epicyst of B is representative of Early Tertiary *Cladopyxidium*; the epicyst of C is representative of Late Cretaceous *Microdinium ornatum*.

D-E - Antapical views corresponding to A and B, showing (in E) the quasisagittal position of **X/Y** when **Iu** is large and **X** is displaced posteriorly.

F - Apical view of *Eyachia prisca*. Dashed line marks approximate position for apparently unexpressed **2/3**, presumed to represent fusion of corresponding thecal plates. Note that **A** clearly meets **B** dorsal to the apical prominence.

G-H - Dorsal views illustrating subdivision of **4** into secondaries **4i** and **4u**, and contrasting positions of intrapostcingular parasutures corresponding to differences in size of **Iu** (compare A and B). G - Paraplate **IV** is middorsal. When combined with ventral view in A, this represents *Glyphanodinium facetum* and *Microdinium opacum*. H - Parasuture **III/IV** is middorsal. This corresponds to the dorsal pattern in *Phanerodinium cayeuxii*. *Microdinium ornatum* is similar, except that **c** and **d** are not fused and additional parasutures are present toward the apex (compare I).

I - Apical details in some species of *Microdinium* (e.g., *M. ornatum*, the type species), with five Kofoidian intercalaries (light stipple) around four apicals (heavy stipple). The entire stippled area forms the archeopyle operculum.

J - Apical view of insert gonyaulacoid pattern with dashed lines to suggest a possible subdivision of three primaries (stippled) that would account for the pattern in I. Reduction of **1i** (ruled) is also required.

K - Apical details as in some species of Early Tertiary *Cladopyxidium*, with four Kofoidian intercalaries and four apicals. Only the two stippled apicals (and the preapical between them) form the archeopyle. Another topologically unusual feature is the apparent shift in **2/3**, causing **2** to be nearly triangular.

L - Similar to J, but with **2/3** shifted relative to its primary position to suggest a possible origin of the pattern shown in K. Markings as in J.

Epicyst features. Paratabulations of the epicyst associated with a partiform hypocyst are more varied than in the case of other gonyaulacoid combinations (Figure 5.18, A-C, F, I-L), and only a few examples can be mentioned here. In Kofoidian terms, the epicyst is composed of apicals and precingulars, with or without anterior intercalaries. Apicals normally number four, but there are only three if **1u** is exsert or metasert.

Intercalaries, if present at all, tend to be relatively large and conspicuous, either symetrically or asymetrically arranged. The presence of intercalaries in some species complicates considerably the interpretation of the pattern, because it appears that they probably represent subdivision of primaries—polars **B** and **C** in some cases (e.g., such small subspherical cysts as *Microdinium ornatum*, *Cladopyxidium saeptum,* and *Histiocysta palla*), possibly preequatorials in others. If this is true, then, an undivided primary in a species without intercalaries is homologous with two or more paraplates in different series in a species with intercalaries. Until the relationship of the anterior intercalaries in species with partiform hypocysts is better understood, it is impossible to give them satisfactory Taylor labels.

Variation in the precingular series is especially interesting, as it appears to show the effect of all three Taylor "processes"—subdivision, fusion, and migration of primaries. The usual number of six gonyaulacoid precingulars may be increased to seven by subdivision of **4** into **4i** and **4u** (*Glyphanodinium facetum* and *Microdinium opacum*, Figure 5.18 G-H), reduced to five by fusion of **2** and **3** (*Eyachia prisca*, Figure 10.8 I), or maintained at six by a combination of the subdivision of **4** and reduction in the size of **1i**, together with its incorporation into the sulcal area (*Microdinium ornatum*, Figure 5.18 I). Other modifications in the precingular series may well be revealed by further studies now underway.

Anterior intercalaries appear not to be present at all in those partiform species whose epicysts are otherwise most similar to the more conventional gonyaulacoid patterns associated with sexiform hypocysts. However, in species in which **1i** is notably reduced and a major preequatorial has been subdivided, intercalaries appear. Their small absolute size and disadvantageous location make them difficult to study. The examples illustrated here (Figure 5.18 I-L) and the suggested origins for the additional plates are only a few of the possibilities.

Midventral Features (Figure 5.18 A-C). Differences in preservation quality and inherently variable clarity of parasutural markings cause considerable variation in the patterns visible in the anterior portion of the parasulcus. As many as six minute units may be discernible here on the cysts (and an even larger number of small plates may have been present in the theca), but commonly the impression is of four units in two pairs. When six areas are differentiated, they appear to be **li**, **lm**, **fu**, **ai**, **1i**, and the flagellar scar. A lower count results when some parasutures are ill defined. The three units considered to be **1i**, **fu**, and **ai** tend to be delineated more strongly and more consistently than the rest. **li**, **lm**, and the flagellar scar are rarely distinct and often appear to merge into a single irregularly shaped area to the left of **fu**. It is possible to assign a variety of Kofoidian labels to these small and inconstant ventral areas, with consequently varied counts of precingulars, sulcals, and cingulars. For example, **1i** may be taken as *7‴* or *8″*, depending upon whether **ai** is called *1″* or *as* and whether **2** is called *1″* or *2″*.

General features. Paratabulation in the fossil cysts with partiform patterns is indicated chiefly by parasutural features. These are expressed in many ways; for example, as simple ridges or septa on the external surface of an uncomplicated autophragm (*Microdinium ornatum*), as narrow linear pericoels (*M. veligerum*), as internal structures within relatively thick and complex walls (*Histocysta palla*), and as interruptions in an ectophragm that is supported on intratabular processes (undescribed species). Many examples range from spherical to ellipsoidal, lack prominent projecting features (horns, processes), and are not cavate; but there are exceptions. Especially in the small subspherical examples, the epicyst is typically smaller than hypocyst, the two separated by a paracingulum that is often conspicuously wide relative to the overall dimensions of the cyst. The ends of the paracingulum are often only slightly offset and, uniquely, the paracingulum may be considerably wider at its beginning (against **ai**) than at its end (against **fu**). In those forms in which **1i** is much reduced and appears to form a part of the parasulcus, another anomalous relationship may result: the first precingular paraplate (*1″* = Taylor **2**) may be smaller than the last (*6″* = Taylor **6**). This contrasts with the normal size relation of *l″* (= **2**) and *6″* (= **1i**) in most sexiform hypocysts.

Comparison of Gonyaulacoid Patterns and Key

Table 5.4, Figure 5.19, and the brief key presented below should be helpful in deciding which of the gonyaulacoid patterns is represented in a cyst under review. Although the corniform pattern, which is restricted to the group we will define as Gc-cysts in Chapter 8, can be distinguished readily from other types, this pattern is so seldom expressed clearly or completely that its presence must usually be inferred from secondary evidence. Conveniently, Gc-cysts are likely to be confused with only one other group (Gv-cysts, also defined in Chapter 8, and included in the key but not the table or figure cited above), whose sexiform paratabulation is also usually obscure. The distinction made between these two groups at the first step of the key draws mainly on the features most frequently seen. The one definitive difference between them in paratabulation that is mentioned will rarely be observed.

Key to Determination of Gonyaulacoid Patterns

1a Outline suggests *Ceratium* in having one (right) or two (right and left) postcingular horns (or right and left cingular horns with postcingular terminations) in addition to one or two antapical horns; cyst in most taxa is lenticular; archeopyle is exclusively apical and parasulcal notch lies distinctly to left of ventral mid-line; paratabulation rarely distinct; **III** (i.e., first postcingular to contact antapical, **Y**) is *2'''* .... corniform

1b Similar to 1a in that cyst is lenticular, parasulcal notch of exclusively apical archeopyle is offset to left of ventral midline, and paratabulation is rarely distinct; but postcingular horns are lacking (or cingular prominences lack postcingular terminations); outline may be rounded at antapex or show one or two lobes; **III** (i.e., first postcingular to contact antapical, **Y**) is *3'''* .... sexiform (Gv-cyst)

1c Not as in 1a or 1b 2

2a Five-sided **Y** contacts **Z**, **X**, **III-V**; **Z** contacts **V**; **Iu** much smaller than **VI** .... quinqueform

2b Six-sided **Y** contacts **Z**, **X**, **III-VI**; **Z** does not contact **V** 3

3a **Im/Z** (= *ls/ps*) abuts against **X** (= *1p*); antapical half or less of parasulcus is composed of **Z**; posterior extremity of parasulcus is along **Y/Z** and usually indents **Y**; **Iu** much smaller than **VI** and **Iu/X** much closer to paracingulum than is **VI/Y** .... sexiform

3b **Im/Z** (= *ls/ps*) abuts against **Iu** (= *1'''*); most of parasulcus posterior to paracingulum is composed of **Z**; **Iu** medium to large, may approach size of **VI**; posterior extremity of parasulcus is **X/Y/Z**, from which **X/Y**, in extending toward **III**, passes much closer to antapex than does **VI/Y** if **Iu** is large .... partiform

INTRASPECIFIC VARIABILITY IN TABULATION

The distinctiveness of tabulation patterns has received much emphasis in this chapter. It should not be understood, however, that tabulation or paratabulation is absolutely constant within a species. Although tabulation, and, especially, paratabulation, tend to show relatively little variation within a species, there are notable exceptions, for these traits of dinoflagellate thecae and cysts are also Variable Variables (Chapter 7). Of course, occasional irregular specimens (teratologic or simply aberrant) can be expected in any species. Observed variations may involve the presence of supernumerary plates, subdivisions of a plate, fusion of two or more plates, unusually large or small plates, and differences in the number of contacts made by a given plate in conjunction with one of the foregoing variables. "Dislocations" of closely spaced triple junctions, resulting in plates that normally are in contact failing to meet (and *vice versa*), are especially common but readily overlooked owing to the small scale of the differences. Examples of some of these variables are illustrated in Figures 1.6 H-T, and 7.8.

SUMMARY

Tabulation and paratabulation are highly varied characters with a key role to play in dinoflagellate taxonomy. Details of paratabulation (including that of the parasulcus) merit diligent exploration and thorough description, despite difficulty of study. Peridinioid and gonyaulacoid patterns (the former represented by living *Peridinium* and *Protoperidinium*; the latter by living *Triadinium*, *Pyrodinium*, *Protogonyaulax*, *Gonyaulax*, and *Ceratium*, and by

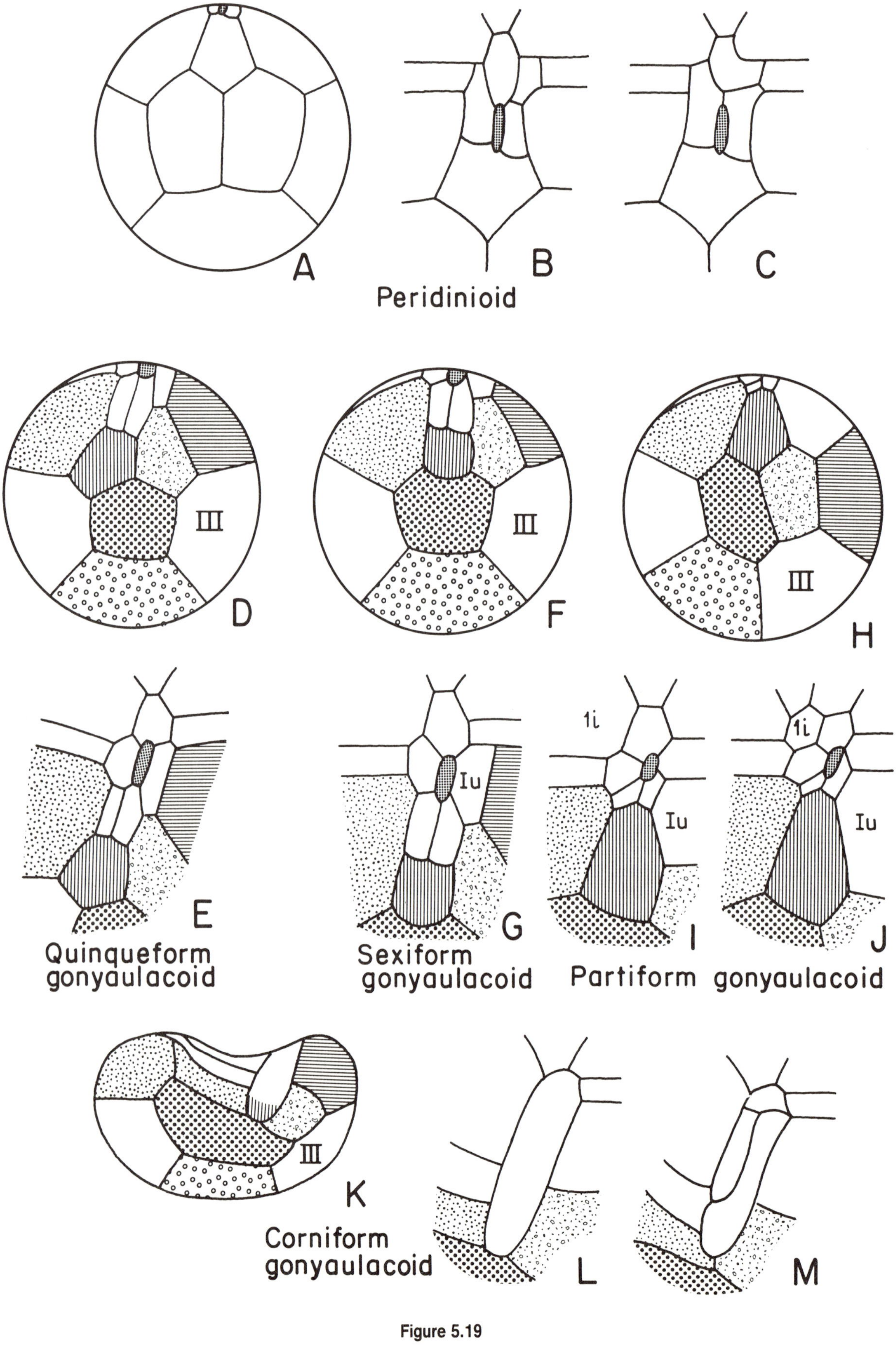
A
B
C
Peridinioid
III
D
III
F
III
H
E
Quinqueform gonyaulacoid
Iu
G
Sexiform gonyaulacoid
1i
Iu
I
1i
Iu
J
Partiform gonyaulacoid
III
K
Corniform gonyaulacoid
L
M

Figure 5.19

Table 5.4 - SUMMARY OF HYPOCYST CHARACTERS FOR GONYAULACOID PATTERNS (See Figure 5.19)

Pattern	Most distinctive features of antapical region	Postcingular features; number of paraplates; and, in parentheses, Kofodian identity of **III** (= first postcingular to contact **Y**)	Posterior limit of sulcus
quinqueform	**Y** is 5-sided, symmetrically between & posterior to two more ventral units: **X** & **Z**; **X/Y** tends to be symmetrical to **Y/Z**	five; **Iu** is entirely within sulcus; (*2'''*)	ill-defined, appears to end at anterior limits of **X** & **Z**; **Z** contacts **V**
sexiform	**Y** is 6-sided, posterior to three more ventral units: **X**, **Z** & **VI**; **Im/Z** meets **X** posterior to **Iu/X**; **X/Y** tends to be symmetrical to **VI/Y**	six, but first (**Iu**) is small, may appear part of sulcus or not; (*3'''* or *2'''*)	**Y/Z**, which makes arcuate, often about symmetrically medial indentation in **Y**; **Z** does not contact **V** and occupies only posterior part of sulcus
corniform	**Y** is transversely elongate; its ventral contacts are unequal: **VIu/Y** long; **X/Y** shorter, sulcal region/ **Y** shortest; horn from **X/Y**	six, but first is **II** (**Iu** not distinguishable, **VI** is divided and **VIu** is modified; (*2'''*); horns from **X/Y** and **V/VIi**, and in some species also **II/III**	not prominent, narrow, on left side; **Z** not defined
partiform	**Im/Z** meets **Iu** anterior to **Iu/X**; when **Iu** is large, **X** shares antapical area with larger, 6-sided **Y**, and **Iu/X** tends to be symmetrical to **VIu/Y**	six, but first (**Iu**) is typically larger than in sexiform and definitely not part of sulcus; (*3'''*)	indentation (often wide, angular, and tending to be symmetrical) between **X** & **Y**; **Z** does not contact **V** but occcupies most or parasulcus posterior to paracingulum

fossil *Cladopyxidium*, *Microdinium* and others) are two major groups of patterns with extensive modern and fossil representation.

Other groups of paratabulate cysts are less well known. They include important fossils, apparently without living representatives, and their paratabulation patterns are in the early stages of study. Examples are two major complexes of patterns (*Rhaetogonyaulax* and *Suessia*) and many distinctive genera from the Upper Triassic and Lower Jurassic.

The much-used Kofoidian system for designating plates according to their position has advantages for objective analysis of individual patterns, but is not satisfactory for comparing patterns when plates that are considered homologous occupy different positions. For this purpose another system, devised by Taylor, is discussed extensively and is applied to those dinoflagellates having gonyaulacoid patterns. The characteristics of Kofoid analysis and Taylor analysis are illustrated. Quinqueform, sexiform, corniform, and partiform gonyaulacoid patterns are distinguished on the basis of differences in hypothecal organization, and various epithecal patterns occur in each of these groups. A key for recognizing the four groups of patterns is presented. L-type and S-type ventral patterns are distinguished among sexiform gonyaulacoid patterns. Details of corniform patterns are infrequently seen in fossils, but partiform patterns characterize a large, varied, and relatively little-known group of taxa. Partiform patterns are discussed in a preliminary fashion using both Kofoid and Taylor analyses.

Figure 5.19 - COMPARISON OF PERIDINIOID AND GONYAULACOID PATTERNS

Schematic antapical views and portions of the ventral surface to show typical patterns in hypothecal and sulcal areas. Transitions from quinqueform to sexiform arrangements occur, but the corniform and partiform patterns stand more distinctly apart. A full transition has not been documented for fossil ceratioid (i.e., corniform) cysts, which rarely show clearly marked paratabulation. Paratabulation is more fully expressed in partiform cysts but few have been studied in detail. The shading on the figures emphasizes the significant homologies thought to exist in the hypocyst; note the constant topologic relationship between **III** and its neighbors. Where two ventral patterns are shown for peridinioid and ceratioid forms, the lefthand one (B and L) is associated with modern marine species and the righthand (C and M) with modern freshwater species. In the case of the two ventral partiform patterns (I and J), the presumed "conservative" condition (hypocyst: **Iu** small; epicyst: **1i** unreduced), is shown at the left, and a "progressive" example (**Iu** large; **1i** about the size of the small sulcals) at the right.

Chapter 6. The Archeopyle

INTRODUCTION

The **archeopyle** (Evitt, 1961, 1967), a feature most authors have considered unique to the cyst (contrast Boltovskoy, 1973), is the opening in the cyst wall through which the protoplast emerges during excystment. The archeopyle is operculate in most species, but it may also be slitlike. The etymological reference of the term to an opening in ancient (i.e., fossil) dinoflagellates is no longer appropriate, for archeopyles are well known in cysts of living species, although this was not so at the time it was proposed. Operculate openings like those already observed in fossil dinoflagellates were first reported in sporopollenin cysts of living marine species by Evitt and Davidson (1964). Slitlike openings in the unfossilizable cysts of modern freshwater *Ceratium hirundinella* had been mentioned much earlier (Entz, 1925). Wall (1965), when observing the induced excystment of *Spiniferites*-type cysts belonging to a modern species of *Gonyaulax*, may have been the first to see a living motile cell emerge through the operculate archeopyle of a cyst compositionally and morphologically comparable to common fossil examples.

Gocht, whose remarks (1983, p. 268-269) are excerpted in a free translation in this paragraph, commented that the archeopyle occupies a special position among the characters of dinoflagellate cysts. In the first place, it cannot be accommodated in the definitions of either thecamorphic or cystomorphic features; it meets the requirements of both groups. Further, the formative process of the archeopyle is different from that of the other cyst structures. On the one hand, the archeopyle is a thecamorphic feature of the cyst in that it most frequently expresses an easily recognizable fraction of the paratabulation. On the other hand, it is typically cystomorphic, because comparable mechanisms are not known on the theca. It must, in fact, be regarded as an "autonomous" feature, functionally critical for the cyst and usually recording in its geometry unmistakable indications of the fundamental tabulation/paratabulation pattern that guides the distribution of many features of cyst and theca. Although it is not designed to fulfill its functional purpose until long after other elements of the cyst wall have been constructed, this "escape hatch" is represented by a "latent" or "preformed" structure formed at the same time as the rest of the wall.

Little is known about the biological processes related to archeopyle formation and function. Normally it is defined by a line of actual or potential separation through the cyst wall, which may be visible as a fine line under the microscope (given optimum conditions) even though the wall may be unruptured. Many archeopyles provide communication directly between the innermost cavity of the cyst and the exterior. Others, in some cavate cysts, provide obvious connection only between the pericoel or the ectocoel and the exterior, whereas the more inward wall layer (i.e., endophragm or autophragm) shows no opening. We do not know how or when an opening for passage of the excysting protoplast forms in the unruptured wall of such species or in the walls of those few living and fossil species in which no structure recognizable as an archeopyle appears to exist in any wall layer. Perhaps penetration of cyst walls without preformed archeopyles is accomplished by a general enzyme-mediated degradation of the wall material, rather than by a highly localized effect along an archeopyle suture.

If the operculum is displaced or missing, the open archeopyle may be a quite conspicuous hole in the cyst wall. Early investigators noticed such openings long before they appreciated the dinoflagellate affinity of some of the specimens in which they occur. Eisenack (1938) used the term "pylome" for openings he had observed in many hystrichospheres—i.e., acritarchs and some chorate dinoflagellate cysts—and this term was widely applied also to openings in other dinoflagellate cysts before "archeopyle" was introduced. He also called attention to the apparently constant position of the archeopyle in any one species, but there was no appreciation then of either the great variety of archeopyles we now know, or of the taxonomic usefulness of detailed observations on archeopyle morphology. In some instances the opening was considered accidental, and even so astute and usually objective an observer as Deflandre published drawings (1936) showing the missing part of the wall "restored," although the opening is clearly visible in the actual specimen.

In general, the simpler archeopyles are widely represented among many species, whereas the larger and more complex types are known from only a small number of species, or even a single species. A statistical summary of archeopyle types at the specific level has never been presented, and current uncertainties and state of flux relative to ideas about the taxonomic significance of archeopyles render a summary at the generic level both difficult to tabulate and of questionable value. Stover and Evitt (1978) listed percentages of various archeopyle types among 279 genera of pre-Pleistocene organic-walled dinoflagellates as defined by taxonomic procedures current to 1977. The following brief extract from their figures shows that, despite the known diversity of archeopyle types, the three basic types are clearly dominant:

Archeopyle Type		Percentage of 279 genera that include species with exclusively the type of archeopyle indicated
Simple apical - Type (tA) or 4A		26 %
Simple intercalary - Type I		22
Simple precingular - Type P		32
	Total	80 %

STRUCTURE AND TERMINOLOGY

As more has been learned about the archeopyle, the concept of its structure and the descriptive terminology applied to it have changed. Means are required for describing its detail explicitly, as well as for summarizing its principal features briefly but unambiguously. In the following synthesis certain aspects of the approaches suggested by Evitt (1967), Lentin and Williams (1976), Norris (1978a), and Davies (1981) are combined, along with a few new suggestions.

Figures 6.1-6.3 illustrate the basic terminology applied to the archeopyle, while Figure 6.4 shows selected aspects of the size and shape of the opening and its relation to the layers of the cyst wall.

The **principal archeopyle suture** is the primary determinant of the archeopyle. This line of structural and probably also compositional discontinuity in the wall apparently originates at the time of wall formation and localizes the opening that develops during excystment. It is often visible in an unopened cyst as a line whose brightness contrasts with that of the rest of the wall (transmitted light), or as a fine linear depression in the wall surface. A difference in ultrastructure of the wall material at the locus of the suture may also be apparent (Jux, 1968). In a modern cyst, the suture may be opened by chemical treatment (acetolysis) as well as by excystment of the living cell.

Most commonly the principal archeopyle suture more or less completely surrounds a portion of the cyst wall, the **archeopyle operculum**, whose release actually forms the opening when the wall ruptures along the suture. It is common for the suture to run approximately along a sequence of parasutures and for the operculum to comprise one or more entire paraplates. In fact, if the angulations in the outline of the opening are clearly defined, its paraplate equivalence and possibly something about the overall paratabulation pattern of the cyst may be evident from the archeopyle alone. However, we know from evidences of paratabulation independent of the archeopyle that the archeopyle suture in many species deviates noticeably from a parasutural course. Then the angulations in its outline tend to be rounded and the opening comprises less than or more than the paraplate(s) to which it chiefly corresponds. Such openings are termed, respectively, **reduced** and **enlarged**. Reduction or enlargement can be assessed readily if parasutures are marked by features other than the archeopyle suture, and rounding alone will suggest that the archeopyle suture does not follow paraplate boundaries closely. However, in the absence of parasutural markings, the relative size of a rounded opening usually is not a satisfactory basis for judging reduction or enlargement, although observations on rounded archeopyles may still provide useful information for comparison of species.

In cysts of a few modern peridinioid species, archeopyle shapes and positions have been reported that are deceptively different from those of the anterior intercalary plates of the theca to which the opening appears to correspond (Harland, 1982). By inference based on closely similar cyst morphology, some fossil cysts (e.g., *Selenopemphix*, see Chapter 9) have been implicated in a similar anomalous relationship. However, archeopyles of most fossil peridinioid cysts are conspicuously regular and symmetrical, consistent with all other indica-

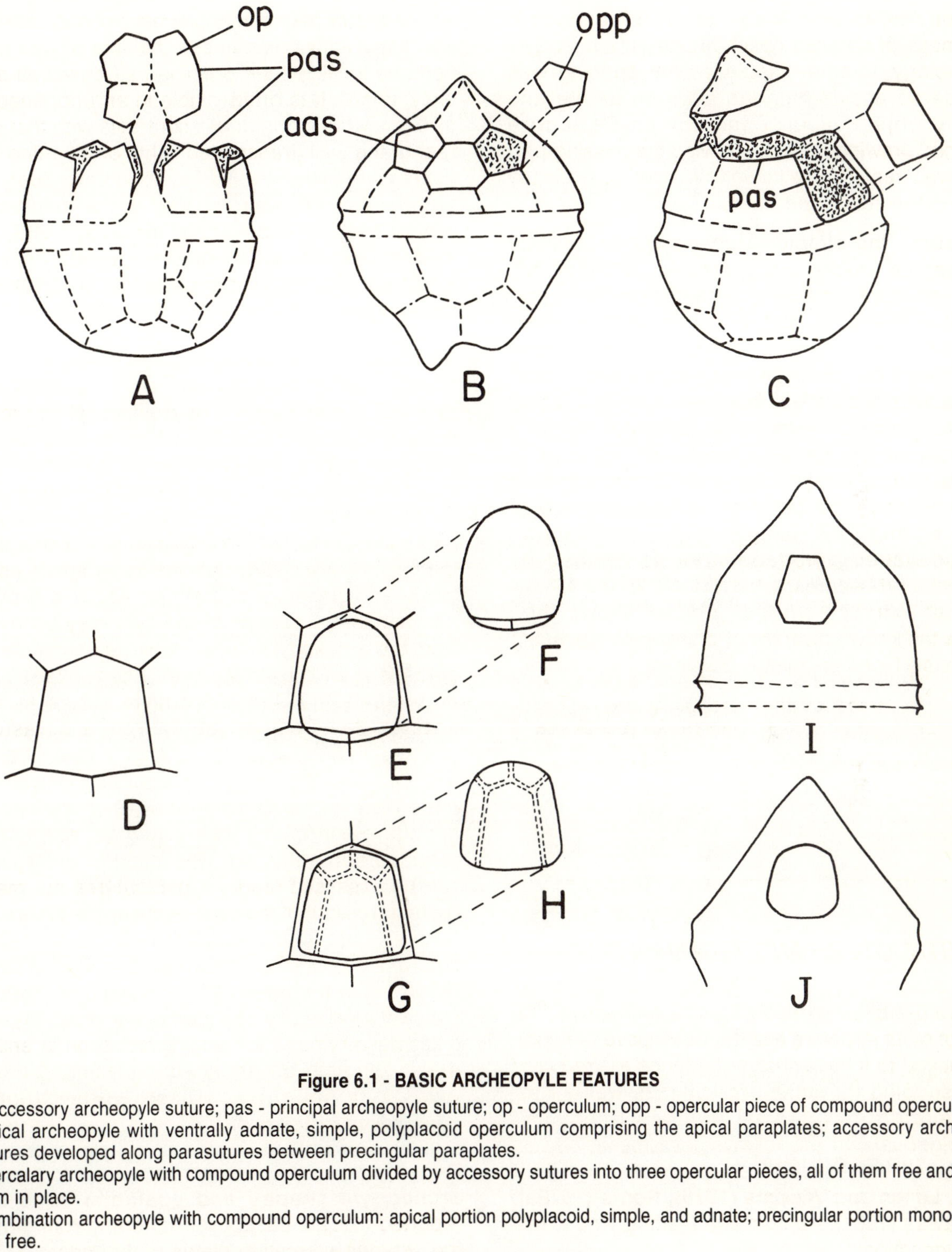

Figure 6.1 - BASIC ARCHEOPYLE FEATURES

aas - accessory archeopyle suture; pas - principal archeopyle suture; op - operculum; opp - opercular piece of compound operculum

A - Apical archeopyle with ventrally adnate, simple, polyplacoid operculum comprising the apical paraplates; accessory archeopyle sutures developed along parasutures between precingular paraplates.

B - Intercalary archeopyle with compound operculum divided by accessory sutures into three opercular pieces, all of them free and two of them in place.

C - Combination archeopyle with compound operculum: apical portion polyplacoid, simple, and adnate; precingular portion monoplacoid and free.

D - Complete congruence of archeopyle and the paraplate to which it corresponds.

E-F - Archeopyle not congruent with paraplate as outlined by parasuture. E -Opening reduced along lateral and apical margins, enlarged posteriorly to include part of paracingulum. F - Operculum of E.

G-H - Reduced archeopyle involving a paraplate whose surface bears accessory ridges related to overlap and growth of thecal plates (e.g., *Cribroperidinium*). H - Operculum of G. Note that accessory ridges on operculum falsely suggest that extensive enlargement has incorporated portions of adjacent paraplates.

I-J - Angularity of the opening. I - Shape and position of the opening are sufficient to indicate equivalence of the archeopyle to paraplate *2a* of a peridinioid cyst, despite absence of definitive paratabulation. J - Rounded opening whose shape and position are ambiguous indicators of paraplate equivalence.

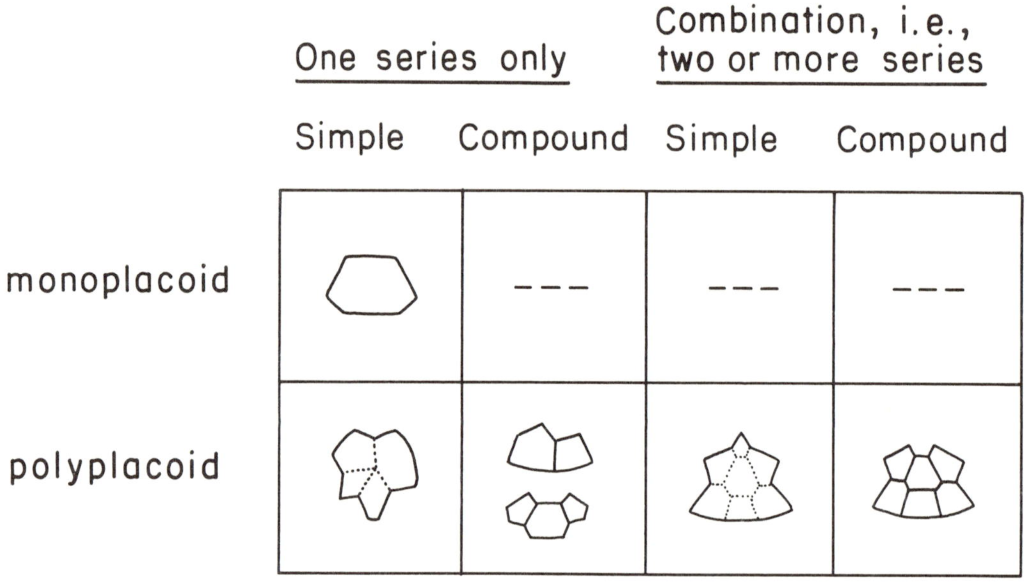

Figure 6.2 - TERMS TO DESCRIBE OPERCULA

Monoplacoid opercula are necessarily simple and represent only one series of paraplates. Polyplacoid opercula may represent only one series or may combine paraplates from two or more series, and may be either simple or compound.
Solid lines = principal and accessory archeopyle sutures
Dotted lines = parasutures along which simple operculum is not divided

tions of paratabulation. There seems to be no good reason to question that archeopyles, virtually without exception, record the approximate shape and position of the respective thecal plates.

Both the opening and the operculum may be described as **monoplacoid** if they represent a single paraplate, and **biplacoid** or **polyplacoid** if they represent more than one paraplate from the same

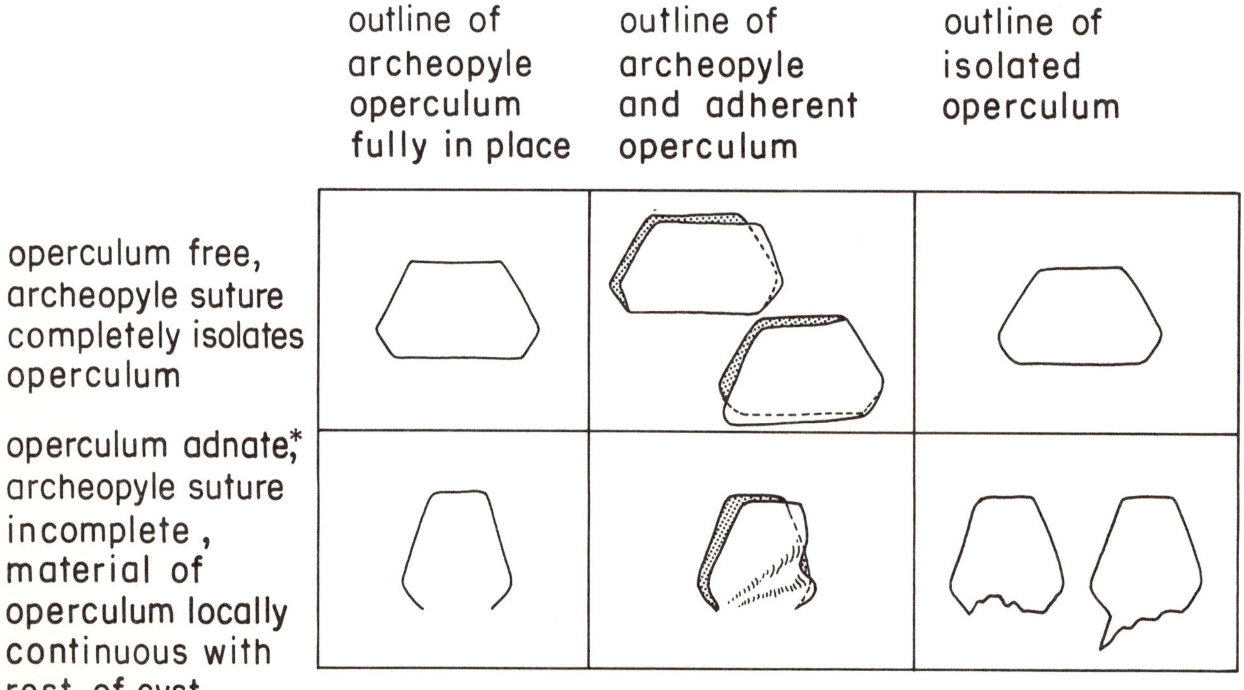

Figure 6.3 - FREE, ADNATE, AND ADHERENT OPERCULA
*replacement for "attached" of Evitt, 1967

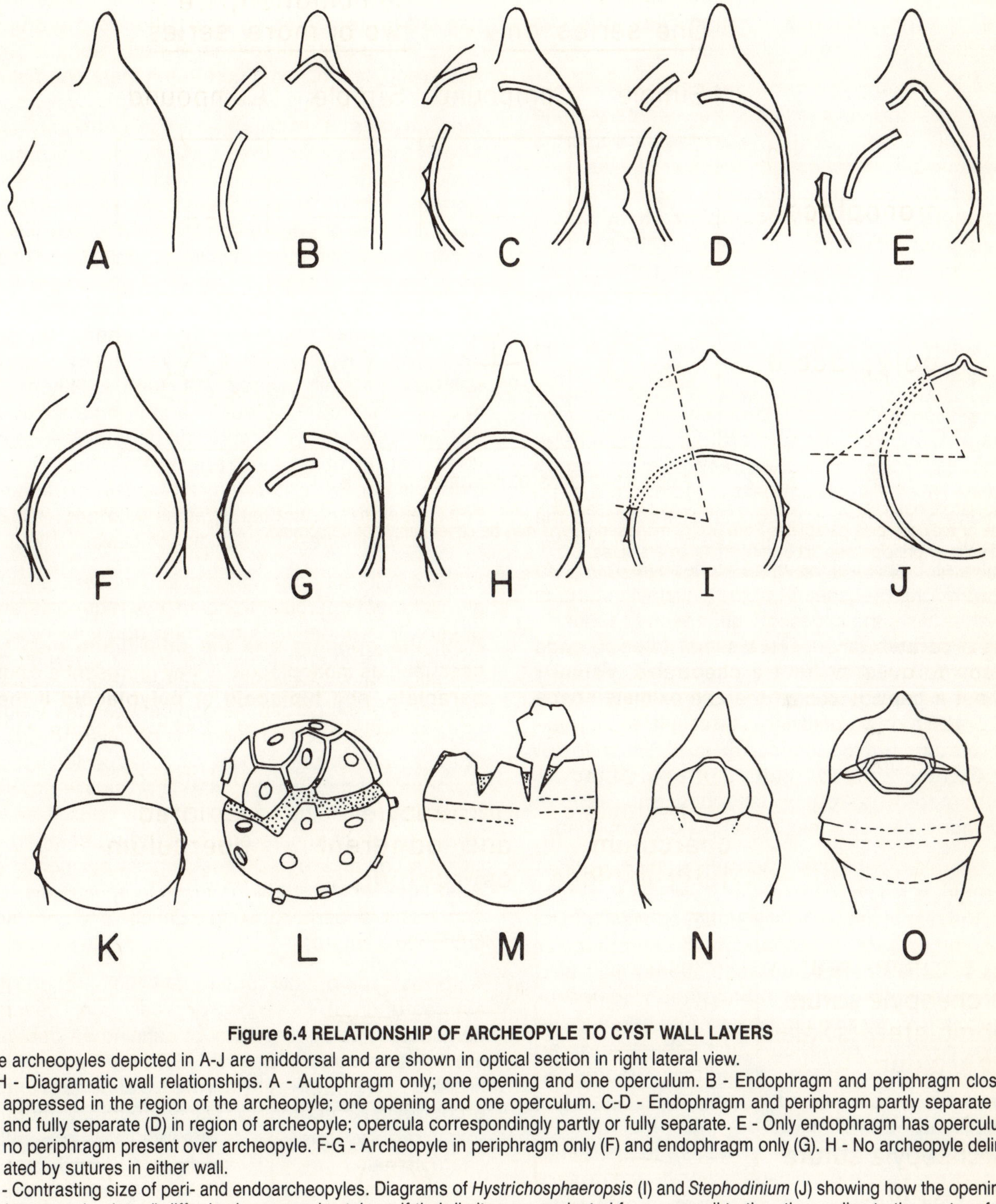

Figure 6.4 RELATIONSHIP OF ARCHEOPYLE TO CYST WALL LAYERS

The archeopyles depicted in A-J are middorsal and are shown in optical section in right lateral view.

A-H - Diagramatic wall relationships. A - Autophragm only; one opening and one operculum. B - Endophragm and periphragm closely appressed in the region of the archeopyle; one opening and one operculum. C-D - Endophragm and periphragm partly separate (C) and fully separate (D) in region of archeopyle; opercula correspondingly partly or fully separate. E - Only endophragm has operculum; no periphragm present over archeopyle. F-G - Archeopyle in periphragm only (F) and endophragm only (G). H - No archeopyle delineated by sutures in either wall.

I-J - Contrasting size of peri- and endoarcheopyles. Diagrams of *Hystrichosphaeropsis* (I) and *Stephodinium* (J) showing how the openings (or opercula, dotted) differ in size approximately as if their limits were projected from one wall to the other on line to the center of the cyst.

K-O - Size of opening. K - Small angular opening in periphragm of cavate peridinioid cyst of *Isabelidinium*. L - Large epicystal archeopyle in chorate cyst (only process bases indicated) of *Homotryblium*, vv; opercular pieces shown separating along accessory sutures. M - *Kallosphaeridium*, vv, showing adnate apical operculum and long accessory sutures between precingulars. Note disparity between effective size of opening and size of apical operculum. N - *Isabelidinium* showing rounded periarcheopyle, Type I, and larger adnate operculum of endoarcheopyle (shown fully open and rotated into axial plane), Type (tAtI)a. O - Omegaform Type I periarcheopyle and Type 3I endoarcheopyle in *Chatangiella*.

series or adjacent series. A **combination archeopyle** is a polyplacoid opening that involves paraplates from more than one series.

The first archeopyles to be described were three basic types distinguished as apical, intercalary, and precingular (Evitt, 1961). These types are still the ones known from the greatest number of examples. The typical apical archeopyle is polyplacoid, equivalent to the entire complement of apical paraplates, whereas the basic intercalary and precingular types include just the single middorsal paraplate in each series. Many other and more elaborate types are now also known (Evitt, 1967, and subsequent authors), and the number and variety continue to grow.

The principal archeopyle suture may be unbranched (straight, curved, or angled), or it may have branches in the form of **accessory archeopyle sutures** which extend onto either the operculum (in the case of polyplacoid opercula) or the rest of the cyst, or both. Accessory sutures may range from barely detectable knicks in the archeopyle margin to relatively long branches from the principal suture. In the latter case, the accessory sutures may separate flaplike portions of the cyst wall, or fully subdivide the operculum into two or more separable **opercular pieces**. If the operculum is entire or only incompletely divided by accessory sutures, it is a **simple operculum**; but if accessory sutures divide it completely, the operculum is **compound**. A monoplacoid operculum is necessarily simple; polyplacoid opercula may be simple or compound.

An operculum completely surrounded by the principal suture is said to be **free**, and it will separate fully from the rest of the cyst if the suture opens along its entire length. A free operculum (or free opercular pieces, if the operculum is compound) may also remain locally **adherent** (connotation: joined or united, although usually separate) wherever the suture does not open, but such adherence is an incidental condition and cannot be recognized unless the suture is partly open elsewhere. The locus of adherence is usually variable within a species. Just as opercula may adhere to the rest of the cyst where the principal archeopyle suture fails to open, so also the opercular pieces of compound opercula may adhere to one another in varying degrees and combinations. Contrasting with a free operculum (whether adherent or not) is an **adnate** operculum, which is incompletely surrounded by the archeopyle suture. This term, having the biological connotation of congenitally grown together, is proposed here to replace "attached" (*sensu* Evitt, 1967), a word whose broad vernacular connotation has led to its confusing application to the archeopyle in the special senses of both "adnate" and "adherent" as employed here. An adnate operculum is locally continuous with the rest of the cyst where the suture is interrupted and this continuity can often be recognized, even if the suture is nowhere open. An adnate operculum will not separate fully from the rest of the cyst unless the wall is actually torn. The free or adnate condition of the operculum is constant within most species and has been used as the principal criterion for distinguishing a few genera. Adnate opercula are usually attached along the adcingular margin of the archeopyle, but adnation is adapical when it occurs in the *Wetzeliella* complex. Laterally adnate opercula possibly occur in cysts of some Pleistocene and living species (if the distinction between adnation and adherence as used herein has been reported correctly); for example, in *Protoperidinium* (*Archaeperidinium*) *compressum* and similar species (Harland, 1982), and *Diplopsalis lenticula* (Bujak and Davies, 1983).

While it is important to distinguish between adnate and adherent opercula, to do so may require careful observation under high magnification with critical lighting. Persistent presence of the operculum, combined with apparently complete confluence of the wall surface across the gap between the visible ends of the archeopyle suture, is positive indication of the adnate condition. However, in species where the zone of attachment is very short relative to the operculum size, the operculum may be torn away readily. Loss of adnate opercula through tearing is indicated by an irregular break whose position is at least somewhat variable in multiple specimens, as opposed to separation along a definite and constant archeopyle suture.

A special type of wall rupture, in which archeopyle sutures open up to varying degrees along several neighboring paraplates without defining an operculum or opercular pieces of constant design or number, has been aptly termed a **disintegration archeopyle** (Dörhöfer and Davies, 1980). Although this irregular opening of the wall is especially characteristic of certain Triassic and Early Jurassic species, similar tendencies toward irregular archeopyle formation observed in cysts of Holocene freshwater species (Norris and McAndrews, 1972) may represent geographic or ecologic variants (Evitt and Wall, 1968) of forms in which the archeopyle is simpler and more regular.

The fact that the archeopyle in fresh cysts of many living species is effectively opened by acetolysis treatment shows that the archeopyle suture is in some measure compositionally different from the rest of the wall. It seems reasonable that decay—and, ultimately, fossilization—may accomplish the same result. Therefore, open archeopyles in fossil cysts do not necessarily bear witness to naturally completed excystment. On the other hand, fossil cysts with truly unopened archeopyles must represent unexcysted cells. In at least one instance, an apparent correlation has been reported (Wille and Gocht, 1979) between sediment type and the open or closed state of the archeopyle suture.

Opercula or opercular pieces can often be observed in the interior of fossil cysts. Special explanations of this situation were once sought because it had seemed an unlikely result of the initially outward displacement of an operculum that presumably accompanied emergence of the protoplast. According to one imaginative, if mechanically implausible suggestion, when the emerging protoplast broke free of the tightly fitting opening through which it had just squeezed, the vacuum within the vacated cyst was filled by an inrush of water that carried the operculum along with it. A simpler explanation is that cysts with opercula inside represent cells that died before excysting, the opercula having been displaced inward when the protoplasts deteriorated and the archeopyle sutures opened.

The characters of the archeopyle and its operculum relate closely to the wall structure at the locus of the opening (Figure 6.4 A-H). If the wall is an autophragm or is composed of two layers closely linked in the archeopyle area, only a single opening (with a single operculum) is developed. However, if wall layers are separated in the area of the archeopyle (e.g., in some cavate cysts), two distinct openings (peri- and endoarcheopyle, or ecto- and autoarcheopyle) may occur. In most cases the openings in different wall layers of one specimen are of the same type; rarely they show a substantial and constant difference.

So far as known, archeopyles in cavate peridinioid cysts are always operculate, whether or not the wall layers are markedly separated in the area where the openings occur. However, in some cavate gonyaulacoid cysts in which periphragm and endophragm are separated in the area occupied by a monoplacoid precingular archeopyle, only the endoarcheopyle appears normally to have an operculum (Eaton, pers. comm.). *Gonyaulacysta jurassica*, *Tubotuberella apatela*, and *Stephodinium coronatum* are examples. Endopercula of such cysts are familiar objects commonly observed in a variety of positions—essentially in place, within the endocoel, or entirely apart from the specimen to which they belong. While the existence of periopercula for the often conspicuous openings in the outer wall of these cysts had been taken for granted, such opercula seem never to have been positively identified.

An ectophragm, if present, normally is either incomplete in the region of the archeopyle (e.g., *Kisselovia*) or opens as part of the periphragm or autophragm on which it is supported (e.g., *Gardodinium*). Therefore,the features of ectoarcheopyle and ectoperculum in such cysts usually need not be detailed separately.

The relative sizes and shapes of archeopyles through the separated walls of cavate cysts are strongly influenced by the contours and spacing of the two walls, even if both archeopyles are of the same type (Figure 6.4 I-J). The effect is especially pronounced if the two wall layers are close together along one edge of the archeopyle but widely separated at the opposite edge. Areas and outlines of archeopyle and operculum in the two wall layers are then related roughly as they would be if they were projected from one layer to the other along lines from the center of the cyst (Chapter 7, Principal of Centripetal Projection). The resulting outline shapes of archeopyles in the two walls may be markedly different (Davies, 1981).

Archeopyle dimensions as a fraction of cyst dimensions vary greatly, but the significance of this is unknown (Figure 6.4 K-O). At one extreme, exemplified by some peridinioid cysts with relatively small intercalary archeopyles, the opening is so small that only a severely distorted or subdivided protoplast could pass through it. At the other extreme, an archeopyle whose principal suture runs approximately along the equator of a spherical cyst would allow the entire protoplast to leave the cyst as one body without distortion. In some cases the arrangement of accessory archeopyle sutures appears as if designed to assure that the potential dimensions of the opening itself will exceed those of the operculum. In fact, free opercula may appear much too small to cover the openings observed in associated specimens of the rest of the cyst. In such cases, the conspecific identity of the separated parts may be easily overlooked, but the adequacy of the fit is undeniable if the operculum is adnate (e.g., *Kallosphaeridium*, Figure 6.4 M). In a number of peridinioid species with cavate wall structure (e.g., some *Chatangiella*, Figure 6.4 0), approximate equality in the sizes of openings in inner and outer walls appears to be achieved by matching a periarcheopyle composed of one intercalary paraplate with an endoarcheopyle comprising three paraplates. It is as if the smaller size of the single paraplate when projected centripetally onto the inner wall were thus compensated for. In other peridinioid species (e.g., some *Isabelidinium*, Figure 6.4 N), a periarcheopyle composed of just the medial intercalary area is combined with an apical endoarcheopyle that involves both apicals and intercalaries, suggesting another modification for increasing the size of the inner opening. However, such inferences are speculative, and it is clear from the observed excystment of *Spiniferites* (Wall, 1965) that the protoplast can undergo considerable distortion as it passes through the archeopyle.

Archeopyles apparently may be situated on any part of the cyst. Although a location somewhere on the epicyst is the most common, cingular archeopyles occur in the unusual cysts of *Nannoceratopsis* (Chapter 12), and, in a few instances, archeopyles are located in the hypocyst (involving antapical paraplates in *Tuberculodinium* and postcingular paraplates in the siliceous cysts of *Peridinites*). Note that the term **epicystal archeopyle** has the special connotation of a large combination archeopyle that involves *all* (not just *any*) paraplates of the epicyst, the archeopyle suture running approximately along the paracingulum.

The term **archeopyle type** applies to the sum of the characteristics of the excystment aperture in a single specimen. Differences in archeopyle type chiefly concern the particular location and paraplate equivalence of the opening, and certain features of the operculum, especially whether it is free or adnate, simple or compound. The size and proportions of the opening and details of its relation to certain styles of paratabulation (e.g., quadra and hexa in some peridinioid cysts) are also important.

Normally the significant features of the archeopyle will be covered in a taxonomic description. However, an **archeopyle formula**, in which symbols stand for the most important morphologic features, can also be useful as an abbreviated statement of the archeopyle type. Such a formula alone may suffice for certain purposes (e.g., in a diagnosis or in comparative tables) and it is often a helpful supplement to verbal description. However, it should not be used alone where clarity or necessary detail would be sacrificed. Also, for those archeopyle types that are poorly understood or anomalous, a careful description is far superior to an ambiguous formula.

The symbols used in archeopyle formulae are explained in the paragraphs that follow. Besides letters and numerals, the symbols include parentheses (or square brackets) and the hyphen, slash, and prime marks. The examples given immediately below illustrate both the appearance of the basic formulae, in which symbols appear only on the main line, and expanded versions, which incorporate more explicit information in subscripts.

Three most common types:	Type (4A)	Type I	Type P
Other examples:	Type I/-	Type ?/3I	Type (tA)aP
Comparing basic formula:	Type A(3A)6P		
with explicit expansions:	Type $A_{1'}(3A_{2\text{-}4'})6P_{1\text{-}6''}$		(Kofoidian)
	Type $A_{\mathbf{1u}}(3A_{\mathbf{A\text{-}C}})6P_{\mathbf{1i,2\text{-}6}}$		(Taylorian)

Upper case letters on the main line identify the one or more paraplate series in which the opening is primarily situated. The letters are based on Kofoid's names for the series, an "H" being added to distinguish those on the hypocyst (Evitt, 1961; Norris,1978), as follows:

A	=	apical	HP	=	postcingular
I	=	anterior intercalary	HI	=	posterior intercalary
P	=	precingular	HA	=	antapical
C	=	cingular	S	=	sulcal

If not preceded by a numeral or qualified by other symbols, only a single paraplate is involved and the operculum is free.

Examples: Type P - monoplacoid archeopyle formed by complete release of the area corresponding to a precingular paraplate. Type I - monoplacoid archeopyle involving one intercalary paraplate.

An arabic numeral or lower case "t" on the main line before a capital letter refers to the number of paraplates of the indicated series that are involved in the archeopyle. The numeral gives the number explicitly; the "t" (for "total") may replace the numeral when it is clear that all the paraplates of a particular series are involved, but when the number of such paraplates is uncertain, variable, or irrelevant to the immediate context.

Examples: Type 3I - a polyplacoid archeopyle representing three intercalary paraplates; the operculum is compound and the three opercular pieces are all free and separate. Type tI - an opening involving the total number of intercalary paraplates present, but the number is unspecified owing to its uncertainty or irrelevance. Type 3I3P - a polyplacoid combination archeopyle in which the compound operculum consists of separate opercular pieces representing three intercalary and three precingular paraplates; all the opercular pieces are free.

Parentheses or square brackets on the main line signify that the paraplates identified by the enclosed symbols are united into a single opercular piece; that is, this part of the operculum is simple. These symbols (Norris, 1978) pose significantly less typesetting difficulty than the horizontal bar above the series designation (Evitt, 1967).

Examples: Type (4A) - an opening corresponding to four apical paraplates and closed by a free, polyplacoid operculum which is simple (i.e., accessory sutures, if present, do not divide it into individual opercular pieces). Type (3I3P) - a combination archeopyle in which the simple and free polyplacoid operculum includes three intercalary and three precingular paraplates.

Note: Since separation of opercular pieces is indicated wherever parentheses are not present, the use of a plus sign (+) to denote separation of opercular pieces derived from different paraplate series (Evitt, 1967) is superfluous.

A lower case "a" on the main line indicates that the opercular piece(s) specified by the preceding notation is (are) adnate to the rest of the cyst.

Examples: Type (4A)a - opening closed by an adnate operculum which is simple and polyplacoid, representing four apical paraplates. Type 3I3Pa - combination archeopyle involving six paraplates: three intercalaries, each represented by a free opercular piece, and three precingulars separated from one another by secondary sutures but individually adnate to the rest of the cyst. Type (tA)aP - combination archeopyle representing a total of five paraplates and having two opercular pieces: an adnate, simple polyplacoid piece including all apical paraplates and a free piece representing a single precingular.

A slash mark (/) on the main line indicates that archeopyles in more than one layer of the wall are being described. The symbols for the openings are given in sequence beginning with that in the outer wall and are separated by the slash.

Examples: Type Ia/I - each wall layer has a monoplacoid archeopyle involving an intercalary paraplate; the operculum of the outer archeopyle is adnate but the inner one is free. Type I/3I - the opening in the outer wall is monoplacoid and has a free operculum; whereas three free opercular pieces close the polyplacoid opening in the inner wall.

A hyphen (-) or query (?) on the main line signals, on the one hand, the apparent absence of an archeopyle in the particular wall of a multiwalled cyst or, on the other, an uncertainty as to the existence or composition of the opening.

Examples: Type I/— - opening in the outer wall closed by a free operculum formed by an intercalary paraplate, but inner wall apparently lacks an archeopyle. Type ?/3I - existence or composition of an opening in the outer wall is uncertain (e.g., because the outer wall is typically thin and severely deformed), but the character of the opening in the inner wall is clear.

Table 6.1 - MEANING OF LETTERS ON MAIN LINE AND IN SUBSCRIPTS

	Meaning	Meaning in subscripts	
Letter	on the main line	Kofoid notation	Taylor notation
A	apical	—	paraplate **A** of polar series
C	cingular	cingular	paraplate **C** of polar series
a	adnate	anterior intercalary	paraplate **a** of equatorial series
p	—	posterior intercalary	—

Subscript notations following the upper case letters entered on the main line provide an optional means of conveying information to supplement that in the basic formula. These subscripts specify the particular paraplates of the series identified by the preceding letter as participating in the archeopyle. Either Kofoidian or Taylorian terminology can be used in the subscript line for gonyaulacoid cysts, but prime marks should always be used in conjunction with Kofoidian notation in subscripts, for they give the most obvious signal that Kofoid's system is being followed. This is especially important when specifying precingular archeopyles in gonyaulacoid forms, since elements of this series are normally indicated by different arabic numerals in the two systems (*3″* of Kofoid usually applying to the same paraplate as **4** of Taylor, etc.). So long as the prime marks are recorded as part of the Kofoidian notation, overlap among the symbols for the Kofoidian and Taylorian systems is unlikely to cause difficulty. Note, however, that some letters have different meanings depending upon whether they appear on the main line or as subscripts (Table 6.1).

SURVEY OF ARCHEOPYLE TYPES

Archeopyles in Peridinioid Cysts

General. Among the very large number of sporopollenin peridinioid cysts, the archeopyle is consistently restricted to the epicyst, but otherwise shows great diversity. The opening ordinarily involves the second of three anterior intercalary paraplates and may consist of this area alone or supplemented by the lateral intercalaries, the apicals, or the precingulars. If the archeopyle is polyplacoid, the operculum may be simple or compound, depending upon the species.

Since all available evidence suggests that the great majority of fossil sporopollenin *Peridinium*-like cysts were derived from thecae with symmetrical hexa or quadra style paratabulation, it follows that the archeopyles in these cysts are normally symmetrical in themselves and also symmetrically positioned. For similar reasons, if the principal or accessory archeopyle sutures lie along the boundary between the precingular and intercalary series, this boundary is middorsally rectilinear (i.e., *4″* is planate) and not geniculate as it is in those non-peridinioid cysts that have intercalary archeopyles. Exceptions to this general symmetry of shape and position have recently been emphasized for certain Pleistocene and Holocene peridinioid cysts (Harland, 1982), and are treated by Bujak and Davies (1983).

Figure 6.5, which summarizes the archeopyle types observed in peridinioid cysts, should be consulted in conjunction with the following numbered paragraphs.

1. Type I Archeopyles. These archeopyles, which involve only paraplate *2a*, are the most common archeopyle types in peridinioid cysts. Therefore, it is convenient that the opening shows many differences in size, shape, and position, but it is less ideal that many of these differences are difficult to describe and are subject to considerable intraspecific variability.

The opening, which may be developed in one or more wall layers is, with rare exceptions, middorsal in position and bilaterally symmetrical. Even within a single species, archeopyle shape may range from sharply angular to strongly rounded, and rounding may be sufficient to completely obliterate the shape to be inferred for thecal plate *2a*. Any degree of rounding implies that some portion of the archeopyle suture deviates from the corresponding parasuture, but clear parasutural lines against which to assess the amount of suspected deviation are infrequently developed in this group of cysts.

Both the hexa and quadra styles of dorsal paratabulation are known to occur in fossil cysts in combination with Type I archeopyles. However, the few examples of the quadra style known occur exclusively among species having a distinctive morphology and limited geologic range (the Paleogene *Wetzeliella* and *Apectodinium* complexes).

In conjunction with the distinction between hexa- and quadra-style paratabulation, the effect of rounding on the basic polygonality of the archeopyle associated with either style of paratabulation must be remembered. Actually, the precise shape of the archeopyle is a variable relatively independent of the style of dorsal paratabulation. The terms "hexa *2a* archeopyle" and "quadra *2a* archeopyle" have been used to convey information about both paratabulation style and archeopyle type simultaneously (Lentin and Williams, 1976), but these phrases do not mean that the openings themselves are geometrically hexagonal and quadrilateral, respectively (Evitt, 1978). Slightly longer but less ambiguous descriptors would be in the form: paratabulation style hexa; archeopyle Type I_{2a}. Actually, Type I archeopyles *2a* that are geometrically either rounded-hexagonal or rounded-quadrangular, as well as openings that approach circularity, may occur in association with either a hexa-style or a quadra-style dorsal paratabulation.

If parasuture intersections are not directly visible, the two styles can be distinguished by comparing the relative widths of paraplates *2a* and *4″* (Figure 6.6 D). Of greater practical value is the knowledge that the quadra style has been documented only in cysts of *Wetzeliella* type.

The dominance of strong bilateral symmetry in peridinioid cysts limits somewhat the variety of sizes and shapes of the second anterior intercalary that are reflected in Type I archeopyles, but the variety is still great enough to have inspired efforts to standardize the expression of these properties in both qualitative and quantitative terms. Lentin and Williams (1976) applied the terms "standard", "attenuated", "broad", and "omegaform" to four of the shapes of the *2a* paraplate that recur with greater frequency than others in the hexa-style setting. These authors and others (Bujak, 1980; Davies, 1981; Bujak and Davies, 1983) have also defined several ratios and indices to express quantitatively (in all cases as decimal quotients) aspects of the relationship of Type I archeopyles to other features

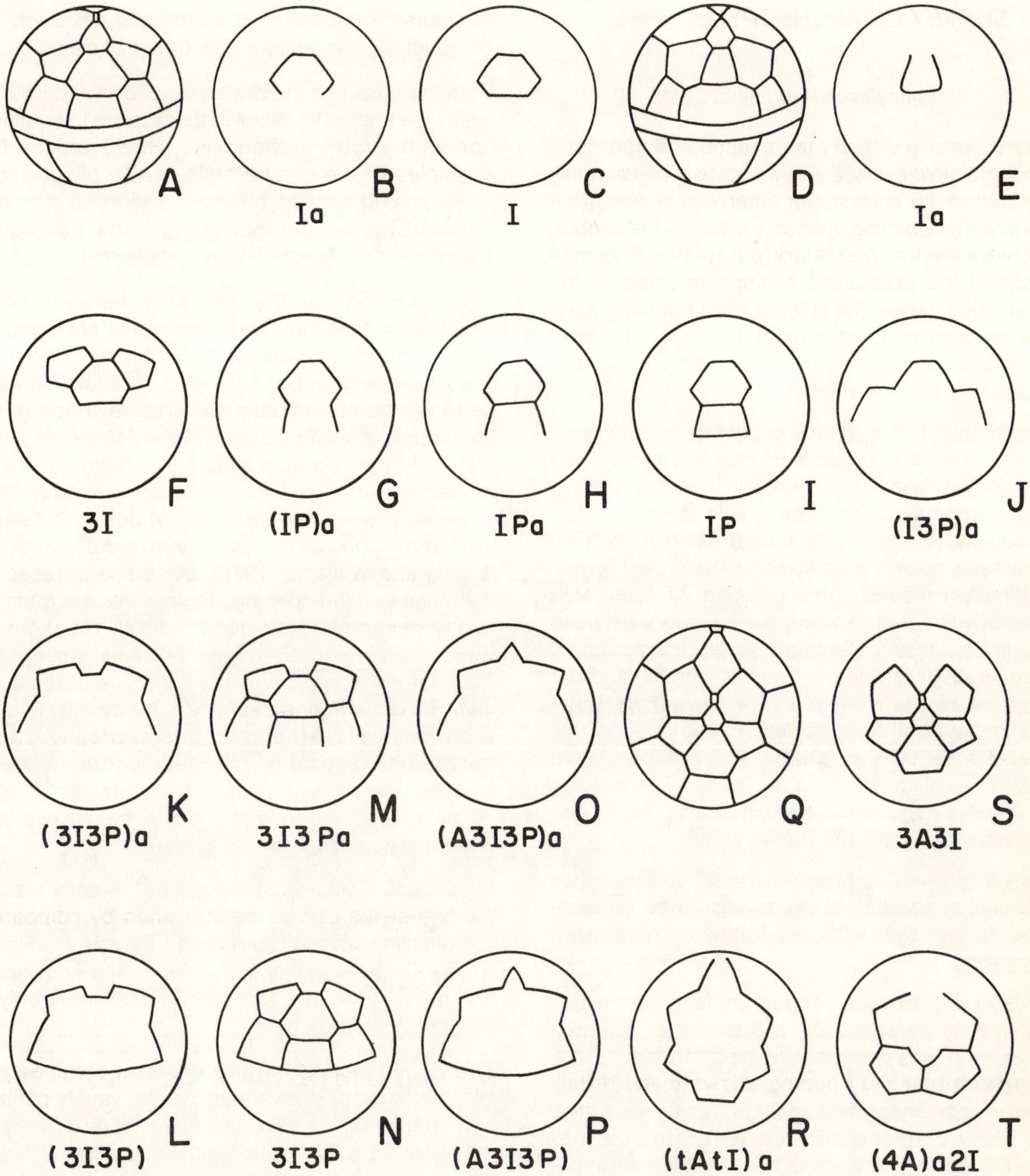

Figure 6.5 - ARCHEOPYLE TYPES - PERIDINIOID CYSTS

Schematic outlines of archeopyle types known from fossil cysts. The epicyst is seen in oblique views normal to 2a (Figures A-P) or in apical view (Figures Q-T). Figures A, D, and Q are orientation diagrams. All except T assume three intercalaries as in *Protoperidinium* subgen. *Protoperidinium*, and all except D, E, and T assume hexa style dorsal paratabulation. T shows the two intercalaries of *Protoperidinium* subgen. *Archaeperidinium*. The formula (short style) for the archeopyle type is given directly below each diagram.

B - *Isabelidinium*, C - *Alterbia*, E - *Rhombodinium*, F - *Trithyrodinium*, G - *Diconodinium*, H - *Spinidinium*, I - *Spinidinium*, J - *Morkallacysta*, K - *Laciniadinium*, L - *Luxadinium*, M - *Ginginodinium*, N - *Luxadinium*, O - *Saeptodinium*, P - *Palaeoperidinium*, R - *Ascodinium*, S - "*Canningia*" *turrita*, T - *Umbodinium*.

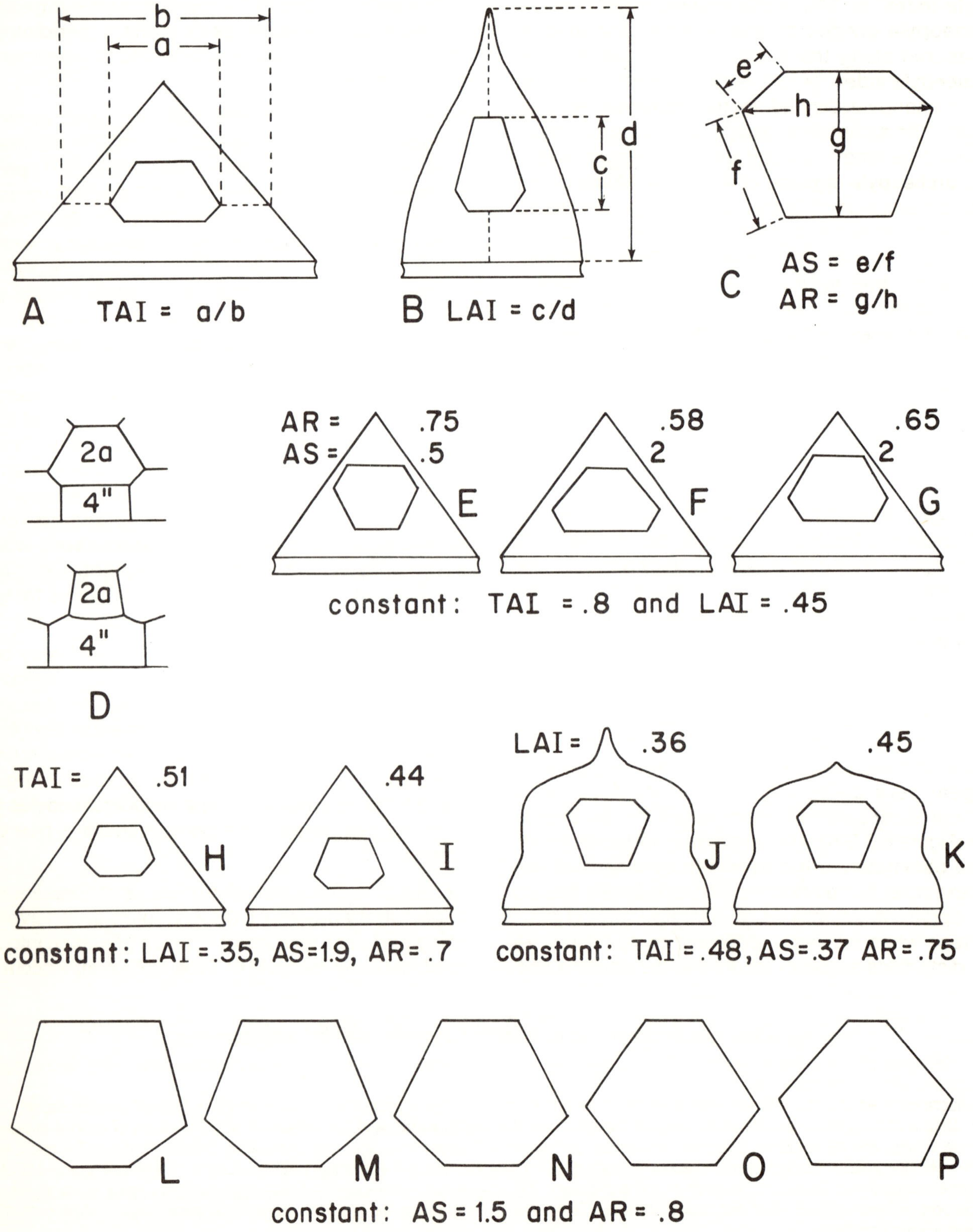

b
a
A TAI = a/b
d
c
B LAI = c/d
e
h
g
f
C
AS = e/f
AR = g/h
2a
4"
2a
4"
D
AR = .75
AS = .5
E
.58
2
F
.65
2
G
constant: TAI = .8 and LAI = .45
TAI = .51
H
.44
I
constant: LAI = .35, AS=1.9, AR= .7
LAI = .36
J
.45
K
constant: TAI = .48, AS=.37 AR=.75
L
M
N
O
P
constant: AS = 1.5 and AR = .8

Figure 6.6

of the epicyst (Figure 6.6). The **transverse archeopyle index**, or TAI, is the maximum width of the archeopyle divided by the width of the epicyst measured along the same line. The **longitudinal archeopyle index**, or LAI, is the maximum length of the archeopyle divided by the maximum epicyst length. The **archeopyle ratio**, or AR, is the maximum archeopyle length divided by its maximum width. The **archeopyle signum**, or AS, is the length of the anterolateral side of *2a* (i.e., *1a/2a* or *2a/3a*) divided by the length of the posterolateral side (i.e., *2a/3″* or *2a/5″*). Lentin and Williams (1976) proposed special symbols for the several sides of the critical *2a* paraplate, but these symbols are not introduced here. The sides of this paraplate can be specified unambiguously (as was done within parentheses two sentences back) by using the plate-slash-plate symbolism that enables one to designate any particular side of any particular plate/paraplate on theca or cyst.

Although these parameters may be helpful, they do not, even in combination, completely characterize archeopyle shape and size. Comparison of the diagrams in Figure 6.6 will bring out some of their advantages and shortcomings. While the AR and AS are helpful for indicating archeopyle shape, rather substantial differences in shape are possible without change in either (Figure 6.6 L-P). LAI and TAI must be used with awareness that they are significantly affected by factors relatively unrelated to the archeopyle; conversely, the archeopyle may appear very differently in different specimens without significantly different values for TAI and/or LAI (Figure 6.6 E-K). The usefulness and reliability of such parameters are also limited by other factors; for example: 1) the effect that foreshortening has on the apparent dimensions of a feature on a curved surface when viewed in different perspectives, 2) the distortions so easily induced in the opening and the cyst by postdepositional folding or compression of flexible walls, and 3) the rounding of the angles in the archeopyle outline seen in many peridinioid cysts, often to a variable extent among specimens of the same species.

Bujak and Davies (1983), building upon earlier proposals by Lentin and Williams (1976), defined nine broadly quantitative categories of Type I archeopyle shape by combining AR and AS values in a two-dimensional matrix (Figure 6.7). These categories, labeled with terms constructed of word stems and prefixes expressing the shapes involved, are descriptively useful if defined broadly enough to avoid the pitfall of false precision from calculated decimal values. Prefixes **steno-**, **iso-**, and **eury-*** are used for AR values decidedly more than, about equal to, and decidedly less than 1; and word stems **omegaform**, **thetaform**, and **deltaform** are used for AS values in the same progression. Lentin and Williams' earlier four terms for archeopyle shape equate to terms in this more comprehensive scheme approximately as follows: **standard** with isothetaform or isodeltaform, **attenuated** with stenodeltaform, **broad** with eurythetaform or eurydeltaform, and **omegaform** with the three omegaform possibilities.

The perioperculum in Type I archeopyles may be free or adnate, and either condition is normally constant within a species. Adnation, when it occurs, is always along the adcingular margin (*2a/4″*) except in the *Wetzeliella* and *Apectodinium* complexes, where it is along the adapical margin (*2a/3′*). The operculum of Type I endoarcheopyles is normally free, whether the perioperculum is free or adnate.

Accessory archeopyle sutures are rarely present in Type I archeopyles. If they develop extensively along the lateral margins of *4″*, then gradation into a combination archeopyle of Type $I_{2a}P_{4''}$ a may occur.

Figure 6.6 - QUANTITATIVE AND QUALITATIVE ASPECTS OF TYPE I ARCHEOPYLES

A-C - Definition of four ratios and indices. TAI = transverse archeopyle index, LAI = longitudinal archeopyle index, AS = archeopyle signum, AR = archeopyle ratio.

D - Diagrams to emphasize different relative widths of *2a* and *4″* in hexa-style (upper) and quadra-style (lower) dorsal paratabulation.

E-P - Diagrams to illustrate coordination—or lack thereof—between the appearance of the archeopyle and the values of various archeopyle indices and ratios. E-G - Diagrams to suggest pronounced differences in appearance possible despite constant TAI and LAI. H-I - Archeopyles of same size and shape in epicyst of same size and shape (i.e., constant AS, AR, LAI) but significantly different TAI dependent only on "latitudinal" position of the archeopyle. J-K - Significant difference in LAI dependent only on apical horn; all other properties constant. L-P - Outlines of five archeopyles to indicate contrasting appearance possible in archeopyles for which both AS and AR are constant.

*Bujak and Davies used "lati-", but otherwise entirely Greek origin of stems and prefixes suggests the greater appropriateness of "eury-", which is adopted here.

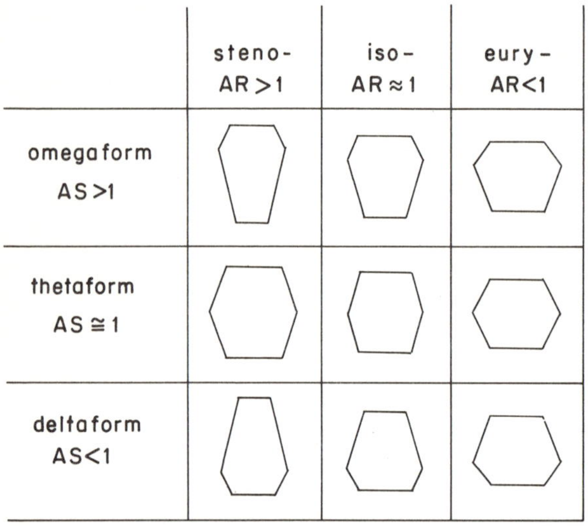

Figure 6.7 - MATRIX OF TYPE I ARCHEOPYLE SHAPES
Outlines of archeopyles having values of archeopyle ratio (AR) and archeopyle signum (AS) less than, about equal to, and greater than 1. Based on Bujak and Davies (1983).

A few fossils, referred to *Selenopemphix*, have Type I archeopyles that are distinguished by their eccentric position. Instead of being middorsal, the opening is off to the left of the dorsal midline. Similarly placed archeopyles are known in modern cysts, but these have been attributed to species with symmetrically disposed intercalary plates (Harland, 1982). Verification of this seemingly unlikely association, which combines a conspicuously asymmetrical paratabulation (as indicated by the eccentric archeopyle) with a symmetrical arrangement of corresponding thecal features, is needed before it can be accepted without reservation.

A Type I archeopyle can be easily mistaken for a Type P archeopyle if paraplate *4″*, which separates the adcingular edge of the opening from the paracingulum is unusually short in a meridional sense. Folding along the paracingulum often adds to the problem of recognizing the intercalary nature of the opening in such cases.

Examples with hexa style paratabulation: Type I - *Hexagonifera* and *Lejeunecysta*. Type Ia/- - *Nelsoniella* and some *Isabelidinium* spp. Type I/I - *Deflandrea phosphoritica* (eurydeltaform archeopyle, Frontispiece B), *Chatangiella* spp. (omegaform archeopyle), *Palaeocystodinium* (stenodeltaform), *Phelodinium* (isodeltaform). Opercula adnate along the adcingular margin occur in some species of *Isabelidinium*, *Chatangiella*, and *Spinidinium*.

Examples with quadra style paratabulation: Type I/I - *Wetzeliella* and *Wilsonidium*. Type Ia/I - *Gochtodinium* (operculum adnate along adapical margin).

A few modern examples of Type I archeopyles occur in organic-walled cysts of species with a penta style epithecal pattern, and similar cysts occur in Quaternary sediments (Wall and Dale, 1968b; Harland, 1982), but they are uncommon in the pre-Quaternary fossil record.

2. Type 3I archeopyles. All three intercalary paraplates participate in forming this type of archeopyle which has a many-angled and highly characteristic outline. In all known examples, the operculum is compound (i.e., opercular pieces separated by accessory sutures) and, almost always, the opercular pieces are free (but they are adnate, Type 3Ia, in some species attributed to *Subtilisphaera* by Bujak and Davies, 1983). The Type 3I openings are easy to identify if specimens are ideally oriented, but, in unfavorably oriented or distorted specimens, they can be mistaken readily for irregular breaks or for apical archeopyles of the type common in cysts with gonyaulacoid features (Figure 6.8 A-C).

The characteristic forms of the adapical and adcingular margins of the opening, which resemble one another in basic outline but differ decidedly in scale, are the most useful clues for correct identification. The most diagnostic feature of the outline is the **middorsal tab**, actually part of paraplate *3′* that projects into the opening from its adapical margin. Straight lines drawn from the angles of this tab to the angles along the adcingular archeopyle margin will approximate the course of accessory sutures and provide a basis for reconstructing the shapes of all three opercular pieces. Often one or more of the opercular pieces will be found inside the cyst or adhering to one edge of the opening. Each piece has a distinctive shape and a clear view of any one of them may help to identify the archeopyle type even if the outline of the opening itself is severely distorted.

The periphragm in some cavate cysts with Type 3I archeopyles is exceedingly thin and may even be torn away without a trace, making it difficult or impossible to identify the type of opening in the outer wall if only a few specimens of less than ideal quality are available. Type 3I openings also occur in some species characterized by having an autophragm only, although these single-walled

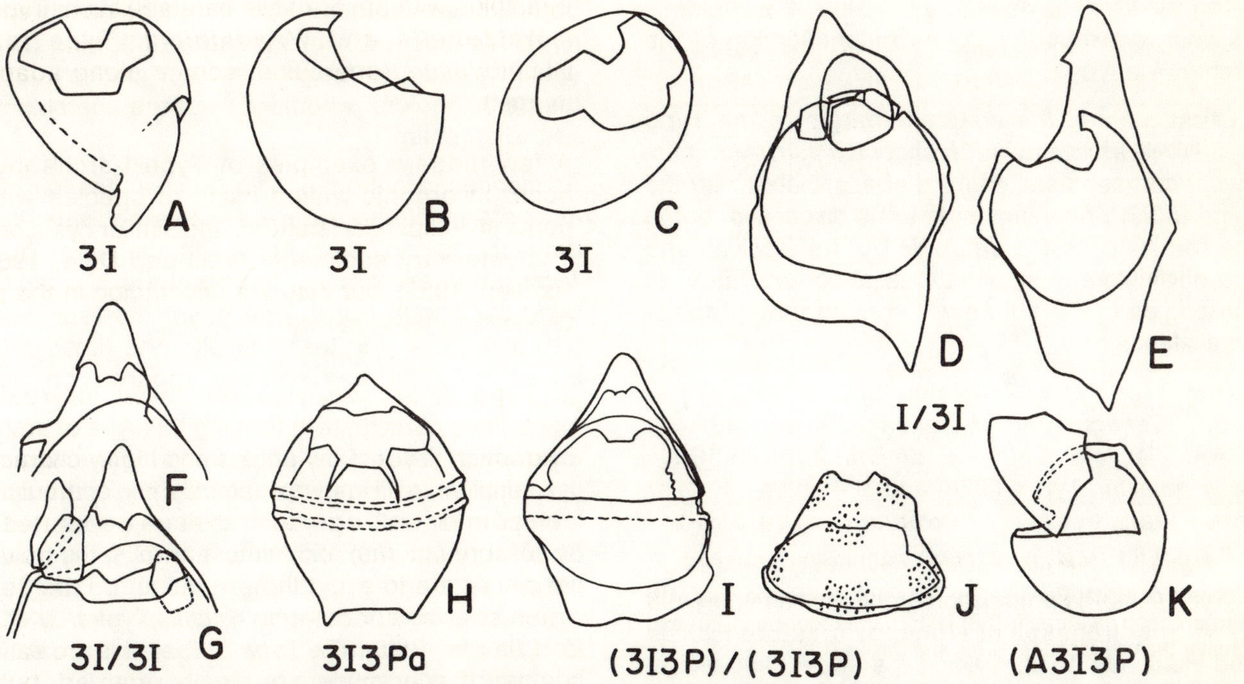

Figure 6.8 - ARCHEOPYLES INVOLVING THREE INTERCALARIES

Drawn from actual specimens; archeopyle types are indicated below the figures.

A-B - *Trithyrodinium* sp. Autocyst (=endocyst?). A - Open, clearly trilobate opening (dv). B - Superposed edges of opening in distorted cyst suggest zigzag margin of apical archeopyle (llatv). Late Cretaceous, Pakistan.

C - *Trithyrodinium suspectum*. Oblique view of open archeopyle in isolated endocyst (thin pericyst missing). Late Cretaceous, arctic Canada.

D-E - *Chatangiella* sp. D - Cyst with periarcheopyle open and pieces of endoperculum dislodged (vv). E - Specimen showing distinctive outlines of Type I periarcheopyle and Type 3I endoarcheopyle (rlatv). Late Cretaceous, arctic Canada.

F-G - *Trithyrodinium* sp. showing distinctive middorsal tab on periphragm (dv). Free opercular pieces visible in G confirm division of both endo- and perioperculum by accessory sutures. Late Cretaceous, arctic Canada.

H - *Ginginodinium* sp. Precingular opercular pieces adnate along paracingulum (dv). Late Cretaceous, arctic Canada.

I-J - *Luxadinium propatulum*. Specimen with open archeopyle, and isolated operculum (both dv), showing apical outlines of opening and operculum in both walls. Surface relief on perioperculum indicates paraplates. Early Cretaceous, western Canada.

K - *Saeptodinium eurypylum*. Endocyst only (dv), showing distinctive apical outlines of archeopyle and operculum. Late Cretaceous, arctic Canada.

bodies may, in fact, be isolated endocysts with no traces of associated periphragm.

A Type 3I opening in the endophragm of a cavate cyst may occur together with either a Type I or a Type 3I opening in the periphragm. A species with a Type I/3I archeopyle illustrates well the presence of the distinctive middorsal tab on the inner wall and its contrasting absence from the outer wall—critical features for distinguishing these two kinds of openings (Figure 6.8 D-E).

Examples: Type 3I and 3I/3I - *Trithyrodinium*. Type I/3I - some *Chatangiella*.

3. Combination archeopyles involving intercalary and precingular paraplates. These types can be viewed as elaborated from Types I and 3I by involvement of the subjacent precingular areas, and they may intergrade with the simpler types.

In Types $I_{2a}P_{4''}$ and IPa the middorsal paraplates of both series participate in the opening.

In Types $3I_{1\text{-}3a}3P_{3\text{-}5''}$ and (3I3P) (Figure 6.8 H-J), the apical margin has the same distinctive shape as the Type 3I opening, including the middorsal tab, but the opening is larger (with distinctively indented lateral margins) and the cingular margin is contrastingly straight. Isolated opercula are easily recognized by their shape, and patterns of surface ornament may individually distinguish the six paraplates. Since the adnate operculum of a Type (3I3P)a archeopyle may lie nearly in its original position, and the sutures may run close to the lateral margins, the existence of this type of opening can be overlooked easily. However, when the archeopyle suture is visible, the angles in its course are useful clues to identification of the archeopyle type.

At first glance, the adapical margin of the Type ($I_{2a}3P_{3-5''}$)a archeopyle (*Chichaouadinium*, *Morkallacysta*) can be mistaken for that associated with the Type (A3I3P) openings next to be discussed, but in the former case the suture is not transapical, and the operculum is smaller. Critical consideration of the angles in the suture will confirm the paraplate equivalence.

Examples: Type IP - some *Spinidinium*. Type (IP)a - *Senegalinium dilwynense*. Type ($I_{2a}3P_{3-5''}$) - *Morkallacysta*, *Chichaouadinium*. Type (3I3P)a - *Laciniadinium*. Types (3I3P) and 3I3P in one, two, or three wall layers - *Luxadinium*. Type 3I3Pa - *Ginginodinium*.

4. Type (A3I3P) combination archeopyles (Figure 6.8 K). A truly **transapical archeopyle suture** (Norris and Hedlund, 1972), dividing the epicyst into virtual dorsal and ventral halves, is characteristic of such openings. The suture extends adapically from the paracingulum on one side, traverses the apex itself, and returns to the paracingulum on the other side. If the operculum is free, a dorsal equatorial portion of the suture extends along the paracingulum between the ends of the transapical portion, and the type can be represented by the formula: ($A_{3'}3I_{1-3a}3P_{3-5''}$). If, on the other hand, the operculum is adnate, the archeopyle, which then is Type (A3I3P)a, is defined by this transapical suture alone, and the operculum, as in the case of Type (3I3P)a, tends to lie more or less in place, even when the wall has separated along the suture. Type (A3I3P) has been termed a **hemiepicystal archeopyle** by Bujak and Davies (1983).

At first consideration, it may seem that these types should be difficult to distinguish from Types (3I3P) and (3I3P)a, which they approximate in composition and exceed only slightly in size. However, they are easy to distinguish if the position of paraplate *3′* relative to the archeopyle suture is noted. In the case of Type (A3I3P), the middorsal tab characteristic of Type (3I3P) is replaced by a V-shaped **middorsal notch** that extends to the apex. Correspondingly, the broad indentation in the adapical margin of the operculum is replaced by an angular projection. Although these differences may seem slight, their visual effect is pronounced and identification is not difficult when the critical adapical outline of the archeopyle can be examined.

In the dorsoventral view most frequently favored by specimen shape, the principal archeopyle suture in both Types (A3I3P) and (3I3P) may be obscured by near coincidence with the lateral margins of the epicyst and by confusion with other markings along the paracingulum, whether or not the operculum is actually present.

As mentioned in the preceding section, the transapical portion of the suture associated with these archeopyles can be confused with the suture in the Type (I3P)a archeopyle. In both there is a median adapical notch, but in one case this represents paraplate *3′* and in the other *2a*, and other differences in outline and size are apparent on close inspection.

Examples: Type (A3I3P) - In *Palaeoperidinium pyrophorum*, an opening occurs only in the outer wall; an inner wall, if present, appears unruptured. In *Saeptodinium hansonianum* and *S. eurypylum*, periphragm and endophragm are in contact and similar openings occur in both walls. Type (A3I3P)a - *Peridinium limbatum*, *P. willei*, and probably a number of other modern freshwater species.

5. Type (tAtI) and 3A3I combination archeopyles. These archeopyles are apical in position, although they involve a combination of intercalaries and apicals. If the angles along the adcingular margin are clearly visible, the course of the suture can be seen to match that of the series of thecal sutures along the adapical margins of the precingular plates in many modern *Protoperidinium*. Conversely, these angulations define the distinctive form of the adcingular margins of all three intercalaries and of two or three of the apicals (*2′* and *4′*; or *1′*, *2′*, and *4′*). Presumably a total of four apicals were present, but this cannot be proven if paraplate *3′* is not defined independently by parasutural markings of some kind.

Examples: Type (tAtI) is best known in *Ascodinium*, the operculum being simple and either free or adnate ventrally. When periphragm and endophragm are closely appressed all along the archeopyle suture, only a single opening is present. For discussion of a remarkably archeopyle-like ventral opening in *Ascodinium verrucosum*, see the comments under Claustra in Chapter 4.

Many species of cavate cysts referable to *Isabelidinium*, *Nelsoniella*, and other genera having Type I periarcheopyles appear to lack openings through the endophragm, but occasional specimens suggest the presence of a Type (tAtI)a endoarcheopyle. The archeopyle suture of this endocyst opening may appear slitlike or may seem to isolate an apical "lid" whose outline shows a middorsal salient suggesting the adcingular margin of paraplate *2a*, as well as an unusually broad area of ventral adnation. However, rounding of angles in the archeopyle outline and lack of indications of paratabulation on the endocyst usually limit the certainty of this interpretation.

An archeopyle unusual in combining the intercalaries with only three of the four apicals, the first apical being excluded and forming an angular ventral projection into the opening, occurs in "*Canningia*" *turrita* (Brideaux, 1977; transferred to *Ovoidinium*? by Below, 1981). Although some uncertainty exists, the opercular pieces are probably all free, and the opening could be represented as Type $3A_{2\text{-}4'}3I_{1\text{-}3a}$.

6. An unusual archeopyle of Type (4A)a2I observed in specimens from Lower Cretaceous strata (Figure 6.5 T; Bint, 1983) appears to give the first good evidence for representation of *Protoperidinium* subgenus *Archaeperidinium* in the fossil record. Although visible paratabulation is restricted to the archeopyle sutures, these clearly indicate the presence of two symmetrical anterior intercalaries and four apicals in the arrangement diagnostic of this subgenus.

7. Hypocystal archeopyles. Archeopyles in the hypocyst are documented only in the case of Tertiary species referred to *Peridinites*, a genus also unique among fossil dinoflagellates for the siliceous composition of its cysts and unusual in the presence of a penta style paratabulation. Deflandre (1933) considered the openings to be in the epicyst, but Dale (1978) has identified openings of Type $HP_{3'''}$ and $3HP_{2\text{-}4'''}$ in specimens from Miocene sediments of the Norwegian Sea.

8. Epicystal archeopyles. Type E archeopyles with simple opercula (openings representing the total of epicyst paraplates) occur in sporopollenin cysts of modern marine species referred to *Diplopeltopsis* and *Diplopsalis*. The opened cyst consists of two hemispheres attached ventrally by a tenuous connection at the parasulcus. In fossil sporopollenin cysts, comparable openings are known only among non-peridinioid taxa. Dale (1978) reports this type of archeopyle in one species of Eocene siliceous cysts from the Norwegian Sea referred to *Peridinites*.

9. Miscellaneous archeopyle types. This category includes those archeopyles whose paraplate equivalence cannot be recognized because requisite traces of parasutures (including distinctive angulations in the archeopyle suture) are not present.

Examples: The organic-walled cysts of two widely distributed modern marine species (*Protoperidinium claudicans* and *P. oblongum*) have archeopyles that are apical in position and are defined by an archeopyle suture that is relatively lower on the dorsal surface than on the ventral one. Possibly this represents a Type (tAtI)a opening modified by severe rounding (Evitt, 1967; Wall and Dale, 1968b; Harland, 1982).

10. No archeopyle present. The consistent absence of an opening in the cyst can hardly be called an archeopyle type, but the condition is highly characteristic of some peridinioid fossils. It is the principal feature that distinguishes the genus *Subtilisphaera* from genera with otherwise similar features. It is also certainly a highly typical feature of *Palaeohystrichophora*, although a few individual specimens among many thousands referable to this genus appear to show a Type (A3I3P)a archeopyle (Harker 1979).

Archeopyles in Non-Peridinioid Cysts

General. The taxonomic diversity of the dinoflagellates represented here means that few generalizations can be made about these archeopyles as a group.

The operculum for an archeopyle involving more than one paraplate may be simple or compound, according to the species, but, with the exception of the Type 4A archeopyle (e.g., *Lithodinia jurassica*), all known apical archeopyles have simple opercula.

If an archeopyle has a compound operculum, the intersections of principal and accessory archeopyle sutures commonly lie at the apices of minute cusps along the archeopyle margins. Since the visibility of these cusps is increased by absence of the operculum, they are excellent clues to the existence and position of accessory sutures in an open archeopyle for which the operculum is not at hand.

1. Apical archeopyles involving all apicals (Figure 6.9 A-J). These are one of the two types of opening (the other being Type P) most commonly found among cysts with gonyaulacoid paratabulation, including all fossil cysts of ceratioid design. Since the principal archeopyle suture in these types follows the adapical boundaries of the precingular paraplates and parasulcus, the margin of the opening and its operculum exhibit the distinctive features of this line as discussed in Chapter 5 (Figure 5.9). The angulations vary greatly in prominence from species to species, and are generally

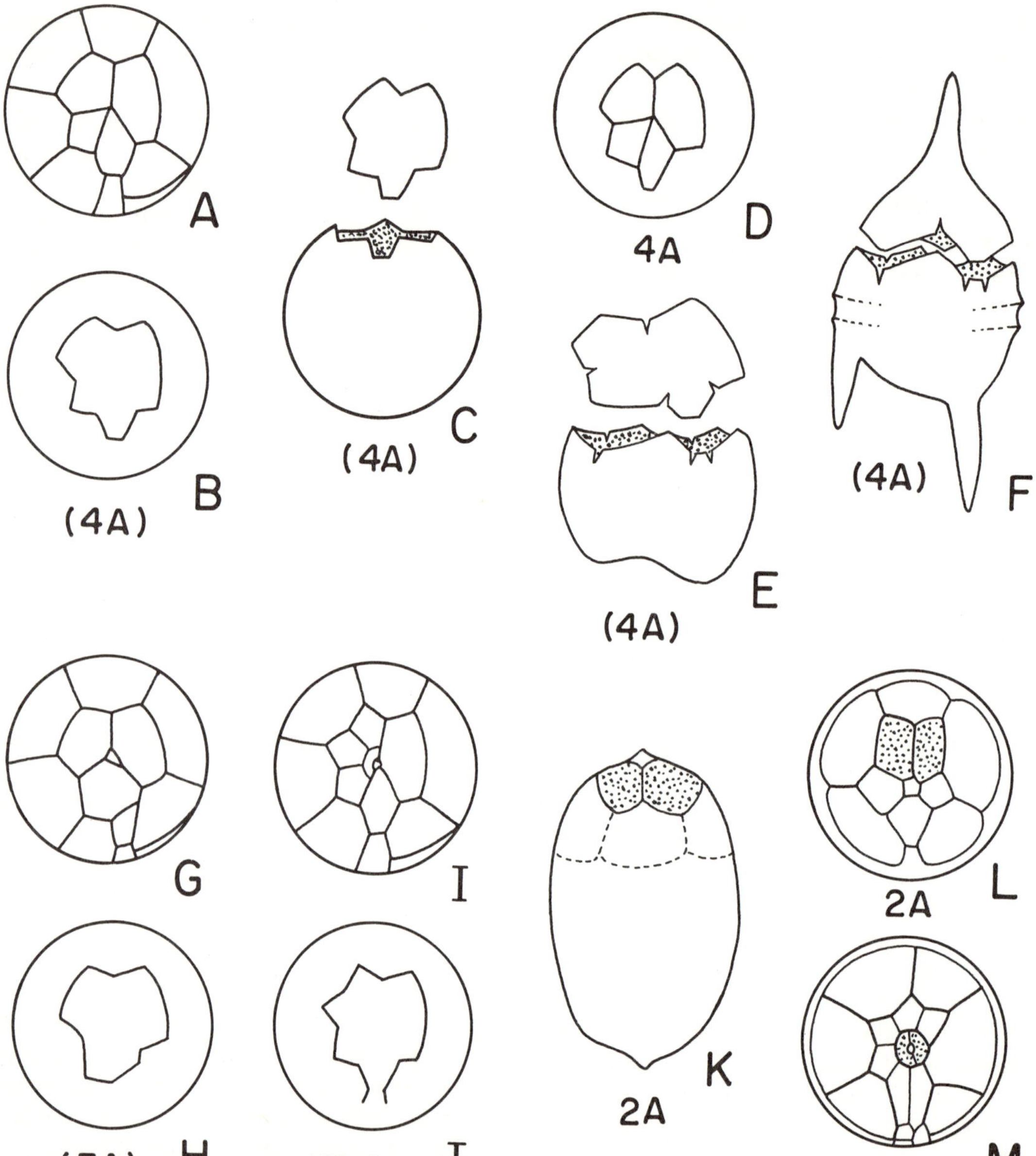

Figure 6.9 - ARCHEOPYLE TYPES - APICAL IN NON-PERIDINIOID CYSTS

Figures are schematic. Archeopyle types are indicated below figures.

A - Orientation diagram of insert gonyaulacoid epithecal plate pattern (av).

B - Outline of archeopyle.

C - Main part of cyst (vv) and free operculum (av).

D - Full division of operculum by accessory sutures, as in *Lithodinia jurassica*.

E - Dorsoventrally compressed cyst with offset sulcal notch (below, vv) and transversely elongate operculum (above, av).

F - Configuration of archeopyle sutures and operculum characteristic of ceratioid cysts (vv); consistently with offset sulcal notch and usually with significant dorsoventral compression.

G-H - Archeopyle associated with exsert gonyaulacoid paratabulation in undescribed Eocene cyst (D. K. Goodman, pers. comm.). G - Orientation diagram of epithecal plate pattern (av). H - Outline of opening (av).

I-J - Archeopyle indicating camerate **5** (*4″*). I - Orientation diagram of epithecal pattern (av) in modern *Gonyaulax polyedra*, showing apical series including subdivided **C** (**Ci** and **Cu**). J - Archeopyle suture outlining adnate operculum of *Kallosphaeridium* sp. consistent with tabulation in I, but details unconfirmable in the absence of visible parasutures on the operculum.

K-L - Two dorsal apicals forming archeopyle in *Comparodinium*; opercular pieces patterned. K - Entire cyst (dv). L - Epicyst paratabulation (av), showing preapical surrounded by five apicals and five intercalaries.

M - Archeopyle in *Cladopyxidium*. Epicyst paratabulation (av), showing four apicals and four intercalaries. Archeopyle suture surrounds simple operculum (patterned).

less prominent in smaller cysts. A **sulcal notch**, formed where **1u** extends toward the paracingulum to meet **ai**, *and a corresponding* **sulcal tongue** on the operculum are typical features along this archeopyle suture. Exceptionally, paraplate **1i** is shorter than **ai** and the relations of the notch and tongue to adjoining areas then reflect this difference.

Some cysts that have Type (tA) archeopyles are approximately spherical or have nearly circular transverse cross-sections with negligible depression of the ventral surface. Others show local excavation of the parasulcal area or a more extensive flattening or concavity of the ventral surface that may be sufficient to account for obvious dorsoventral compression of the cyst as a whole. With few exceptions (chiefly among very small cysts having partiform paratabulation on the hypocyst), pronounced dorsoventral compression is associated with displacement of the parasulcus toward the left of the ventral midline and the consequent appearance in the archeopyle margin of an **offset sulcal notch**. In these dorsoventrally compressed cysts, paraplate **1i** tends to be especially broad, as if in compensation for the shift in position of the notch. It acquires an approximately midventral position, and projects as a uniquely large and conspicuous angulation in the archeopyle outline adjacent to the notch. Reciprocal features are found on the operculum.

The amount of transverse elongation of the operculum (i.e., its departure from approximately equidimensional proportions) is a sensitive and easily perceived indicator of dorsoventral compression. Therefore, polar views of isolated opercula (which are likely to be common orientations in strew mounts) may provide a more practical and reliable basis for judging the amount of original compression than the main part of the cyst, in which assessment of compression may be difficult without polar views of undistorted specimens.

Accessory archeopyle sutures are commonly developed to some extent between the precingular paraplates, and they may extend a third—rarely more—of the distance to the paracingulum. When estimating the size of the operculum likely to fit such an archeopyle, allowance must be made for the considerable effect of the accessory sutures in enlarging the actual opening (Figure 6.4 M).

Type (tA) is a suitable designation for the archeopyle if the opening is obviously apical but lacks typical angularity, and other indications of paratabulation are also absent. This category includes openings in many fossils that are not demonstrably dinoflagellates, but which are, nevertheless, treated with them for convenience or on the strength of suspected affinities.

In contrast to Type (4A) and Type (tA) archeopyles, Type 4A archeopyles appear to be quite rare (Gocht, 1975; Below, 1981). Identification of this type, of course, requires the presence of the operculum, since the opening has the same shape as that for Type (4A). The opercular pieces, being individually quite small, can be overlooked easily if they are dissociated.

Examples: Type (4A) - *Meiourogonyaulax*, *Gardodinium*, *Odontochitina*, *Endoceratium*, *Hystrichosphaeridium*, some *Microdinium*. Type 4A - *Lithodinia*. Type (4A)a and (5A)a - *Kallosphaeridium* spp. Type (tA) - *Fromea*, *Trigonopyxidia*, *Palaeotetradinium*.

2. Apical archeopyles involving only certain apicals (Figure 6.9 K-M). In addition to the preceding examples, in which the apical opening includes all apical paraplates, archeopyles are now known which include some, but not all, of the apicals. Such archeopyles occur in the partiform gonyaulacoid genus *Cladopyxidium* and in *Comparodinium*, whose relationships are uncertain. Both genera are characterized by relatively large anterior intercalaries between the archeopyle and the precingulars, and, in both cases, other aspects of the paratabulation are somewhat unusual or as yet imperfectly understood. In *Cladopyxidium*, the minute polyplacoid operculum of the Type (2A) opening lies within a collarlike ring around the apex and comprises apicals *2′* and *3′* (out of four apicals), together with a small preapical between them (Goodman and Stover, 1981). In *Comparodinium* (Wille and Gocht, 1979), the two opercular pieces of the Type 2A archeopyle are not centrally located at the apex, but lie on the dorsal slope where they might be taken at first glance for intercalaries. However, they approximate middorsal paraplates *3′* and *4′* (out of six apicals), although the archeopyle suture is not fully congruent with the paratabulation indicated by parasutural features. Individual specimens of one species of *Comparodinium* have also been observed to lose paraplates *1′*, *2′*, and *5′* in addition to *3′* and *4′*, but the extent and significance of this variability is unknown.

Perhaps other examples of "selective exclusion" of certain apical paraplates from apical archeopyles will be discovered. Certainly these examples, while somewhat similar, do not suggest close relationship of the taxa in which they occur.

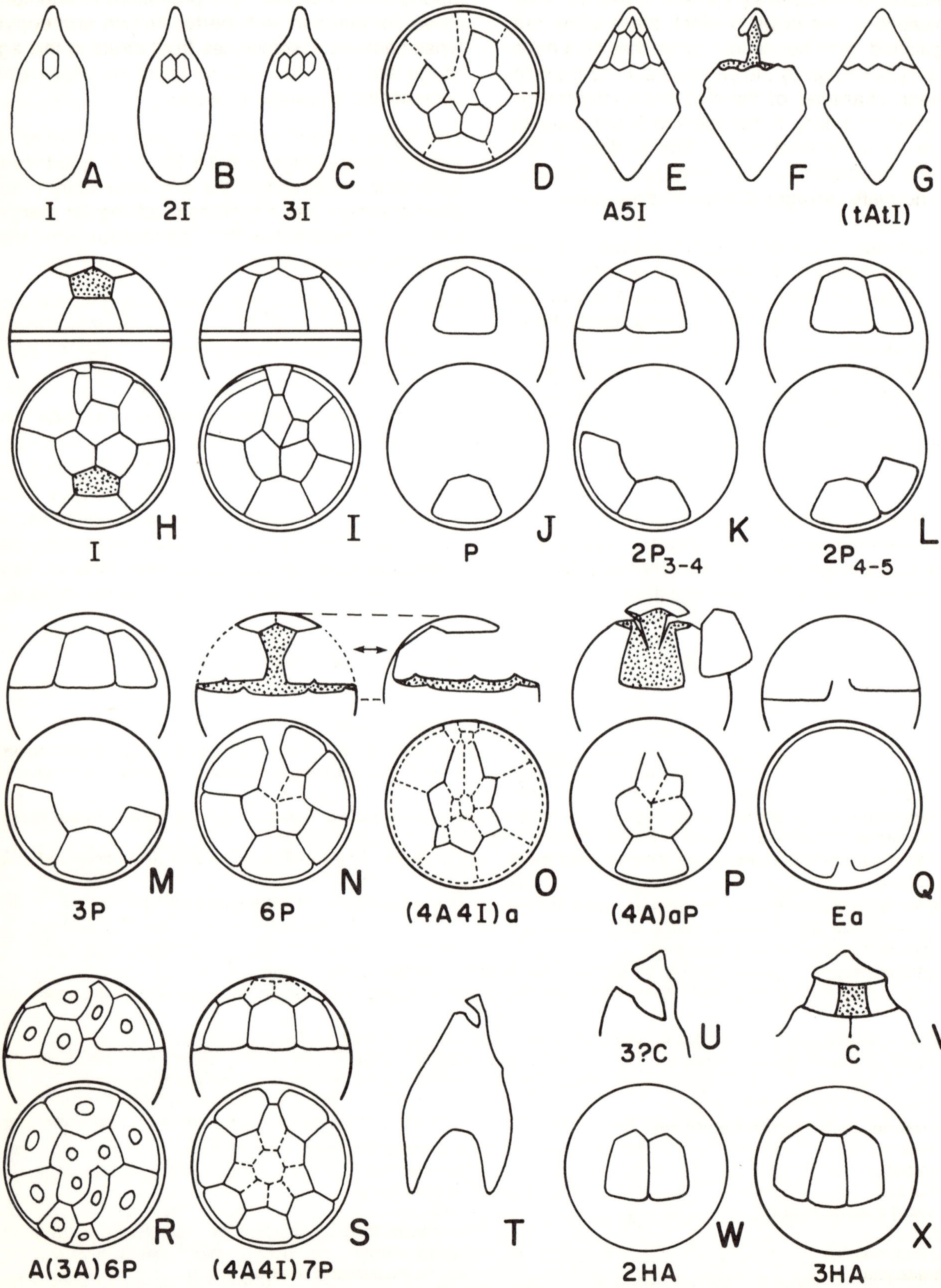
A
I
B
2I
C
3I
D
E
A5I
F
G
(tAtI)
H
I
I
J
P
K
2P3-4
L
2P4-5
M
3P
N
6P
O
(4A4I)a
P
(4A)aP
Q
Ea
R
A(3A)6P
S
(4A4I)7P
T
U
3?C
V
C
W
2HA
X
3HA

Figure 6.10

3. Intercalary archeopyles (Figure 6.10 A-C, H). With few exceptions, the cysts in which these occur are distinguished from peridinioid cysts that also have intercalary openings by the geniculate, in contrast to rectilinear, character of the middorsal intercalary-precingular boundary. If the opening is polyplacoid, the operculum is compound. The outline of the middorsal intercalary (or middorsal pair of intercalaries) is normally hexagonal when clearly visible.

There is much uncertainty about details of these archeopyles owing to a combination of factors: the cysts are seldom dorsoventrally compressed, so orientations that are best for viewing the opening are not favored; the cysts are commonly pyriform, so that dorsoventral views, even when available, do not necessarily display the opening to advantage; other traces of paratabulation are rare or obscure, and independent clues to the paraplate equivalence of the opening are therefore lacking.

These types of openings appear to be especially well represented among Late Triassic (Carnian) to Middle Jurassic species. Great diversity in overall morphology and in archeopyle structure is apparent among them. However, it is premature to attempt a close comparison with better known archeopyle types because assemblages from strata in this age range are only now beginning to be thoroughly documented in taxonomic works.

The Type I archeopyle of "*Lacrymodinium*" is extraordinary (Figure 6.10 H). It is middorsal, symmetrical, represents the only intercalary in the paratabulation, has a rectilinear adcingular margin, and is associated with a paratabulation that suggests modern *Triadinium* (see Chapter 10).

Examples: Type I - *Broomea*, "*Lacrymodinium*," *Moesiodinium*. Type 2I - *Cantulodinium*, *Gochteodinia*, *Pareodinia*, *Valvaeodinium*. Type 3I - some *Pareodinia*.

4. Precingular archeopyles (Figure 6.10 I-O). The archeopyle that involves just the third Kofoidian precingular paraplate (paraplate **4** of Taylor) is the second of the two types of archeopyle most frequently encountered in cysts with gonyaulacoid paratabulation. In the previous literature, this archeopyle type has commonly been cited as "Type P (*3″* only)." In the style of notation favored here, it

Figure 6.10 - ARCHEOPYLE TYPES - OTHER THAN APICAL IN NON-PERIDINIOID CYSTS

Figures are schematic. Archeopyle types are indicated below figures. Solid lines (except in orientation diagram, I) indicate archeopyle sutures; patterning indicates open archeopyle or interior surface.

A-C - Intercalary types in cysts with middorsally geniculate intercalary-precingular boundary (dv).

D-G - Diagrams of combination apical and intercalary types in *Rhaetogonyaulax*. D - Apical view (parasulcus at top), dashed lines to indicate precingulars. E-F - Dorsal views with opercular pieces in place (E) and with open archeopyle (F) to show apical cone and ventral connection. G - *R. rhaetica* (dv) with operculum in place.

H - "*Lacrymodinium*" sp. (dv and av), showing epicyst paratabulation pattern comparable to *Triadinium*, but with Taylor **4** divided into two secondaries: **4v**, which forms the archeopyle (patterned), and **4h**, adjacent to the paracingulum.

I - Orientation diagram for J-Q (except O), showing typical insert gonyaulacoid epicyst paratabulation in dorsal view (above) and apical view (below).

J-N - Various types of precingular archeopyles. N - Upper right figure shows left lateral view of same archeopyle type.

O -Combination apical and intercalary opening in *Microdinium* (av only).

P - Combination apical and precingular opening.

Q - Epicystal archeopyle in cyst having no visible paratabulation except archeopyle suture.

R - Epicystal archeopyle in chorate cyst with intratabular processes (bases indicated by ellipses); vv above, av below with parasulcus at top.

S - Precingular archeopyle in a cyst with both apicals and anterior intercalaries; dv above, av below with parasulcus at top.

T-V - Cingular archeopyles in *Nannoceratopsis*. T-U - Overall view and detail (rlatv) showing large opening resulting from loss of more than one cingular paraplate. V - Detail of anterior part of specimen (dv), showing archeopyle formed by middorsal cingular only.

W-X -Archeopyles in the hypocyst (aav).

Taxa represented by the diagrammatic figures:

A - *Broomea*
B - *Gochteodinia*
C - *Pareodinia*
D-F - *Rhaetogonyaulax testacea*
G - *R. rhaetica*
H - "*Lacrymodinium*"
J - *Gonyaulacysta*
K - *Occisucysta*
L - *Exiguisphaera*
M - *Lingulodinium*
N - *Isthmocystis*
O - *Microdinium*
P - *Florentinia ferox*
Q - *Wanaea*
R - *Homotryblium*
S - *Mancodinium*
T-U - *Nannoceratopsis pellucida*
V - *N. senex*
W - *Tuberculodinium rossignoliae*
X - *T. wallii*

would be Type $P_{3''}$ following Kofoid or Type P_4 in Taylor's system. The obvious similarities and contrasts between the two notations for the same opening illustrate the critical need to include the distinctive prime symbols when identifying paraplates in Kofoidian terms.

The Type P_4 opening tends to be approximately middorsal and the operculum is free (Frontispiece D). Rounding and enlargement or reduction are common. Enlargement may incorporate a portion of the paracingulum into the opening, a condition that is obvious if the archeopyle suture obliquely transects a line of surface ornament associated with the paracingulum (Figure 6.1 E-F).

As mentioned earlier in this chapter, the periarcheopyle associated with a cavate gonyaulacoid cyst in which the wall layers are separated in the region of the archeopyle appears to be non-operculate, whereas the endoarcheopyle is clearly operculate. Failure to recognize this situation has led to the mistaken impression that the archeopyle in such species as *Gonyaulacysta jurassica* is of the reduced type. The error lies in contrasting the position of the endoarcheopyle suture with the parasutural features that bound the opening in the periphragm. The endoarcheopyle is definitely smaller than the outer opening, but no parasutures are visible on the endophragm to suggest that the opening is reduced, and the opening in the periphragm extends quite to the bases of the bounding parasutural ridges.

Accessory ridges and other elevations on the operculum of some cysts with Type P archeopyles (Fig. 6.1 G-H) may suggest enlargement of the opening at the expense of portions of neighboring paraplates. These features have also been interpreted (Dörhöfer and Davies, 1980) as evidence for the evolutionary origin of paraplate **4** through fusion of what were several separate plates in ancestral species. The correct explanation apparently lies in a third direction: Gocht (1979) showed that these patterns of surface relief are almost certainly related to bands of incremental growth on thecal plates. For further discussion, see Chapter 9 under *Cribroperidinium*.

In more complex openings—Types $2P_{3\text{-}4}$, $2P_{4\text{-}5}$, $3P_{3\text{-}5}$, and $6P_{1i,2\text{-}6}$, the operculum is compound and the intersections of the principal and accessory sutures are marked by small but usually distinct angulations in the archeopyle margin. The opercular pieces are free.

Examples: Type P - *Gonyaulacysta*, *Pentadinium*, *Spiniferites*, *Leptodinium*. Type $2P_{3\text{-}4}$ - *Occisucysta*. Type $2P_{4\text{-}5}$ - *Exiguisphaera*. Types 3P, 4P, and 5P - *Lingulodinium*. Type 6P - *Dissiliodinium*, *Isthmocystis*.

Examples with operculate Type P endoarcheopyle and non-operculate periarcheopyle: *Gonyaulacysta jurassica*, *Tubotuberella apatela*, *Hystrichosphaeropsis* spp., *Stephodinium coronatum*.

5. Cingular archeopyles (Figure 6.10 T-V). Archeopyles that chiefly involve cingular paraplates have been observed only in the Jurassic genus, *Nannoceratopsis* (Piel and Evitt, 1980), in which both Types C_{3c} and $2C_{3\text{-}4c}$ occur. In this genus the third cingular paraplate is middorsal.

The term "cingular archeopyle" has been applied with a different connotation (Downie and Sarjeant, 1966) to those epicystal openings for which the archeopyle suture runs definitely (or arguably) within the limits of the paracingulum. In this usage, "cingular" refers to the position of the archeopyle suture, rather than to the paraplates (which are chiefly precingular and apical) that account for most of the opening. Since this use of "cingular" is inconsistent with the method used for naming other archeopyle types, and since archeopyles that are cingular in the sense of involving just cingular paraplates are now known, the connotation of "cingular archeopyle" is restricted here to openings of the latter type. (See comment under 8. Epicystal archeopyles.)

6. Archeopyles in the hypocyst (Figure 6.10 W-X). Openings in the hypocyst are rare, but are represented by archeopyles of Types 2HA and 3HA in modern and fossil cysts of *Pyrophacus* (= *Tuberculodinium*). The thecal tabulation in living species of this genus shows high intraspecific variability, with from three to seven plates in the antapical series. Although the full pattern of paraplates is difficult to determine in the cysts, archeopyles with three to five free opercular pieces have been observed (Wall and Dale, 1971). These authors suggest that the archeopyle in *Caligodinium*, a genus of fossils with incompletely known paratabulation, may also be formed at the antapex.

7. Combination archeopyles that are not epicystal. These involve apicals together with either intercalaries or precingulars.

Type (tA)P_4 archeopyles (Figure 6.10 P). Such openings are more or less keyhole-shaped, with an approximately equidimensional apical portion and a precingular extension that reaches to the paracingulum on the dorsal side. The opercular piece for either portion of the opening, or for both portions, may be adnate (ventrally along the parasulcus in the case of the apical piece and dorsally along the paracingulum in the case of the precingular piece).

Examples: Type (tA)aPa - *Sirmiodinium*. Type (tA)aP and Type (tA)P - *Florentinia*, some *Diphyes*.

Types (tAtl) and A5I. Biconical cysts that have been referred to *Rhaetogonyaulax* have many traits in common (including that the adcingular middorsal margin of the archeopyle is geniculate), but the openings are of at least two quite different kinds in different species (Figure 6.10 D-G). The simple operculum of the Type (tAtl) opening is a large conical structure, nearly as large as the rest of the cyst, whereas the Type A5I archeopyle is characterized by a small apical cone connected with the rest of the cyst by a narrow ventral isthmus of autophragm. An acutely angular indentation in the middorsal margin of this small cone apparently results from separation of a single apical opercular piece; the large intercalaries account for most of the opening.

A Type (tAtl) archeopyle of somewhat different kind occurs in some cysts referred to *Microdinium* (Figure 6.10 O) and similar genera. At first glance this may appear to be Type (tA), especially if only the opening is seen. However, parasutural markings on the operculum establish that up to five intercalary paraplates are present. The paraplates can be distinguished in only the most ideal specimens and the homologies of the smallest paraplates are debatable (see discussion of partiform gonyaulacoid patterns and Figure 5.18 in Chapter 5).

8. Epicystal archeopyles. Type E (Figure 6.10 Q). This is a combination archeopyle in which all paraplates of the epicyst remain associated in a simple operculum; it might also be designated Type (tAtP) or Type (tAtltP). The operculum may account for up to half of the surface area of the cyst and cover an opening whose diameter may equal that of the entire specimen. Alternatively, the paracingulum may be located so that the operculum is decidedly the smaller portion of the cyst wall. Intercalary paraplates, if present, are usually small and more or less "incidental" to the major plan of the paratabulation; they rarely approach being the large, symmetrical and conspicuous elements present in many peridinioid cysts (exception: *Luehndea* with two or three large and rather *Peridinium*-like intercalaries and a partiform hypocyst). The designation "E" emphasizes the unitary and chiefly epicystal composition and avoids the sometimes irresolvable issue of the identity of the incorporated paraplates.

The archeopyle suture runs close to the anterior limit of the paracingulum, but its exact position relative to this limit is not always clear. An archeopyle whose suture lies definitely within the paracingulum (e.g., in *Dinopterygium* and *Heteraulacacysta*) has been termed **hemicystal** (Bujak et al., 1981; a term equivalent to the original "cingular" archeopyle of Downie and Sarjeant, 1966), but this will be treated here as a special case of an epicystal opening. The expression of paratabulation on the operculum varies in kind and degree. At one extreme, paraplates are clearly indicated by parasutural ridges or by large intratabular processes; at the other, the opercula are smooth and virtually featureless. Frequently—perhaps in a majority of instances—the operculum is adnate at the parasulcus. However, the connection is so small relative to the size of the operculum in large spherical cysts with subequatorial archeopyle sutures that it is easily torn. Consequently, opercula of such openings are customarily found isolated from the rest of the cyst.

Examples: With hemispherical or subconical epicyst forming essentially half the area of the cyst surface - *Ctenidodinium*, *Dinopterygium*, *Heteraulacacysta*, *Luehndea*, *Mendicodinium*, *Tubidermodinium*, and *Wanaea*; with epicyst much smaller than the rest of the cyst - *Biorbifera*; a chorate cyst - *Callaiosphaeridium*.

Type (4A4I)7P (Figure 6.10 S). The epicystal archeopyle in *Mancodinium* involves eight free opercular pieces, seven of which represent the precingulars individually, while the eighth comprises four apicals and four intercalaries as a polyplacoid unit.

Type A(3A)6P (Figure 6.4 L, 6.10 R). This is an especially complex epicystal archeopyle that occurs in some cysts with quinqueform gonyaulacoid paratabulation. An expanded formula specifying the eight opercular pieces would be: Type $A_{1u}(3A_{A\text{-}c})$ $6P_{1i,2\text{-}6}$, or its Kofoidian equivalent: Type $A_{1'}(3A_{2\text{-}4'})$ $6P_{1\text{-}6''}$. A small but definite preapical area not recorded in the formula may also be visible in the polyplacoid opercular piece.

The principal archeopyle suture shows diagnostic ventral angulations related to the anterior sulcal paraplate (**ai**) and to the offset in the paracingulum. Observation of the distinctively shaped individual opercular pieces may indicate the presence of specimens with this archeopyle type, even before examples of the hypocyst have been recognized.

Examples: Living *Pyrodinium bahamense* (Wall and Dale, 1969) and fossils either known or thought to be closely related to it (*Homotryblium*, *Eocladopyxis*, *Polysphaeridium*).

In both of the two preceding types, separation of the wall along the principal archeopyle suture leaves a distinctive sulcal tab projecting from the edge of the opening on the main part of the cyst (Figure 6.4 L; cf. Figure 6.10 R). This small projection may be recognizable in specimens too poorly preserved to permit identification of other archeopyle details.

9. Miscellaneous archeopyle types. In many non-peridinioid dinoflagellates the exact nature of the archeopyle suture is not known. Small size of specimens, scarcity of individuals, delicacy of the wall material, poor visibility of the opening in preferentially oriented specimens, absence of independent indications of paratabulation, and unusual location or composition of the opening may all contribute to this situation. Many examples that might be listed here are relatively nondescript or have been described from poorly preserved material only. Among the more distinctive genera that may be cited as examples are: *Codoniella*, *Dapcodinium*, *Dinogymnium*, *Gillinia*, *Noricysta*, *Phanerodinium*, and *Suessia.*

Note on Alternative Archeopyle Formulae

Certain archeopyle types can be represented by more than one formula, although one of the alternatives may seem preferable to others. Take, for example, the archeopyle in *Rhaetogonyaulax testacea* (Figure 6.10 D-F), in which release of six free opercular pieces (five anterior intercalaries and one apical) nearly isolates a conical apical structure attached by a narrow connection to the parasulcus. This archeopyle can be characterized as either Type (6A)aA5I or Type A5I, depending upon whether the apical cone is considered an adnate opercular piece or is not included as a part of the archeopyle.

An example similar in principle but different in detail is the archeopyle in *Isthmocystis* (Figure 6.10 N). Here also release of six opercular pieces (in this case all precingulars) leaves an apical cone that is connected to the parasulcus, a situation expressed with equal accuracy as Type 6P or Type (tA)a6P.

Even three alternative interpretations are possible in several species of living freshwater *Peridinium*, whose archeopyle was discussed earlier in this chapter as Type (A3I3P)a. The obvious rupture in these cysts (Figure 6.5 O) is a transapical suture whose ends are not interconnected by a suture along the paracingulum. In two alternatives this archeopyle is viewed as an opening with an operculum adnate along its adcingular edge. These interpretations can be represented by formulae that are equally descriptive: Type (3A4P)a, in other words Type (3$A_{1',2',4'}$4$P_{1-2',6-7''}$)a, and Type (A3I3P)a, or Type ($A_{4'}$3I_{1-3a}3$P_{3-5''}$)a. The first of these represents the archeopyle as having a ventral position (a condition otherwise unknown), and seems less desirable than the second, which permits easier comparison with other types of archeopyles in peridinioid cysts where the dorsal paraplates play a dominant role. The second formula suggests at once the similarity to the Type ($A_{3'}$3I_{1-3a}3$P_{3-5''}$) archeopyle (Figure 6.8 K) and the slight but significant differences from the Type (3I_{1-3a}3$P_{3-5''}$)a archeopyle (Figure 6.8 I-J). The third possible interpretation of the opening is as a slitlike suture not involving an operculum. Although this concept cannot be represented by a formula that uses the symbols employed in this chapter, note that it is the only one of the three interpretations for which the term "transapical archeopyle" would be appropriate. The other two are combination archeopyles partly defined by a transapical suture.

TAXONOMIC SIGNIFICANCE

The archeopyle is clearly one of the typical features of most dinoflagellate cysts—fossil and modern. Its many differences in shape, size, position, and operculum construction—features which, in turn, are related to paratabulation and wall structure—combine with its frequently easy visibility as an obvious opening in the wall to give it high appeal as a discriminating taxonomic character. Thus, Evitt (1967) strongly recommended that species differing in archeopyle type should be assigned to different genera, and procedure in taxonomic works over a period of about a decade was largely consistent with this view. To be sure, Wall observed early on (1967) that several archeopyle types occur in cysts that he considered to represent a single species, *Lingulodinium machaerophorum* (the cyst now known to be that of the living thecate dinoflagellate known as *Gonyaulax polyedra*). Nevertheless, most generic diagnoses of fossil dinoflagellates written through 1978 tolerate only slight variation in archeopyle morphology, and the elaborate suprageneric classification of Norris (1978), which is based on earlier ones by Downie and Sarjeant (1966), Sarjeant and Downie (1974) and Sarjeant (1974), applies at the generic level a principle that might be termed "the taxonomic integrity of archeopyle type."

Today, however, that principle seems unacceptably simplistic as we see increasing evidence for archeopyle variability within genera and even species, judging these taxa on the basis of morphological features aside from the archeopyle (for sporopollenin cysts: Stover and Evitt, 1978; Goodman and Witmer, 1980; Islam, 1982; Goodman, 1983—for siliceous cysts: Dale, 1978).

In other words, the variability of the archeopyle itself now appears to be variable. More explicitly: within many genera and most species of dinoflagellates (as these taxa generally have been defined), archeopyle variability is sufficiently slight to be encompassed by what we have called a single archeopyle type. However, in quite a number of genera and a few species, archeopyle variability exceeds the limits of a single type. Obviously it would be possible to define all genera and species so that archeopyle type is invariable within each, but in a growing number of cases this would require crediting the archeopyle with considerably more importance than any other taxonomic criterion, while denying the validity of similarities in other features as indicators of taxonomic proximity.

Table 6.2 - SOME GENERA INCORPORATING MORE THAN ONE ARCHEOPYLE TYPE

Genera in the first column include species with the different archeopyles indicated by the formulae in the second column:

Genus	Archeopyles	References
Ceratiopsis	I/I, I/3I	pers. observ.
Chatangiella	I/I, I/3I	Lentin & Williams, 1976
Coronifera	P, 2P, (tA), (tA)P, (tAtP)	Norvik, 1976; May, 1980
Diphyes	P, 2P, (tA), (tA)P	Goodman & Witmer, 1980; Goodman, 1983
Exochosphaeridium	P, 2P	
Florentinia	P, (tAtP)	Davey & Verdier, 1973; Davey, 1976
Hebecysta	2I, 3I	Bujak & Fisher, 1976
Heibergella	I, 2I, 3I	Bujak & Fisher, 1976
Lingulodinium	P, 2P, 3P, 4P, 5P, (tAtP)	Wall, 1967
Luxadinium	3I3P, (3I3P)	Brideaux & McIntyre, 1975
Phthanoperidinium	I, 3I, IP, 3I3P	Edwards & Bebout, 1981; Islam, 1982
Rhaetogonyaulax	(6A5I)a, 5I, A5I	Fisher & van Helden, 1979
Spiniferites	P, 2P	Dobell & Taylor, 1981
Sverdrupiella	I, 2I, 3I	Bujak & Fisher, 1976

A few examples illustrating the range of current uncertainties about the taxonomic significance of the archeopyle at the generic level are given in Tables 6.2 and 6.3. Archeopyle diversity acceptable within a genus is indicated in Table 6.2. In contrast, Table 6.3 shows how archeopyle differences have been used at the generic level to separate otherwise similar species. Only a more comprehensive and fundamental revision of generic circumscriptions than any yet undertaken can be expected ultimately to alleviate the increasingly undesirable situation illustrated by these examples, but such a task seems unlikely to be undertaken in the near future.

Table 6.3 - EXAMPLES OF GENERIC SEPARATION ON ARCHEOPYLE TYPE

The following pairs of species, despite other similarities, have been referred to different genera almost solely on the basis of archeopyle differences which are no greater than those encompassed within single genera in Table 6.2:

Type	Species	Remarks
P	*Millioudodinium pallum*	—
2P	*Occisucysta evittii*	
P	*Tectatodinium pellitum*	—
2P	*Bitectatodinium tepikiense*	
(4A)	*Meiourogonyaulax bulloidea*	*L. jurassica* is the only species known with Type 4A
4A	*Lithodinia jurassica*	
Ea	*Ctenidodinium continuum*	Identical in respect to distinctive tabulation details (2pr, 2a)
P	*Leptodinium mirabile*	
(4A)	*Hystrichosphaeridium tubiferum*	—
P	*Kleithriasphaeridium loffrense*	
(tA)	*Hystrichokolpoma cinctum*	—
P	*Achilleodinium biformoides*	

Part III - Interpretation of the Cyst

After one or more readily visible features suggest that a specimen under study is a dinoflagellate cyst, microscopic analysis will usually proceed in predictable stages: a survey of basic morphology confirms the suspicion and indicates general identity; scrutiny of wall structure, archeopyle, and paratabulation provides critical information; finally, an identification is made, often with the aid of published or unpublished records. While the length of the process will vary with the experience of the microscopist and the complexity of the specimen, uneven progress will inevitably result from the unpredictable interplay of objective observation, subjective interpretation, and occasional insightful perception. A certain configuration of lines first suggests the identity of a feature which, in turn, implies the presence of certain other features in accord with some conceptual model. Periodically the growing concept of the whole is tested more comprehensively against similar models until the microscopist determines that he has seen enough to reach the conclusion toward which he has aimed—or, in less successful cases, until he decides that his goal is not attainable.

With increasing experience, each investigator tends to develop his subconscious mental pathways for integrating the observations, hypotheses, testing, and conclusions into an efficient and productive operation. However, regardless of experience and efficiency, the conclusions will be correct only if the observations have been adequate, the hypotheses based on sound principles, and the testing valid.

In Part II, our focus was on the individual features that provide the observational data on which the rest of the procedure of analysis depends. In Part III, we turn to the subsequent trio of steps: hypothesis, testing, and conclusion—in other words, interpretation.

Chapter 7 sets the stage with a discussion of certain principles of interpretation. Chapter 8 opens with the outline of an analytical procedure and a checklist of important morphological features, continues with a guide to 17 major morphological categories of cysts, and closes with a key for identifying these categories.

In Chapters 9-12, selected taxa representative of the 17 morphological categories are discussed. While the general objective is always to facilitate accurate interpretation of the morphology, the means selected to accomplish this varies from group to group. In one it may be a straightforward description of features and a guide to their recognition; in another, an exposé of a "theme-and-variations" relationship among several taxa; in still others, explorations of especially critical or potentially misleading features. In every case, however, the subjects treated should be taken merely as representative of the kinds of considerations that can be applied profitably to a wide assortment of other groups. In this way specific examples can have general value, and the limited information presented here can guide the study of many morphologically interesting and stratigraphically important examples too numerous to include, and many more that remain to be discovered.

The discussions under the 17 morphologic categories in Chapters 9-12 are organized by reference to informal "complexes" of genera. The name of a prominent genus is used to name each complex. The first part of each of these chapters includes a key to the complexes, followed by a more detailed statement of the morphological features associated with each complex, emphasizing those features useful for identification and pointing out problems commonly encountered in its pursuit. A list of representative genera and a note on stratigraphic occurrence are also given.

The comments on complexes have two objectives. First, to point out the general features of a large number of dinoflagellate cysts in such a way as to stress the features that should be carefully evaluated in descriptions or identification. Second, to indicate some of the prominent lacunae in our current knowledge and, thereby, to suggest opportunities for future research.

The second part of each chapter includes a discussion of short topics illustrative of the problems or potentials of morphological analysis within the category under consideration.

The taxa discussed in Chapters 9-12 are not placed in a formal suprageneric classification. The concern here is with comparative morphology, not affinities and formal taxonomy, and a "framework" more flexible than the usual systematic hierarchy seemed desirable. It is emphatically to be noted that the categories and complexes are not to be thought of as families and subfamilies. They are purely morphological groupings, created for organizational convenience. If some of them have scopes similar to those of formal taxonomic units in previously published classifications, this merely reflects the obvious reliance of any scheme for the classification of fossils on morphology, and the well-documented tendency for many closely related organisms to look alike.

The coverage in these chapters is not intended to be encyclopedic, either in the enumeration of complexes or the discussion of selected topics. Although the numerous examples given include many important or distinctive groups of cysts, other groups which some readers may rate as equally deserving of attention have been omitted. Certainly, the space devoted to R-cysts and to Triassic and Early Jurassic cysts in general is small in proportion to their merit, but this deficiency will have to be remedied when more is known about these intriguing taxa.

The 17 morphological categories compare very roughly with the lineages discussed by Wall and Dale (1968) and Williams (1977). Also, the complexes approximate in concept the sublineages or groups of Williams (1977), although many of them differ significantly in scope. However, even when the same name used by one of those authors is applied to a category or complex, it is not the intent to suggest precise equivalence. The scheme of categories and complexes presented here involves many arbitrary choices in which strictly "biological thinking" has played a minimally important role.

Chapter 13, which brings Part III to a close, addresses the much broader topics of classification at levels above the genus, and the general evolutionary patterns that may be suggested by fossil dinoflagellates. It is concerned chiefly with some of the ideas expressed in a suite of recent publications.

Chapter 7. Guides to Interpretation

When the fundamental question raised in the mind of a microscopist by a dinoflagellate cyst under the microscope is "What cyst is this?", then the microscopist faces the familiar problem of identification. This is usually attacked through a routine of observation, analysis, and comparison, and is considered solved when it becomes possible to agree upon what some might facetiously call the systematist's alias for knowledge, a specific name. Sequential operational questions during this routine are:

1. What are the visible characters of the cyst under study?
2. What is their relative importance?
3. How are they to be used in deciding the proper assignment to one genus and species?

Progress is normally from the relatively simple to the more difficult, from the relatively obvious to the more obscure, from the relatively objective to the more subjective.

In this chapter we will explore some principles of interpretation and reasoning that should facilitate progress from the initial examination of morphology to the final evaluation of similarity and difference. While all the matters will be discussed in a dinoflagellate context, some pertain peculiarly to dinoflagellates and others are more broadly biological or are derived from other disciplines, including simple common sense. This miscellany of ideas, which are really no more than "things to bear in mind" as one studies a dinoflagellate cyst, is presented under 15 headings, after which the chapter concludes with a section on interpreting the literature.

1. SELECTIVE REPRESENTATION

Principle: The dinoflagellate fossil record is a selective one dependent upon geologically preservable materials produced chiefly for a special structure associated with a resting stage in the life cycle in a limited number of taxa at certain geological times.

Consequence 1: The absence of dinoflagellate fossils from sediments deposited in an evidently suitable environment may mean that these organisms were not present near the site of deposition; alternatively, it may mean only that those species that were present, although possibly numerous and represented by abundant individuals, did not produce preservable materials.

Consequence 2: The date of the first and last appearances of a dinoflagellate species in the fossil record may accurately record the range of that species, but closely related species with no fossil record may have been present in undeterminable numbers before, during, or after that individual range.

Consequence 3: No matter how distinctive or otherwise significant a particular group of fossil dinoflagellates may be, its occurrence does not mean that lineages either ancestral to or descendent from it are necessarily represented by fossil cysts.

Consequence 4: Statistics involving counts of dinoflagellate cyst species (or individuals) are useless as the basis for reliable estimates of the numbers of dinoflagellate species (or individuals) that were living at the time and site represented by the sample.

Discussion: Refer to Chapter 2 for a discussion of this important topic and its ramifications.

2. SPHERICAL CONCEALMENT

Principle: In any single view of a strongly convex object, a significant fraction of the surface is concealed from reasonably direct inspection.

Discussion: See Figure 7.1 A-F. A reasonably direct view of the surface of a transparent sphere is limited to regions within about 60 degrees of the center of its circular outline. The peripheral zone not easily subject to study occupies less than one eighth the radius of the visible circle but includes half the surface area of the sphere. On an undeformed

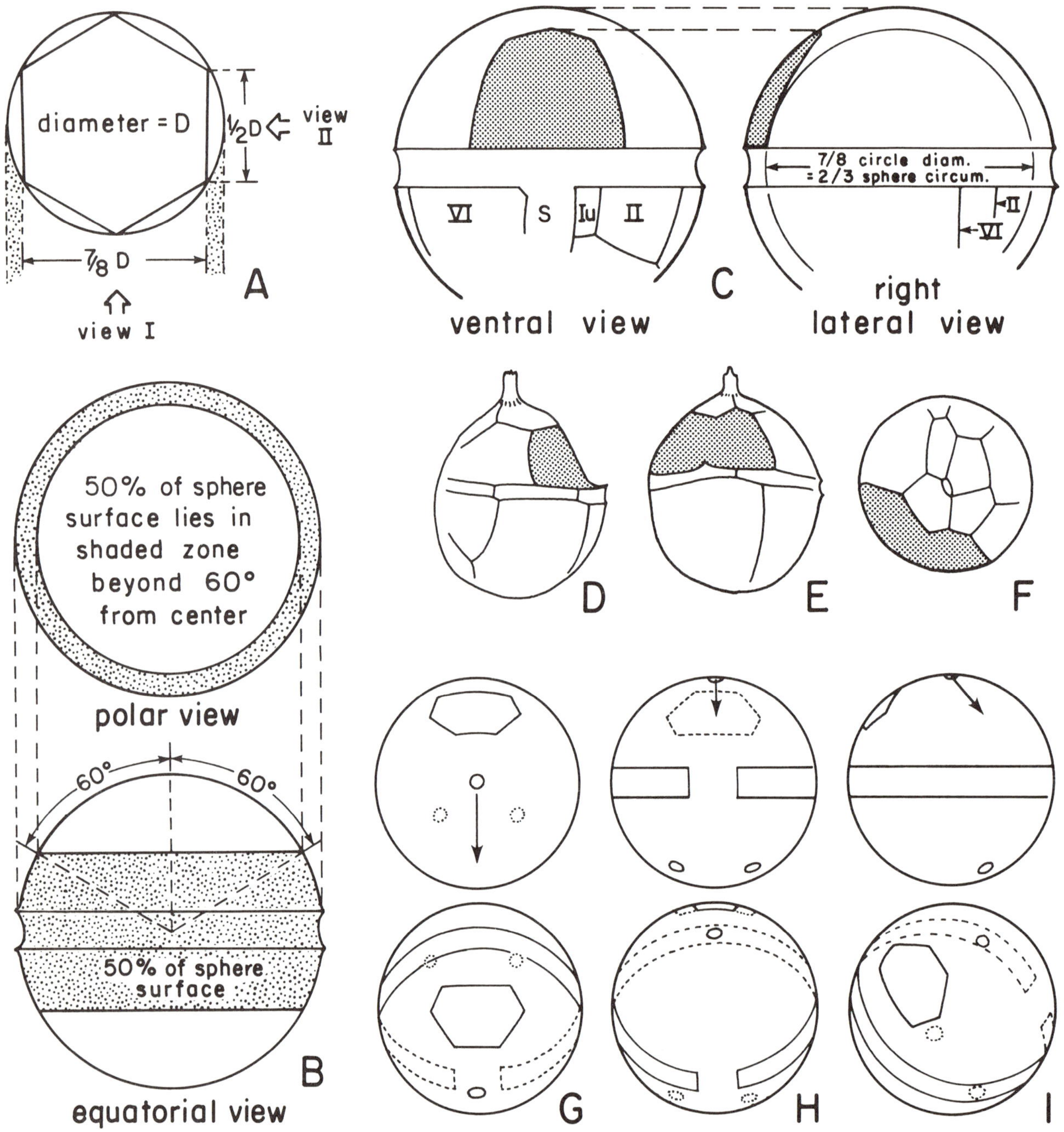

Figure 7.1 - EFFECTS OF SPHERICAL CONCEALMENT AND COORDINATED DISPLACEMENT

A-F - Effects of Spherical Concealment.

A - The circle represents the equatorial section of a spherical dinoflagellate. The inscribed regular hexagon divides the periphery into segments approximately comparable in extent to the largest pre- and postcingular plates or paraplates. The meaning of "spherical concealment" becomes apparent if one considers such a specimen as being viewed from directions I and II. In view I, two of the six segments are visible more or less directly at the equivalent of high focus, and two more, on the other side of the specimen, at the equivalent of low focus. The remaining two segments, accounting for one third the circumference of the section, are hidden in the two shaded regions, which account (together) for only an eighth of the transverse dimension of the image. In view II, two superposed segments (one at high and one at low focus), whose combined extent equals only one third of the specimen circumference, account for half the apparent specimen width, whereas the remaining two thirds of the circumference lie in the other half of the image width.

B - Polar view (above) and equatorial view (below) of a hypothetically spherical cyst, to show how a peripheral "shadow zone" bounded by a line 60 degrees from the center of the circular outline of the sphere includes half the sphere surface.

spherical cyst under the microscope, the dimensions of the obscured region are sufficient to conceal: a) about a third of the paracingulum in an equatorial view, b) the entire parasulcus and substantial portions of adjoining plates in a lateral view, or c) any single paraplate except for the largest (normally **2**, **III**, **IV**, and **V** in gonyaulacoids and *3‴* in peridinioids).

As applied to dinoflagellate cysts, the effect of spherical concealment is aggravated by other factors (in addition to the strongly oblique view of the surface), including optical interference from projecting ornament or internal wall structure, and low illumination owing to light absorption on the long light path obliquely through the wall material.

In the microscopic interpretation of dinoflagellate cysts, the difficulties associated with the effect of spherical concealment increase with decreasing cyst size. Three factors (in addition to the general problem of lost resolution at higher magnifications) are primarily responsible for this. First, wall strength and thickness do not decrease in proportion to size. Therefore, smaller cysts tend to retain initial shape, whereas larger ones tend to show more varied degrees of compression. Any compression (so long as it does not involve severe folding or rupture) tends to expose more of the surface to direct view. Second, the greater relative thickness of the wall in smaller cysts increases problems related to optical interference by the wall material. Third, initial sphericity is more frequently approximated by small cysts than by larger ones, which include many more varied shapes.

3. COORDINATED DISPLACEMENT

Principle: If a specimen is rotated, all its visible features will appear to be displaced in a coordinated way.

Discussion: See Figure 7.1 G-I. This is a statement of the obvious if one has in mind actually rotating a moveable specimen as it is being observed under the microscope. However, important implications of the statement are not always self-evident. For example, when comparing two immoveable specimens that are differently oriented, the "rotation" referred to amounts to the movement necessary to change the orientation of one specimen to match that of the other. The observed shifts in the apparent positions of all correctly identified features in the two specimens must be consistent with this movement.

Or, in the still more common case of studying a single fixed specimen, the principle has a high interpretive value if a procedure similar to the following is employed:

1) Locate some feature on the specimen that can be identified with confidence (or hypothesize its identity if only that is possible).

2) Imagine the position which that feature would occupy if the specimen were viewed from an ideal or "cardinal" direction; that is, in dorsal, ventral, apical, antapical, or direct lateral view.

3) Consider the rotational movement needed to displace the selected feature from its position in the cardinal view to its actual position in the observed specimen.

4) Consider the positions on the observed specimen into which other significant features would be brought by the same rotational displacement.

5) Look for these features in the predicted positions or, as will usually be more appropriate, check hypothesized identifications of these features against observed positions.

C - The effect of spherical concealment is shown by schematic ventral and right lateral views of a transparent spherical cyst. The lefthand figure combines a 60 degree-wide Type P archeopyle on the dorsal epicyst with a plausible distribution of paraplates on the ventral hypocyst. The righthand figure shows how the "shadow zone" that lies beyond 60 degrees from the center of the figure completely encloses the archeopyle and one third of the equatorial circumference along the paracingulum. The ventral paraplates of the hypocyst appear foreshortened at the lower right.

D-F - Outlines of a single specimen (*Occisucysta* sp., Early Cretaceous) in three orientations (llatv, dv, av). The considerable width of the Type 2P archeopyle, which is large and obvious in the dorsal and apical views, is effectively disguised in the lateral view.

G-I - Effects of Coordinated Displacement. An apical horn, two antapical horns, a paracingulum, and a Type I, hexa style intercalary archeopyle are shown on a stylized spherical dinoflagellate in different orientations. The upper figures show the features as they would appear to be located in three "cardinal" views: apical, ventral, and right lateral. The lower figures show them in various oblique orientations. The arrows in the upper figures indicate the apparent directions of displacement of the features that accompanies the change from the upper to the lower orientation.

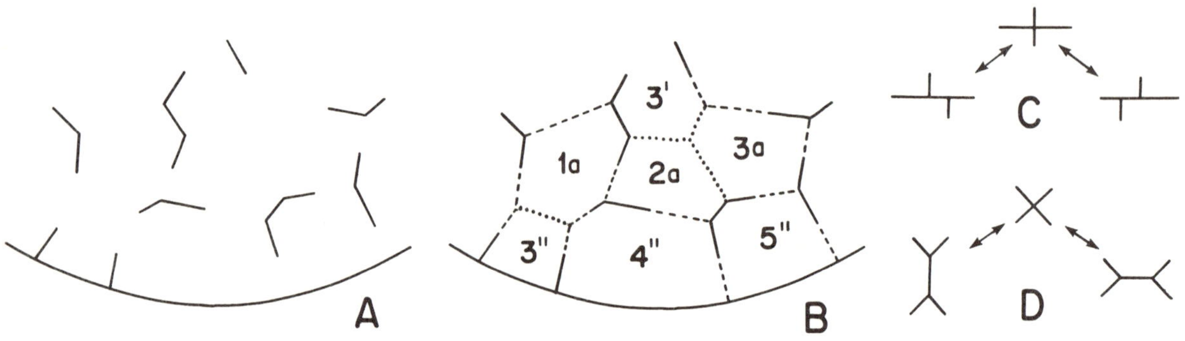

Figure 7.2 - TRIPLE JUNCTIONS AS AIDS TO INTERPRETATION

A - Hypothetical fraction of a cyst surface with traces of parasutural lines too incomplete for a pattern to be recognized; arcuate line represents edge of paracingulum. B - The first figure has been modified on the assumption that triple junctions prevail. Dashed lines, which are rectilinear extensions of visible lines, were drawn first. Dotted lines were then added to connect apparent triple junctions. At this point, the tentative interpretation, indicated by labeling as a dorsal epithecal peridinioid pattern, is ready for testing. C, D - Diagrams showing the relationship of two closely spaced triple junctions to a quadruple junction in configurations likely to be encountered along the cingulum (C) and elsewhere (D).

4. TRIPLE JUNCTIONS

Principle: The great majority of plates meet at triple junctions.

Discussion: Inspection of a variety of tabulation patterns will quickly show this to be true, particularly of the larger plates. In most cases, all three sutures radiating from a junction point are separated by obtuse angles, but angles of about 90, 90, and 180 degrees are typical along the cingulum and occur occasionally elsewhere. Acute angles occur rarely and usually in areas of locally strong surface curvature. A rectilinear crossing of sutures (i.e., a quadruple junction) at major plate contacts is unusual, but not unknown as once claimed (Barrows, 1918).

The value of this principle lies in the guidance it provides when analyzing paratabulation patterns in which parasutural markings are weak or incomplete (Figure 7.2 A-B). It enables one to predict, from the observation of two parasutures convergent toward an intersection, where another boundary between paraplates is likely to be found, or to test the reasonableness of a suspected but unconfirmed parasuture.

Variation of the principle: A quadruple junction can be viewed as the limiting case of two triple junctions that are increasingly closely spaced.

Discussion: See Figure 7.2 C-D. Quadruple junctions are most frequently encountered along the cingulum, where sutures between cingular plates may line up with those between plates in the adjacent areas. Even here close inspection will often show that two deceptively close triple junctions are the explanation of an apparent quadruple junction. A true quadruple junction is essentially an accident.

5. INTERDEPENDENT PROPERTIES

Principle: In a given tabulation pattern, a change in the number of plates contacted by any particular plate results in predictable changes in certain other plates.

Discussion: Those thecate dinoflagellates and their paratabulate cysts with which we are chiefly concerned can be viewed as solid geometrical bodies covered by polygonal fields (i.e., plates and paraplates). Simple geometric constraints determine that no single factor (e.g., the number of plates, or the contact relationship between two particular plates) can be changed without affecting one or more other factors. Familiarity with some of the geometric considerations involved offers both an avenue of insight and a potential measure of control in the interpretation of patterns.

If "change" in the preceding lines is construed as an observed difference between two dinoflagellates being compared, the principle then tells us that the tabulation patterns of two dinoflagellates cannot differ in *only the number of plates* or in *only the shape of a particular plate*. If one such difference is observed, certain other differences must also be present, whether they have been noticed or not.

A rigorous evaluation of observed and suspected differences in tabulation can be guided by this principle. It allows us to reason: If difference A is observed (or reported!), then differences B and C must also exist. Confirmation that differences B and C do, in fact, exist will support the accuracy of the first observation, but failure to confirm B and C may only leave the case unsettled. On the other hand, confirmation that differences B and C do not exist will utterly negate the accuracy of observation A and require a different interpretation, no matter how strongly general appearances may suggest contrariwise.

The potency of the principle can be illustrated best in terms of several examples phrased in quite specific terms.

Example 1: If four plates are associated at two triple junctions, a set of closely coordinated relationships exists between several factors: the relative sizes of the plates, the position of the triple junctions, the orientation of the suture between them, the identity of the plates that make contact along that suture, and the number of contacts made by each plate with its neighbors.

Discussion: Consideration of Figure 7.3 A-C will show that no one of the cited factors can change alone, and changes in some of them will involve a change in all the others, altering the visual effect of the pattern to a degree that may seem out of proportion to the magnitude of the change. The differences among the ventral tabulation styles of *Peridinium* known as ortho, meta, and para, and among the dorsal tabulation styles known as quadra, penta, and hexa, illustrate the point (see Figures 5.2 and 7.3 D-I).

If attention is focused solely on the number of contacts made by each of the four plates with its neighbors, then the relationship determines that a change in the number of sides on any one plate has the particular consequences stated in Example 2.

Example 2: If only triple junctions are present and the number of plates is constant, a change in the number of plates contacted by a particular plate requires a specific change in contacts made by three other plates, as follows:

An observed gain of a contact by one plate means the loss of a contact by each of the plates on either side of the new contact line, and a gain by the plate opposite.

An observed loss of a contact by one plate means the gain of a contact by each of the plates on either side of the lost contact line, and a loss by the plate opposite.

Discussion: Figure 7.3 J-K show a portion of an unusually variable paratabulation encountered in cysts of a few Late Cretaceous species. The variability can be viewed in terms of the contacts among four paraplates associated at two triple junctions. Figure 7.3 D-F also illustrate the systematic interdependence that exists in the number of plate contacts in a particular pattern. Note that the total number of contacts is constant despite the different numbers of contacts made by particular plates. These three diagrams also suggest how a simple count of the number of sides on a paraplate may serve as a useful control on the interpretation of a pattern: in this example of a peridinioid dorsal epicyst, if it can be established that *1a* or *3a* has six contacts, or that *3″* or *5″* has four, one can be sure that the pattern is not hexa, even without seeing the outline of *2a*.

Example 3: If a new plate is inserted in a pattern of polygonal fields in which only triple junctions are permitted, the number of plate contacts made by three (or, in one case, only two) other plates is changed to a degree that depends upon the number of plate contacts made by the new plate.

Discussion: Figure 7.4 A-F illustrate the example diagrammatically by showing the gain and loss in numbers of contacts which fall in the range of values likely to be encountered among dinoflagellates. These can be summarized as follows: If the "new" plate contacts-

-3 plates, then each of the three adjoining plates gains one contact.

-4 plates, then those two of the four plates which did not contact one another before the insertion gain one contact each. (This is the only case in which the number of plates whose contacts are affected is other than three.)

-5 or more plates, then the end members of a sequence of three plates gain one contact each, and the intermediate plate loses contacts equal in number to the number of sides on the "new" plate, minus four.

It is immaterial whether the "new" plate is perceived as being inserted among existing plates, or as originating by subdivision of an existing plate; and, in the case of subdivision, it is immaterial which portion of the subdivided plate is considered "new".

Figure 7.4 G-H illustrate the dorsal epicyst in two similar species from the Early Jurassic whose paratabulation differs in the number of anterior

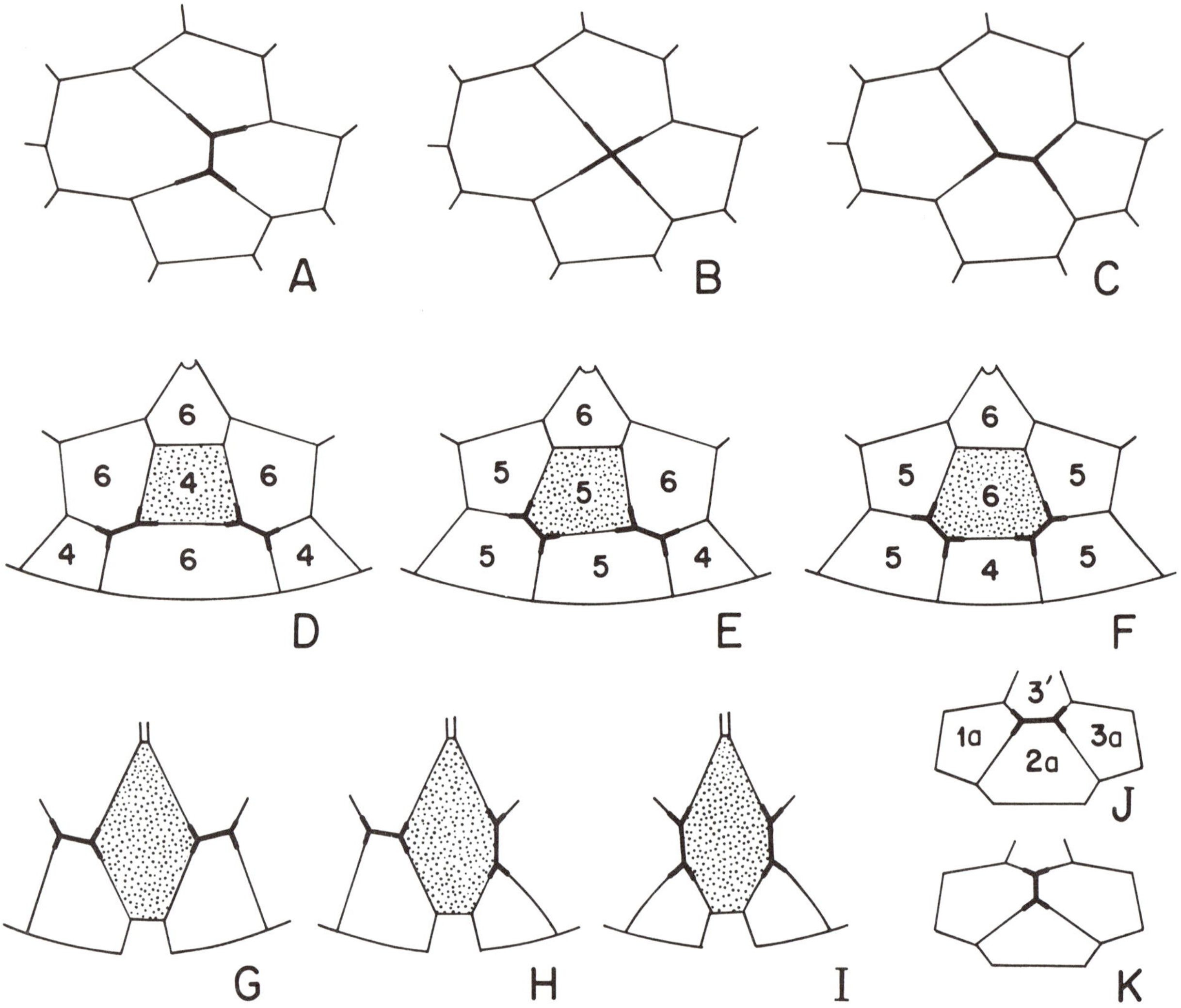

Figure 7.3 - INTERDEPENDENT PROPERTIES: EXAMPLES 1 AND 2

(See discussion in text)

A-C - Example 1. Altering the plate pattern from that shown in A, through intermediate stage B, to C involves a set of closely coordinated changes. These include changes in the relative sizes of plates, the position of triple junctions, the orientation of the suture between triple junctions, the identity of the plates in contact along that suture, and the number of contacts made by each plate with its neighbors.

D-I - Diagrams of the standard variations in dorsal and ventral epithecal patterns in *Protoperidinium* (D-F: quadra, penta, hexa; G-I: ortho, meta, para). The "critical" plates, *2a* and *1′* are patterned. Heavy sutures emphasize the contrasting plate contacts, in each case involving four plates at two triple junctions. Numerals on the plates in D-F record the contacts with surrounding plates; in all three cases, the seven plates outlined make a total of 36 contacts.

J-K - Example 2. Portions of two specimens of *Trithyrodinium striatum* (Late Cretaceous) showing the three opercular pieces of a Type 3I archeopyle. J - The middle intercalary piece has the 6-sided outline typical of the hexa style of peridinioid paratabulation. K - In this common variant, *2a* is distinctively 5-sided. The two triple junctions at the top of these figures which are involved in the contrast are emphasized by the heavier parasutures. Comparing K with J, plate *2a* can be said to have lost its contact with *3′*. As noted in the discussion of triple junctions, this "change" involves a difference in the position of the two triple junctions and in the orientation of the parasuture between them. As the figures show, loss of one contact by *2a* also involves a gain of one contact by the plate on each side (*1a* and *3a*), and a loss of one contact by the opposite plate (*3′*).

intercalaries. The differences inevitably associated with this difference are analyzed in the figure caption. Figure 7.5 I-J relate to differences in certain gonyaulacoid patterns associated with the presence or absence of an incidental paraplate (**K**) in the epicyst.

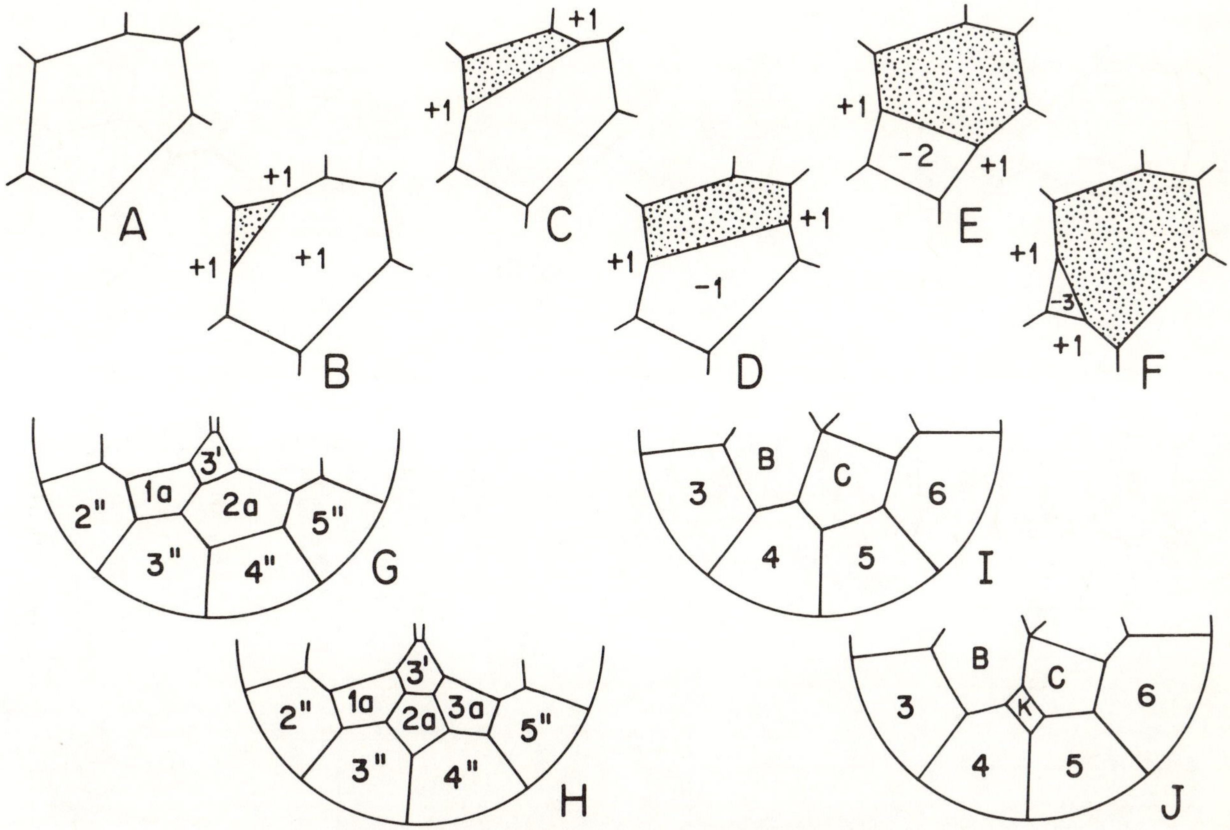

Figure 7.4 - INTERDEPENDENT PROPERTIES: EXAMPLE 3
(See discussion in text)

A-F - In a portion of a tabulation pattern in which seven plates meet at triple junctions (A), a new plate (stippled) is inserted so that it contacts an increasing number of original plates. Changes in the number of plate contacts are labeled on the affected plates.

G-H - Schematic diagrams of portions of the epicystal paratabulation in *Luehndea spinosa* (H) and an undescribed species of the same genus (G), both from the Early Jurassic. The presence of an "additional" paraplate in *L. spinosa* is immediately apparent, although it is open to choice whether *2a* or *3a* is considered "new". Less obvious—certainly when examining a small and spiny specimen under the microscope—but no less important, is the gain in one plate contact by *3′* and by *4″* (i.e., the end members of the trio of other affected paraplates) and the loss of one or two sides by the intermediate member, depending upon whether paraplate *2a* or *3a* of *L. spinosa* is considered to have been the "added" one. If the "new" plate is taken to be the five-sided *2a*, then it follows that there are five sides on *3a*: 5 = 6 - (5 - 4). If this five-sided *3a* is taken to be "new", then the number of sides on the "residual" *2a* is also predicted: 6 - (5 - 4) = 5.

I-J - Schematic diagrams of the dorsal epicyst in generalized gonyaulacoid cysts of two designs, differing by the presence in J of an incidental paraplate, which contacts four other paraplates. The addition of **K** requires that two non-touching paraplates (**B** and **5**)—and no others!—have an additional side in J as compared with I.

6. UNCOMMON CONCAVITY

Principle: The sides of most major plates normally appear straight or nearly so.

Discussion: The single important exception is in conjunction with the posterior end of the sulcus (Figure 7.5 A-B), where two plates (*1′′′′* and *2′′′′*) in the peridinioid pattern and one (**Y**) in the sexiform gonyaulacoid pattern typically have a concave side. A concave margin on a major plate, therefore, immediately suggests the antapical end of the sulcus. The concavity in the outline of **Y** in the gonyaulacoid pattern is commonly the more conspicuous because the end of the excavated sulcus meets the convex surface of the hypotheca entirely along this plate, whereas the corresponding arcuate intersection in the peridinioid pattern is divided between two antapicals. Other concave plate margins may occur on small plates bounding the flagellar pore (flagellar scar), and on small plates (preapicals) or the short edges of larger plates (apicals) at the apex. The gentle concavities in the

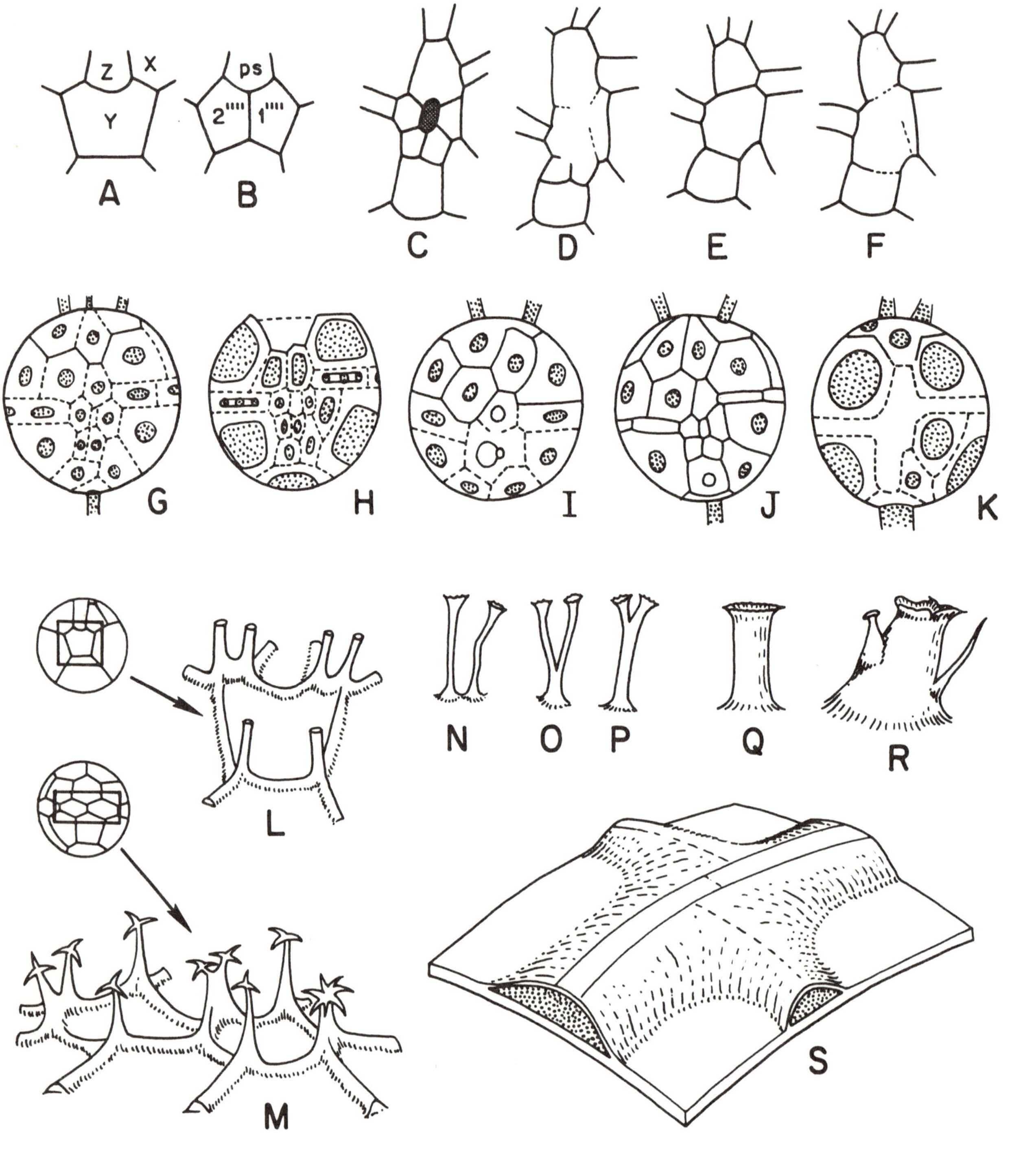

Figure 7.5 - EXAMPLES OF UNCOMMON CONCAVITY AND MINORITY SUPPRESSION

A-B - Concave margins of antapical plates in sexiform gonyaulacoid pattern (A) and peridinioid pattern (B).

C-F - Suppression of parasutures among small sulcal paraplates. Complete parasulcal pattern in Late Jurassic *Leptodinium mirabile* (C) serves for orientation. D-F - Selective suppression in three Eocene species of *Impagidinium* (D - *I. californiense*, E - *I.* sp., F - *I. maculatum*).

G-K - Diagrams showing differential expression of intratabular processes in gonyaulacoid chorate cysts (vv). Processes consistently present are stippled; those variably present or absent are shown by open circles. Parasutures shown by solid lines indicate archeopyle sutures (or, in J, parasutural markings on one species of the genus). Process representation on the dorsal surface (not shown) is consistent with that on the ventral surface. Apical and antapical processes whose bases are not clearly visible are shown by silhouettes. G - *Hystrichosphaeridium tubiferum* (Late Cretaceous-Early Tertiary). Each paraplate is represented by a process sides of plates that form large horns are seldom conspicuous.

A second exception is presented by the *2a/4″* parasuture in peridinioids with quadra-style arrangement on the dorsal epicyst. Since this organization is rare among fossil cysts, the concavity in the outline of *4″* is unlikely to be observed frequently (Figure 9.3).

7. MINORITY SUPPRESSION

Principle: The clarity with which paratabular features (other than the archeopyle) are expressed is roughly proportional to the relative sizes of the paraplates involved.

Discussion: To a certain extent, it is obvious that features associated with larger paraplates will tend to be more readily visible and more fully separated from one another. However, a distinct tendency toward both the suppression and greater variability of paratabular features associated with small paraplates is also apparent.

In many gonyaulacoid cysts with parasutural markings, boundaries within the sulcal region are often poorly expressed (Figure 7.5 C-F). When only some parasutures in this area are visible, they are most likely to be those that separate the larger paraplates (**ai** and **Z**) from the midsulcal area. In cysts with intratabular features (Figure 7.5 G-K), the smaller ventral paraplates and also the cingulars may be represented by smaller and less elaborate processes, by processes more variable in form and position (e.g., fusion between varied combinations of sulcal processes), or by progressive absence of processes, as is shown in Table 7.1.

Table 7.1 - PARAPLATES NOT REPRESENTED BY PROCESSES IN SELECTED SKOLOCHORATE CYSTS

Hystrichosphaeridium tubiferum and *Hystrichokolpoma* spp.	none (i.e., all paraplates represented)
some *Homotryblium* spp.	**li**, **lm**, **lu**, **fu**, **ai**
most *Oligosphaeridium* spp.	**li**, **lm**, **lu**, **fu**, **au-fi**, **ai** (also **Z** in a few species)
Litosphaeridium siphoniphorum	**li**, **lm**, **lu**, **fu**, **au-fi**, **ai**, **Z**

Only in special circumstances is a tendency contrary to this progression (in other words, toward the suppression of processes representing the larger paraplates) evident, but this occurs in a variety of taxa in association with pronounced dorsoventral compression. In some species of *Areosphaeridium* (Figure 3.3) and *Areoligera*, for example, processes representing some of the larger paraplates and occupying the central parts of dorsal and ventral surfaces (**4**, **1i**, **IV**, and **VI**) are reduced or absent, whereas processes representing such relatively small paraplates as the lateral cingulars may be present.

Variation: Closely spaced features on the cyst tend to lose their individuality.

Discussion: This could be taken as accurately describing the progressive loss of features related to the small ventral paraplates just discussed, but it has more specific applications.

One example involves the gonal processes in *Spiniferites* (Figure 7.5 L-M). Where these stand close together owing to the shortness of the parasutures between them, the processes tend to fuse, but where they are more widely separated, they remain as independent processes. Pairs of processes at the ends of the short parasutures between cingular paraplates are typically fused at their bases, part

(total = 30) in this quinqueform species. H - *Hystrichokolpoma* spp. (Early Tertiary). All paraplates are represented by processes, but these are of variable size and form: diminutive in the parasulcus, commonly clustered on the cingular paraplates. I - *Homotryblium* spp. (Early Tertiary). Interspecific variables include the presence of the **ai** process and the presence and form of an apparently compound process in the mid-sulcal area (see figures Q-R). J - *Oligosphaeridium* spp. (Cretaceous-Tertiary). Pattern of parasutures (directly evident only in *O. abaculum*) confirms absence of all cingular processes and most sulcals (**Z** = *ps* is present in most species). K - *Litosphaeridium siphoniphorum* (mid-Cretaceous) and closely similar species. Presence of a fourth apical process is variable. No sulcal processes (**ai**, **fu**, **li**, **lm**, **lu**, **Z**) or cingular processes present.

L-P - Fusion of bases of closely spaced processes. L-M - Variable fusion of closely spaced gonal processes in *Spiniferites* (schematic) at ventral corners of antapical paraplate (L) and along paracingulum (M). Orientation diagrams at left; distal parts of processes omitted in L. N-P - Variable fusion of processes **li** and **lm** in *Hystrichosphaeridium tubiferum*.

Q-R - A simple process representing one of the principal paraplates in *Homotryblium* sp. (Q), compared to a larger and irregular process in some species that may be the collective representation of several paraplates (**li-lu** and possibly **fu**).

S - Schematic illustration of wall structure in *Pentadinium*. Generally, a narrow pericoel underlies each parasuture. However, where parasutures are relatively close together along the paracingulum, two pericoels are not separately resolved and there is only a single broader cavity. (Layering not indicated where pericoel is absent.)

way up the shafts, or quite to the tops of the shafts. Contrasting with these are the discrete single processes that stand along the paracingulum at intersections with parasutures between the major pre- and postcingular paraplates. Similarly, processes at the following triple junctions tend to fuse at their bases while remaining distinct at their tips: **X/Y/Z** with **X/III/Y**, also **V/VI/Y** with **VI/Y/Z**. In the epicyst a compound process is apt to mark the closely spaced triple junctions near the apical end of 6″, and an elaborate apical process may result from the fusion of a cluster of processes at the angles between apicals and preapicals.

Other examples are provided by fused or otherwise atypical processes in the parasulcus. In *Hystrichosphaeridium tubiferum* (Figure 7.5 N-O), in which a full complement of processes is normally present, the two minute and closely spaced processes representing **li** and **lm** (i.e., *rs* and *ls*) can be found in varying states of fusion from base to tip. A peculiarly irregular process on the ventral hypocyst of some *Homotryblium* species (Figure 7.5 Q-R) appears to represent several small sulcal paraplates (**fu**, **li**, **lm**, **lu**), although **ai** and **Z** are individually represented.

Finally, the same tendency finds a different expression in the typically suturocavate species of *Pentadinium* (Figure 7.5 S). Where the parasutures in such forms are widely separated, as between the major plates of the pre- or postcingular series, each is marked by its own narrow pericoel. However, where parasutures are close together, they tend to merge. The most prominent effect of this appears along the paracingulum, which is underlain by a single broad, equatorial "tunnel," rather than by one pericoel beneath each of its margins (Edwards, 1982).

8. CENTRIPETAL PROJECTION

Principle: Corresponding areas on two different layers of a cyst wall (or on a theca and the wall of an enclosed cyst) compare in size and shape about as they would if they were projected from one layer to the other along lines to the center of the cyst.

Discussion: See Figure 7.6 and also Figure 6.4 I-J. If the surfaces (i.e., walls, layers) under consideration are closely spaced, corresponding features on them will have quite similar outlines. But if parts of the surfaces are substantially separated and of different configuration (as in the sphere-inside-a-cone situation approximated by many peridinioid cysts), then a certain contrast in the shapes and sizes of corresponding features will be inevitable and will depend upon the position of the features on the two surfaces. The contrast in appearance occasionally is sufficient to disguise the fact that separated endo- and periopercula represent a single species.

9. INTERNAL CONSISTENCY

Principle: If the paratabulation pattern hypothesized for a particular cyst species is to be considered reasonable, observed relationships among all features visible on the cyst must be compatible with the requirements of the hypothesized pattern.

Discussion: This fundamentally important and almost self-evident principle is easily overlooked when interpretation must rest on isolated and variably preserved specimens, no one of which reveals details of more than a portion of the surface. In the course of interpreting a cyst, each successive observation will suggest the possibility of certain relationships and preclude others. Each of these must be explored and tested until the implications of the hypothesized pattern are consistent with all the confirmed features of the cyst. For example, tentative impressions of orientation, location of visible features, identification of processes and paraplates, and recognition of archeopyle type should always be checked *against one another* and the implications of whatever other evidence regarding paratabulation may be available. While no interpretation with internal inconsistencies can be accepted as satisfactory, a consistent relationship among observed features may both justify and require acceptance of even a startlingly unexpected interpretation.

Corollary: If equivocal evidence permits multiple interpretations, choice among them should be guided by likelihood of occurrence.

Discussion: Any interpretation is subjective. Ideally, all items of evidence are mutually consistent with a single interpretation, but if evidence is conflicting, the best interpretation will result from the most astute weighting of the evidence. For example, in the case of a chorate cyst with intratabular processes and an apical archeopyle, there are likely to be three items of evidence for the number of apical and precingular paraplates: the processes, the outline of the archeopyle, and the outline of isolated opercula. Although the implications of all

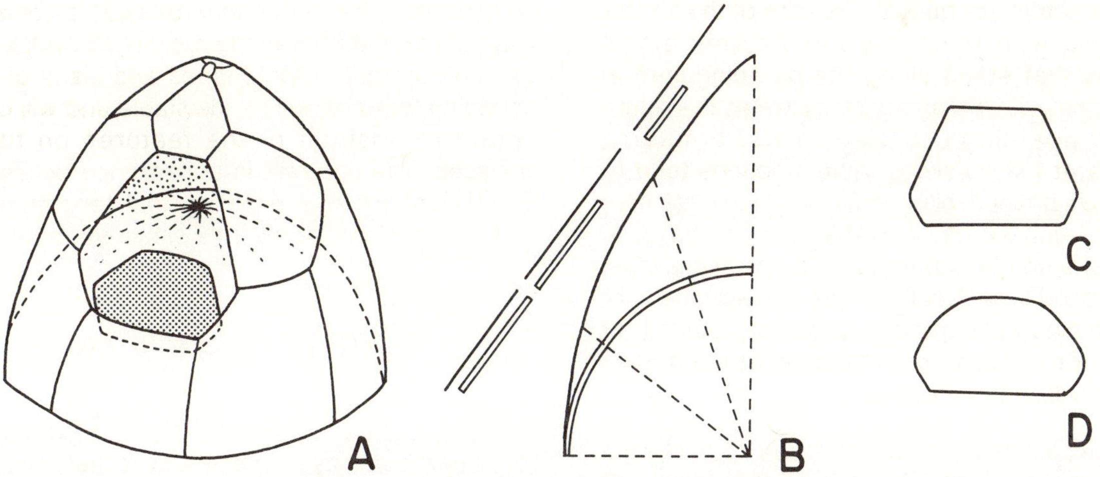

Figure 7.6 - CENTRIPETAL PROJECTION AND ITS EFFECTS

A-B - Perspective diagram and sagittal section, which can be taken to represent either part of a peridinioid theca (oblique dv) and an enclosed cyst, or outer and inner cyst walls. Apical, anterior intercalary, and precingular areas are shown on the outer surface and the medial intercalary area as "projected" onto the inner surface, to suggest the different sizes and shapes of corresponding areas. Straight line segments at left in B compare meridional dimensions along the outer surface (single line) and inner surface (double line) between intersects with the radial projection lines (dashed).

C-D - Contrasting outlines of perioperculum and endoperculum in Early Tertiary *Deflandrea* cf. *D. phosphoritica*, showing the effect of centripetal projection (and also rounding of the archeopyle).

three should be consistent, in case of conflict or uncertainty, the outline of the operculum should be the most reliable. This is because orientation of the operculum is likely to favor an unambiguous view of the course of the archeopyle suture; whereas bent, broken, or suppressed processes may reduce process count, while folds, tears, or oblique views along the edge of the archeopyle can easily lead to an inaccurate count of paraplates there.

10. SIMPLEST SOLUTIONS AND THE PARABLE OF "ONEIROHIPPUS PENTAPUS"

Principle: Normally one should favor the explanation that involves the least departure from conditions already known to exist.

Discussion: This is Occam's Razor applied to the interpretation of paratabulation in dinoflagellate cysts. Although new observations occasionally will require radical explanations, these should be advanced only after all less radical ones that might suffice have been considered and shown to be inadequate.

Variation 1: In analyzing a new situation, one should give special attention to factors most like those known in more familiar situations.

Discussion: Circumstantial evidence is strong that the dinoflagellate hypotheca is evolutionarily more conservative than the epitheca, and that the dorsal thecal surface is more conservative than the ventral area in and around the sulcus. The organizational patterns in the more conservative areas provide a more reasonable framework for analyzing an unfamiliar pattern than more strikingly distinctive features of other areas.

For example, the interpretation of the dorsal tabulation patterns in *Gonyaulax* and *Ceratium* followed here (Chapter 5) implies that the first postcingular of *Ceratium* is homologous with the second postcingular of *Gonyaulax*. Although this conclusion requires hypothesizing the subdivision of a higher-numbered postcingular to account for the total of six postcingulars in *Ceratium*, that explanation seems preferable to considering the first postcingulars in the two genera to be homologous. The latter explanation would require severe contrasts between the patterns of the two genera in the more conservative areas. That is, the dorsal postcingulars in *Ceratium* would be the second through fourth, and they would have grossly different relationships to the dorsal precingulars than do the corresponding postcingular plates in *Gonyaulax*.

Variation 2: Despite the high probability that new paratabulation patterns will continue to be discovered, the apparent uniqueness of a prospectively new pattern should be tested rigorously before it is accepted.

Discussion: Patterns indisputably different from those previously known among living dinoflagellates have been found in the fossil record, and, no doubt, others will be discovered. However, it is already clear that a few patterns are predominant, whereas the rest are relatively uncommon. This makes meticulous interpretation and documentation of prospectively new patterns essential. When analyzing such patterns, we are challenged to avoid forceably molding perceptions to fit our admittedly incomplete experience, while simultaneously taking full advantage of the fundamental lessons that that experience has already taught us. Emphasis of this important point is the objective of the following fanciful tale of "*Oneirohippus pentapus*":

Parable: A bird-watching biologist was driving one day through a countryside of pasturelands. Passing a field dotted with grazing animals, his attention was caught by the singular appearance of a horse at the far edge of the field (Figure 7.7). The immediate impression given by this animal was so absurd that it had nearly traversed the observer's mind without registering, and his first conscious reaction was also to dismiss the thought and proceed on his way. But, some trivial movement by the horse fixed his glance for a further critical instant and he realized that the animal *really did appear to have five legs*.

How should a biologist react to this perception? Should he believe the message of his eyes without further concern and proceed to his birding objective without seeking another explanation? Of course not. And a likely and immeasurably more acceptable alternative was obvious: some fortuitously positioned stump or post stood in the grass just beyond the horse, suggesting a fifth leg.

Still, the impression was so extraordinary that the biologist stopped his car in order to confirm the suspected explanation with his binoculars. Since what he saw with this examination was distinctly more leglike than postlike, he now got out of the car and walked a distance along the road, so that his changing viewpoint might bring the leglike object into full view and reveal its identity. Astonishingly, however, under this test and further scrutiny with binoculars, the would-be leg was not stripped of its disguise but appeared still more convincingly a part of the horse. In fact, within the limits of observation possible at that distance, there could be no doubt that the horse truly *appeared* to have five legs.

What should he do next? Should he now accept the judgment of his senses, return to his car and drive off, convinced of the existence of a five-legged horse? Or should he investigate the situation further, considering the magnitude of the conflict between the appearance of this animal and common knowledge of horses and related vertebrates? Certainly, the latter!

So, he climbed the fence and walked toward the distant animal. The closer he approached, the more convincing the "fifth leg" appeared. At 100 feet the anomalous limb still was not resolved into a recognizable feature apart from the horse. But a partial circuit of the animal at this range would surely settle the matter. It did, more or less, but in a startling way. The limb was definitely attached to the horse! Obviously closer approach was necessary. Finally, an inspection from a distance of 10 feet and a further circuit of the placid animal (which, by now, had even walked a few feet in the direction of taller grass) made it completely clear that the fifth leg was not only attached but apparently functioned in effective coordination with the normal four.

Now there was no longer room for doubt: the animal was remarkable, indeed! And, what was a reasonable explanation? Quite evidently a freak, appropriate for a circus sideshow and worth careful anatomical study by someone but, surely, no longer justifying the distraction of a birder from his avian objectives. However, not even complete conviction of the reality of this equine pentapod was sufficient to engender a conscious thought that standing there was the type specimen of a new taxon! That would require quite another major enterprise—a successful breeding stable of the animals.

Comment: Certainly, the perspective in which we view paratabulation patterns among dinoflagellate cysts differs in many respects from that in which we view such a familiar animal as a horse. In the world of terrestrial vertebrates, our experience is extensive enough to justify complete confidence that viable species of five-legged horses do not exist and never have existed. In the world of fossil dinoflagellate cysts, however, our sampling of the total number of species is vastly more limited. Accordingly, we can expect to discover a reasonable number of patterns as distinctive as a five-legged horse when measured against our previous experience; that is, species in which features of

Figure 7.7 - THE CHALLENGE OF SIMPLEST SOLUTIONS

"*Oneirohippus pentapus*" seen from a distance.

paratabulation depart significantly from any previously known example. Nevertheless, apparent novelties should invite special care.

The situation with fossil dinoflagellates is aggravated by the ease with which incidental or accidental features (such as folds, tears, crystal molds, debris, optical artifacts) on imperfectly preserved specimens of previously unknown species can be mistaken for significant morphological features. For example, folds or lines of ornament mistaken for parasutures can easily give impressions of paraplates in unusual number or position (witness *Cribroperidinium*, for instance). On the other hand, a variety of deficiencies (such as faint parasutures, or paraplates at the margin of a specimen hidden by "spherical concealment" or otherwise obscured) can as easily lead to an abnormally low paraplate count.

In summary, one should strive for an openmindedness unprejudiced in principle against five-legged horses, but accompanied by a skepticism and critical judgment that insist upon acutely testing the legitimacy of each serious prospect. Such openmindedness is not to be confused with an empty-headedness that would accept any suggestively positioned post or shadow as a fifth leg without check or reservation. Only when all reasonable efforts fail to account for apparently unusual features by usual means, when exhaustive checks confirm internal consistency of all available evidence, when multiple specimens are reasonably similar, and *still* the fifth leg remains, should we conclude—and then it will be with Holmesian logic and confidence—that the most improbable is real. In the words of an old saying on a similarly equine theme, "When you hear hoofbeats, think of horses, not zebras," —and one might add: "and, if horses must be excluded, think of zebras before unicorns."

11. DISTINCTIVE DEFORMATION

Principle: The location, character, and extent of postdepositional compression or other deformation may suggest or confirm important properties of undistorted specimens, even when these properties cannot be observed directly.

Discussion: Some *postmortem* deformational features (folds, tears, etc.) are random and truly accidental, but many are related to important aspects of the original specimens. Probably the most obvious and best known example is the selective orientation of specimens in a strew mount, reflecting original cyst shape as accentuated by sediment compaction following original preferred orientation on the depositional surface. In contrast, random orientation of specimens in the same preparations can be interpreted to mean essentially spherical original shapes. Some other examples:

1) Tears in the archeopyle margin may indicate the presence of less conspicuous accessory archeopyle sutures.

2) Consistent patterns of folds, wrinkles, or tears may indicate the location of local changes in the convexity or thickness of the wall.

3) Presence of opercular pieces inside the cyst cavity implies burial of cysts before excystment (contrary to earlier suggestions), for the "implosion" seems a reasonable consequence of wall collapse following disintegration of cell contacts.

12. UNCERTAIN SIGNIFICANCE

Principle: The taxonomic significance of a morphological feature is a) independent of the ease with which it can be observed, and b) subjectively determined, easily modified, and never provable.

Discussion: a) In terms of impact on the human perception, the features of organisms might be categorized crudely as ranging from subtle to obvious, and in terms of signficance for indicating biological affinitiy, as ranging from trivial to fundamental. Relatively few features of a given organism will be both obvious and fundamentally significant in these terms, especially since the most obvious features are often those more expressive of adaptation to particular habits and habitats than of biologic origin. However, it is the more obvious features that inevitably form the initial basis for classification. As in other human activities, only experience can impart an ability to perceive the significance of the more subtle.

In the case of dinoflagellates, the progression has been from major taxonomic categories defined on gross features (e.g., only two decades ago there was one group of fossil dinoflagellates and another group of fossil hystrichospheres, and many descriptions dating from the 1960s mention the archeopyle only in vague terms of shape and position) to a general awareness of the importance of many morphological details (e.g., wall structure, paratabulation from intratabular processes, and the archeopyle, in addition to the more obvious horns and parasutural patterns). Future refinements remain uncertain but seem likely to include greater emphasis on details of paratabulation, some of which have been completely ignored heretofore.

b) Dinoflagellates, like most organisms, evolved without the aid of taxonomic classifications or phylogenetic trees. These inventions are arbitrary, subjective, and unreal, however essential they may be in making possible a census of organisms and discussions of their similarities, differences, and possible relationships. Even theoretically, a taxonomy can only be satisfactory retrospectively. That is to say, so long as we are discovering new organisms, observing new features in them, and evaluating these features in new ways, our taxonomy will be temporary, incomplete, and more or less contentious.

13. VARIABLE VARIABLES

Principle: All characters of living organisms vary, including character variability.

Discussion: Within each taxon, every character exhibits a certain variability. The range of this character variability is also a character, genetically determined and variable from one taxon to another.

No intrinsic aspect of a character indicates what its range of variability may be, but an assessment of this variability is critical for sound taxonomy. The ubiquity of variability and the unpredictability of its extent limit severely the taxonomic significance of a single specimen, even when preservation is excellent. Given the uncertainties presented by many fossil specimens (e.g., resulting from unfavorable orientation, distortion, corrosion, or local concealment by debris), the number of specimens necessary to "fix" the characters of a species and to assess variability may be large.

Variations that occur at especially low frequency can be considered individual anomalies. Although examination of multiple specimens makes it easy to discount the taxonomic significance of such erratic individuals, they can play a grossly exaggerated role in a taxonomy that includes species based on single specimens.

Corollary: Within a diverse group of organisms, the taxonomic importance attributed to a given character may vary from one subgroup to another.

Discussion: If characters once thought to have low variability and used in constructing a taxonomy are shown by subsequent observations to have higher

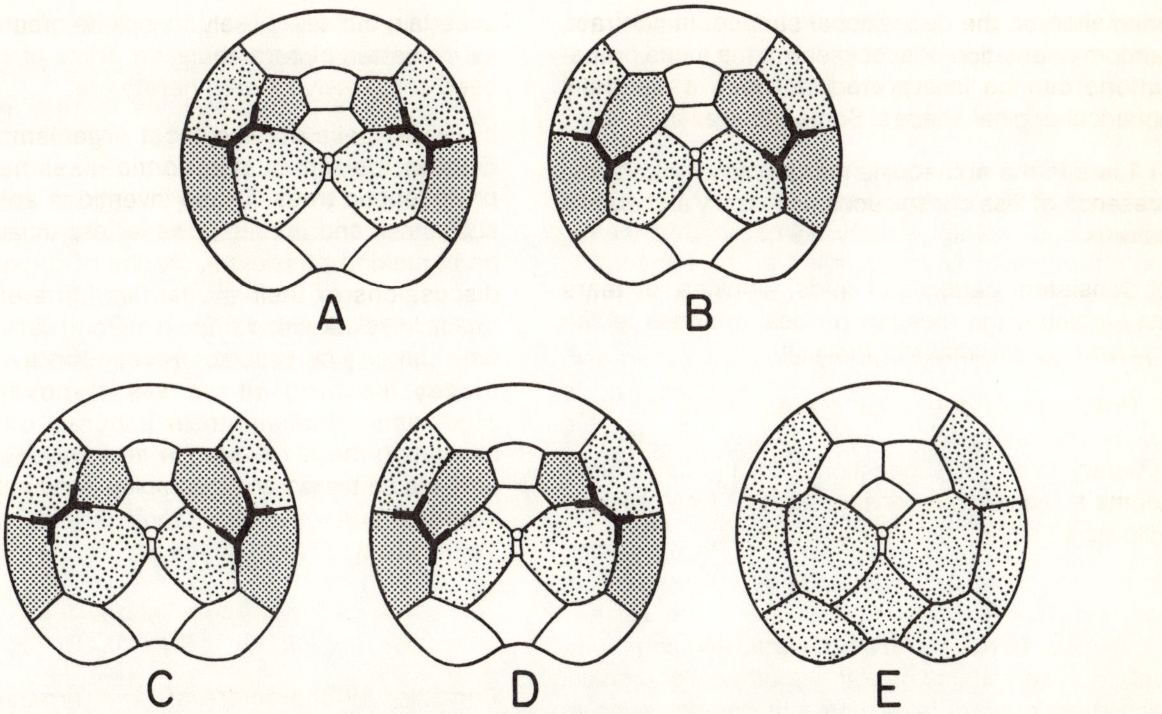

Figure 7.8 - EXAMPLES OF VARIABLE VARIABLES

A-E - Schematic representation of variable epithecal tabulation in *Protoperidinium* (*Protoperidinium*) *depressum*, a modern species with basically ortho quadra organization. A-D - Four variants that are all ortho quadra. Plate contacts within the unpatterned area are constant. Patterning emphasizes variable contacts laterally among four plates on each side, which result in two symmetrical arrangements (A-B) and two asymmetrical ones (C-D). Heavy lines emphasize plate contacts. Most other species of the subgenus show the lateral contact relationships seen in B. E - Contacts in the patterned area of this variant are as in A, but in the unpatterned dorsal area only two intercalaries are present and, as a consequence of the median suture between them, the middorsal apical and precingular plates have shapes quite different from those in A-D. Such an arrangement of two dorsal intercalaries is normally diagnostic of *Protoperidinium* subgen. *Archaeperidinium*.

variability, their role as taxonomic criteria may need corresponding adjustment. For example, the taxonomic importance originally ascribed to the archeopyle is currently being reconsidered (Stover and Evitt, 1978; Goodman, 1983) because the variability of this feature now appears to be greater in some groups than was previously appreciated.

Characters that are taxonomically discriminating at a certain level are generally ones that show a limited variability within taxa at that level, but relative contrasts between taxa. A character may have low variability and therefore be distinctive at a certain level in one group (e.g., hair color, at the generic level in *Camelus*), whereas the same character may have high variability and little diagnostic value at the same level in a related group (e.g., hair color in *Homo*). Characters considered to distinguish genera or subgenera in some circumstances are encompassed within a single genus or single species in others. An example among dinoflagellate cysts is discussed in Chapter 10 (*Areosphaeridium*). An especially instructive example among dinoflagellate thecae concerns the tabulation pattern in *Protoperidinium* (*Protoperidinium*) *depressum*, a large, frequently abundant, and geographically widespread modern marine species having a basically ortho quadra tabulation. Graham (1942) discussed the extreme variability that affects the epithecal tabulation of this species in areas where the tabulation is otherwise conspicuously stable from species to species virtually throughout the subgenus.

The extensive variability analyzed by Graham, which was evident even in single collections, involves intergradations among four distinct patterns in specimens having three intercalaries, and a further pattern in specimens having only two intercalaries (Figure 7.8). Comparison with Figure 7.3 will show that the principal variation here is a matter of how four particular plates on each side of the epitheca (the lateral apical, the lateral intercalary, and the two subjacent precingulars) meet at two triple junctions. *P. depressum* is distinctive in having high variability in the topologic relationship of these plates, which is normally low among *Protoperidinium* species (the contacts being typically between the lateral intercalary and the penultimate precingular, as in Figure 7.8 B).

The variations in *P. depressum* occur in significant fractions of population samples. Even the highly "anomalous" examples with only two intercalaries—the character ordinarily applied at the subgeneric level to distinguish *Protoperidinium* subgenus *Archaeperidinium* from *P.* subgenus *Protoperidinium*–had a frequency of one to two percent. Graham concluded (1942, p. 17) that the epithecal plate pattern "in this group of species is not a stable feature . . . and that between two patterns there is a continuous line of intergrades." Therefore, "the epithecal plate pattern alone cannot be used for the classification of the species related to *P. depressum*, and . . . must be used with considerable caution in other parts of the genus as well." He further concluded that this epithecal variability, combined with marked stability of both hypothecal and sulcal tabulation, renders tabulation a useless criterion at the specific level in the *P. depressum* group. Only by statistical analysis of thecal shape could he satisfactorily categorize the taxa.

Although the case of *P. depressum* provides a dramatic example of variability, it should not be taken to vitiate the contention that tabulation/paratabulation is of extreme importance in dinoflagellates. Empirical studies show that such a high level of variability is unusual. What this example does emphasize is the importance of determining the range of variability to whatever extent a sample permits and then applying variability with care and understanding as a criterion for distinguishing taxa.

14. DECEPTIVE DISTINCTION

Principle: Highly visible distinctive features shared by two taxa tend to divert attention from taxonomically more discriminating but less striking features.

Discusson: Identification based on only the more obvious features will often be incorrect because few features are restricted to only one species. Even seemingly intricate combinations of features are known to recur in different species. When intraspecific variability is great, it may be especially difficult to look past such superficial similarities and recognize underlying differences that can be used to advantage. The best guard against this difficulty is a systematic analysis of all morphological evidence available, including such consideration of variability as the material permits.

The example of two species of Tertiary chorate cysts with extreme superficial similarity, *Operculodinium divergens* and *Cordosphaeridium funiculatum*, will serve to illustrate the point. Specimens of both these species are large, similar in general size and shape but frequently fragmented, and have as their most striking feature an intricate and characteristic surface reticulum that merges in similar ways with the bases of solid, rather fibrous-appearing processes. Although the two species have been confused, differences in process distribution (nontabular in *O. divergens* and intratabular in *C. funiculatum*) enable the species to be recognized and their different stratigraphic ranges in the Early Tertiary to be utilized effectively.

15. PREDICTABLE PROGRESSION

Principle: Impressions during the systematic study of any large number of generally similar organisms usually progress through three stages: an initial simplistic perception of sweeping similarity, followed by a bewildering hyperawareness of individual distinctions, and, eventually, but only after considerable experience, a feeling of reasonable confidence in evaluating similarities and differences.

Discussion: Few experienced systematists have escaped acute and humbling awareness of the feelings described, and most of them find that the same succession is repeated with each new group studied. Awareness of the phenomenon may make no practical contribution toward improving the interpretation of dinoflagellate cysts, but it may lift morale during the inevitable frustrations associated with the study of assemblages rich in variable species.

INTERPRETING THE LITERATURE

Since the mid-1950s a veritable flood of publications has presented us with a vast array of new taxa of dinoflagellate cysts and revisions of older taxa, often based on materials derived from previously unsampled areas or recovered by improved laboratory techniques and studied at levels of detail possible only with modern light and electron microscopes. With this flow of new observations, descriptions of species and genera that had seemed adequate and reasonable have become obsolete after a few years and we are assured that concepts of dinoflagellate taxonomy at higher levels will remain highly fluid in the foreseeable future.

The contemporary author wishing to describe a species also faces a special problem. The continuing rapid expansion of our knowledge and the excellence of modern equipment and techniques impose upon him an obligation to produce comprehensive descriptions and effective illustrations. Conciseness is desirable, but, for our purposes, brevity easily becomes counterproductive. The details required to assure the identifiability of a species cannot be presented in a short description, and, although a species can be over-illustrated, more likely faults are figures that are too few and of inferior quality. Unquestionably, the most valuable older species descriptions are those accompanied by the best illustrations. Good illustrations may be nearly as useful as original specimens, as long as one knows what to look for.

The microscopist inexperienced in the study and identification of dinoflagellate cysts may assume that the specimen he has under the microscope presents the questions to which the taxonomic literature has the ultimate answers, but he will soon find that the situation is not so simple. Although the literature on dinoflagellate cysts is an invaluable resource, in some respects the species described and illustrated in it present today's reader some of the same challenges and raise some of the same questions as do the "real" specimens under his microscope. This is inevitable in a field marked by such recent and rapid expansion as ours, for one must remember that the literature records ideas and concepts in the process of their development. At any stage, it is incomplete and partly in error through omission, inadequate understanding, and faulty communication. Currently with regard to fossil dinoflagellates, techniques and concepts are outmoded nearly as fast as they are published. Under these circumstances, no descriptive taxonomic work—nor any synthesis, such as this, based on such works—can be more than a subjective and imperfect attempt to distill an essence of information out of a virtually infinite amount of raw data, with the further handicap of having to communicate this information using, in effect, a language whose vocabulary and syntax are constantly evolving.

Accordingly, one should not merely accept the published literature at face value. To do so is to perpetuate its shortcomings. One must be prepared to evaluate figures and descriptions for oneself, applying to them the most rigorous of currently accepted criteria, reading between the lines what is not said in the descriptions, and seeing in the figures the significant features that are not mentioned in the captions. One must be ready to translate expressions and concepts into current terms, alert to innocent oversights and mistakes no matter how emphatic or convincing an author's original words or how esteemed his reputation.

Let us explore this point with a few examples dealing with paratabulation patterns, the topic that dominates so much of this volume. While the examples were taken directly from published works, the sources are not cited, because the intent here in no way is to hold up for criticism what particular authors have said or thought, but to emphasize that a reader with a different perspective—such as may have been influenced by studying the previous pages of this volume—can often go beyond what has been written. Surely, progress in our field will be furthered if he recognizes and utilizes this ability.

In regard to paratabulation as treated in the literature, one is commonly faced with two closely related but different questions: 1. Am I able to accept the description or illustration as a plausible record of the actual pattern? 2. Am I able to accept the author's interpretation of the pattern as indicated by his labeling or descriptive remarks? Many of the principles discussed in the foregoing pages of this chapter can be drawn upon effectively in answering these questions.

A practical approach to a concrete example is, at first, to disregard the author's labeling and descriptive comments when examining the pattern presented by a published photomicrograph or paratabulation diagram. At this point, one might consider questions along the following line: What would I consider the most reasonable interpretation of this pattern? Is it one of the conventional "standard models"? If not, how does it differ? Are the apparent differences likely to be "legitimate" or might they express extraneous influences (e.g., folds, debris)? To what familiar pattern is this one most similar? What alterations would be necessary to convert the familiar one into the reported one?

Having pursued such questions to the limit of productiveness, one might next consider the author's labeling and text in asking, essentially: Did he see it the way I do? If not, what are the specific differences in our impressions? How can they be accounted for? Which ones am I prepared to maintain as defensible improvements on his interpretations, and on which points do I recognize that

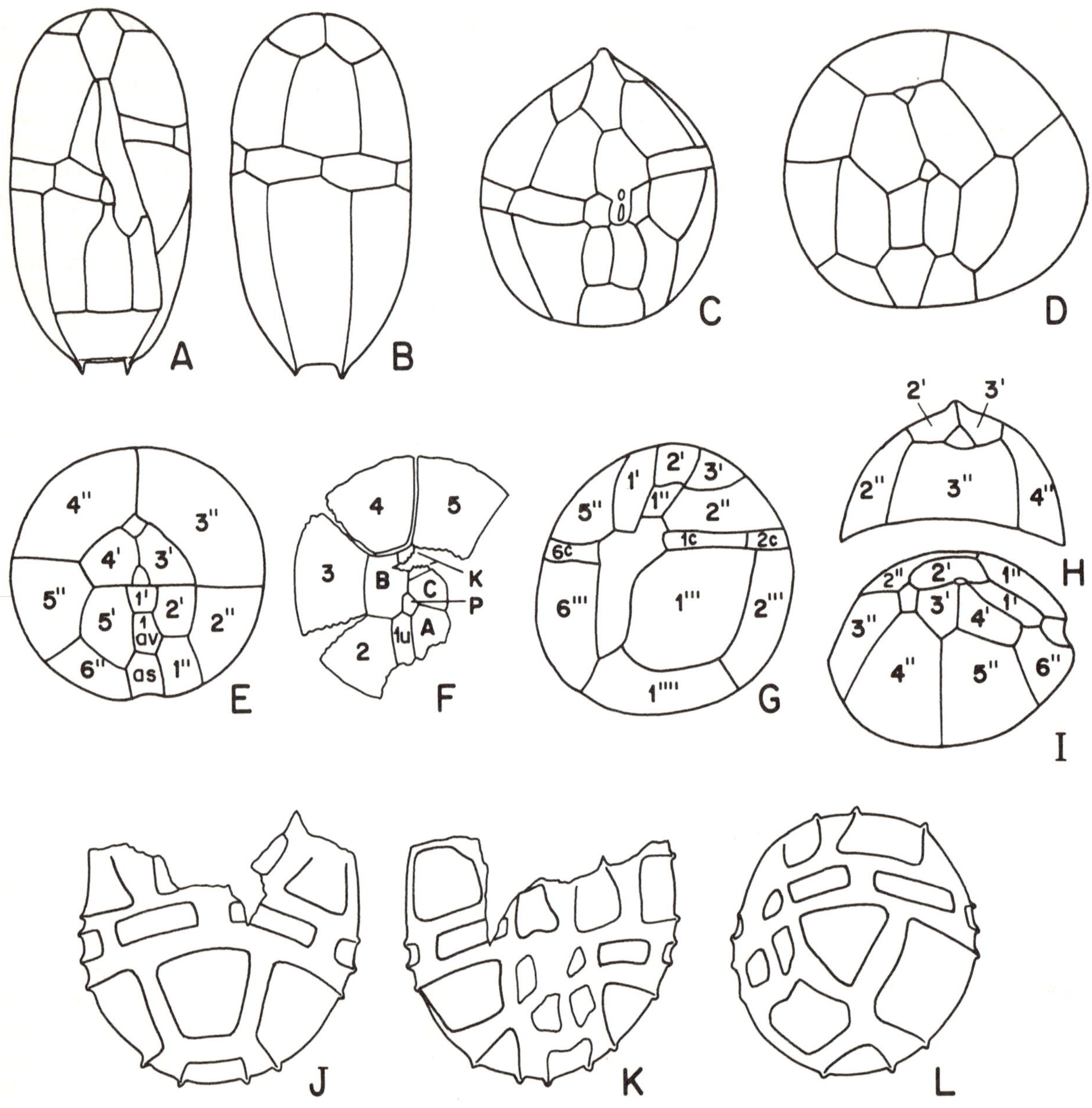

Figure 7.9 - EXAMPLES OF PATTERN INTERPRETATION, I.

A-B - Example 1 (vv and dv). C - Example 2 (vv). D - Example 3 (av). E-F - Example 4 (av and aav of epicyst). G - Example 5 (vv). H-I - Example 6 (dv and oblique av). J-L - Example 7 (dv at high and low focus, and oblique anterodorsal view at low focus).

his interpretations were more reasonable than mine? If the author has claimed a novelty, does it really appear to be one, is it adequately documented, or may this be a case of a post appearing to be a horse's fifth leg?

Turning to the microscope and one's own specimens, whose identification may have inspired consultation with the literature in the first place, one might consider: What features in my material would be consistent with or in conflict with the author's interpretation? Do my specimens provide an opportunity for more plausible interpretation of the morphology or for extracting information of importance that was originally overlooked or misinterpreted?

Throughout, of course, the purpose is not to "second guess" the author, but only to confirm, extend, or modify his interpretations to the extent—which will always be limited—that the published text or figures (or the specimen under the microscope) provide the information to justify an independent analysis in which one can have confidence.

With this approach in mind, let us consider, one by one, a few examples. The line drawings on Figures 7.9 and 7.10 represent as faithfully as possible the patterns shown in the authors' original illustrations. Each is discussed in the following paragraphs. The diagrams for four of the examples are left unlabeled on Figure 7.9 to encourage readers to develop their own interpretations of these patterns before reading this author's commentary. After studying the unlabeled diagrams, reference to Figure 7.10 will show these figures repeated twice, once with the original author's labeling and once with an alternative labeling (using Taylor symbols) consistent with the interpretation advanced in the discussion here.

Example 1 (Figure 7.9 A-B, 7.10 A-C). The figures are modified in shape but not in topology from the original paraplate diagrams showing ventral and dorsal views. Although the paratabulation has definitely distinctive details, the basic pattern is clear and all the defined paraplates have "conventional" topologic relationships. The symmetrical group of three apicals, strongly offset paracingulum, triangular **1i**, and mostly undivided parasulcus are all traits of the gonyaulacoid pattern with S-type parasulcus, but definition of **fu** (here labeled *7c* although conforming to the customary Kofoidian *ras*) in the parasulcus is unusual. A good "landmark" is afforded by the "step" in the outline of the parasulcus representing **lu/X**. The most distinctive feature of the pattern is the fully ventral position of **Y** and its strikingly transverse and rectilinear boundary against **VI**, **Z**, and **X**.

The stated formula is *3′, 6″, 7c, 6‴, 1⁗*. Suppression of parasuture **A/1u** commonly associated with the S-type ventral arrangement, easily accounts for the presence of only three apicals. The six precingulars are "conventional." Citation of *7c*, while an obviously plausible interpretation in this setting, suggests a non-existent contrast to the "normal" gonyaulacoid pattern. On the other hand, the listing of *6‴*, although appearing to record the "normal" gonyaulacoid postcingulars, here omits **lu** while including **Z**. Thus, all numerals for the postcingular elements are shifted by one relative to most species having six Kofoidian postcingulars (*1‴-6‴* = **lu**, **II-VI**). Interpretation of **X** as the posterior part of the parasulcus removes the usual *1p* from the formula and suggests a non-existent fundamental contrast with the gonyaulacoid standard. The dorsal surface presents no problems; all relationships shown there are typical for a large number of gonyaulacoids.

Example 2 (Figure 7.9 C, 7.10 D-E). The figure shows a ventral view and represents a published line drawing. The pattern is "conventional" sexiform gonyaulacoid throughout. Seen without labels, it presents no problems, but some of the labels and the formula that would represent them require "translation." The epicyst interpretation is standard. The labeling of the paracingulum, while indicating the "conventional" total of six paraplates, incorporates **fu** (= Kofoidian *ras*) as *6c*. That means the paraplates labeled *4c* and *5c* occupy the positions of the "normal" Kofoidian *5c* and *6c* (= **e** and **fi**). This gives a strong suggestion that one cingular has been overlooked, perhaps "spherically concealed" at the sides or undetected on the dorsal surface because the defining parasuture was obscured. The hypocyst paratabulation, while topologically consistent with the usual gonyaulacoid pattern, has been interpreted in a way that suggests significant contrasts with that pattern. At the latitude of the postcingular series, *7‴* is the usual Kofoidian *rs* (**li**) and *ps* is the usual left sulcal (**lm**). Finally, the paraplate labeled *1pv* is the usual posterior sulcal (**Z**).

One might protest that the reader is in a poor position to "improve upon" labeling that may have been applied quite objectively by the person who studied the actual specimen—counting as paraplates only those that could be clearly seen, and numbering them in Kofoidian sequence without "interpretive prejudice". This contention has a certain merit, but objectivity untempered by judgment that incorporates what we know about the relatively stable properties of a number of "standard" patterns is like treating each specimen as if it "existed in a vacuum." With five legs or four, a horse is a horse, and that should be the starting point.

Example 3 (Figure 7.9 D, 7.10 F-G). The figure, which originally appeared as a line drawing, shows the paratabulation of the epicyst (= operculum of epicystal archeopyle) in apical view. The sequences of paraplates around the periphery and about the apex in the center of the figure are "conventional" for the precingulars and apicals in an insert gonyaulacoid pattern. The small dorsal intercalary is distinctive but not surprising. Novel elements are the three large intercalaries in the ventral half. These seemingly anomalous features would disappear if some of the would-be parasutures were, in fact, folds, as has been suggested by the dashed lines in Figure 7.10 G.

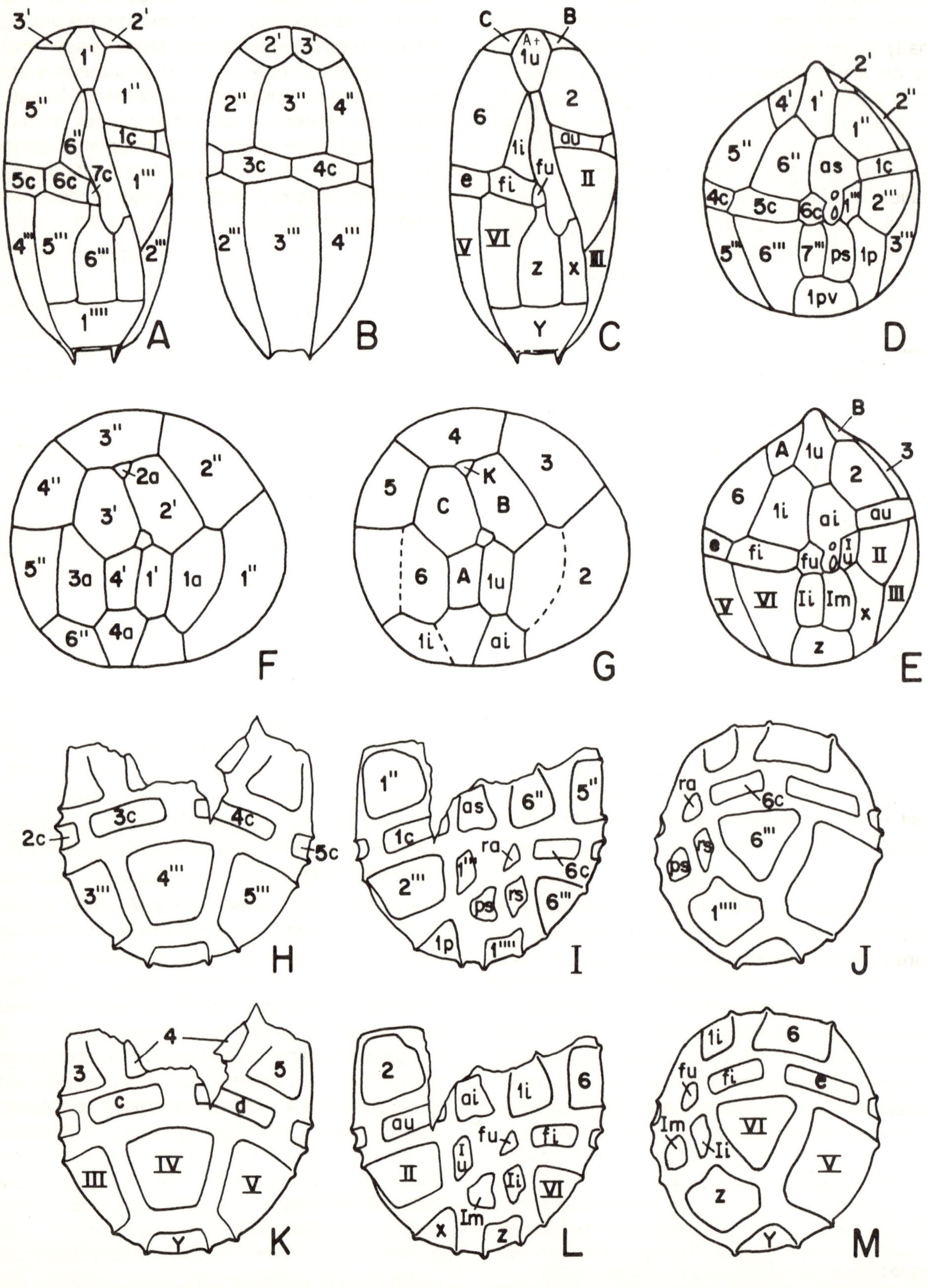
3' 2' 1' 5'' 1'' 6'' 1c 5c 6c 7c 1''' 4''' 5''' 6''' 2''' 1'''' A
2' 3' 2'' 3'' 4'' 3c 4c 2''' 3''' 4''' B
C B A+ 1u 6 2 1i au fu e fi II V VI z x III Y C
2' 4' 1' 2'' 1'' 5'' 6'' as 1c 4c 5c 6c 1''' 2''' 5''' 6''' 7''' ps 1p 3''' 1pv D
3'' 2'' 4'' 2a 3' 2' 5'' 3a 4' 1' 1a 1'' 6'' 4a F
4 3 5 K C B 6 A 1u 2 1i ai G
B A 1u 3 2 6 1i ai au e fi fu I u II V VI Ii Im x III z E
3c 4c 2c 5c 3''' 4''' 5''' H
1'' as 6'' 5'' 1c ra 6c 2''' 1''' ps rs 6''' 1p 1'''' I
ra 6c rs 6''' ps 1'''' J
4 3 5 c d III IV V Y K
2 ai 1i 6 au fu fi I u II Ii VI Im x z L
1i 6 fu fi e VI Im Ii V z Y M

Figure 7.10

Example 4 (Figure 7.9 E-F). The operculum of an epicystal archeopyle is shown in apical view (E), originally a line drawing; the second figure (F) is based on a photomicrograph without paraplate labels. The overall appearance (E) vaguely suggests a gonyaulacoid epicyst pattern, but the diagram presents difficulties which cannot be resolved without reference to the specimen. Several of the anomalies plausibly result from folds being mistaken for parasutures at the same time that some true parasutures have been overlooked. Several prominent anomalies are associated with the "suspicious" rectilinear sequence of parasutures extending transversely through the apex. These include: *3″* unusually wide relative to *2″*, *2′* and *3′* together occupy the area topologically customary for *2′* (**B**) alone, topologic relations of *4′* and *5′* are mostly those customary for conventional insert gonyaulacoid *3′* and *4′* (**C** and **A**). Also, the ventral meridional sequence of *1′-1av-as* is strange; *av* is perhaps an artifact defined by folds.

Another epicyst (obviously torn and incomplete) was attributed to this species by the same author in a subsequent publication. This specimen, illustrated by a photomicrograph rendered here as a paratabulation diagram (Figure 7.9 F), appears to show a "conventional" insert epicyst pattern (with the addition of **P** and **K**) in interior view.

Note that the patterns shown by the two illustrations are mutually incompatible, although they were referred to the same species. The paraplate topology in Figure 7.9 F (labels ignored) appears acceptable at face value, but that in Figure 7.9 E is spectacularly novel, if real—which it almost certainly is not.

Example 5 (Figure 7.9 G). The figure represents a line drawing of the type species of a new genus. If they are correctly portrayed—and the author notes that parasutures are obscured in many specimens —the topologic relationships in the visible areas are highly unusual. One might guess that artifacts of deformation have disguised a "conventional" gonyaulacoid pattern. Here one has only unsatisfactory alternatives: to accept the author's interpretation without being able to really see any independently supportive evidence for it, or to reject it outright as too bizarre to be reasonable but, again, without supportive evidence. More specimens and more illustrations are required, and authors should be expected to provide them in documenting such a distinctive pattern. If an author's work portrays a five-legged horse, the burden of proving its legitimacy falls on him; the critical reader is justified in remaining skeptical until the case is clear, and a "snapshot from the distance" will not do.

Example 6 (Figure 7.9 H-I). The two views shown are based on line drawings of the epicyst, showing the holotype (dorsal view, above) and paratype (oblique view, below) of a new species. Note the prominent internal inconsistency: the anterior intercalary occupies a fraction of the width of *3″* in the holotype, but extends from *2″* to *4″* in the paratype, spanning *3″* entirely. It is highly unlikely that this difference results from intraspecific variation. The condition in the holotype diagram is not unusual; the second figure is most easily explained as a misinterpretation in detail, although the basic organizationis clearly an insert gonyaulacoid pattern with a dorsal anterior intercalary.

Example 7 (Figure 7.9 J-L, 7.10 H-M). The figures, drawn after labeled photomicrographs, show dorsal views of one specimen (in both high and low focus) and an oblique anterodorsal view of another (low focus). The paratabulation is clearly shown by penitabular septa. In the first specimen, the operculum of the apical archeopyle is missing and the archeopyle margin shows some tearing. The author's descriptive remarks portray a generally gonyaulacoid pattern, and he calls attention to the triangular shape of *6‴*.

The pattern is "conventional" quinqueform gonyaulacoid; all paraplates are present in their usual relationship. The contact between **Z** and **V**, combined with the opposing angles on **VI** and **Y** (indicating their lack of mutual contact) clearly documents the quinqueform hypocyst. The Kofoidian labels offered by the author present problems only in the posterior part of the parasulcus. Here, possibly "spherical concealment" suggested that the outlines seen at the different

Figure 7.10 - EXAMPLES OF PATTERN INTERPRETATION, II.

See text for discussion; see Figure 7.9 caption for directions of view.

In each case, the first labeling of paraplates presented is that of the original author, using Kofoid symbols; the second uses Taylor symbols to suggest the homologies of the paraplates shown.

A-C - Example 1. The pattern on the dorsal surface is not repeated, as it matches exactly the "standard" pattern (compare Figure 5.10 E). D-E - Example 2. F-G - Example 3. In the second figure, suspected folds are dashed. H-M -Example 7. In H-J, the complete labeling given by the author is shown.

focus levels at the antapex pertain to a single paraplate. In fact, they must represent both **Y** and **Z**, not just the antapical (*1''''* = **Y**), a point clarified by reference to the oblique figure. In this figure also, however, the misleading effects of "spherical concealment" may have confused the issue by fostering the impression that the unlabeled outline at the bottom (actually **Y**) is part of a postcingular showing "from over the horizon." Confusion of **Z** for **Y** in the oblique view, in turn, occasioned an anomalous interpretation of Im as the posterior sulcal. The species exhibits the traits of a quinqueform hypocyst in almost diagrammatic perfection.

To summarize: Hopefully, it is not necessary to emphasize that the intent of this section is constructively to encourage a critical and analytical, rather than a passive, approach to printed works that deal with the descriptive taxonomy of dinoflagellate cysts. The intent certainly is not to call for a challenge to every author's word. Even at the "accidental level," which we have not discussed, one must be alert to occasional irregularities with high potential for introducing confusion, such as reversed photographic images and inadvertent "slips of the pen" or proofing oversights that may record "apical" for "antapical" or "epicyst" for "hypocyst." At a more significant level, the foregoing examples should have made clear how necessary it is to "translate" another author's work carefully into one's own language and perspective—especially as regards identification of paraplates and statements made in the form of paratabulation formulae. One can speak quite accurately here of an author's style and the molding of his ideas by the "climate" of his times. Individual competence, of course, also is a factor. Style and competence vary greatly, and climate changes rapidly, but, no matter what an author's views or competence or how they change, his expressions lie fixed and dated on the printed page, from where they must be repeatedly recalled and amended by successive readers.

Chapter 8. Analytical Procedure; Major Morphological Categories

INTRODUCTION

Identifying a species through comparison with the published literature, describing a species thought to be new, and surveying an assemblage of specimens as a step toward dating and correlation, all depend on analysis of morphology and comparison of one specimen with another. Equipped with a basic knowledge of morphological features (from Part II), and familiar with certain principles for interpreting them (Chapter 7), we proceed in the first part of this chapter to the outline of a simple analytical procedure for studying a specimen and a checklist of features to be observed. In the larger part of the chapter, which follows, diagnoses of 17 major morphological categories of sporopollenin dinoflagellate cysts are presented, together with a key for their identification.

Most practicing palynologists working with dinoflagellates are not primarily concerned with formal suprageneric taxonomy, but they have a real need for a convenient framework for organizing voluminous and diverse descriptive material. For a well-known group of organisms with an established and widely accepted classification, it is logical to look to the suprageneric hierarchy for such a framework. However, for a group about which information is expanding rapidly and for which the criteria of suprageneric classification are not yet agreed upon, another means to serve this limited purpose may be more appropriate. The morphological categories devised here have that objective. They provide a means for grouping similar cysts independent of a suprageneric hierarchy. They will also serve as the basis for organizing the additional discussions of morphology which constitute Chapters 9 to 12.

AN ANALYTICAL PROCEDURE

Probably the most efficient way to begin the analysis of a specimen under the microscope is with the question: How is the specimen oriented? The answer will provide the frame of reference for other observations and, on the way to it, attention will be directed early to key features and areas (e.g., paracingulum, parasulcus, apex, antapex, and archeopyle). If indications of paratabulation are obvious and reasonably extensive, the analysis can proceed at once to considerable detail on this point, including observation of the mutual relationships of the paraplates and identification of the particular paratabulation pattern displayed. If, on the other hand, the surface is nearly smooth or shows only nontabular relief, the archeopyle may give important clues to both orientation and paratabulation. The shape of a specimen free from postdepositional compression may also assist in determining orientation. The indentation of the parasulcus that commonly marks the ventral surface is unlikely to be confused with any other feature except the antapical depression between two horns or prominences. Since primary compression is most likely to be dorsoventral, specimens with noncircular equatorial sections will probably present a dorsoventral view. Remember also that elongation is nearly always in a polar direction and that, if the outline is angular or extended by horns, a single apical prominence and two symmetrical or asymmetrical antapical prominences are the most common.

With specimen orientation established and the presence and character of some related features noted, attention can be turned to such details as wall layers and cavation, the archeopyle, processes and other projecting structures, and small-scale surface ornament. Whenever possible, observations should be extended and refined by reference to other specimens of the same species.

Isolated opercula will help confirm archeopyle details and, in the case of apical or epicystal archeopyles, should give an independent indication of the amount of dorsoventral compression. Torn or broken specimens should not be overlooked. They may be especially helpful in two respects. First, they offer the best opportunity for viewing cross-sections of the wall, often providing a sharpness of detail unlikely to be seen in a complete specimen. Second, they commonly make it possible to view directly parts of the surface normally hidden or seen only obliquely in whole specimens whose orientation is selectively determined by overall shape.

If specimens being examined represent an unfamiliar species, there is no substitute for photomicrographs and drawings to help record critical features for later reference and comparison. While some photomicrographs are essential for documentation and publication, the merits of drawings are commonly underappreciated. Modern improvements in the ancient camera lucida bring accurate drawings of features ranging from outline shapes of whole specimens to fine details at high magnification equally and easily within the competence of any skilled microscopist, leaving virtually nothing to artistic ability. Diagrammatic records of process tips, surface ornament, and, especially, local details of paratabulation are also highly useful. As a basis for later review and comparison, such graphic notations can save time far beyond that consumed in their making. They are quicker and cheaper than photographs and, for some features, definitely superior. Also, unlike the memory, which is the alternative commonly chosen, they remain indefinitely accurate.

A complete paratabulation or process arrangement can seldom be determined (as opposed to recognized after once having been determined) from a single specimen fixed in position as is customary in most conventional permanent mounts. This limiting circumstance can be overcome by preparing a synthetic diagram based on either multiple specimens or multiple views of a single specimen individually manipulated. The hunt-till-you-find-the-right-view technique is familiar to all; the more direct approach of specimen manipulation probably is not. It is not so demanding in either time or skill as might be imagined and its reward—the guarantee that all features observed in one completely visible cyst are attributes of the species under study—is important enough to be worth careful consideration whenever specimen quality is appropriately high.

While drawings and photographs can help in preparing a synthetic diagram, whether based on one or many specimens, an especially effective technique for assembling the fragmental pattern pieces (as noted already in Chapter 5) is to draw them with a marking pen on a transparent glass ball (e.g., a 250ml roundbottomed evaporating flask). Corrections in interpretation can be made easily by erasing with acetone, and details can be plotted from specimens viewed in either high or low focus by drawing on either the near or far surface of the clear ball—not to mention the potential of insights derived from gazing deeply into the ball.

A Note on Image Reversal

In most compound microscopes, the image perceived by the viewer is inverted (but not reversed), but in some instruments there is reversal, with or without inversion. Similarly, the image recorded on a photographic negative may be reversed by some optical systems, but not by others. For example, in one model of the Zeiss Photomicroscope the following situation obtains:

- –The viewer sees the specimen through the oculars as upright but reversed; that is, left and right are interchanged.
- –But, when the photo slide is inserted for photography, the reversal is removed and the image is both upright and normal.
- –The image on film exposed in the 35mm automatic camera is reversed and must be printed through the "back" of the film if the photographic print is to match the specimen.
- –The image is not reversed on a film exposed in a flat film holder mounted atop the microscope (e.g., for Polaroid™ photomicrographs).

However, these same optical effects are not observed in all Zeiss Photomicroscopes. The optics differ from model to model, and in some models can be altered by accessory lenses available upon special order.

Image reversal is of little consequence for microscopists who study such symmetrical objects as spores, pollen, diatoms, and silicoflagellates; but with asymmetrical dinoflagellates, it is critical to know if, and precisely at what stage in the course of examination and documentation, the image may be reversed. Particular care must be taken to properly instruct photographic technicians if any special procedures (e.g., printing with the emulsion surface of the film toward the light instead of toward the paper) are required in order to produce a finished print that shows the specimen in correct orientation.

A quick and simple test of image orientation is to examine under low power a capital letter R typed on a thin sheet of paper. The eye easily recognizes the inverted or reversed image of this familiar form, allowing one to determine at once what unsuspected optical "tricks" a given microscope may be playing.

CHECKLIST OF FEATURES

Whether used as an adjunct to identifying, describing, or simply studying a specimen, a checklist can help in developing a consistent routine of observation and also in assuring that important points are not overlooked. Moreover, reference to such a list may assist in bringing to light deficiencies in a published description, after which the missing information may be sought in the illustrations.

The topics in the following outline are arranged in the sequence one might follow when examining a specimen, although, of course, some of the points will be inappropriate or irrelevant to certain specimens.

- Orientation and direction of view (final judgment may depend upon other observations)
- Overall character
    - Size
    - Major features of shape, wall structure and surface relief
    - General definition of paracingulum, parasulcus, apex, antapex
- Details of shape
    - Elongation or compression
    - Outline of main body relative to that subtended by process tips
    - Relative conformity of pericyst and epicyst, etc.
    - Horns - location, number, gross form
    - Processes unique in size or form at particular locations
- Surface relief and paratabulation
    - Form of surface relief
        - General identity: processes, septa, minor elements
        - Details of processes and septa
            - Size, number
            - Basal, intermediate, and distal parts
            - Linkages between bases, shafts and tips of processes
        - Details of minor elements
- Wall structure
    - Number, thickness, prominence, and continuity of layers
    - Internal structure and surface features of each
    - Distribution of cavation and contacts between layers
    - Openings in the wall besides the archeopyle (claustra)
    - Distribution of features
        - General: parasutural, intratabular, penitabular, nontabular
        - Pattern type (summary formula if appropriate)
            - peridinioid - ortho, meta, para; quadra, penta, hexa
            - gonyaulacoid - quinqueform, sexiform (S-type, L-type), corniform, partiform
            - suessioid, rhaetogonyaulacoid, other
        - Details of distribution
        - In critical areas: ventral surface, apex, antapex, paracingulum, parasulcus
        - Location of distinctively differentiated (or missing) elements
- Archeopyle
    - Type, including number and identity of paraplates involved
    - Shape and size
        - Presence and extent of accessory sutures
        - Reduced or enlarged
        - Special data as appropriate (e.g., TAI, LAI, archeopyle ratio)
        - Distinctive features of outline (e.g., sulcal notch, middorsal tab)
    - Operculum
        - Free, adnate; simple, compound
        - Notable features of individual parts
    - Differences, if any, between openings in different wall layers
- Variability of all features observed

MAJOR CYST CATEGORIES

The number of described species of sporopollenin dinoflagellate cysts now approaches 2500 in over 500 genera, and the figures are continually growing. A comprehensive biological classification of these taxa would be desirable, but contemporary views differ widely on how it should be structured and even whether it is currently useful to attempt its construction. Long before reaching this page, the reader will have perceived that formal classification of dinoflagellate cysts is not a principal concern of this volume. Given the current diversity of views and the rate of new discoveries relating to fossil dinoflagellates and their formal taxonomy, no stable system of classification above the generic level that will find support from a majority of specialists can be expected in the near future. Meanwhile, an informal system based on morphology and avoiding some of the inflexibilities of formal taxonomic nomenclature, although having limited attractiveness in the long term, may have short term advantages where systematics is of secondary importance and the principal objective is to compare species.

The scheme of morphological categories explained on the following pages is not meant to be a classification defensible on biological grounds, although close biological relationship is certainly responsible for many of the morphological similarities among cysts in some of the categories. These categories are intended principally as a convenience for organizing extensive observational or descriptive data that have been reduced to identifications at the generic and specific level *without* requiring a commitment to any formal suprageneric scheme. While paratabulation plays a dominant role in the definition of most of the categories, the limits of some are drawn using features of trivial biological significance, or even are fashioned arbitrarily to accomodate distinctive cysts with confusing combinations of features or simple cysts with nondescript features. Nothing in the scheme is intended to suggest or to favor any particular system of formal classification above the generic level. Thoughts on that matter are entirely reserved for Chapter 13. Where terms like "peridinioid" and "gonyaulacoid" are used in the scheme, reference is to morphological similarity, not suspected biological affinity.

The categories are identified by letters that serve merely as labels, although most of these labels have some relevance either to the name of an included entity or to one of its important morphological features. Thirteen of the categories are designated by double letters (e.g., Pq-cysts, Gc-cysts, Rp-cysts) to indicate association into three broader morphological groupings (P-cysts, G-cysts, and R-cysts). Distinctive features are summarized for each category, followed by comments on its characteristics, suspected affinities, and stratigraphic range. Finally, some genera representative of the category are listed. These lists are far from exhaustive for the larger categories, but they include examples of well-known genera displaying contrasting form. The scheme is summarized in a key and its concepts are reinforced by a discussion of the choices at each step of the key.

Although most well-described genera fall entirely in one or another of the categories, in a few cases species of a single genus (see *Areosphaeridium* Complex, Chapter 11) fall into different categories. This suggests that the circumscriptions of such genera are somewhat unusual, and also makes quite obvious the failure of this scheme to solve all the evident problems!

No attempt has been made to "plan for the future" by imagining morphologies that have not yet been observed. Therefore, the scheme will grow increasingly incomplete as new types of cysts continue to be discovered. Nevertheless, a large number of currently familiar, stratigraphically important, or morphologically interesting genera and species can be accommodated, and some of the "non-conforming" new discoveries will find temporary "residence" in one of three catchall categories (Px-M-, and Gx-cysts).

Although no formal suprageneric classification of organic-walled dinoflagellate cysts is suggested here, Table 8.1 shows the coordination believed to exist between 15 of the 17 morphological categories about to be defined and orders and families of the modern classification. Comments about family allocation of representative morphological types will be found under appropriate headings in Chapters 9 to 12.

D-Tests

Note. The term "test" is used in this exceptional case in preference to "cyst" because quite possibly these fossils are not the remains of resting cysts, but of sporopollenin coverings of cells in the motile stage (May, 1976).

Distinguishing Features. See Frontispiece F and Figure 5.3 G. The outline is similar to motile cells of

Family living today?	yes	yes	yes	?	yes	no	no	yes	yes	no
Families (Orders) / Cyst Categories	Peridiniaceae Protoperidiniaceae (Peri.)	Gonyaulacaceae Triadiniaceae (Peri.)	Ceratiaceae (Peri.)	Microdiniaceae ?=Cladopyxidaceae (Peri.)	Pyrophacaceae (Peri.)	Broomeaceae (Peri.)	n. fam. *(Suessia)* (Gymno.)	Gymnodiniaceae (Gymno.)	Polykrikaceae (Gymno.)	Nannoceratopsiaceae (Nanno.)
Pp	●									
Pq	●									
Gc			●							
Gv		●	○							
Gq		●								
Gs		●								
Gi		●		○	○					
Gn		●		○		○	○	○	●	
Gp				●		○				
Rp	●					○				
Rr	○					●				
M	○				○					
S							●	○		
D								●		
N										●

Table 8.1 - CORRESPONDENCE OF CYST MORPHOLOGICAL CATEGORIES WITH FAMILIES AND ORDERS

Solid circles represent the principal coordination believed to exist between morphological cyst category and family. Open circles indicate other families probably or possibly represented in the category owing to uncertainty of criteria or limited expression of distinctive features. The modern family Cladopyxidaceae may, in part, equate with the fossil family Microdiniaceae. Ordinal abbreviations: Peri. - Peridiniales, Gymno. - Gymnodiniales, Nanno. - Nannoceratopsiales. The categories of Px- and Gx-cysts are not listed since they include morphologically generalized cysts deficient in diagnostic characteristics.

living *Gymnodinium*; that is, subspherical to biconical or fusiform, without horns, and usually with well-marked cingulum and sulcus. The single-layered wall does not bear processes or septa, but is minutely perforate and is frequently distinguished by meridional folds. A small apical opening (which may be only analogous to an archeopyle although it has been referred to by this term) is formed by two short arcuate sutures at the apex. The surface is smooth or has ornament of low relief.

Comment. The gymnodinioid shape of D-tests makes them easy to recognize, although in a few species the cingulum and, especially, the sulcus are not pronounced.

Affinity. The biological integrity of the group is probably high, affinity lying with living Gymnodiniaceae, more specifically, perhaps, with *Balechina* (a living form with unusually thick and perforate pellicle at least superficially resembling *Dinogymnium*) and its allies.

Range. Restricted to Late Cretaceous.

Sole Genus. *Dinogymnium*.

N-Cysts

Distinguishing Features. Laterally compressed cysts with a minute epicyst and large hypocyst separated by a near-apical paracingulum. In lateral view, a narrow sagittal band margins the hypocyst, whose outline is also marked by one or more horns or prominent angles. The archeopyle is cingular and minute, but often associated with a partial displacement of the epicyst that may be readily visible. Details of wall and surface are distinctive but, at best, are barely visible in the light microscope. In known examples the surface is smooth to granular or reticulate, without processes or septa.

Comment. See Frontispiece H, Figure 5.3 F. These cysts are not easily confused with any others. The scanning electron microscope (Piel and Evitt, 1980) shows that their paratabulation combines an essentially peridinialean pattern on the epicyst with a dinophysialean pattern on the hypocyst.

Affinity. N-cysts give every appearance of being a closely interrelated group without living representatives. They are now referred to the order Nannoceratopsiales.

Range. Jurassic.

Sole Genus. *Nannoceratopsis*.

S-Cysts

Distinguishing Features. Cysts having a paratabulation comprising a large number of paraplates (well over 30), more or less similar in size and shape and variably defined by parasutural or intratabular features.

Comment. See Figure 5.3 E. These cysts are highly distinctive but as yet poorly known. Both the number of paraplate series (generally more than seven) and the number of paraplates (over 60) exceed those known from other dinoflagellate cysts. The composition of the archeopyle is uncertain but it includes the apical area and, in at least some examples, it involves paraplates from more than one series.

Affinity. Unknown. Superficial similarity of surface pattern is with some living species originally placed in alliance with *Gymnodinium* but now known to contain thin plates in the numerous amphiesmal vesicles (e.g., *Woloszynskia*, *Amphidinium carterae*; Dodge and Crawford, 1970b).

Range. Late Triassic to Middle Jurassic.

Representative Taxa. *Suessia*; *Dapsilidinium*? *langii* and undescribed forms with intratabular verrucae (under study by W. Wille, pers. comm.).

P-Cysts

Distinguishing Features. Either some indications of paratabulation (other than paracingulum and parasulcus) are present and are compatible with a peridinioid pattern represented by the formula *0-2pr, 4′, 2-3a, 7″, 5‴, 2⁗*; or a) no features incompatible with such a pattern are present, AND b) the outline is peridinioid.

Comment. This large group of cysts includes most of those that have been considered "typically peridinioid." Wall structure, surface features, and shape are highly varied, although a peridinioid outline is common. Paratabulation is rarely prominent. While parasutural and penitabular features do occur, long parasutural or penitabular projections are rare (some *Phthanoperidinium*) and prominent intratabular processes are unknown.

A characteristic peridinioid paratabulation may be fully developed, only partially discernible, or not visible at all. Details in the paracingulum and parasulcus, although critical in taxonomy of living peridinioid species, are usually too unclear in fossil cysts to be practically useful. The hypocyst paratabulation is exclusively *5‴, 2⁗*, whereas the epicyst paratabulation may match variations represented today in *Peridinium*, *Protoperidinium* subgenus *Protoperidinium* (i.e., with *4′, 3a, 7″*) or *P.* subgenus *Archaeperidinium* (i.e., with *4′, 2a, 7″).* The pattern with four apicals and three anterior intercalaries is most abundantly represented and can be considered "standard". Among fossil cysts, the most common variation of this pattern is the ortho hexa combination, but a para quadra combination is present in one important group (*Wetzeliella*). The archeopyle may involve *2a* alone; *1a*, *2a*, and *3a*; or precingulars or apicals in combination with *2a* or additional intercalaries. The adcingular margin of *2a* is typically rectilinear. The penta arrangement is as yet unknown among sporopollenin cysts of Paleogene or greater age.

The diagnosis is intended to encompass, in addition to distinctively peridinioid cysts, relatively generalized examples apparently lacking an archeopyle or other significant traces of paratabulation, providing the outline is peridinioid and no features are present that would place the example in another category.

Cysts which have a hypocyst paratabulation of *5‴, 2⁗*, but epicyst patterns other than *0-2pr, 4′, 2-3a, 7″*, fall into the category of Rp-cysts.

Affinity. Cysts of living *Peridinium* and *Protoperidinium* include many features well represented among fossil P-cysts. The paratabulation, when it is discernible, is so similar to the thecal patterns of living examples that biological affinity in these instances seems assured. However, cysts with much less definitive features are also included here. Most of them are probably from thecae that would be placed in the Peridiniaceae or Protoperidiniaceae, but this cannot be claimed for all of them, and taxa only remotely related to these families may be represented.

Range. Uncertainly Early Jurassic (*Liasidium*) to Holocene, but most abundant and varied from the Aptian on, and especially characteristic of the Late Cretaceous and Paleogene.

Representative Genera. See individual categories that follow.

Pq-Cysts

Distinguishing Features. P-cysts whose discernible paratabulation (disregarding the paracingulum and parasulcus) matches the tabulation of *Protoperidinium* subgenus *Protoperidinium* and whose dorsal epicyst organization is quadra (whenever determinable), accompanied by other features in various combinations. The ventral epicyst organization, once considered ortho, is now believed to be para, but this can seldom be established with certainty owing to the infrequency with which a clear paratabulation of the critical area is visible. Most specimens are relatively large (over 100µm) and wide (L:W under 1.25) cysts with pronounced dorsoventral compression and distinct peridinioid outline modified by prominent cingular angles or horns. The surface ranges from smooth to highly ornamented with spines or short processes, and an ectophragm may be suspended on process tips. The wall is cornucavate to circumcavate. The archeopyle, Type I, is often rounded in outline and the operculum is adnate in some species (along *2a/3′*).

Comment. This category includes the *Wetzeliella* and *Apectodinium* complexes. Most examples share many of the features mentioned, but significant exceptions require the extended characterization above and the qualifying phrases in the key. For example, *Rhombodinium* is divergent in its smooth exterior, and *Apectodinium* in the rounded outline and poorly visible archeopyle in some species, as well as in the close appression of wall layers and a thin endophragm.

Affinity. The cysts in this category appear to be a compact group with close affinities to living Peridiniaceae.

Range. Late Paleocene to early Miocene.

Representative Genera. *Apectodinium*, *Dracodinium*, *Gochtodinium*, *Kisselovia*, *Rhombodinium*, *Wetzeliella*, *Wilsonidium*.

Pp-Cysts

Distinguishing Features. P-cysts whose discernible paratabulation (aside from paracingulum and parasulcus) matches the tabulation of those living species of *Peridinium* or *Protoperidinium* (or genera created from then by subdivision) that have penta or hexa arrangements of three intercalaries, or only two intercalaries (such as species of *Protoperidinium* subgenus *Archaeperidinium*).

Comment. Cysts in this category are highly varied in all features. *Deflandrea* (Frontispiece B) and similar genera are typical but not exclusively representative. The external surface is normally smooth or with features of low relief, but prominent projections occur in some taxa. Cavation is often conspicuous. Clearly and extensively expressed paratabulation is uncommon but is notable in some (e.g., *Phthanoperidinium* spp. and *Palaeoperidinium pyrophorum*).

Affinity. Many appear to be close to living Peridiniaceae and Protoperidiniaceae, but relationship of forms with less distinctive morphology is correspondingly uncertain.

Range. Uncertainly from the Early Jurassic (*Liasidium*); Early Cretaceous to Holocene.

Representative Genera. *Alterbia*, *Ceratiopsis*, *Chatangiella*, *Deflandrea*, *Geiselodinium*, *Ginginodinium*, *Isabelidinium*, *Lejeunecysta*, *Manumiella*, *Nelsoniella*, *Palaeocystodinium*, *Palaeoperidinium*, *Phelodinium*, *Phthanoperidinium*, *Selenopemphix*, *Spinidinium*, *Umbodinium*; some living species of *Protoperidinium* and *Peridinium*.

Px-Cysts

Distinguishing Features. P-cysts which cannot be placed in either of the two preceding categories. The archeopyle or paratabulation may be absent, incomplete, or indeterminate; or the segments of the archeopyle suture may not be identifiable with particular sutures in *Protoperidinium*.

Range. Early Cretaceous to Holocene.

Representative Genera. *Diconodinium*, *Palaeohystrichophora*, *Subtilisphaera*.

R-Cysts

Distinguishing Features. The archeopyle or other indications of paratabulation identify three or more large anterior intercalaries, but these occur in an epicyst pattern that is not compatible with those in P-cysts. Hypocyst patterns may be obscure or otherwise, and may include *5‴*, *2⁗*, as in P-cysts; but a partiform hypocyst pattern excludes a cyst from these categories.

Comment. In all R-cysts, the anterior intercalaries number three or more (even if only one or two of them are involved in archeopyle formation). Typically, they appear to be about symmetrically disposed and the medial intercalaries are fastigiate, but exceptions may occur.

The epicyst pattern may be incompatible with that of P-cysts for a variety of reasons, such as having more than four apicals, other than seven precingulars, or fastigiate medial intercalaries.

Wall structure is varied, and surface ornament, although mostly of low relief, may include intratabular bulges. The archeopyle may involve one or more of the intercalary paraplates, alone or in various combinations with apicals and precingulars. Paratabulation, except for the archeopyle and traces elsewhere, is seldom fully indicated, although it has been determined for a few genera. Careful analysis of excellently preserved material is needed to further illuminate this interesting group.

Affinity. Almost certainly, several major biological groups are represented. Paratabulation patterns of most R-cysts are unknown among living dinoflagellates, but some are close to or apparently identical with thecal tabulation in living non-cyst-forming *Heterocapsa* (Goodman and Evitt, 1981), a genus included in the Peridiniaceae by Dodge (1981) but excluded from the Peridiniineae by Bujak and Davies (1983).

Range. ?Late Silurian (*Arpylorus*); Late Triassic (Carnian) to Early Cretaceous (Valanginian and Albian).

Rp-Cysts

Distinguishing Features. R-cysts having a determinable hypocyst paratabulation that is peridinioid (*5‴, 2⁗*) and an epicyst pattern that is compatible with *0-2pr, 5′, 3a, 7″* insofar as it can be determined.

Comment. This is a group of cysts of which few examples have yet been described, but paratabulations suggestive of living *Heterocapsa* (=*Cachonina*) occur in a number of taxa now under study. However, one species of *Heterocapsa*—the type species, *H. triquetra*—has a posterior intercalary not present in the better known *H. niei* (Morrill and Loeblich, 1981). A comparable intercalary has not been described among fossil cysts. Anterior intercalaries are generally involved in archeopyle formation and these paraplates may be fastigiate or linteloid. If hypocyst paratabulation in such cysts were not recognizably peridinioid, those having fastigiate anterior intercalaries would fall in the category of Rr-cysts.

Affinity. Unknown, except as suggested by remark in preceding paragraph.

Range. Early Jurassic-Early Cretaceous (Albian).

Representative Genera. *Angustidinium*, *Susadinium*.

Rr-Cysts

Distinguishing Features. R-cysts in which the hypocyst pattern is: a) not decipherable, or b) decipherable but not 5‴, 2⁗.

Comment. There are at least three important morphological components to this category. One includes distinctive autocysts and some cavate cysts (mostly more or less unornamented) from the Jurassic and Early Cretaceous, which are typically pyriform and have the archeopyle as about the only indication of paratabulation. *Broomea*, *Pareodinia*, and *Cantulodinium* are examples; another commonly encountered representative of this group is *Gochteodinia*, but it is atypical in having conspicuous nontabular processes.

A second component is a possibly heterogeneous assortment of stratigraphically older types (Late Triassic to Middle Jurassic) whose morphology is quite varied. The shape may be spherical to biconical, or even roughly peridinioid; the wall cavate or consisting of an autophragm only; the ornament subdued, parasutural or nontabular; the archeopyle simple Type I or more complex, including vaguely defined types; the paratabulation ranging from an indication by archeopyle alone to complete delineation and, for the most part, as yet very imperfectly known. Examples are *Dapcodinium*, *Heibergella*, *Moesiodinium*, and *Rhaetogonyaulax*.

A third component of even less fully explored cysts includes *Comparodinium*, recently interpreted (Wille and Gocht, 1980) to have an archeopyle composed of two out of the series of apical paraplates, which, owing to their position on the dorsal slope somewhat away from the topographic apex, give the impression of being intercalary. The intercalaries below these apicals are poorly set off from the precingulars. Other indications of paratabulation are usually incomplete or difficult to follow, but a row of posterior intercalaries may be indicated in the hypocyst.

Transfers from Rr-cysts to Rp-cysts can be expected if hypocyst patterns in currently uncertain cases can be resolved to *5‴, 2⁗*.

Affinity. Probably varied; uncertain.

Range. Silurian?; Late Triassic (Carnian) to Early Cretaceous (Valanginian and Aptian). Sarjeant's (1978) interpretation of the paratabulation of *Arpylorus antiquus* places it among the Rr-cysts, but analyses of additional and better preserved examples of this exceptional species are needed before the interpretation of its pattern can be considered free of uncertainties.

Representative Genera. *Broomea*, *Cantulodinium*, *Dapcodinium*, *Gochteodinia*, *Kalyptea*, *Pareodinia*, *Rhaetogonyaulax*.

G-Cysts

Distinguishing Features. Either the paratabulation corresponds to one of the gonyaulacoid patterns, or, if the pattern is not definitive, any traces of paratabulation present are compatible with such a pattern, insofar as they are expressed, and do not indicate any other.

Comment. The large set of G-cysts is apportioned among eight categories and distinguishing features of each are stated. It should be understood that, in each case, additional features not mentioned do not include any that are diagnostic of another G-cyst category. If a cyst under consideration is taken to be a Gp-, Gq-, or Gs-cyst, but cannot be placed unambiguously owing to uncertainties in critical details of hypocyst paratabulation, it should be considered a Gs-cyst.

In the long run it is unlikely that sharp boundaries between these subsets of G-cysts will be reinforced by new observations. More likely they will become blurred if, as seems likely, the categories are but facets of a single major group. However, practical distinctions between them seem possible, even if a few individual forms must be placed more or less arbitrarily.

Gc-Cysts

Distinguishing Features. G-cysts in which one apical, one antapical, and one or two postcingular horns are present and usually prominent, resulting in a distinctly ceratioid outline. The archeopyle is apical and is usually associated with both distinct dorsoventral compression and clear offset of the sulcal notch to the left of the ventral surface.

Comment. Only rarely is the diagnostic corniform pattern of parasutures on the hypocyst clearly expressed, and frequently the multiangular margin of the archeopyle is the only indication of paratabulation. Wall structure and ornament are varied, but distinctly developed processes occur in only a few taxa. Nevertheless, the conspicuous horns usually make positive identification easy, providing their positions are correctly identified. If short and broad horns (e.g., *Aptea*) can be confused with the low prominences in the outline of some Gv-cysts, one of the hypocyst horns (often with traces of the paracingulum close to or across its base) is distinctly more remote from the polar axis than the other. In some species of *Phoberocysta* and *Xenascus* (Frontispiece J) the smaller horn(s) may be masked by elaborate or irregular processes. In *Muderongia* the bases of two nearly equal postcingular horns are essentially equatorial, and only the orientation of the distal portion may suggest that they are related to postcingular horns, as is confirmed by paratabulation in occasional species.

Affinity. Almost certainly close to the the modern family Ceratiaceae.

Range. Latest Jurassic (*Muderongia*) through Cretaceous, with greatest diversity from mid-Early to mid-Late Cretaceous.

Representative Genera. *Aptea*, *Endoceratium*, *Muderongia*, *Odontochitina*, *Phoberocysta*, *Pseudoceratium*, *Xenascus*.

Gv-Cysts

Distinguishing Features. G-cysts which lack a distinctly ceratioid outline but have a primary dorsoventral compression normally sufficient to establish a strongly preferred orientation of settled specimens. As in Gc-cysts, the archeopyle is apical and the sulcal notch is offset to the left of the ventral surface.

Comment. The hypocyst paratabulation is sexiform gonyaulacoid in the uncommon cases where it is apparent; large anterior intercalaries are not indicated. Wall structure and surface features are varied. Usually they are modified over the central dorsal and central ventral areas. This effect (most commonly expressed as a suppression of ornament and fusion of wall layers) may be conspicuous.

This category is defined separately out of purely practical considerations. Its representatives are often abundant and not infrequently are dominant elements in assemblages. Although full intergradation between Gv-cysts and Gs-cysts probably occurs, the strongly preferred orientation determined by the lenticular shape of Gv-cysts is a convenient basis for separating them. Actually, the only distinctions between Gv-cysts and Gs-cysts appear to relate to this primary dorsoventral compression which may be suspected first from discovering the transversely elongate operculum.

The apical archeopyle and dorsoventral compression may also suggest Gc-cysts, causing uncertainty in a few cases; however, the one or two hornlike protrusions that may occur in the hypocyst outline of a Gv-cyst are antapical, not one antapical and one postcingular as in a Gc-cyst.

Affinity. Almost certainly this lies (as for Gs-cysts) with living Gonyaulacaceae.

Range. Latest Jurassic to Miocene; especially abundant and varied in Maastrichtian to Oligocene strata.

Representative Genera. *Areoligera*, *Canningia*, *Canninginopsis*, *Chiropteridium*, *Cyclonephelium*, *Glaphyrocysta*, *Membranophoridium*, *Palynodinium*, *Renidinium*, *Senoniasphaera*.

Gq-Cysts

Distinguishing Features. G-cysts with quinqueform hypocyst paratabulation indicated by parasutural or penitabular features; or a) a hypocyst paratabulation that, while not demonstrably quinqueform, is compatible with that pattern insofar as it is indicated, AND b) the archeopyle operculum includes two separate pieces for **1u** (=*1′*) and **A-C** (=*2′-4′*); for example, as in Type A(3A)6P.

Comment. Both proximate and chorate examples are known, with varied surface features and wall structure. The archeopyle most frequently involves the entire epicyst, the complex Type A(3A)6P being the most distinctive example. Since the only hypocystal paratabulation known to occur in combination with the Type A(3A)6P archeopyle is the quinqueform arrangement, the presence of this type of archeopyle is taken to indicate a Gq-cyst, even when the paratabulation of the hypocyst may not determinable independently.

This category includes important and varied morphological types whose distinction from Gs-cysts has seldom been specified in published works because the critical paraplate relationships in the antapical region have been overlooked.

Affinity. The quinqueform gonyaulacoid pattern on the hypocyst points to affinity with living *Triadinium* and those genera of the Gonyaulacaceae having a similar hypothecal tabulation (e.g, *Protogonyaulax* and *Pyrodinium*). The cyst of living *Pyrodinium bahamense* is morphologically a *Polysphaeridium*.

Range. Early Cretaceous to Holocene, with greatest diversity and numbers in the Early Tertiary.

Representative Genera. *Alisocysta*, *Dinopterygium*, *Eisenackia*, *Eocladopyxis*, *Heteraulacacysta*, *Homotryblium*, *Polysphaeridium* (=*Hemicystodinium* of many authors), *Tubidermodinium*.

Gp-Cysts

Distinguishing Features. G-cysts which have a partiform hypocyst paratabulation; or a) have a hypocyst paratabulation that, while not demonstrably partiform, is compatible with that pattern insofar as it is indicated, AND b) are small (under 50µm), essentially spherical, and have an indication of relatively large dorsal anterior intercalaries.

Comment. One difficulty in defining this category arises from the overlap that appears between those examples with large dorsal anterior intercalaries and R-and P-cysts, especially if other indications of paratabulation are vague. Another problem is that this category currently cannot be delineated as satisfactorily as some of the others, simply for lack of information. Few of the many observed species that apparently fall here have been described and details of paratabulation for most of them are uncertain. Two groups within the category appear distinguishable.

1. *Microdinium* complex. This includes small (under 50µm), mostly spherical to ellipsoidal cysts, usually with relatively wide paracingulum, epicyst smaller than hypocyst, and surface marked by ridges or other ornament of low relief. The walls are varied and may be relatively thick, refractive, and internally structured, making analysis of paratabulation difficult. Anterior intercalaries may be absent or may number up to four or five. Several types of archeopyles have been observed, including Type (2A),Types 4A and (4A), Type P or 2P, Type E.

2. *Phanerodinium* complex. This includes equally small cysts distinguished from the *Microdinium* group by dorsoventral compression, reduction of ornament and paratabulation in the central ventral region, and lack of anterior intercalaries. In addition, the wall may be single layered or cavate with selectively localized pericoels. The archeopyle is apical or precingular, but details are difficult to decipher owing to the very small size of the epicyst.

In addition to these two groups, several isolated taxa with properties similar in general but distinct in detail can be referred here.

Affinity. Uncertain and possibly mixed. Paratabulation of some species of the *Microdinium* complex suggests living Cladopyxidaceae. Dorsal intercalaries of other taxa (e.g., *Luehndea*, *Carpathodinium*) at first suggest ties with living Peridiniaceae, but this appears to be a misleading superficial resemblance inconsistent with other details of paratabulation.

Range. Early Jurassic (*Eyachia* and *Luehndea*) to Miocene; especially varied among small cysts in Late Cretaceous-Miocene, with occasional earlier and, in part, much larger representatives.

Representative Genera. *Microdinium* complex: *Cladopyxidium*, *Fibradinium*, *Hystiocysta*, *Microdinium*, *Subtilidinium*. *Phanerodinium* complex: *Druggidium*, *Gillinia*, *Phanerodinium*. Miscellaneous: *Carpathodinium*, *Eyachia*, "*Lacrymodinium*", *Luehndea*, *Paragonyaulacysta*.

Gs-Cysts

Distinguishing Features. G-cysts typically with a hypocyst paratabulation that is indicated by parasutural features and that is either sexiform or a) while not demonstrably sexiform, is compatible with that pattern insofar as it is indicated, AND b) cannot be distinguished as quinqueform or partiform. Prominent intratabular features are not present.

Comment. The diagnoses of Gq-, Gp-, and Gs-cysts are phrased so that equivocal types will be considered Gs-cysts. This category includes most fossils that generally have been considered "typically gonyaulacoid" on the basis of well-defined parasutural pattern. Wall structure and surface ornament vary greatly; shape is undiagnostic. Both S-type and L-type ventral organizations occur, with variations. Archeopyle Types (4A) and P are most common. Large dorsal anterior intercalaries do not occur.

Affinity. Some living forms with the morphology of *Nematosphaeropsis* and *Spiniferites* are known to represent modern Gonyaulacaceae. The same relationship is also suggested by the sexiform gonyaulacoid pattern on the hypocyst of many others and is supported by the close similarity of the epicystal patterns of some fossils to modern epithecal patterns.

Representative Genera. *Achomosphaera*, *Cannosphaeropsis*, *Ctenidodinium*, *Evittosphaerula*, *Gonyaulacysta*, *Hystrichodinium*, *Hystrichostrogylon*, *Impagidinium*, *Isthmocystis*, *Leptodinium* (Frontispiece D, G), *Meiourogonyaulax*, *Muratodinium*, *Nematosphaeropsis*, *Rottnestia*, *Spiniferites*.

Gi-Cysts

Distinguishing Features. G-cysts whose paratabulation is indicated by relatively few (normally 30 or fewer) intratabular features. These may include single processes, process groups defined by spacing or interconnections, or (less commonly) arcuate septa.

Comment: Parasutural features normally are not present (but note *Oligosphaeridium abaculum* and some species of *Hystrichokolpoma*; see Chapter 10). In the *Callaiosphaeridium* complex, intratabular features are limited to the paracingulum and parasutural features may be conspicuous toward the poles. Archeopyle varied, but Types (4A) and P are the most common.

Affinity. Most are clearly gonyaulacoid but the intratabular features leave details uncertain. A sexiform pattern of parasutures in *Oligosphaeridium abaculum* suggests living *Gonyaulax*, whereas an insert **1u**, combined with quinqueform hypocyst documented by shapes of process tips in specimens of *Hystrichosphaeridium tubiferum* from some localities, suggests living *Protogonyaulax*, and this association is also suggested in other cases by the arrangement of intratabular processes.

Representative Genera. *Amphorula*, *Areosphaeridium* (most species), *Callaiosphaeridium*, *Conosphaeridium*, *Cordosphaeridium*, *Florentinia*, *Hystrichokolpoma*, *Hystrichosphaeridium*, *Kleithriasphaeridium*, *Litosphaeridium*, *Oligosphaeridium* (Frontispiece E), *Perisseiasphaeridium*, *Polystephanephorus*, *Schematophora*, *Silicisphaera*, *Surculosphaeridium*, *Systematophora*, *Turbiosphaera*.

Gn-Cysts

Distinguishing Features. Skolochorate G-cysts which have numerous (usually well over 30) processes in nontabular distribution or have parasutural alignment indicated only locally and with insufficient clarity for the paratabulation to be resolved.

Comment. The cysts in this varied category share the combination of spherical to moderately ellipsoidal or lobate shape, conspicuous processes, and absence of a well-defined paratabulation. Most have an autophragm or closely appressed periphragm and endophragm; conspicuous cavation is rare. Type (4A) and Type P archeopyles are the most common.

Affinity. Probably varied; some clearly have links to living Gonyaulacaceae. Cysts with the morphology of *Lingulodinium* and *Operculodinium* occur in living *Gonyaulax polyedra* and *G. grindleyi*, respectively.

Representative Genera. *Cleistosphaeridium* (of many authors), *Coronifera*, *Diphyes*, *Exochosphaeridium*, *Kiokansium*, *Lingulodinium*, *Operculodinium*.

Gx-Cysts

Distinguishing Features. Not like any of the other categories of G-cysts.

Comment. Although this category is a catchall defined by a negative criterion, not all Gx-cysts are nondescript. Some are, indeed, relatively featureless, but others are morphologically distinctive types whose features simply do not match those selected as discriminators for the other categories. Additional categories might be designed to receive them, but these would be minimally useful because many would contain only one or two genera. This circumstance, combined with the universally obscure paratabulation, suggests gross "lumping" of these varied cysts as the most satisfactory treatment.

Affinity. Some of the taxa that fall here are probably close to living Gonyaulacaceae, but obvious heterogeneity of the group renders meaningless any discussion of relationship on an inclusive basis.

Representative Genera. *Aldorfia*, *Apteodinium*, *Bitectatodinium*, *Chlamydophorella*, *Eodinia*, *Fromea*, *Gardodinium*, *Kallosphaeridium*, *Palaeotetradinium*, *Prolixosphaeridium*, *Samlandia*, *Tectatodinium*, *Trigonopyxidia*, *Wallodinium*.

M-Cysts

Distinguishing Features. Cysts which have clearly defined properties of paratabulation that, while not necessarily complete, are sufficient to preclude identification as a gonyaulacoid pattern.

Comment. This is a catchall for cysts with paratabulation that is not compatible with those patterns accomodated in the other categories.

Affinity. Likely to be mixed and of varied certainty.

Representative Genus. *Tuberculodinium* (= cyst of *Pyrophacus*, with two or three antapicals indicated by archeopyle); living *Diplopsalis, Zygabikodinium.*

KEY TO 17 MORPHOLOGICAL CATEGORIES OF SPOROPOLLENIN DINOFLAGELLATE CYSTS

The key discussed in the following paragraphs is presented in Figure 8.1 in both textual and graphic form. This commentary is intended to increase the understanding of the key and simplify its use with species whose morphology lies within the range represented by currently described taxa. With this understanding, it should be relatively easy to determine the cyst category for most species. However, some uncertainties and ambiguities are inevitable and they are bound to increase as new taxa, which could not be considered when the key was designed, are encountered.

The first step in considering an "unknown" dinoflagellate should be to assess its morphology (see the Checklist early in this chapter) to the fullest extent permitted by preservation and orientation. Naturally, the degree of conscious effort which this requires and the speed with which it is accomplished will vary with the experience of the observer. The same applies to all later steps in the process, as well. Also, the completeness of the information that can be garnered about the species will be affected critically by the number of specimens available, the quality of their preservation, and their orientation. When observations on the specimen(s) are complete, one can commence with the keying-out process.

At Step 1, three distinctive groups of fossils–D-tests, N-, and S-cysts—can either be eliminated from further consideration or one of them quickly selected as the sought-for category.

D-tests (all currently referable to *Dinogymnium*) have a single-layered wall and are most readily recognized by a gymnodinioid outline (subconical to subhemispherical epi- and hypotest, indented cingulum), lack of tabulation, subdued surface ornament, and, very often, longitudinal folds of the wall. The small apical opening of distinctive form is a confirming feature if it is visible. Species with other outline shapes, lacking an indented cingulum, or with unfolded wall may require further consideration. The most common shapes of D-tests favor equatorial views that are otherwise random in orientation, but the highly flexible wall permits depositional compression in many directions, with corresponding variety in the appearance of flattened specimens. Refer to Chapter 12.

N-cysts (all currently referable to *Nannoceratopsis*), although apparently autocysts, may have walls that appear reticulate or spongy. Their diagnostic feature is a strong compression (lateral), combined with absence of a paracingulum anywhere close to the midlength of the long axis. Most specimens will show a paracingulum and minute conical epicyst close to one end of the cyst, and one to three pointed to rounded lobes at the opposite end. Any orientation other than a lateral view is exceptional. Most species are conspicuously resistant to staining or coloring by acetolysis. Specimens with the anterior part of the cyst partly or completely separated from the much larger hypocyst are common; probably this is related to archeopyle development but, ordinarily, neither this opening nor any other trace of paratabulation is distinct in the light microscope. Refer to Chapter 12.

S-cysts, in contrast to D-tests and N-cysts, have a paratabulation evidenced by parasutural ridges or septa, or by intratabular processes. They are distinguished from peridinioid, gonyaulacoid, or any similar types by the large number of small paraplates, which are of similar size and are arranged in more than the usual five or six latitudinal series (up to 10). Superficial resemblance of *Suessia* (with parasutural ridges/septa) to the acritarch *Cymatiosphaera* or a reticulate lycopod spore may be striking if paracingulum or parasulcus is not visible. Original shape is subspherical to subbiconical and orientation is correspondingly nearly random. Refer to Chapter 12.

At Step 2, either of two observations may be critical. The more important one relates to the paratabulation, the other to the outline shape. If one or more dorsal paraplates of relatively large size are indicated (by the archeopyle or other paratabulation) in an apparently anterior intercalary position, OR if the cyst has a peridinioid outline and no features that would exclude the possibility of a peridinioid tabulation, then one should proceed to Step 3. If neither of these conditions fits, go to Step 8. Intercalary paraplates that are relatively minute compared to adjacent precingulars and apicals, such as occur in some gonyaulacoid patterns (Figure 5.16 B-D), should not be considered. For purposes of the choice at Step 2, a peridinioid outline should be construed to encompass the range shown in Figure 9.2 and to include extreme (but rarely encountered) examples in which the usual second antapical horn is unrepresented.

At Step 3, the decision concerns exclusively the hypocyst. If this area shows the gonyaulacoid pattern we have defined as partiform, then the cyst is a Gp-cyst. If the paratabulation is not sufficiently clearly expressed to be identified as partiform, or if it is clear and is definitely not partiform, then proceed to Step 4. Gp-cysts in which **X** is nearly as large as **Y** and occupies an approximately antapical position can be distinguished from superficially similar Pp- and Rp-cysts by considering that the partiform hypocyst comprises six postcingulars, of which the third from the last (**IV**) contacts only one antapical (**Y**), whereas the hypocyst in the other two categories has five postcingulars, of which the third from the last (*3‴*) contacts both antapicals.

Gp-cysts, may lack anterior intercalaries, or one or more (up to five are known) may be symmetrically or asymmetrically arranged, and may be part of the archeopyle or not. Size, wall structure, and archeopyle type vary. Shape also varies and includes some pyriform cysts that are confusingly suggestive of R-cysts, but can be distinguished by the gonyaulacoid hypocyst. Examples are: *Cladopyxidium*, "*Lacrymodinium*", *Microdinium*, and *Paragonyaulacysta*.

At Step 4, the choice will be easy if the archeopyle or parasutural features give positive indication of the presence of relatively large dorsal anterior intercalaries. Step 5 should follow if the form of the intercalaries and the rest of the paratabulation (disregarding the paracingulum and parasulcus), to whatever extent it is evident, are compatible with the pattern of tabulation in modern marine *Protoperidinium* subgenus *Protoperidinium* or freshwater *Peridinium—4′, 3a, 7″, 5‴, 2⁗*—or modern marine *Protoperidinium* subgenus *Archaeperidinium—4′, 2a, 7″, 5‴, 2⁗*—or genera created from them by subdivison. Step 5 should also follow in the less ideal case where the outline of the cyst is peridinioid but intercalaries are not indicated, providing that the paratabulation, insofar as it can be made out, is compatible with the presence of intercalaries as in the examples just specified. Note that satisfactory evidence for such anterior intercalaries need not include complete delineation of the paraplates. For example, in cysts whose only indication of paratabulation is provided by an archeopyle that includes paraplates other than the intercalaries, the intercalaries may be reliably indicated (although not necessarily obviously) by the archeopyle margins (e.g., Figures 6.5, 6.10 A-I). Cysts which do not fit these conditions are R-cysts and should be considered at Step 7.

At Step 5, Pq-cysts, a category comprising chiefly cysts of the *Wetzeliella* complex, are distinguished. For placing the remainder, proceed to Step 6. Paratabulation on the dorsal epicyst in the *Wetzeliella* complex is quadra style when fully expressed, but it is often indistinct. All cysts with generally peridinioid features and a quadra style paratabulation fall here.

For practical purposes, the quite distinctive shapes of cysts within the *Wetzeliella* complex are sufficient basis for placing them here. Typically, these circumcavate to cornucavate cysts are strongly compressed dorsoventrally and show distinct periphragm and endophragm. The conspicuously peridinioid outline of the pericyst commonly is modified laterally by prominent angulations at the paracingulum, which may be extended into definitely hornlike projections. The subcircular to rounded-polygonal outline of the endocyst is usually distinct. An intercalary archeopyle (only Type I and Ia known) is almost always obvious, but its outline is commonly rounded. A quadra organization is rarely distinct and is evidenced more often by the relative widths of the archeopyle (*2a*) and paraplate *4″* than by clearly expressed outlines of the critical paraplates. The surface may be smooth to granulate, or provided with few to many short processes with or without ectophragmal connections. Some members of the complex exhibit less compression or an otherwise atypical combination of features,

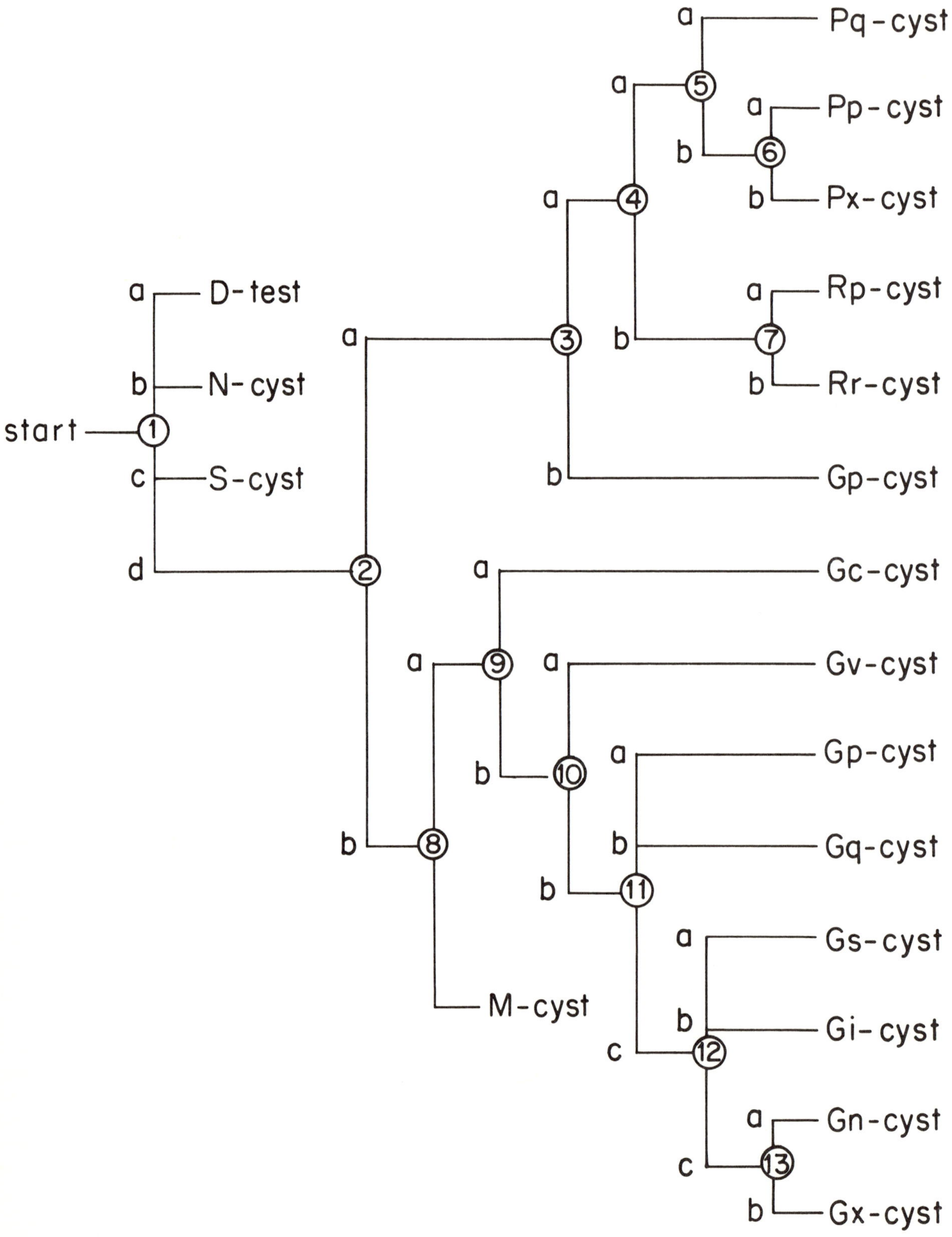
start
D-test
N-cyst
S-cyst
Pq-cyst
Pp-cyst
Px-cyst
Rp-cyst
Rr-cyst
Gp-cyst
Gc-cyst
Gv-cyst
Gp-cyst
Gq-cyst
M-cyst
Gs-cyst
Gi-cyst
Gn-cyst
Gx-cyst

Figure 8.1

Figure 8.1 - KEY TO 17 MORPHOLOGICAL CATEGORIES

The objective of the key is to facilitate placing a dinofagellate with a sporopollenin wall into one of the 17 categories on the basis of morphological features that can be determined as easily and positively as possible. The design and use of the key are discussed in the text. General familiarity with the characteristics of the categories set forth in the earlier part of this chapter is assumed.

1a Fossil without traces of paratabulation, small opening at extreme apex formed by two short, arcuate slits; wall of one layer and perforated by minute canals; surface smooth to granulose, without processes or septa, but commonly with longitudinal folds; shape suggests *Gymnodinium* ... D-test

1b Cyst strongly compressed laterally and paracingulum close to apex; epicyst minute relative to hypocyst; paratabulation not normally visible in light microscope; archeopyle cingular and minute, but entire epicyst may be displaced ... N-cyst

1c Cyst with well over 30 paraplates of about similar size and shape defined by parasutural or other features and arranged in more than 6 latitudinal rows and more numerous in each row than normal for a peridinialean; archeopyle apical or otherwise, may be of disintegration type ... S-cyst

1d All other sporopollenin cysts ... 2

2a Archeopyle or other paratabulation indicates one or more relatively large dorsal anterior intercalaries; OR both the following apply: a) paratabulation, insofar as it is indicated, is compatible with the presence of such intercalaries, and b) cyst outline is peridinioid ... 3

2b Not as in 2a ... 8

3a Hypocyst paratabulation is not clearly partiform gonyaulacoid (see next step for clarification) ... 4

3b Hypocyst paratabulation is clearly partiform gonyaulacoid (*6''', 1p, 1'''' or 6''', 2''''* in Kofoid terms) ... Gp-cyst

4a Insofar as it is indicated, paratabulation (disregarding the paracingulum and parasulcus) is compatible with the tabulation of modern marine *Protoperidinium* subgenus *Protoperidinium*, *Protoperidinium* subgenus *Archaeperidinium*, freshwater *Peridinium*, or genera created by subdivision of them ... (P-cysts) 5

4b Not as in 4a ... (R-cysts) 7

5a Dorsal epicyst paratabulation is quadra style; OR all the following apply:

1) Paratabulation, insofar as indicated, is compatible with quadra style, AND
2) Archeopyle is Type I_{2a} or Type I_{2a}a, AND
3) At least two of the following apply:
   —L:W not greater than 1.25
   —moderate to severe dorsoventral compression (most forms have prominent peridinioid outline and conspicuous cingular angles or horns),
   —ornament of spines or short processes (tips often T-shaped or expanded into ectophragm),
   —cornucavate to circumcavate wall structure ... Pq-cyst

5b Not as in 5a ... 6

6a Archeopyle (in at least one wall) is Type I_{2a} or I_{2a}a; OR the archeopyle or other traces of paratabulation indicate a pattern of parasutures in accord with the tabulation of modern *Protoperidinium* subgenera *Protoperidinium* (3 intercalaries) or *Archaeperidinium* (2 intercalaries) ... Pp-cyst

6b Not as in 6a ... Px-cyst

7a Hypocyst paratabulation is definitely peridinioid (*5''', 2''''*) ... Rp-cyst

7b Not as in 7a ... Rr-cyst

8a Paratabulation is not indicated; OR, insofar as it is indicated, it is compatible with a gonyaulacoid pattern and not definitive of any other ... (G-cysts) 9

8b Paratabulation is indicated sufficiently to show that it is incompatible with a gonyaulacoid pattern ... M-cyst

9a Outline ceratioid; that is, with one apical, one antapical (rarely two), and one or two postcingular horns or prominent angles ... Gc-cyst

9b Not as in 9a ... 10

10a Dorsoventrally compressed cysts having sulcal notch offset to left side of ventral surface (except cysts with intratabular processes having expanded tips characteristic of the *Areosphaeridium* complex) ... Gv-cyst

10b Not as in 10a (including cysts with processes as in the *Areosphaeridium* complex) ... 11

11a Hypocyst paratabulation is partiform ... Gp-cyst

11b A quinqueform hypocyst paratabulation is indicated by parasutural or penitabular features in a cyst that is not skolochorate with intratabular processes; OR b) the archeopyle is Type A(3A)6P and the paratabulation, insofar as it is indicated, is compatible with a quinqueform pattern ... Gq-cyst

11c Not as in 11a or 11b ... 12

12a Hypocyst paratabulation indicated by parasutural features (including gonal or intergonal processes, or ectophragmal interconnections), or closely spaced penitabular features, BUT prominent intratabular features are absent ... Gs-cyst

12b Paratabulation indicated by intratabular features (e.g., single processes, process groups, or arcuate septa) ... Gi-cyst

12c Not as in 12a or 12b ... 13

13a Skolochorate cysts that are spherical to moderately ellipsoidal (L:W less than 1.5) or irregularly shaped (e.g., lobate) and whose process distribution is nontabular (excepting traces of alignment along paracingulum or elsewhere) ... Gn-cyst

13b Not as in 13a ... Gx-cyst

requiring the rather lengthy specification of characters at Step 5a3) in the key. The most important "misfit" in the category is *Apectodinium homomorphum*, which merits special attention. Refer to Chapter 9.

At Step 6, Pp-cysts, the major group of P-cysts, are distinguished from a more generalized category, Px-cysts. Pp-cysts include peridinioid cysts in which there is no indication that the paratabulation of the dorsal epicyst is quadra style and in which the archeopyle consists of *2a* alone or in combination with other paraplates of the epicyst. The category includes *Deflandrea*, *Palaeoperidinium*, *Phthanoperidinium*, *Trithyrodinium*, and a host of more or less similar forms. Collectively, they show a great range in shape (although an approximately peridinioid outline and some dorsoventral compression are usual), wall structure (normally autophragm, or periphragm and endophragm, but occasionally also mesophragm), surface ornament (typically smooth or having nontabular elements of low relief, but a few taxa have parasutural to penitabular spines, processes, or ridges), and archeopyle (compare Figure 6.5).

To distinguish between Pq- and Pp-cysts in specimens without paratabulation sufficiently clear to indicate whether the epicyst pattern is quadra or hexa style, consider carefully the characters under 5a2) and 5a3) in the key. Unless contraindicated by clear paratabulation, a Type I archeopyle with operculum adnate along *2a/3′* is presumptive evidence of a Pq-cyst. Pp-cysts with spines or short processes do not normally have strong dorsoventral compression and a *Wetzeliella*-like outline. Of currently known P-cysts, all having pronouncedly bicavate wall structure are Pp- or Px-cysts, and all with archeopyles other than Type I are Pp-cysts. P-cysts that either have no archeopyle or have an archeopyle suture that does not recognizably correlate with parasutures of a normal peridinioid paratabulation are Px-cysts. For further discussion of P-cysts, refer to Chapter 9.

At Step 7, a distinction is made between two categories of R-cysts. Rp-cysts have a peridinioid hypocyst paratabulation (*5‴, 2⁗*) which is combined with an epicyst pattern that, as already has been registered by the choice made at Step 4b, is incompatible with that of P-cysts. See comment under Step 3 regarding distinction between hypocysts of Rp- and Gp-cysts. In Rr-cysts, if the hypocyst paratabulation is determinable, it is neither peridinioid nor partiform gonyaulacoid; epicyst patterns are diverse. For further discussion of R-cysts, refer to Chapter 12.

At Step 8, the dichotomy that divides the remaining types of cysts is based on relatively simple and obvious criteria. In the great group of G-cysts, to whatever extent a paratabulation is indicated, it expresses or is compatible with a gonyaulacoid pattern; in some cases no paratabulation is recognizable. In the miscellaneous category of M-cysts, at least traces of a paratabulation that is definitely incompatible with a gonyaulacoid pattern are indicated. *Tuberculodinium*, with two or three large antapical plates indicated by the archeopyle, is an example. Refer to Chapter 12.

G-cysts are exceedingly diverse and are divided into eight categories. Some of these are quite precisely defined on the basis of paratabulation, archeopyle type, or shape; others are more generalized, and one is a heterogeneous catchall category. More or less close affinity of some G-cyst categories with allies of living *Gonyaulax* is suggested by similarities in inferred tabulation, but some of the categories almost certainly include forms that have nothing closely to do with that genus.

At Steps 9 and 10, Gc-cysts and Gv-cysts, two categories which have some conspicuous traits in common, are separated off. In both cases, cysts free from postdepositional compression, as well as those that have been flattened secondarily, exhibit a strongly preferred dorsoventral orientation in strew preparations. In the great majority, this is a consequence of substantial primary dorsoventral compression (lenticular shape); in a few that show no primary compression, it results from a nearly coplanar placement of horns. Without exception, the archeopyle in both Gc- and Gv-cysts is apical and, if angulations indicate the paraplates along its margin (as they usually do), the sulcal notch is offset to the left of the ventral surface and the portion of the margin bounding **1i** (*6″*) forms a broad, approximately midventral angle. Wall structure and surface features are highly varied, but in both categories (probably in conjunction with the lenticular shape) surface relief tends to be less pronounced (even to the point of absence) in midventral and middorsal regions than it is peripherally. Indications of paratabulation are usually local and faint. Intraspecific variability may be extreme.

The distinction between the two categories depends upon outline shape and horn development. Gc-cysts have genuinely ceratioid outlines, with one horn or prominent angle at the apex, one (rarely two) at the antapex, and one (right) or two (right and left) in postcingular position. Bases of the postcingular horns may be essentially equatorial in some cases. The hypocyst outline typically is strongly asymmetrical (e.g., *Odontochitina*, *Pseudoceratium*, *Xenascus*), but is nearly symmetrical in some (e.g., *Muderongia*, *Phoberocysta*). Refer to Chapter 10.

Gv-cysts lack long *Ceratium*-like horns, and the outline may be nearly circular. Frequently, however, rounded lobes in approximately the positions of horns in Gc-cysts are present and these may cause confusion. It is critical for the distinction that, in Gv-cysts, the two prominences on the hypocyst outline are both antapical, rather than one antapical and one postcingular. This may be indicated by relative symmetry of the antapical outline, relatively greater distance of the righthand element (i.e., the one that might be confused with the right postcingular horn of a Gc-cyst) from the paracingulum than in a Gc-cyst, or by traces of paratabulation indicating that the antapical prominences both arise from the antapical paraplate (*1''''* = **Y**). Typical examples are *Areoligera* and *Cyclonephelium*.

Note that special provision is made here for the group of cysts assembled in the *Areosphaeridium* complex. Although a few species in this group have the dorsoventral compression and offset sulcal notch otherwise characteristic of Gv-cysts, the prominent processes are intratabular and have the distinctive ectophragmal elaborations at their distal ends that are the subtle but critical indicators of this complex.

Gc-cysts are known from latest Jurassic and Cretaceous sediments; none have been reported from the Tertiary. Gv-cysts appear in the Late Jurassic and occur commonly from the Early Cretaceous until the Miocene; they are especially abundant and varied in many Tertiary samples. For further discussion of both Gc- and Gv-cysts, refer to Chapter 10.

At Step 11, a threefold division is made. Two categories, Gq- and Gp-cysts, are precisely defined by factors of paratabulation, whereas all cysts that do not have these particular features are further treated at Step 12.

Gp-cysts include some forms with and others without anterior intercalaries. The former have been considered already at Step 3. At this point we encounter those that do not have these paraplates. The critical feature distinguishing Gp-cysts from remaining groups of G-cysts is the presence of a partiform hypocyst pattern and, when it is visible, a characteristic arrangement of paraplates along the ventral meridian, both indicated by parasutural features. Most species are small (under 50μm) and subspherical to lenticular. Wall structures and surface ornament vary, but the latter typically includes only features of low relief. Archeopyle types vary but are mostly apical. Refer to Chapter 10.

Gq-cysts either have a quinqueform gonyaulacoid hypocyst paratabulation indicated by parasutural or penitabular features, or have a highly distinctive archeopyle, Type A(3A)6P, combined with a paratabulation that, to whatever extent it is indicated, is compatible with a quinqueform pattern. Chorate cysts with intratabular processes are included here only if the archeopyle is Type A(3A)6P and any traces of paratabulation visible are consistent with the quinqueform pattern. The category includes both proximate and chorate cysts with varied wall structures (rarely with pronounced cavation) and more or less equidimensional shape (subspherical to polyhedral). Archeopyle types vary. The effect of the characterization of Gq-cysts is to bring together some chorate forms (e.g., *Eocladopyxis*, *Homotryblium*, *Polysphaeridium*) and some proximate forms (*Dinopterygium*, *Heteraulacacysta*, *Tubidermodinium*) having features strongly suggestive of living *Triadinium*, *Pyrodinium*, and *Protogonyaulax*. Some cysts that may also correspond to thecate stages similar to these living genera either lack clear parasutural patterns or have those patterns combined with prominent intratabular processes and fall elsewhere (e.g., chorate *Hystrichosphaeridium* in Gi-cysts). Refer to Chapter 10.

At Step 12, two further categories are discriminated by combinations of paratabulation pattern and character of surface relief, leaving two final categories of considerably more generalized form and without clear paratabulation patterns for separation at the final stage, Step 13. The distinctive categories defined at Step 12 are Gs-cysts and Gi-cysts.

Gs-cysts are distinguished by a hypocyst pattern indicated by parasutural or penitabular features. By virtue of the discrimination already made at Step 11, this category will include cysts with a distinct sexiform pattern as well as those in which the pattern, if not demonstrably sexiform, is also neither partiform (Gp-cysts) nor quinqueform (Gq-cysts). Parasutural ridges and septa constitute the typical relief (e.g., *Cribroperidinium*, *Gonyaulacysta*, *Impagidinium*, *Leptodinium*, *Pentadinium*). Wall structures range from an autophragm (*Cribroperidinium*) to periphragm and endophragm with varied development of pericoels (*Gonyaulacysta*, *Pentadinium*) and include trabecular ectophragmal connections between processes in some cases (*Cannosphaeropsis*, *Nematosphaeropsis*).

Although prominent intratabular processes are not developed in Gs-cysts, parasutural processes (gonal or intergonal) may be conspicuous and, if ridges between them are suppressed (e.g., *Achomosphaera*), they may at first be taken for intratabular. *Actinotheca* and similar forms (including *Callaiosphaeridium*) require special attention and are discussed in Chapter 11. These genera are treated as Gi-cysts on the basis of their unique paracingular elaborations, which are considered to be intratabular.

Large dorsal intercalaries comparable to those in Gp-cysts do not occur in Gs-cysts. However, small ones, treated here as incidentals, are characteristic of a few taxa. One or two preapicals may also be present. Archeopyle types vary, but Types (4A), P, and E are the most common. Refer to Chapter 10.

Gi-cysts include a large and varied group of skolochorate cysts with subspherical to ellipsoidal endocyst or autocyst, from which project prominently intratabular features arranged to define partly or completely a gonyaulacoid paratabulation. The projecting structures may include processes that are isolated at the base and either free or interconnected distally (e.g., *Achilleodinium*, *Araneosphaera*, *Areosphaeridium*, *Callaiosphaeridium*, *Cordosphaeridium*, *Eatonicysta*, *Hystrichokolpoma*, *Hystrichosphaeridium*, *Litosphaeridium*, *Oligosphaeridium*, *Surculosphaeridium*, *Turbiosphaera*), process groups whose intratabular nature is indicated by spacing, alignment, or interconnections (e.g., *Emmetrocysta*, some *Florentinia* spp.), or penitabular ridges or septa sufficiently removed from the parasutures so that the emphasis is distinctly intratabular rather than parasutural (e.g., *Amphorula*, *Hemiplacophora*, *Histiophora*, *Hystrichosphaerina*, *Schematophora*, *Systematophora*). Typically one unit or one complex corresponds to one paraplate; occasionally one to several paraplates may be marked by more than one unit (e.g., some species of *Areosphaeridium* and *Cordosphaeridium*). Some Gi-cysts show marked selective suppression of processes (e.g., *Litosphaeridium*, *Oligosphaeridium*). Refer to Chapter 10.

At Step 13, a diversity of cysts remain. They exhibit a great variety of shapes, wall structures, surface features, and archeopyles. Traces of paratabulation, if present, are too faint, incomplete, or complicated by other wall structure for analysis under the light microscope to be practical. Some are almost certainly *Gonyaulax* relatives, but, for most, even a tentative suggestion of affinity would be speculative. Intergradation with groups in which paratabulation is more clearly indicated (especially Gs- and Gi-cysts) occurs and is difficult to deal with in a satisfactory manner. Pairs of existing genera that epitomize the problem include *Endoscrinium* and *Scriniodinium*, *Cribroperidinium* and *Apteodinium*, *Turbiosphaera* and *Fibrocysta* (paratabulate form listed first in each case). In borderline cases, generic assignment may be influenced by preservation, preparation, or quality of microscope optics. Only a single, relatively simple and consistently applicable subdivision is proposed here, using a rather gross difference in surface character.

Gn-cysts are spherical to moderately ellipsoidal or lobate skolochorate cysts with nontabular process distribution. Typical examples are *Cleistosphaeridium*, *Coronifera*, *Diphyes*, *Dapsilidinium*, *Exochosphaeridium*, *Fibrocysta*, *Lingulodinium*, and *Operculodinium*. Note that elongate (L:W over 1.5) skolochorate cysts fall into Gx-cysts.

Gx-cysts include all the rest. Cysts with intricate surface structure involving short projections or septa (e.g., *Ellipsoidictyum)* and cysts with conspicuous wall structure beneath the outermost surface (e.g., *Aldorfia*, *Chlamydophorella*, *Eodinia*, *Gardodinium*) are included in Gx-cysts, as well as cavate cysts (e.g., *Scriniodinium*) and acavate forms (e.g., *Apteodinium*) with simpler wall structure. Also included here are decidedly elongate cysts, regardless of wall structure and character of projecting features (so long as they are not parasutural or intratabular; e.g., *Distatodinium*, *Prolixosphaeridium*, *Tanyosphaeridium*, *Wallodinium*).

Chapter 9. Morphology of P-Cysts

Indications of paratabulation (including the archeopyle, supplemented by characteristics of shape and wall structure like those in living species) suffice to establish that the majority of fossil cysts here called P-cysts represent species that would be referable to living *Peridinium*, *Protoperidinium*, or closely similar genera on the basis of thecal morphology—if their thecae were available for study. Three individual categories of P-cysts are recognized: Pp-, Pq-, and Px-cysts.

CHARACTERISTICS OF CATEGORIES AND COMPLEXES

Introduction and Key

The category of Pp-cysts is, perhaps, the largest and most diverse of all the cyst categories discussed in this volume. It includes species that show extremes of almost any feature of cyst morphology that can be mentioned. Such variety renders generalization about the group fruitless and subdivision nearly essential. Accordingly, the following complexes are recognized and discussed individually: *Deflandrea* complex, *Spinidinium* complex, *Selenopemphix* complex, *Palaeoperidinium* complex, *Ascodinium* complex, and *Phthanoperidinium* complex.

Pq-cysts constitute a relatively restricted group of cysts, distinguished by direct or indirect indication of quadra style paratabulation on the dorsal epicyst. Most cysts of this category are also distinguished by relatively large size, dorsoventral compression, and prominent peridinioid outline (often including extensions of the angles at the paracingulum). Many have, in addition, higher or more robust surface relief (as spinelike or T-shaped processes, with or without ectophragmal interconnections) than is common among other P-cysts except for a few examples in the *Phthanoperidinium* and *Spinidinium* complexes of Pp-cysts and in *Palaeohystrichophora* and closely similar forms of Px-cysts. The *Wetzeliella* and *Apectodinium* complexes are recognized.

Px-cysts constitute an undifferentiated category of relatively nondescript cysts lacking the features necessary for placement as Pq- or Pp-cysts. Although some Px-cysts are almost certainly related to these groups, others may well not be, but diagnostic features are lacking or are as yet poorly understood.

Although the varied types of P-cysts are all believed to represent species more or less closely related to living genera of the Peridiniaceae (before emendation by Bujak and Davies, 1983), it does not follow that all cysts of peridiniacean taxa are (or were) morphologically P-cysts. For example, cysts of some living peridiniaceans (*Diplopsalis*, *Zygabikodinium*), with epicystal archeopyles and a paratabulation (when it can be determined) unlike that of P-cysts, fall in the category of M-cysts. Also, mid-Cretaceous *Angustidinium* and some Jurassic species with similar and distinctive epicyst paratabulation (e.g., *Susadinium*) fall in the category of Rp-cysts, despite paratabulations closely similar to the tabulation of the living (but non-cyst-producing) peridiniacean *Heterocapsa*.

Key to Complexes of P-cysts

I. Pp-cysts

   A. Without prominent parasutural ridges or septa that define paratabulation more or less completely, archeopyle characteristics important (Figure 9.1)

      1. Periarcheopyle or autoarcheopyle formed by *2a* alone; *2a* in conjunction with subjacent *4''*; or all three intercalaries only. Types I, Ia, IP, IPa (IP)a, 3I, 3Ia. Endoarcheopyle like periarcheopyle, or Types I/3I, I/(tAtI)a.

         a. Archeopyle middorsal

            1) Outer surface smooth or with relatively inconspicuous features

of low relief (grana, denticles, verrucae, short and delicate pila) ....*Deflandrea* cpx

2) Outer surface bears spines or diminutive processes.................... ....*Spinidinium* cpx

b. Archeopyle not middorsal ....*Selenopemphix* cpx

2. Combination archeopyles (periarcheopyle or autoarcheopyle formed by *2a* in combination with: more than one precingular, one or more apicals, or both precingulars and apicals) and apical archeopyles

a. Archeopyle formed by one or more intercalaries in conjunction with *3-5″*, with or without inclusion of *3′*; operculum either free or adnate, simple or compound. Types (I3P)a, I3Pa, (3I3P)a, (3I3P), 3I3Pa, 3I3P, (A3I3P), (A3I3P)a ....*Palaeoperidinium* cpx

b. Archeopyle formed by apicals alone, or in combination with intercalaries. Types (4A)a, (3A3I), (4A3I), (4A3I)a, (4A)a2I ....*Ascodinium* cpx

B. With prominent parasutural or penitabular ridges or septa defining paratabulation more or less fully................ *Phthanoperidinium* cpx

II. Pq-cysts

A. Endocyst well defined, cornucavate to circumcavate.......................... *Wetzeliella* cpx

B. Endophragm thin and appressed to or near periphragm, generally cornucavate.............. .... *Apectodinium* cpx

III. Px-cysts (undifferentiated)................................. .... Px-miscellaneous cpx

Category of Pp-Cysts

In any absolute sense, it is impossible to subdivide this great category of cysts on morphological grounds in a logically defensible manner, let alone one whose merits can be argued on sound biological grounds. All basic morphological features—shape, surface features, wall structure, archeopyle—are expressed in a profusion of ways and in a seemingly infinite variety of combinations. Although founded on morphologically distinct types, existing genera become increasingly difficult to maintain as

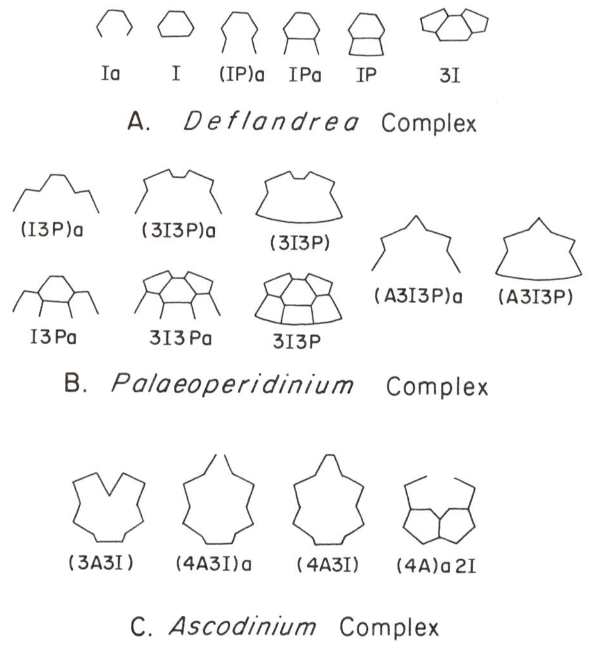

Figure 9.1 - ARCHEOPYLE TYPES IN MAIN COMPLEXES OF P-CYSTS

Principal and accessory archeopyle sutures as seen in approximately dorsal views (A-B) and apical view (C).

new species fill in the gaps originally perceived between genera. Nevertheless, the more conspicuous and more frequently encountered forms tend to cluster around relatively distinct "morphological foci". The complexes enumerated in the key are an attempt to recognize these clusters, with limits drawn on obvious features of the external surface and the archeopyle as the principal criteria. At the level of genus and species, the characters to consider are:

Size

Shape

In preferred orientation

In polar and lateral views whenever possible

Degree of primary dorsoventral compression

Horns and their extent of development

Walls

Number and identity of layers

If more than one: relative shape, thickness and conformation; distribution of cavation

Surface features - ornament

Archeopyle

Precise character in each wall layer, including size and shapes of individual paraplates where possible

Paratabulation, to whatever extent expressed

Paracingulum and parasulcus

Ornament distribution, pandasutural areas

Variability, for all features, which may be extreme.

Postive indications of a paratabulation that does not include three symmetrical anterior intercalaries are extremely rare among P-cysts. Notable exceptions include the asymmetrically placed middle intercalary in *Selenopemphix*, and the pair of symmetrical intercalaries in *Umbodinium*.

See Figure 9.1 for illustrations of most of the archeopyle types known to occur among P-cysts.

Note: The attempt by Harland (1982) to integrate the nomenclatures of Holocene and Quaternary peridinioid taxa based on thecae and cysts may have an important effect on the names applied to P-cysts. In the discussion on these pages, established names based on cysts are used in the generic sense customary before Harland's revision.

Deflandrea Complex

Genera: *Alterbia*, *Andalusiella*, *Ceratiopsis*, *Chatangiella*, *Deflandrea*, *Eucladinium*, *Eurydinium*, *Hexagonifera*, *Isabelidinium*, *Leipokatium*, *Lejeunecysta*, *Liasidium*(?), *Manumiella*, *Nelsoniella*, *Omanodinium*, *Palaeocystodinium*, *Phelodinium*, *Senegalinium*, *Stelladinium*, *Svalbardella*, *Trithyrodinium*, *Xenikoon*.

Occurrence: Possibly represented in the Liassic by *Liasidium* (Drugg, 1978); chiefly Late Cretaceous to Holocene. Relatively short ranges for many particular combinations of morphological features.

This complex is the largest and most varied of all, unified by the subdued surface features and the archeopyle (Figure 9.1 A). Most commonly, the archeopyle involves *2a* alone or all three intercalaries together, and the openings in the two walls are similar if the cysts are cavate. Types I, Ia, 3I, I/I, and 3I/3I are examples. Less common archeopyle types included here are Types IP, IPa, and (IP)a; that is, intercalary-precingular combination archeopyles in which only one precingular (*4″*) is involved. In still less common instances, pericyst and endocyst have distinctly different openings, as in the relatively easily interpreted example of a Type I/3I opening illustrated in Figure 6.8 D-E. A more complex and less readily deciphered example is the Type I/(tAtI)a archeopyle shown in Figure 6.4 N. Interpretation of such openings is virtually impossible if the operculum is not appreciably displaced and only a portion of the slitlike suture is visible middorsally. The range of intraspecific variability in some species of *Nelsoniella* and *Isabelidinium*, extends from some individuals having fully developed endoarcheopyles of this type to others that appear to have no endoarcheopyle at all. Openings such as the endoarcheopyle just discussed fall among what Bujak and Davies (1983) have termed "transverse archeopyles." In cysts with exceedingly thin and finely wrinkled or otherwise deformed walls, it is usually impossible to ascertain the character of the archeopyle.

Almost always, the anterior boundary of *2a* is rectilinear and is formed by *3′/2a*. Unusual examples in which *2a* is camerate and its anterior limit is determined by *1a* and *3a* whose meeting prevents contact between *2a* and *3′*, include individuals toward one extreme of the range of variation in *Trithyrodinium striatum* (Figure 7.3 K).

Features unknown or rare (and, therefore, distinctive) within this complex are:

- Archeopyle not symmetrically centered on midline (see *Selenopemphix* complex),
- Marked polar, in contrast to dorsoventral compression (see *Selenopemphix* complex),
- Distinct mesophragm (further discussion below),
- Ornament of high relief on the periphragm or development of an ectophragm,
- Fully developed paratabulation on periphragm (*Svalbardella*, also visible occasionally on others in SEM),
- Openings other than the archeopyle,
- Adnation of intercalary opercular pieces, except *2a* along adcingular margin,
- Anteriorly displaced endocyst (*Deflandrea phosphoritica*, Frontispiece B),
- Right antapical horn longer or larger than the left in forms where the two are distinctly unequal,
- Endophragm internally structured; even more rarely the periphragm,
- Connecting structures of any kind between periphragm and endophragm, or any other filling of the pericoel.

The characteristics of a "common denominator" for the *Deflandrea* complex might be considered as follows: Cyst bicavate to circumcavate with a peridinioid pericyst (left antapical horn more prominent than the right) and a thicker-walled, subspherical endocyst, each having a Type I archeopyle but of somewhat different shape and size owing to differences in wall contours. Periphragm surface smooth to granulose; endophragm smooth. Traces of paratabulation limited to ridges defining the paracingulum (interrupted ventrally), and the archeopyle. Equatorial cross-section shows moderate dorsoventral compression or ventral concavity which, in conjunction with horns, favors dorsoventral views.

Figure 9.2 shows representative shapes of cysts in the *Deflandrea* complex. Differences in outline shape, wall relations, and archeopyle have been the principal criteria for genera, while finer aspects of the same features have been supplemented by surface detail at the species level. Seemingly infinite interspecific variations in relative shapes of peri- and endocysts, their proximity, and the distribution of pericoel(s) occur. Where these differences distinguish genera, borderline species must be expected and will have to be assigned arbitrarily. Intraspecific variation in these features may also be great and, at least in some species, appears to record stages in cyst development (discussed below).

If the periphragm formed as only a thin and delicate wall surrounding a robust endocyst, it may be torn away in some, or even most, individuals. Multiple specimens should permit recognition of this situation. However, some spheroidal and ellipsoidal cysts (*Hexagonifera*, some *Trithyrodinium*) appear to represent the inner body only, without any trace of an outer wall. Although these may be most conveniently designated autocysts, they obviously differ from other autocysts whose typically peridinioid shape suggests close approximation to the theca and, therefore, equivalence to periphragm alone or to an indistinguishable union of periphragm and endophragm (e.g., *Lejeunecysta*; living *Protoperidinium* species, such as *P. leonis*, *P. obtusum*, and others).

When periphragm and endophragm are differentiated and the walls are in contact or closely spaced, the archeopyle is of the same type in both. When the walls are widely separated, the openings may be of the same type, may only appear to be of different types, or may, in fact, be different. If the endocyst has a nearly circular outline, the precise character of the endoarcheopyle may be difficult to recognize in the normally prevalent dorsoventral orientation. For example, the Type I endoarcheopyle in *Deflandrea phosphoritica* may be mistaken for apical. Similarly, a Type 3I inner opening may also be taken for apical if its shape or accessory sutures are overlooked. If a Type 3I archeopyle is suspected but not completely obvious, it will be important to look for the medial offset in the adapical margin adjacent to *3′* and the larger scale but somewhat similar outline along the adcingular margin. Typically, the operculum of an archeopyle involving just three intercalaries is fully divided by accessory sutures into three free opercular pieces whose individual shapes match the three intercalaries in a hexa style paratabulation. Exceptions are rare but include what was cited in Chapter 6 as Type 3Ia, with three adnate opercular pieces (termed a "transverse archeopyle" by Bujak and Davies, 1983, in Early Cretaceous species they refer to *Subtilisphaera*), and the opening with variably shaped opercular pieces that occurs in *Trithyrodinium striatum* (Benson, 1976; May, 1980; Figure 7.3 K).

One of the few available clues to differences in details of paratabulation in this complex is a conspicuous difference in size, shape, and position of paraplate *2a* as indicated by the archeopyle. These aspects of the archeopyle have an appeal as potentially useful taxonomic criteria (conveniently expressed in terms of transverse and longitudinal archeopyle indices, archeopyle ratio, and archeopyle signum). However, if one considers the great range in shapes assumed by plate *2a* among living species of *Protoperidinium*, it is difficult to attribute great biological or taxonomic significance to the differences in archeopyles that represent that plate.

Aside from the archeopyle and general features of paracingulum and parasulcus, indications of paratabulation are uncommon. A notable exception is the fine detail revealed by a few species under close scrutiny with the light microscope, and dramatically displayed by the SEM. For example, some forms close to *Deflandrea phosphoritica* show polygonal fields of delicate, short, closely spaced pila, separated by broad, pandasutural areas that are featureless except for a finely delineated parasuture. Such examples demonstrate the presence of a small paraplate (*1c*) at the proximal end of the paracingulum, but full details of parasulcus and paracingulum remain to be explored.

A few incidental features are noteworthy. Accumulation bodies are frequent inclusions in cysts of this complex. Specimens of normally cavate species that lack an endocyst may account for a significant

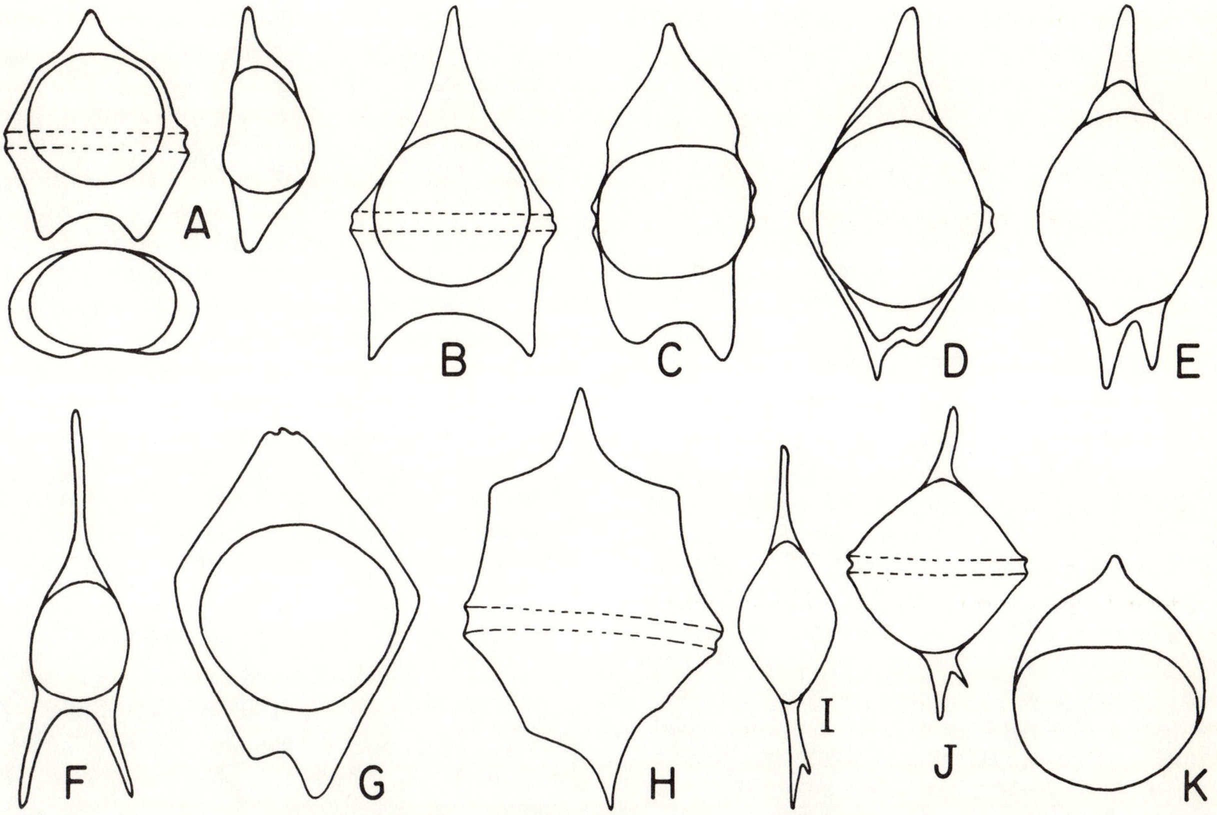

Figure 9.2 - OUTLINES OF REPRESENTATIVE FOSSIL CYSTS OF THE *DEFLANDREA* COMPLEX

All are dorsoventral views except A in which dorsoventral view is supplemented by left lateral (to the right) and apical (below) views of same individual.
Selected species of *Deflandrea* (A, B), *Chatangiella* (C), *Trithyrodinium* (D, E), *Ceratiopsis* (F), *Manumiella* (G), *Liasidium* (H), *Palaeocystodinium* (I), *Andalusiella* (J), *Nelsoniella* (K). Ages: Jurassic (H), Late Cretaceous (C-G, I-K), Tertiary (A-B). Wall structure: periphragm and endophragm (A-C, F-G, I-K); periphragm, mesophragm, and endophragm (D-E); autophragm (H).

percentage of individuals in some species (discussed below). Many species show flagellar scars in the parasulcus, as well as some kind of a modification (often a nipplelike structure) at the apex that probably reflects an apical pore in the theca. Finally, conspicuous rosettes etched in the relatively smooth outer surface of Pp-cysts (like those that also occur on levigate spores) may record fungal attack on the sporopollenin wall.

Spinidinium Complex

Taxa: *Dioxya*, *Epelidosphaeridia*(?), *Ginginodinium*, *Lentinia*, *Spinidinium*, *Sumatradinium*, *Uvatodinium*, *Vozzkennikovia*; some living *Protoperidinium* species.

Occurrence: Early Cretaceous (Aptian-Albian) to Holocene.

Relatively few peridinioid cysts have bold surface relief, and some of those that do are placed elsewhere in recognition of the dominating effect of other features (e.g., complex archeopyle type in the case of living *Protoperidinium claudicans*, and pronounced parasutural features in *Phthanoperidinium*). A general surface granulation, occasionally conspicuous, occurs in some species of the *Deflandrea* complex. In the *Spinidinium* complex, the outer surface carries obvious, if not always prominent, ornament in the form of projecting spines or diminutive processes. Among currently known taxa that are included here, archeopyles in most cases are of the relatively simple Types I and 3I found in the *Deflandrea* complex, but accessory sutures may divide the subjacent precingulars. The wall is an autophragm, or the cysts have a periphragm and an endophragm that are not widely separated, and are acavate to cornucavate or circumcavate, not bicavate. Thus, differences in wall relations tend to go along with the more highly ornamented surface. Distribution of ornament varies and is a critical taxonomic character. *Epelidosphaeridia* is distinct from the rest by having an archeopyle in apical position, but uncertainly Type (tA) or (tAtI). The cyst of living *Protoperidinium divaricatum* is essentially chorate (Matsuoka et al., 1982).

Selenopemphix Complex

Taxa: *Selenopemphix*; living *Protoperidinium subinerme*.

Occurrence: Paleogene to Holocene.

Characteristic of this complex is the asymmetric paratabulation indicated by the position of the archeopyle, which lies distinctly to the left of the middorsal line and is Type I in known examples (Figure 9.3 K). Shortening of the polar axis, which is another unusual feature associated with species in this complex, may be sufficient to favor polar views in compacted specimens, an orientation not recognized when *Selenopemphix* was first described. The theca of living *Protoperidinium subinerme*, which produces a cyst closely similar to some fossil examples, has a similar shape. Some fossil representatives have a conspicuous ornament of spine-like processes.

Palaeoperidinium Complex

Taxa: *Chichaouadinium*, *Laciniadinium*, *Luxadinium*, *Morkallacysta*, *Palaeoperidinium*, *Saeptodinium*; living *Peridinium limbatum* and several other freshwater species of *Peridinium*.

Occurrence: Early Cretaceous to Holocene.

The complex includes a relatively small number of distinct taxa in which the archeopyle (Figure 9.1 B) is a relatively large dorsal opening which includes one or more intercalaries together with the three dorsal precingulars, or all six of these paraplates in combination with 3′. The operculum may be simple, incorporating all the paraplates of the archeopyle in a single piece, and either free or adnate along the paracingulum. Alternatively, the operculum may be compound. In that case, all opercular pieces may be free or variously adnate.

The extent of the archeopyle along the paracingulum and the course of its anterior margin are the most useful features for recognizing the archeopyle. Since three precingulars are involved, the intercepts of the longitudinal portions of the principal archeopyle sutures with the paracingulum are relatively far apart and may lie essentially on the lateral margins. The outline of the medial adapical archeopyle margin will indicate whether the opening includes *2a* as the only intercalary (*Chichaouadinium*, *Morkallacysta*), all three intercalaries (*Laciniadinium*, *Luxadinium*), or the three intercalaries plus the dorsal apical (*Palaeoperidinium*, *Saeptodinium*). Where *3′* is included, a true transapical suture is developed; Norris and Hedlund (1972) discussed its characteristics, and Bujak and Davies (1983) termed this a "hemiepicystal" archeopyle. If the operculum is adnate (*Laciniadinium*, *Saeptodinium*), one will see only a slitlike separation in the wall, rather than a clear opening. In known examples having more than one wall layer, the same type of archeopyle occurs in each. Mesophragms have been observed. The unusual aspects *Palaeoperidinium pyrophorum* of are discussed later in this chapter.

Ascodinium Complex

Taxa: *Ascodinium* (including *Ovoidinium*, see Helenes, 1983; and *"Canningia" turrita*); living *Protoperidinium claudicans*, *P. oblongum*, *Peridinium wisconsinense*.

Occurrence: Late Early Cretaceous (Albian) to Holocene (discontinuous distribution and not closely similar types).

Diagnostic is an archeopyle (Figure 9.1 C) that includes the apex of the cyst and does not reach the paracingulum (or equator), but may extend noticeably closer to it dorsally than ventrally. If the principal archeopyle suture has a neatly angular course, the parasutures which it represents can be recognized, and Types (4A3I)a and (3$A_{2\text{-}4''}$3I) have been identified in this manner. Frequently, however, the archeopyle outline is less clear, and its exact course is conjectural if independent indications of paratabulation are absent. Note that in a few species of the *Deflandrea* complex this type of archeopyle occurs in the endocyst in combination with a Type I_{2a} periarcheopyle.

The ventral opening in the periphragm of *Ascodinium verrucosum* is discussed under Claustra in Chapter 4.

Cysts of living *Protoperidinium claudicans* and *P. oblongum* resemble one another in design (dorsoventrally compressed, roundly peridinioid autocysts), but contrast in surface ornament. The cyst of the latter species is smooth, but the former has a dense cover of nontabular pila. The exact course of the archeopyle suture in both is uncertain, since no indications of paratabulation are present.

Living freshwater *Peridinium wisconsinense* has a cyst which is, so far, unique among peridinioid examples in having a strictly apical archeopyle with an operculum that is broadly adnate ventrally, Type (4A)a (Wall and Dale, 1968). No other living dinoflagellate is known to have this type of opening. Although apical archeopyles are well known among fossil gonyaulacoid species, the type here occurs in a species with a peridinioid tabulation. Accessory sutures are developed in some specimens between the anterior intercalaries, suggesting a variety of the archeopyle formula—Type (4A)a3Ia—but the consistency of these accessory sutures has not been established.

Phthanoperidinium Complex

Taxon: *Phthanoperidinium.*

Occurrence: Paleogene.

Distinctive of the species in this complex is the presence of a paratabulation prominently expressed by parasutural or penitabular ridges or septa, which may be unmodified or extended into spines or processes. The archeopyle is commonly Type I, but Types 3I, IP, and 3I3P are also known. The cysts are usually acavate or cornucavate, and horns are seldom prominent. Some species show complex but as yet unresolved features of wall structure associated with the parasutural ridges and spine bases.

Cysts of this type offer an opportunity to learn more about details of paratabulation in Early Tertiary peridinioids, since the pattern is clearly marked. The paracingulum and parasulcus should be observed carefully. Details of wall structure are prospectively important features in the more elaborate species.

Category of Pq-Cysts

This category, which includes some of the most obvious and distinctive dinoflagellate cysts found in Paleogene samples, encompasses a rather limited range of morphological types. The *Wetzeliella* complex is typical and is widely represented. A second group, the *Apectodinium* complex, is possibly ancestral to the other.

Most Pq-cysts are relatively large (100µm or more) with a conspicuously peridinioid outline commonly accentuated by equatorial hornlike extensions in line with the paracingulum (Figure 4.1 C-D). As a consequence of primary dorsoventral compression, normally only a dorsoventral view is presented to the observer unless specimens are individually manipulated. Although some species have smooth exteriors and a few have fine acicular processes, most have short, often T-shaped processes in various arrangements; ectophragmal interconnections may be prominent. A distinct endocyst is typical (but commonly obscure in representatives of the *Apectodinium* complex), and most species are circumcavate. The archeopyle is exclusively Type I or Ia (adnation along anterior margin).

Paratabulation is rarely obvious except locally, but whenever determinable it has proved to be quadra on the dorsal surface. An ortho ventral arrangement has been assumed in the past and is, indeed, compatible with the large diamond-shaped *1′*. However, Goodman (1983) has determined in some species having a clear paratabulation that *1′* definitely contacts *2″* and *6″*. Barring the discovery of positive evidence for ortho patterns in this group of cysts, it now seems highly probable that their thecae were universally para quadra (Figure 9.3 A-D). However, it is most unlikely that the paratabulation on the fossil cysts will frequently be clear enough for this relationship to be observed directly since the critical areas are generally concealed at the edges of specimens even if parasutures are expressed.

Spinidinium has been suggested (Costa and Downie, 1979) as an ancestral source for the *Wetzeliella* complex, with *Apectodinium* as an intermediate stage. However, after discovery of the para quadra organization of *Wetzeliella*, ancestry of the group in the ortho hexa *Spinidinium* appears unlikely. Bujak and Davies (1983) suggest possible origins in *Manumiella* (Pp-cyst), but paratabulation in this genus is poorly expressed and, while its *2a* (i.e., its archeopyle) may be described as quasiquadrate, the ventral pattern is entirely unknown.

Whereas cyst shape, wall relations, and archeopyle types are the most significant variables among the majority of Pp-cysts, these features are relatively constant in Pq-cysts. Here, the form and distribution of surface features on the periphragm are the most important variables.

Features to observe for identifying or describing taxa include:

- Shape
  - Outline in dorsoventral view
  - Number, form, and prominence of horns and equatorial projections

Degree of dorsoventral compression (significant when not extreme)

Walls

Distribution of cavation and shape of endocyst

Form and structure of spines or processes

Presence of ectophragm and its structure

Paratabulation, to whatever extent expressed

Ornament distribution

Details of ectophragm

Archeopyle

Adnation of the operculum

Differences between peri- and endoarcheopyles

Variation, which is commonly great

Wetzeliella Complex

Taxa: *Dracodinium*, *Gochtodinium*, *Kisselovia*, *Rhombodinium*, *Wetzeliella*, *Wilsonidium*.

Occurrence: Early Eocene to early Miocene.

Cysts in this complex have a distinct endocyst and are typically circumcavate. Taxonomic distinctions rest chiefly on differences in the kind and distribution of surface ornament and projections, and to a certain extent on outline shape. Ideal representatives of the several genera are quite distinct, but variability, both inter- and intraspecifically, is so great that many determinations require more or less arbitrary allocation to both genus and species. Subdivision of the complex at the generic level hardly seems defensible in a biologic sense however useful stratigraphically may be finely drawn taxonomic units.

The most common element of surface relief in these cysts is a relatively short, often T-shaped or capitate, hollow process whose interior communicates with the pericoel. Variation on this theme extends to solid and acicular processes in one direction, to interconnections of process tips through an ectophragmal membrane (Figure 9.3 I-J) or reticulum of trabeculae in another, and to reduction and even complete elimination of processes in a third. Processes may be nontabular, intratabular, penitabular, or parasutural in distribution.

The pericoel tends to surround the endocyst in dorsoventral view, but actual spacing and shapes of the two bodies vary. Commonly the endophragm is elaborated locally at the four "corners" (apex, antapex, and equator) by granular thickenings (Figure 9.3 F) which may be quite conspicuous but are neither uniformly present nor similarly developed, even within a species.

Species with ectophragmal connections between processes (*Kisselovia*) and occasional examples of other morphologies invite careful inspection of paratabulation. Ectophragmal units corresponding to paraplates may be clearly defined by pandasutural spaces between them, and the quadra style dorsal pattern is convincingly demonstrated by the best examples (Figure 9.3 E). Further study of such material with the SEM can be expected to show conclusively the details of parasulcal and paracingular patterns that will permit closer comparison of these fossils with living taxa.

Figure 9.3 - Pq-CYSTS AND OTHER P-CYSTS

A-D - Epicyst paratabulation of *Wetzeliella* (in vv, llv, dv, and av) shown schematically as para quadra. Arrows in A indicate *2′/2″* and *4′/6″* parasutures whose position is critical for distinguishing para and ortho ventral patterns. In the ortho pattern, these parasutures would abut against *1″* and *7″* rather than against *1′* (compare Figure 5.2 C).

E - *Kisselovia* sp. Detail of dorsal epicyst surface, showing quadra paratabulation defined by open archeopyle (*2a*) and ectophragm units separated by pandasutural gaps. Small dots show location of processes supporting ectophragm. Ectophragm strip representing paracingulum continues uninterrupted beyond margins of dorsal surface. Eocene.

F - *Wetzeliella* sp. Optical section of specimen in vv (dashed outline of archeopyle projected to plane of section). Note five horns, graded lengths of peripheral processes from horn to horn, and thickening of endophragm by granular material below horns. Eocene.

G-H - *Apectodinium homomorphum*. Optical (dvv) sections of two specimens showing contrasts in outline, definition of horns, and character of processes; apex uppermost in both specimens. Dashed outline of archeopyle as in F. Paleocene.

I-J - *Kisselovia* sp. Relative positions of three wall layers and typical degree of dorsoventral compression (vv and rlv). Parasutural interruptions in ectophragm and processes connecting it to periphragm are not shown. Dashed outline of periarcheopyle projected in I. ec - ectophragm, p - periphragm, en - endophragm. Eocene.

K - Cyst of living *Protoperidinium subinerme* (av). Parasulcus at bottom and apex in center emphasize asymmetric position of Type I_{2a} archeopyle. Dashed line indicates concavity of the paracingulum. Comparable Early Tertiary cysts have been referred to *Selenopemphix*.

L - *Palaeohystrichophora infusorioides*. Dorsal view showing characteristic hairlike processes in essentially nontabular arrangement except along paracingulum. Late Cretaceous.

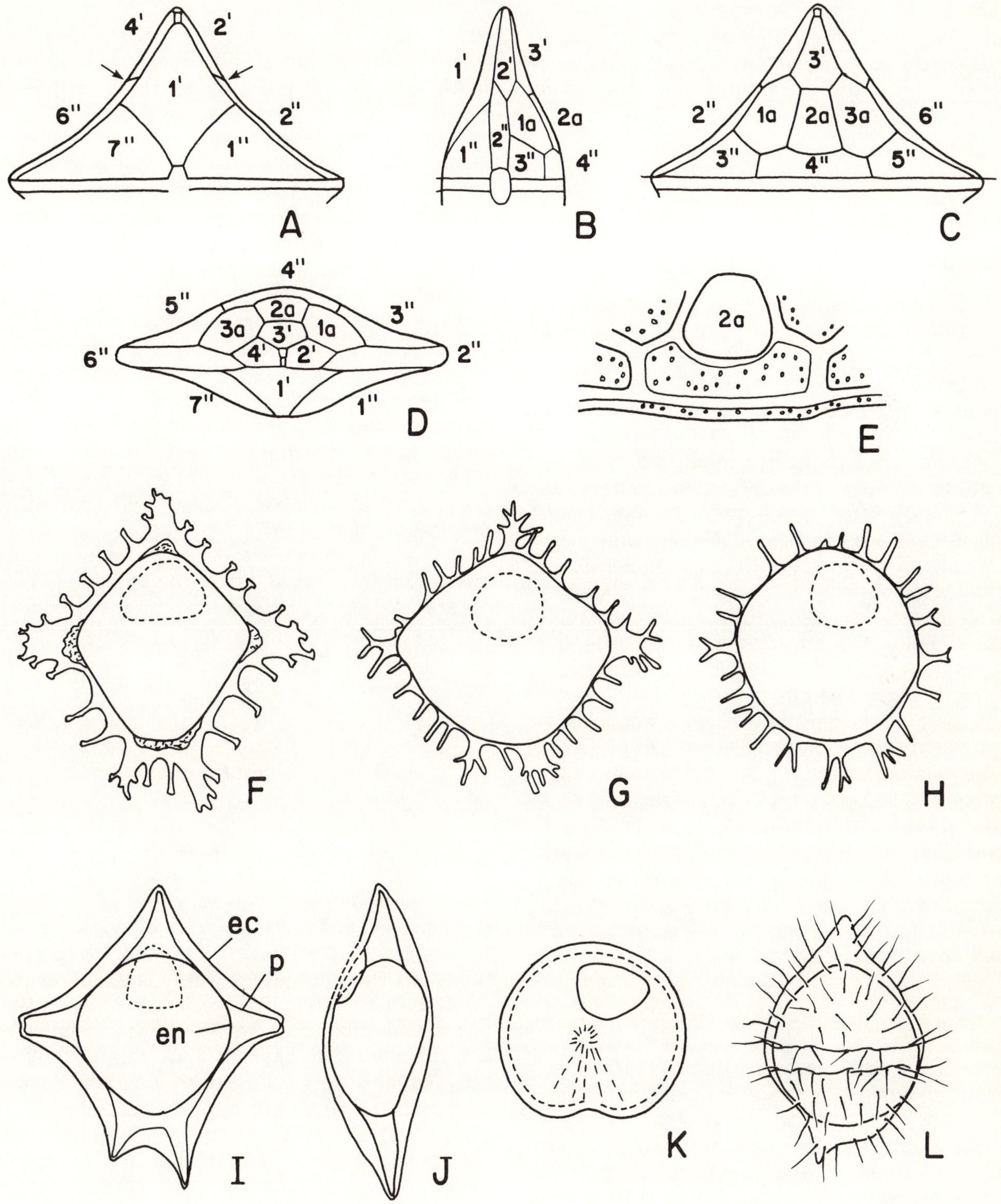
4'
2'
1'
6''
2''
7''
1''
A
1'
3'
2'
1a
2a
1''
2''
3''
4''
B
3'
2''
1a
2a
3a
6''
3''
4''
5''
C
4''
5''
3''
2a
3a
3'
1a
6''
4'
2'
2''
1'
7''
1''
D
2a
E
F
G
H
ec
p
en
I
J
K
L

Figure 9.3

The group has received much attention in the course of studies of its stratigraphic distribution in the Early Tertiary (Costa and Downie, 1976, 1979; Williams, 1977; Châteauneuf and Gruas-Cavagnetto, 1978; and references therein). Variability in form has been explored by Gocht (1969). For a general survey and evaluation, see Goodman (1983).

Apectodinium Complex

Taxon: *Apectodinium*.

Occurrence: Late Paleocene to late Eocene.

This monogeneric complex is, in many respects, intermediate between Pp-cysts and the more distinctive Pq-cysts of the *Wetzeliella* complex. The critical feature of the genus—close approximation of the periphragm and endophragm or near indistinguishability of the two wall layers—serves also for the complex, but there are other less sharply defined contrasts with the *Wetzeliella* complex as well. A typical *Wetzeliella*-like outline is rarely clear, and processes are abundant, generally nontabular, and commonly long relative to overall dimensions. Paratabulation normally is not clear; the archeopyle may be nearly circular and is rarely obvious. However, a quadra arrangement has been observed in *A. longispinsoum* (Wilson, 1968). The principal species, *A. homomorphum* (Harland, 1979), is spectacularly variable (Figure 9.3 G-H), and individual specimens may show virtually no basis for reference to this complex or category. In fact, some specimens could be taken for compressed remains of nearly spherical chorate cysts. However, the species is commonly represented by large numbers of individuals, some of which show more characteristic features of outline and archeopyle. Assignment of some individual specimens to *Apectodinium* or *Wetzeliella* may be purely arbitrary.

Category of Px-Cysts

This is a catchall category for cysts whose peridinioid outline or traces of paratabulation suggest affinities with many peridinioid dinoflagellates but which lack more diagnostic features that would place them in one of the other categories. Archeopyles, if present at all, are difficult to recognize. Most specimens usually encountered clearly lack such openings. Accordingly, taxa have been distinguished on the basis of shape, wall structures, and surface ornament. As with most other P-cysts, intraspecific variability tends to be great.

Taxa: *Diconodinium*, *Geiselodinium*, *Palaeohystrichophora*, *Subtilisphaera*, *Teneridinium*.

Occurrence: Early Cretaceous to Paleogene.

Palaeohystrichophora infusorioides is an especially distinctive species, commonly abundant, and restricted to mid-Cretaceous (Albian to Campanian) strata. *Diconodinium* is a more varied genus (Morgan, 1976), some of whose species have interesting promise as subjects for closer examination when more abundant well-preserved specimens are encountered. Details of paratabulation and affinities of the genus are not yet entirely clear, but sufficient traces of paratabulation have been observed to encourage further study. Species of *Diconodinium* with Type IP and (IP)a archeopyles (Morgan, 1976) would fall in the category of Pp-cysts. Ruptures of the periphragm in rare individuals of *Palaeohystrichophora infusorioides*, a distinctive and well-known species from late Albian and younger Cretaceous sediments (Figure 9.3 L) which typically shows no trace of an archeopyle, suggest an archeopyle involving intercalary and precingular paraplates (Harker, 1979). *Geiselodinium* and *Teneridinium* are apparently freshwater cysts, whose thin and folded walls leave details of morphology quite unclear. *Subtilisphaera* is a more distinctive marine form, suggesting an archeopyleless *Alterbia* (although Bujak and Davies, 1983, attributed some cysts with archeopyles to *Subtilisphaera*).

ADDITIONAL SELECTED TOPICS

Peridinium limbatum

No other living cyst-producing peridinioid dinoflagellate has been so informative in providing information applicable to the interpretation of fossil cysts (Eisenack and Fries, 1965; Norris and McAndrews, 1970; Evitt and Wall, 1968; Evitt, and Pfiester, both unpublished; utilizing materials from Pleistocene and Holocene lake sediments, modern lake plankton, and laboratory cultures). Although the species has already been the subject of comment in Chapters 1 and 4 in connection with introductory remarks on several cyst features, we return to it here in order to summarize more comprehensively the observations it has afforded. Frontispiece A, and Figures 1.5 A-E and 4.2 J in earlier chapters should be consulted, along with Figure 9.4.

Of all known cysts of living dinoflagellates those produced by this species are most similar to a major

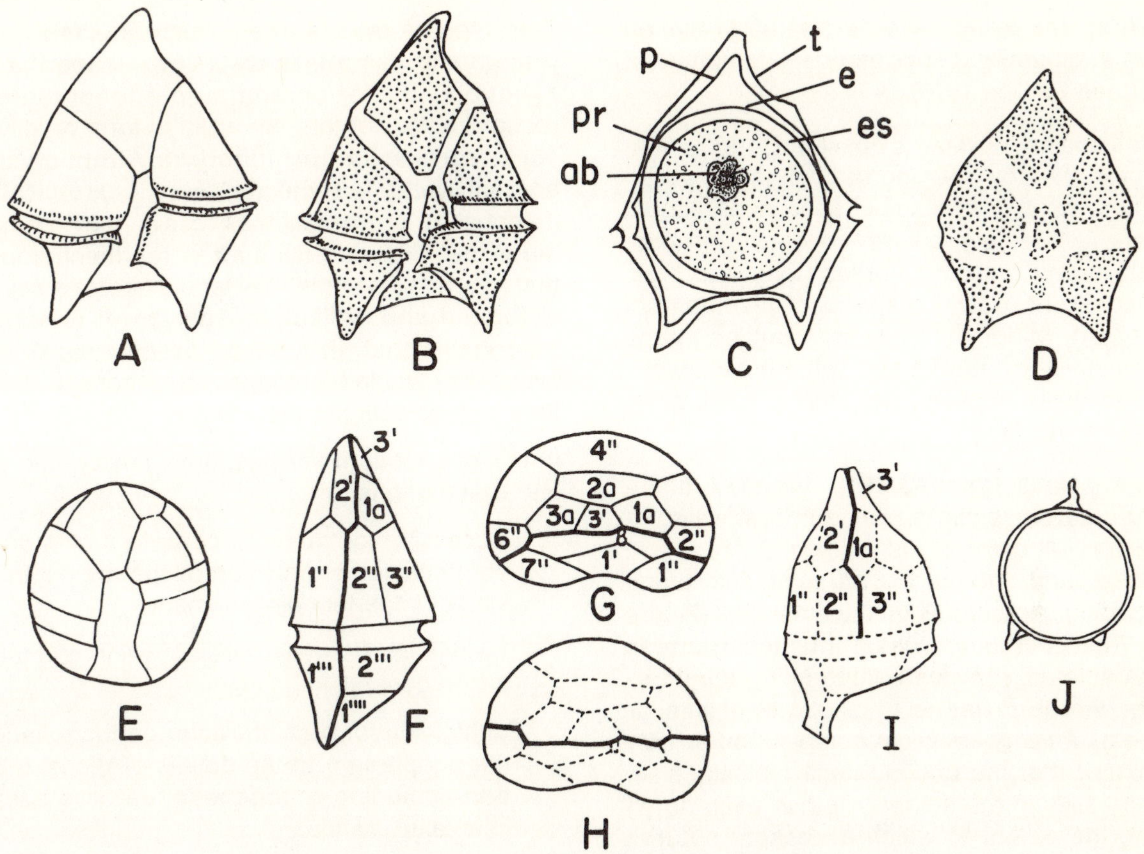

Figure 9.4 - THECA AND CYST OF LIVING *PERIDINIUM LIMBATUM*

A - Vegetative theca or immature zygotic theca, showing ventral tabulation.
B - Zygotic theca in ventral view, showing intercalary bands (plain) between reticulate/perforate areas of original plates (stippled).
C - Section of specimen with fully formed cyst inside theca, ventral view. t = theca, p = pericyst (= mesospore of neobiologists), e = endocyst, es = endospore of neobiologists (nonpreservable), pr = granular protoplast, ab = accumulation body.
D - Zygotic cyst, showing pandasutural areas (plain) and polygonal ornamented fields (stippled) corresponding to reticulate/perforate areas on theca.
E - Isolated endocyst, showing paratabulation of single grooves on inner surface.
F-G - Vegetative theca (llv and av), showing (heavy lines) course of ecdysal suture that divides the theca into dorsal and ventral portions.
H-I - Zygotic cyst (av and llv), showing (heavy lines) course of archeopyle suture, for comparison with F-G. Schematic and simplified, dotted lines replace pandasutural areas.
J - Pellicular or ecdysal cyst derived from a presumably vegetative theca, for comparison with zygotic cyst shown in D.

group of fossils—the peridinioid cysts of general *Deflandrea* type. In this modern species we can see that the peridinioid pericyst closely mimics the theca in shape (Figure 9.4 A-D), with minor modifications in detail. Prominent intercalary growth bands between reticulate areas of original plates in the zygotic theca are matched by ornament-free pandasutural areas between polygonal fields of apiculate grana on the outer periphragm surface. To whatever extent it is apparent, the paratabulation revealed by these surface features is faithful to the thecal tabulation. Acetolyzed cysts normally show an ellipsoidal endocyst within the pericyst. The endophragm (Figure 9.4 E) is externally smooth, but internally it is minutely corrugate with prominent grooves that delineate a paratabulation, like that of the pericyst in pattern but conspicuously different in lacking any trace of pandasutural areas. The significance of this difference was discussed in Chapter 4 in conjunction with controls on the development of cyst features. Similar patterns of internal paratabulation are known in a few fossil peridinioid cysts (e.g., in a Late Cretaceous species of *Trithyrodinium* studied by Stein, 1983, and in an Early Cretaceous *Ascodinium* described by Helenes, 1983). In living cysts, the thick, clear, nonpreservable endospore lines the preservable endophragm, separating it from the granular protoplast within. Accumulation bodies from the living cells persist as shriveled remnants in acetolyzed specimens (= "omphali" of Dörhöfer and Davies, 1979).

In cultures, the endocyst was observed to form after the pericyst, thereby seeming to confirm the suspicion that the small percentage of specimens without endocysts, which have been recovered from both lake waters and bottom samples (Evitt and Wall, 1968), represent individuals whose development was arrested. Minor differences in outline shape often distinguish cysts with and without endocysts. Parallels to these observations among fossil cysts will be mentioned later.

The archeopyle is Type (A3I3P)a, and the archeopyle suture develops similarly in both cyst walls, as a transapical slit along *2″/3″*, *2″/1a*, *2′/1a*, *2′/3′*, to the apex, and back to the paracingulum along corresponding parasutures on the other side (Figure 9.4 H-I). The operculum, which thus incorporates seven paraplates, remains adnate along the paracingulum. Archeopyles in fossil cysts of similar design (e.g., *Palaeoperidinium* complex) differ only to the extent that the operculum is typically free. Some variability in the archeopyle has been noted (involving the extent to which accessory sutures may divide the operculum), but distinctly contrasting archeopyle types have not been recorded. Under some conditions accessory archeopyle sutures develop between constituent paraplates of the archeopyle. The archeopyle suture in living cysts is not apparent in fresh specimens, but is opened by acetolysis.

Rupture of the theca during ecdysis follows a course along thecal sutures identical to the course of the archeopyle suture over the apical region, but consistently different toward the cingulum (Figure 9.4 F-G). Here the thecal separation is along *1″/2″* and *2′/2″* instead of *2″/3″* and *2″/1a*, with comparable differences on the right side. Consequently, the ecdysal rupture can continue smoothly from the epitheca across the cingulum along *2c/3c* and then along *1‴/2‴* (and, finally, *1‴/1⁗* and corresponding sutures on the right side) through the hypotheca, permitting the separation of the theca into dorsal and ventral portions, which is characteristic of the species.

In cultures of *P. limbatum*, two distinct kinds of cysts appear. Discussion up to this point has concerned only those that form in zygotic thecae. The second type of cyst is presumably an ecdysial cyst. It forms in vegetative thecae that are smaller than and lack the pronounced intercalary bands of zygotic thecae. This type of cyst is itself much smaller, thinner walled, and has a less distinctive shape than the zygotic cyst. Periphragm and endophragm are recognizable, but the thin walls deform readily and surface relief is not apparent. Although these ecdysal cysts withstand acetolysis, thus indicating a potential for preservation as fossils, they are hardly more than wrinkled films after treatment, implying that in a fossil sample they would likely be severely deformed and difficult to analyze. It is certainly improbable that, in a mixed association of fossil cysts, they would be recognized as conspecific with their zygotic counterparts.

In this one modern species, then, the zygotic theca and cyst demonstrate:

- Successive formation of pericyst and endocyst, but occasional omission of the endocyst as a kind of imperfect development,
- The relation between paratabulation and tabulation in general,
- Faithfulness of paratabulation on the exterior of the periphragm to the details of thecal tabulation, including ontogenetic features such as intercalary bands,
- Features on cyst walls more remote from the theca that record only the basic paratabulation pattern, without ontogenetic detail of the theca,
- Existence of a complex archeopyle and opening of the archeopyle suture in an ungerminated cyst by chemical treatment,
- Similarity, but not identity, between the lines along which the theca ruptures in ecdysis and those that outline the archeopyle, and
- Preservability of accumulation bodies.

In addition, the species illustrates production of conspecific, potentially preservable cysts of two different types: 1) robust zygotic cysts whose morphology is analogous to familiar fossils of the *Deflandrea* type, and 2) smaller, thin and delicate ecdysial cysts of essentially similar form but unlikely to be well preserved as fossils.

The Paracingulum and Parasulcus

The role in the taxonomy of living peridinioid dinoflagellates played by tabulation details in the cingulum and sulcus (Balech, 1980) was mentioned briefly in Chapter 1. Unfortunately, comparable details of paratabulation are seldom visible in dinoflagellate cysts and only very few examples of complete cingular and sulcal patterns are currently known from the fossil record. The areas are difficult to see directly at the requisite magnifications and, even if they are visible, paratabulation there is seldom clearly expressed. Although what has been seen invites comparison with tabulations in living species, broad generalizations about relationships between living and fossil peridinioids on the basis of cingular and sulcal details are not yet justified. Consequently, it is especially important to record any details that can be observed in these areas.

Figure 9.5 A-C show schematically three patterns of cingular tabulation exemplified today by marine *Protoperidinium* and *Scrippsiella*, and freshwater *Peridinium*. The *Protoperidinium* pattern has a total of four cingular plates, of which the first (*1c = t* or "transitional" of some authors) is much smaller than the second, while the third is very large, extending without interruption across more than the entire dorsal surface. In *Scrippsiella* and *Peridinium*, the cingulum is paved dorsally by three plates which appear to equate with the large one in *Protoperidinium*. Both *Scrippsiella* and *Peridinium* currently include some species having five and others having six cingulars, depending upon whether or not a small initial cingular like that in *Protoperidinium* is present. Since this small *1c* is absent in the type species of *Peridinium* (*P. cinctum*) and present in the type of *Scrippsiella* (*S. sweeneyae*), these generic names provide convenient labels for distinguishing two further types of cingular arrangements. However, the reader must be aware that variation occurs within the genera as currently constituted and also that *Scrippsiella* species form contrasting types of cysts and have an especially complex nomenclatural history and debated biologic relationship (Dodge, 1982, for brief review and references).

The *Protoperidinium* cingular pattern is displayed nicely by the ectophragm connections between short processes in some species of Early Tertiary *Kisselovia*. It seems probable that the same cingular arrangement occurred throughout the *Wetzeliella* and *Apectodinium* complexes, but there is little likelihood that this will be provable in very many instances. The same cingular paratabulation occurs in Eocene species close to *Deflandrea phosphoritica*, evidenced by intratabular clusterings of minute coni and pila barely visible in the light microscope but clearly revealed by the SEM (W.R.E., pers. obs.). How extensively the pattern may occur among other members of the large *Deflandrea* complex and among other Pp-cysts is not known.

The *Peridinium* pattern with five cingulars is clearly expressed in cysts of modern *Peridinium limbatum* from cultures, lake plankton, and Holocene sediments (Evitt and Wall, 1968), but representation of this cingular pattern in the older fossil record has not been established. In contrast, a few excellent fossil examples of the *Scrippsiella* cingular arrangement are known from the Late Cretaceous and Early Tertiary. They include *Palaeoperidinium pyrophorum* (Figure 9.5 D-E; Gocht and Netzel, 1979) and the previously mentioned species of *Trithyrodinium* (Figure 9.5 H-I; Stein, 1983). However, it is noteworthy, perhaps, that the sulcal arrangement in these fossil species is more closely similar to that in modern *Protoperidinium* (Figure 9.5 E-F) than to that in either *Peridinium cinctum* (Figure 9.5 G) or *Scrippsiella* (Balech, 1980). Boltovskoy (1979) called attention to the similarity between such fossil species and those modern freshwater *Peridinium* species having six cingulars, and Bujak and Davies (1983) emphasized this similarity by placing *Palaeoperidinium pyrophorum* as a "stepping stone" on the way from Deflandreaceae to Peridiniaceae (the latter in their emended sense, excluding Protoperidiniaceae), the two families which they recognized as containing the dominant fossil marine and modern freshwater peridinioids.

A distinctive feature associated with the paracingulum occurs as a prominent specific trait in a variety of Late Cretaceous species (especially of *Chatangiella*). This has been referred to as a "segmented paracingulum", but, in fact, it gives no direct evidence of division within the paracingulum. Rather, it results from a prominence of the adcingular edges of the pre- and postcingular paraplates, interrupted at points that would correspond to the junctions with interprecingular and interpostcingular parasutures in the pre- and postcingular areas (Figure 9.5 J). If complete, there are seven ridges along the apical edge of the paracingulum, and five along the antapical edge (McIntyre, 1975). However, in many species only those ridges on the dorsal surface are easily seen. Since the lateral limits of *4″* and *3‴* are nearly in line with one another across the paracingulum, a quick impression may be that the interruptions between ridges are an expression of paratabulation within the paracingulum. Actually details of that paratabulation are unknown.

The small size, unpredictable number, varied and irregular shape, and location within the sulcal depression which make the sulcal plates difficult to analyze in modern thecae also make analysis of the

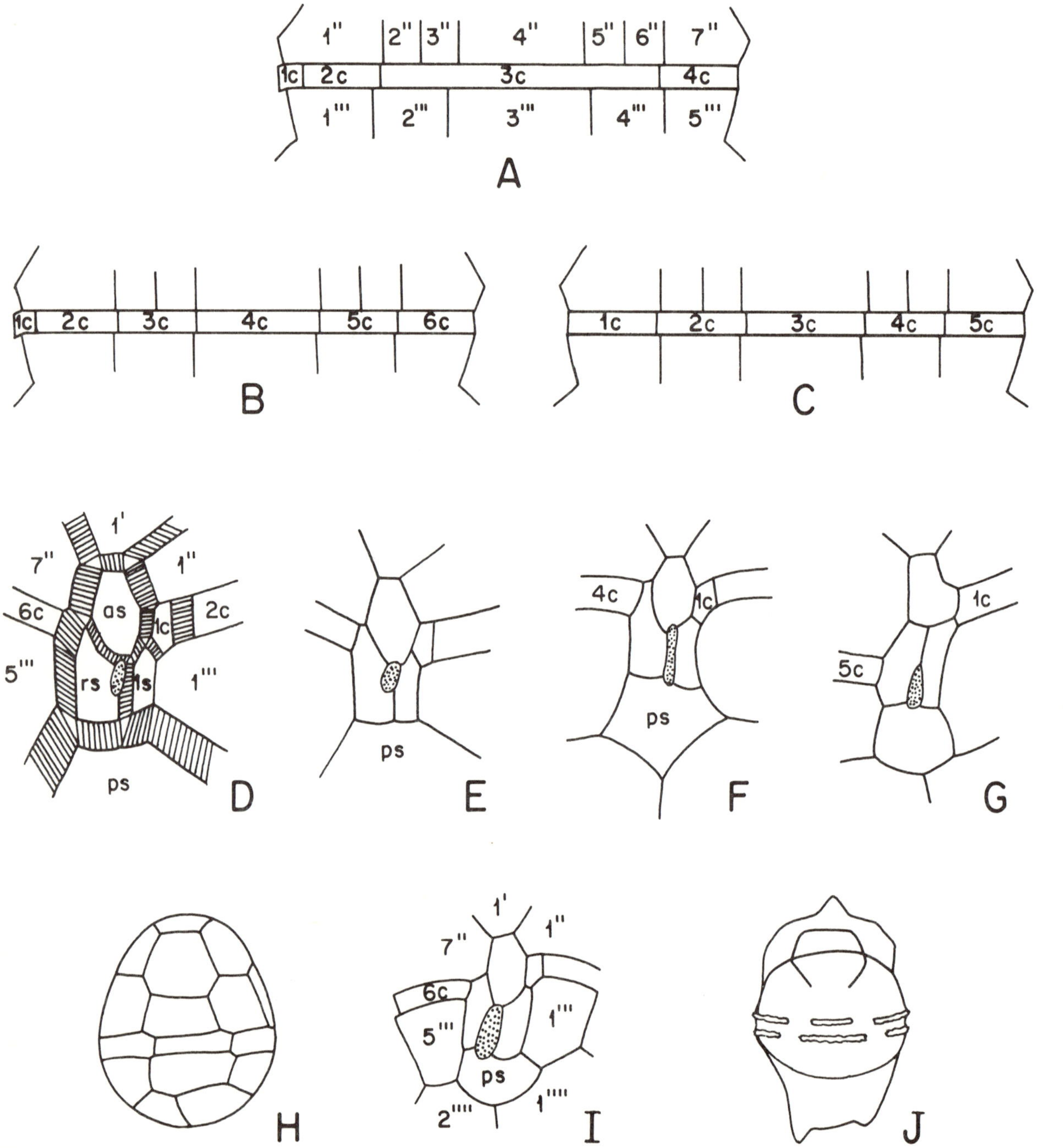

Figure 9.5 - PATTERNS IN CINGULAR AND SULCAL AREAS

A-C - Schematic patterns in the cingulum/paracingulum and immediately adjacent areas. A - Four cingulars include small *1c* and very large *3c*, as in living *Protoperidium* and Eocene *Kisselovia*. B - Six cingulars include small *1c* and three dorsal cingulars, as in some species of living *Scrippsiella* and Late Cretaceous *Palaeoperidinium pyrophorum* and *Trithyrodinium* sp. C - Five cingulars are apparently homologs of those in B except for the small initial unit, as in living freshwater *Peridinium*. Actual offsets of longitudinal (vertical) sutures at the cingulum may be so slight that intercalated growth bands appear continuous from epicyst onto hypocyst.

D-E - Sulcal paratabulation in Late Cretaceous *Palaeoperidinium pyrophorum*. D - The pattern as normally expressed in the fossils. The left sulcal, shown here as a single plate, was interpreted as two by Gocht and Netzel (1976). E - The same pattern "reduced" by removal of striate pandasutural areas that are normally conspicuous in actual cysts, to facilitate comparison with F and G.

F-G - Generalized sulcal tabulation for living marine *Protoperidinium* (F) and the specific tabulation for living freshwater *Peridinium cinctum* (G after Boltovskoy, 1976).

paratabulation of this region in cysts especially difficult. When parasutures are clear, four principal sulcal paraplates are indicated, but additional ones may be suggested by markings whose significance is difficult to evaluate without multiple, suitably preserved specimens. Two examples from the fossil record are shown in Figure 9.5 D-E and I, and these can be compared with the patterns typical for marine *Protoperidinium* and living freshwater *Peridinium* (Figure 9.5 F-G).

The Mesophragm and Traces of Cyst Ontogeny

Mesophragms appear in some Pp-cysts as a wall between layers that are plausibly identified as periphragm and endophragm because of their similarity to these layers in more common two-walled specimens. Mesophragms have been observed occasionally in species of *Chatangiella*, *Isabelidinium*, *Trithyrodinium*, and *Luxadinium* (Figures 4.1 I, 9.2 D-E, 9.6 A-E). Comparable structures have not been reported in cysts of living species, and their precise nature and significance are, therefore, uncertain.

In some species, the mesophragm occurs in conjunction with intraspecific morphological variation among hundreds to thousands of specimens derived from single preparations (Evitt, 1961 and unpublished; May, 1980; Stein, 1983), suggesting the following sequence in wall development (Figure 9.6 A-D):

1. Formation of the pericyst. This stage is represented by undamaged specimens in which no internal wall is visible.

2. Formation within the pericyst of a (future) mesocyst, which at first lies close against the pericyst but gradually contracts, pulling away from it, especially in the apical and antapical areas, so that a bicavate to circumcavate cyst with increasingly large pericoels develops.

3. Eventual formation of an innermost body, the endocyst, like that typical of other species of the genera.

4. Continued contraction of the mesocyst until it forms mostly a "cap" over each polar surface of the endocyst. This "cap" may appear thickened or wrinkled where its contraction has been greatest, but of unaltered thickness to the sides, where its continuity across the equator may be discernible in only occasional individuals.

Alternatively to these four steps, development may be limited to step 1 and step 3. This is indicated by individuals, which occur in the same preparation with specimens having fully developed endocysts and shrunken mesocysts, but differ from them only in showing no traces of mesophragm whatsoever, although the endophragm is strongly developed.

In species having a Type I or Ia archeopyle in the pericyst, that opening appears also to penetrate the mesocyst if the latter fills most of the interior and is closely appressed to the pericyst in the anterior intercalary area. In specimens judged to be more "advanced" because the mesocyst is more extensively free from the pericyst, no discrete opening has been observed in this intermediate wall. A Type (tAtI)a opening is occasionally distinct in the endocyst, but commonly this innermost body appears to be without an opening. The feature that Evitt (1967) reported as the operculum of the mesoarcheopyle is not an operculum; rather, it appears to be the nipplelike remnant of the extreme apex on the contracted mesocyst (Figure 9.6 D).

During the apparent sequence of developmental stages, the pericyst exhibits nothing like the change in shape that affects the mesocyst. However, outlines of cysts composed of periphragm alone tend to differ from those of cysts in which the endophragm is fully developed, especially by having less marked independent convexity in the equatorial region where the endocyst is centered, and less pronounced anterolateral "shoulders" in the epicyst outline.

The occurrence of occasional "empty" pericysts having more variable outlines than typical for cysts with fully formed endocysts matches observations on living *Peridinium limbatum*, as already discussed. The comparison even extends to the presence within such "incomplete" specimens of granular matter dispersed through the interior instead of the

H-I - Late Cretaceous *Trithyrodinium* sp. of Stein (1983). H - Paratabulation on dorsal surface as indicated by grooves on interior of cyst wall. Note that here, as in endocysts of living freshwater *Peridinium limbatum* (Figure 4.2 J), these internal grooves give no suggestion of intercalary growth bands that may well have been present on the zygotic theca. I - Detail of sulcal paratabulation, topologically like D but with much smaller posterior sulcal.

J - Late Cretaceous *Chatangiella* sp. (dv), showing interrupted ridges along dorsal paracingulum which represent adcingular edges of pre- and postcingular paraplates.

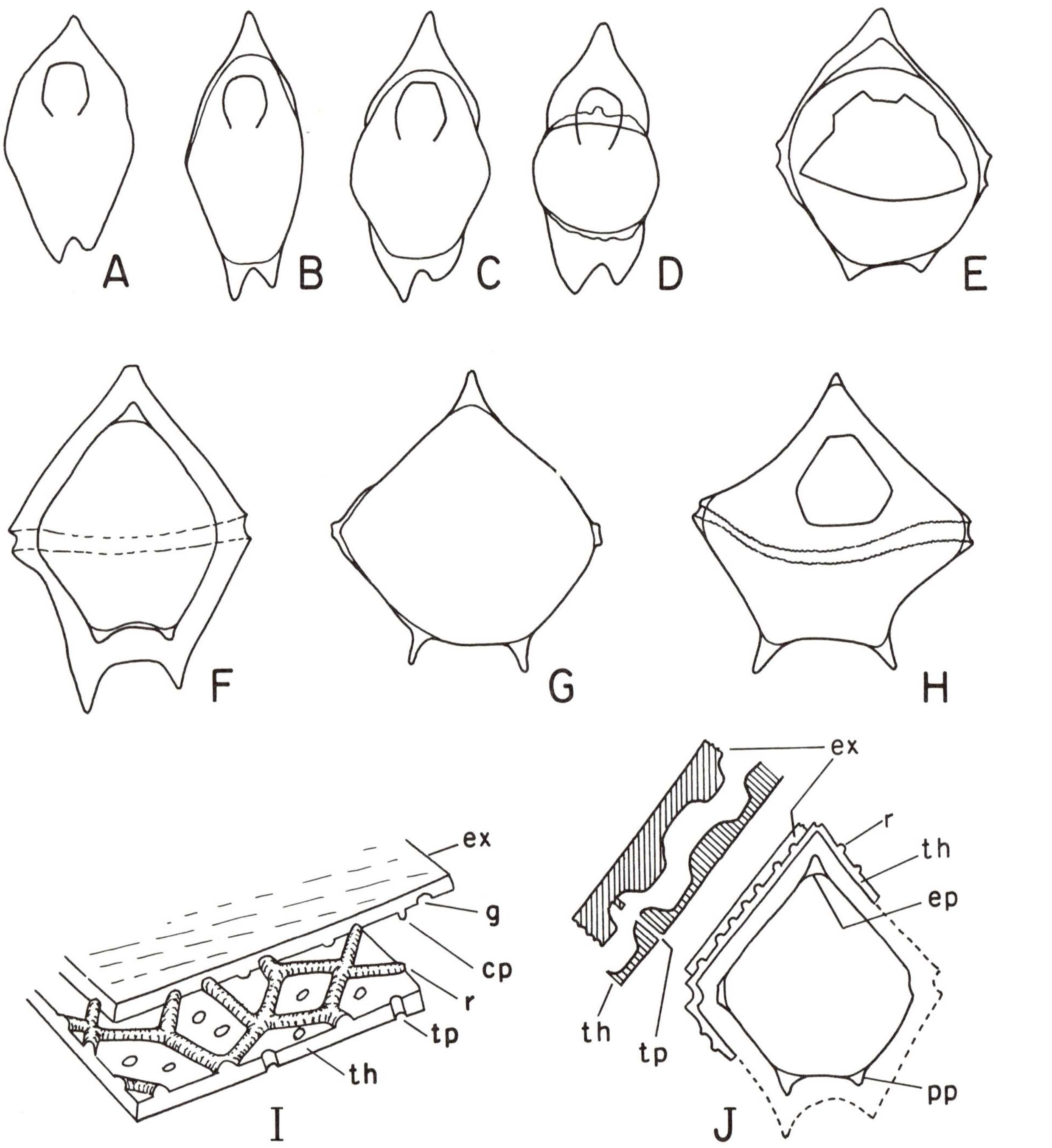

Figure 9.6 - THREE-WALLED CYSTS (WITH MESOPHRAGMS) AND EXTRATHECAL SPOROPOLLENIN BODIES

A-D - Apparently progressive development of wall layers shown by four specimens of Late Cretaceous *Chatangiella* sp. cf. *C. victoriensis* (dv). Inside the initially formed pericyst (A) appears the future mesophragm (B), which contracts progressively from the poles (C) until it eventually forms only wrinkled polar "caps" on the late-forming subspheroidal endocyst (D). The archeopyle suture is also shown.

E - Mid-Cretaceous *Luxadinium* sp. Optical section (dvv) showing distinct mesophragm between peri- and endophragm at apex, and outline of displaced operculum of Type (3I3P) archeopyle.

F-G - "Conventional" cornucavate peridinioid cyst of *Palaeoperidinium pyrophorum*. F - Cyst inside the extrathecal layer (from which the distinctive inner surface pattern is omitted). G - Cornucavate cyst from another specimen, shown without its enclosing paratabulate wall for comparison with H.

H - Cornucavate peridinioid cyst of Late Cretaceous *Phelodinium magnificum* (with Type I archeopyle).

more compacted accumulation bodies found in most of the typical double-walled cysts.

Aside from the biological interest that may attach to this case, the palynologist should be aware of the amount of substantive difference in morphology (i.e., not just degrees of difference in shape, ornament, and the like, but in the basic structure of the specimen) that may occur within some species of P-cysts. In an example like the one described, character contrasts among a small number of individuals may appear so great as to suggest several distinct species, and a substantial number of specimens may be needed to demonstrate the intergradation that actually exists.

Palaeoperidinium pyrophorum

Of all organic-walled microfossils described by Ehrenberg (1838), there can be no question that those with most obvious similarity to modern dinoflagellates were the ones he named *Peridinium pyrophorum*, a large species (to 136 x 100μm) having ortho hexa paratabulation, which we today attribute to *Palaeoperidinium*. Experience has shown this to be a geographically widespread species, often represented abundantly in sediments ranging from late Campanian to Paleocene in age. It is appropriately symbolic of the slowness of progress toward real understanding of a complex group of organisms that this long-known microfossil and obvious dinoflagellate should still be a challenge to interpretation.

Ehrenberg, knowing nothing about resting cysts, accepted it tacitly as a fossil theca on the basis of the strong similarity with thecae of such living species as *Protoperidinium* (then *Peridinium*) *conicum*. Nearly a century later Deflandre encountered the fossil in Parisian cherts and, after first (1935) referring to it as *Palaeoperidinium* cf. *pyrophorum*, described it later (1936) in more detail also under the name of *Peridinium conicum*. Lejeune (1938) and Sarjeant (1967a) restudied Ehrenberg's original preparations, which consist of several chert flakes containing numerous individuals, and added precision to his very incomplete remarks and figures. Sarjeant designated the species as type of the previously untypified *Palaeoperidinium*. Evitt (1975d) discussed the archeopyle and the nomenclatural status of the genus, and Stover and Evitt (1978) listed synonyms for species. Examination for the first time with the SEM led to a meticulous description of morphologic detail by Gocht and Netzel (1976), which includes many hitherto unobserved features.

Although the interpretation to which Gocht and Netzel were inclined was plausible in light of earlier observations on other fossil cysts, the evidence they revealed leads this author to a very different conclusion—one they considered but rejected—namely, that the fossil we know as *P. pyrophorum* is, as we usually see it, a structure formed outside the theca, not inside it like more ordinary cysts. By this interpretation, *P. pyrophorum* is truly unique among all dinoflagellates known to date, living as well as fossil. The unusual nature of the case justifies a brief review.

The obviously dinoflagellate-like features of the fossil (Frontispiece C) are its peridinioid shape and, when well preserved, a fully revealed peridinioid paratabulation (ortho hexa) evidenced by polygonal regions with fine, reticuloid texture, normally separated by prominent pandasutural areas bearing transverse striations. Curiously, the holotype, alone among specimens in the type material restudied by Sarjeant (1967a), has simple parasutural lines separating the textured paraplates. Three-dimensional specimens show strong dorsoventral compression, and it is typical for specimens from compacted sediments to have dorsal and ventral surfaces fully appressed.

The archeopyle is Type (A3I3P), and the size and distribution of paraplates is such that the transapical archeopyle suture nearly coincides with the lateral margins of the specimen. Presence or absence of the operculum is most easily determined by observing whether the paratabulation visible on the epicyst presents the pattern of the ventral surface alone or a rather confusing superposition of patterns of both dorsal and ventral surfaces. Both isolated opercula and specimens of the rest of the cyst without the operculum attached are rare.

I-J - Perspective (I) and sectional (J) diagrams showing relationship of internal sporopollenin cyst and cellulosic theca (th) to extrathecal sporopollenin layer (ex) inferred for Late Cretaceous *Palaeoperidinium pyrophorum*. Grooves on the inner surface (g) of the extrathecal layer represent the ridges of the reticulate thecal surface (r), and short cylindrical projections (cp) appear to have originated as incomplete fillings of the trichocyst pores (tp). Sporopollenin layers of inner cyst include periphragm (pp) and endophragm (ep) in cornucavate relationship. In the detail (upper left of J), a separate small portion of the extrathecal layer is shown slightly separated from a portion of the underlying theca.

The first surprise discovered by Gocht and Netzel with the SEM was that the relief expressing the paratabulation pattern that is the hallmark of the species is located on the inner surface of the wall, the outer surface being either smooth or showing only faint suggestions of the inner detail. Although detectable with care under the light microscope, the location of the surface relief was not determined in that manner.

The second surprise was that this internal relief matches with high fidelity, but in negative form, the relief typical for the external surface of the theca of a modern *Peridinium*, "as if the inner surface were a cast of the outside of a theca" (W.R.E. translation; Figure 9.6 I-J). Specifically, the textured surface of the polygonal areas in the fossils resolves into a reticulum of rounded grooves along which short, peglike projections stand at irregular intervals, negative analogs to the network of ridges on the thecal plates in living *Peridinium* and to the trichocyst pores distributed along these ridges in some species (or, between them in other species). Similarly, the transverse ridges of the pandasutural bands lie on the inner surface of the fossil, while the corresponding thecal pattern is expressed in exterior relief.

While acknowledging an inclination to regard this relief at first glance as an impression of the external thecal surface, Gocht and Netzel reasoned that observations on living species have firmly established the basic cyst-inside-theca relationship and that evidence is also clear for the development of thecalike patterns on inner cyst surfaces (e.g., *Peridinium limbatum*, as already discussed in this chapter). Accordingly, they did not question the fundamental nature of *Palaeoperidinium pyrophorum* as a cyst that formed inside the theca. However, in this extraordinary case, their conclusion seems untenable for at least the following reasons:

Firstly, whereas most specimens of *P. pyrophorum* appear to be constructed of a single layer of material and have, therefore, been considered autocysts, a significant number of specimens in some samples contain internal bodies (Figure 9.6 F-G). On close inspection, these thin and wrinkled structures are clearly double walled—cornucavate, in the terminology we are accustomed to applying. Excepting the apparent absence of an archeopyle, they closely resemble the periphragm and endophragm of a P-cyst similar to fossil *Phelodinium* of the *Deflandrea* complex (Figure 9.6 H). In other words, this fossil structure, whose peculiarity lies in its possibly extrathecal origin, sometimes befittingly surrounds what appears to be a reasonably "conventional" intrathecal cyst.

Secondly, it does not seem reasonable that a surface independent of the mature theca should exhibit features (the pandasutural areas) that are so obviously the counterparts of thecal features (the intercalary bands) that are, in turn, phenotypic expressions of individual ontogeny. Specimens of *P. pyrophorum* show the same sort of variation in relative width of pandasutural areas that is typical for intercalary bands in thecae of living species at different stages of cell growth (Boltovskoy, 1979). As if to emphasize this point, the holotype of the species has only simple parasutures, apparently recording a particularly early stage in the growth of a presumably zygotic cell.

Finally, although the thecalike patterns on both fossil *P. pyrophorum* and the endocyst of living *Peridinium limbatum* are expressed by internal relief, the form and location of this relief strongly suggest significantly different origins. While the structure of cyst wall (i.e., typical pericyst and endocyst) in the modern species precludes the possibility that the simplified pattern on the endocyst was determined by direct contact with the exterior thecal surface, nothing appears to preclude the possibility of such contact in the case of the characteristic wall of *P. pyrophorum*. Furthermore, this contact would provide a simple mechanical explanation for the unusually detailed and distinctive topography of that wall (Figure 9.6 I-J). The situation in *P. pyrophorum* is evidently unique—in this case, the parsimonious explanation consistent with Occam's Razor leads to apparent authentication of a "five-legged horse" (Chapter 7).

If the fossils we normally consider to represent *P. pyrophorum* formed outside the theca, by what mechanism did they form, and can we assume that they are zygotic hypnocysts as we now take most other fossil dinoflagellates to be? In respect to the first question, some dinoflagellates today form a gelatinous sheath external to the cell wall, suggesting at least the possibility that other materials might also be laid down there. However, it is clear that the pellicle could not have served as a precursor for such an exterior layer of sporopollenin, as it may do for "conventional" hypnocysts. In respect to the second question, the answer is unclear. Variation in width of pandasutural areas from specimen to specimen appears to record different amounts of cell growth. It will be interesting to discover if living species, such as *P. limbatum*, occasionally produce zygotic cysts in a theca that has not developed intercalary bands. Also, the occurrence of double-walled bodies inside the usual fossil suggests the possibility that these bodies may be the equivalent of the more customary dinoflagellate resting cyst and that the extrathecal structure is of quite a different nature, albeit, in that event, one within which formation of a "normal" zygotic cyst may have proceeded.

However one chooses to read the evidence of *P. pyrophorum* at this point, the case provides emphatic endorsement of Gocht and Netzel's suspicion that we still stand at the beginning of our knowledge of *Peridinium* cysts!

No other fossil cysts with features just like those of *P. pyrophorum* are now known. *Palaeoperidinium cretaceum* from Early Cretaceous sediments is closely similar but excellently preserved specimens of that species have not been studied. Among other species, although relatively smooth strips between intratabular fields of ornament or penitabular rows of relief are frequently observed (as in some *Deflandrea*, *Spinidinium*, *Alisocysta*), pandasutural areas closely resembling striate thecal intercalary bands are rare and appear to be on the exterior of the periphragm, as in *Saeptodinium eurypylum* (Evitt, 1975d). The fossil whose structure is, perhaps, most suggestive of *P. pyrophorum* is the gonyaulacoid cyst, *Sirmiodinium grossii* (Warren, 1973). It would be interesting to know how closely details of the inner surface of the patterned wall in this species resemble those in *P. pyrophorum*, but specimens have yet to be studied with the SEM.

Chapter 10. Morphology of G-Cysts. I.
Gc-, Gv-, Gp-, Gq-, and Gs-Cysts

G-cysts include morphologies that range widely, from those for which a fully and clearly developed paratabulation directly indicates a gonyaulacoid pattern to those which are barely determinable as dinoflagellates and whose affinities within the class are not demonstrably close to *Gonyaulax*. The only character common to all G-cysts besides their sporopollenin composition is that such traces of paratabulation as are present are either diagnostic of, or at least compatible with, the gonyaulacoid pattern.

In the interest of convenient organization of the large number of taxa that express this diversity, the group is subdivided here into eight categories, five of which are discussed in this chapter and three in Chapter 11 (along with several other groups).The first two to be discussed, Gc- and Gv-cysts, illustrate well the biological inconsistencies of the morphological system followed here. Gc-cysts are almost certainly a closely knit group in a biological sense, with close affinity to modern *Ceratium*. On the other hand, Gv-cysts, although probably related to *Gonyaulax*, are brought together by morphological effects associated with pronounced dorsoventral compression, a feature that may well have developed in multiple lineages that also included uncompressed cysts excluded from this category. Gc-cysts and Gv-cysts have much in common and the dividing line between them requires careful definition. The characteristics of the categories are set forth in the first two sections of the chapter, and their distinctions are discussed further as a special topic later in this chapter.

Gp-, Gq-, and Gs-cysts include primarily cysts with, respectively, partiform, quinqueform, and sexiform hypocyst patterns expressed by parasutural features.

CHARACTERISTICS OF CATEGORIES AND COMPLEXES

Category of Gc-Cysts

Introduction

The most obvious features distinguishing cysts in this category are prominent horns in positions comparable to those in living *Ceratium* (Figure 10.1 A-B); that is, one at the apex, one (rarely two) at the opposite end, and one (or two) at the right (or right and left) side of the cyst with its base at or posterior to the paracingulum and its tip deflected posteriorly. Detailed comparison of cyst morphology of modern species is not possible however, because cysts are not known to be produced by the many living marine species, and because cysts of the few freshwater species are not geologically preservable.

The three horns of Gc-cysts are essentially coplanar and, in most examples, the main part of the cyst shows primary dorsoventral compression. Both factors strongly favor the same preferred orientation of settled specimens, with the consequence that the great majority of individuals present a dorsoventral view. When paratabulation is fully visible, it has the diagnostic characters of the corniform gonyaulacoid pattern. Usually only traces of paratabulation are indicated and they are often confined to angulations in the margin of the exclusively apical archeopyle. The sulcal notch, which is visible in many cases adjacent to the broad angular margin of midventrally located **1i** (= *6″*), is consistently offset to the left of the ventral surface. Wall structure variations include autophragms and both periphragm-endophragm and ectophragm-autophragm combinations.

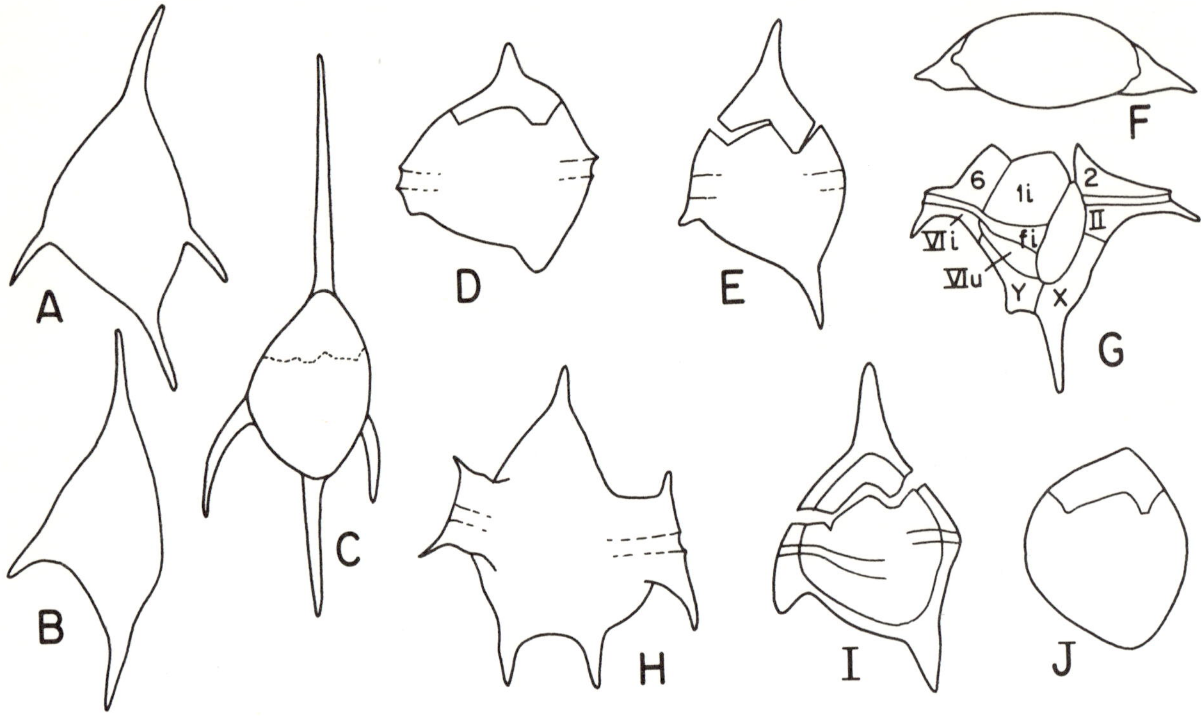

Figure 10.1 - CYSTS OF MODERN *CERATIUM* AND FOSSIL Gc-CYSTS

All are fossils of Early Cretaceous age except A-B, which are modern freshwater cysts from California.

A-B - *Ceratium furcoides*. Geologically unpreservable cysts without sporopollenin (vv). Upper figure shows second postcingular horn but both variants are from same sample.

C - *Muderongia crucis*. This cornucavate cyst (vv) lacks the symmetry of normal *Muderongia* and suggests an *Odontochitina* with a second postcingular horn. Compare outline with A. Dotted line indicates archeopyle suture.

D-E - *Aptea* sp. and *Pseudoceratium* sp. Acavate cysts (vv) with reduced postcingular horns.

F-G - *Muderongia* sp. cf. *M. mcwhaei*. Cornucavate cyst (endophragm shown only in F) with operculum to apical archeopyle missing. F - Optical section (av) through cingular prominences. G - Ventral paratabulation (vv).

H - "*Muderongia*" sp. Somewhat distorted specimen (vv). Unusual for the genus in having anterior projections on lateral prominences.

I-J - *Endoceratium ludbrookiae*. Complete cyst with dislodged operculum (I), and endocyst only, drawn without pericyst for comparison with Gv-cysts (both vv).

Many fossil cysts show at least a trace of the paracingulum where it crosses the base of the postcingular horn, and this feature may make a prominent indentation in the lateral margin of the horn or the locally expanded portion of the main part of the cyst into which the base of the horn merges.

Key to Complexes of Gc-cysts

I. Apical, antapical, and right postcingular horn determine a typical ceratioid outline; left postcingular horn, if present, smaller than right, so that outline remains distinctly asymmetrical.

  A. Equatorial cross section essentially circular .... *Odontochitina* cpx

  B. Dorsoventral compression distinct................ .... *Pseudoceratium* cpx

II. Right and left postcingular horns nearly or quite symmetrical, their basal portions about in line with paracingulum; one or two antapical horns ... . *Muderongia* cpx

Common morphological variables within the category are suggested by the following outline of points to consider in identification or description:

Shape

Length, number, prominence, precise position, detailed shape, and orientation of horns

Degree of dorsoventral compression of main cyst body

Walls

Number of wall layers and their relationship, including distribution of cavation

Surface ornament: character and distribution

Paratabulation

Distinctive details, if any, of archeopyle margin

Additional traces, if any

Regarding paratabulation, it is important to note that the author of this work does not concur with the opinion of Dörhöfer and Davies (1980) that more than four apical paraplates are represented in the operculum of Gc-cysts. In the rare examples where paratabulation is fully and clearly expressed, it shows the characters typical for modern *Ceratium* (Wall and Evitt, 1975).

Prominent horns in a non-peridinioid arrangement, combined with the apical archeopyle and offset parasulcal notch, make almost all Gc-cysts easily distinguishable from those in any other category. The currently recognized genera are simple morphological constructs on the basis of differences in shape, surface ornament, and wall structure. Therefore, transitions are likely to be found and limits of the genera may become obscure. If the horns are severely reduced (e.g., *Aptea*), distinction from Gv-cysts may be difficult, resting on the matter of whether one of the hypocystal prominences is postcingular, rather than antapical. Isolated endocysts from cavate species would probably be considered Gv-cysts unless traces of periphragm on some specimens were sufficient to reveal the distinctive shape of the pericyst. For recent discussion of morphology, see Helby (in press).

Odontochitina Complex

Taxa: *Odontochitina*, *Xenascus*.

Occurrence: Cretaceous (mid-Early to Maastrichtian).

The presence of a single postcingular horn, or distinct inequality in horn size if two are present, imparts a clear asymmetry to the cyst in the usual dorsoventral view (Figure 10.1 C-E, I-J). *Odontochitina* is the more commonly encountered genus; both are typically cornucavate. *Xenascus* (Frontispiece J) is, in effect, an *Odontochitina* with processes, and the two may intergrade. They show no dorsoventral compression and seldom any angularities along the archeopyle margin. Analysis of shape and horn structure seems likely to justify more species than are currently recognized. Multiple small perforations in rows or a network may mark the horns of some species. There may be no suggestion of paratabulation, or parasutures may be expressed as folds or ridges; gonal and intergonal positions of processes may be recognizable in *Xenascus*. The paracingulum, when traceable, may cross the base of the right postcingular horn or this projection may arise definitely posterior to the paracingulum.

Pseudoceratium Complex

Taxa: *Aptea*, *Endoceratium*, *Pseudoceratium*.

Occurrence: Cretaceous (mid-Early to mid-Late).

In many respects, these forms are similar to those in the *Odontochitina* complex, but wall structure is more diverse, dorsoventral compression is pronounced, and angulations along the archeopyle margin typically give clear indications of adjacent apical and precingular paraplates. The sulcal notch is offset to the left. The surface may be smooth or ornamented with features of low to moderate relief, and these may show some reduction or modification in the central dorsal and ventral areas. Paratabulation may not be apparent or may be expressed by faint markings on an otherwise smooth surface or by parasutural interruptions in distribution of ornament. Species with numerous short processes should be examined closely for ectophragm. Circumcavate forms (*Endoceratium* of authors, but not Helby, in press) may exhibit occasional (especially marginal) interconnections between periphragm and endophragm. Isolated endocysts probably would be indistinguishable from Gv-cysts (Figure 10.1 J; see comment under Gv-cysts).

Muderongia Complex

Taxa: *Muderongia*, *Phoberocysta*.

Occurrence: Late Jurassic to early Albian.

Approximate symmetry in horn development distinguishes these cysts from the *Odontochitina* and *Pseudoceratium* complexes (Figure 10.1 F-H). In addition, right and left postcingular horns tend to have a basal portion that projects laterally (carrying an extension of the paracingulum on its surface) and a distal portion that is directed posteriorly. The origin of this feature is suggested by the thecal structure of some modern freshwater *Ceratium* (discussed later in this chapter). In fossil cysts, the proportions of the horns and the relative prominence of the two portions are interspecific variables. Another distinctive feature is a second antapical horn in some species. This varies interspecifically from only a spur at the base of the principal horn to a free-standing projection of nearly equal size; no equivalent features are known in any modern species of *Ceratium*.

In terms of general morphology, *Muderongia* (without processes) and *Phoberocysta* (with processes) can be regarded roughly as symmetrical equivalents to *Odontochitina* and *Xenascus*, respectively.

Category of Gv-Cysts

Introduction

This category includes cysts having some of the characters of Gc-cysts, but lacking the horns in recognizably postcingular position that are diagnostic of that group (Figure 4.5, 10.2 A-J). As in the case of Gc-cysts, dorsoventral compression and an apical archeopyle with free operculum and sulcal notch offset to the left are common to all examples. Local but incomplete indications of paratabulation are typical for some taxa. In the few instances where the paratabulation is complete (Figure 10.2 B), the pattern is sexiform gonyaulacoid, in contrast to the corniform variety of the gonyaulacoid plan evidenced by a few Gc-cysts.

The category of Gv-cysts can be defended as eminently convenient on the grounds that it is an easily distinguished grouping of morphologically diverse taxa, some of which are significant stratigraphic markers. However, there can be little doubt that this category, quite in contrast to the Gc-cysts, lacks biological justification. On the basis of wall structure and outline shape, some Gv-cysts cannot be distinguished from Gc-cysts, and in some instances may actually be isolated endocysts without well-developed horns, such as those that occur in certain Gc-cysts (e.g., *Endoceratium ludbrookiae*, Figure 10.1 J). Other Gv-cysts are distinguished from counterparts elsewhere among categories of G-cysts only in having primary dorsoventral compression and the other features associated with this. Examples of such similar pairs among Gv-cysts and other categories are: *Areoligera* (Gv) and *Systematophora* (Gi) (Figure 10.2 C-F), *Glaphyrocysta* (Gv) and *Adnatosphaeridium* (Gi), *Cyclonephelium* (Gv) and *Sentusidinium* (Gn), *Canninginopsis* (Gv) and some Gs-cysts.

Dorsal and ventral surfaces of Gv-cysts can be distinguished easily and reliably by noting the position of the offset parasulcal notch. The extent of primary dorsoventral compression, while typical for a species, is interspecifically variable. When it is slight, the choice between this category and some other (as represented, for example, by the pairs of genera named above) may be arbitrary. The degree of primary compression in three-dimensionally undistorted specimens can be assessed from accidental polar or lateral views, or from manipulated specimens. Alternatively, transverse elongation of an operculum seen in apical view is a sensitive indicator of primary dorsoventral compression. Outline shape of the cyst often includes rounded (rarely subangular and prominent) lobes, one at the apex and one or two at the antapex. Paratabulation, when expressed, shows the antapical lobes to involve the antapical paraplate in the vicinity of the **X/Y/Z** and **VI/Y/Z** triple junctions, in contrast to the arrangement in Gc-cysts.

Two aspects of surface ornament and wall relations that are associated with primary compression are apparent in all but the simplest Gv-cysts. One is the marked tendency for reduction or suppression of features in the central ventral and, to a somewhat lesser degree, the central dorsal parts of the cyst. The other is the related tendency for elaboration of features, both surface ornamentation and wall relations, in the marginal areas (i.e., the peripheral regions: apical, antapical, right and left lateral). Both tendencies are apparent in the diagrams of Gv-cysts in Figures 4.5 and 10.2.

Although many living peridinioid dinoflagellates show various amounts of dorsoventral compression, and this is reflected in the shapes of the sporopollenin cysts produced by a few of them, dorsoventral compression does not appear commonly today among gonyaulacoids other than *Ceratium*. Cysts with the characteristics of Gv-cysts have not been reported from post-Miocene sediments or from any living species.

As these remarks suggest, a much more extensive variation in wall structure and surface ornament exists among Gv-cysts than among Gc-cysts, and this is the principal basis for distinguishing four complexes. Features meriting close attention in individual cases are:

- Shape
  - Outline, especially at poles
  - Amount of dorsoventral compression
- Walls
  - Identity of layers and distribution of cavation
- Surface relief: scale, structure, and distribution

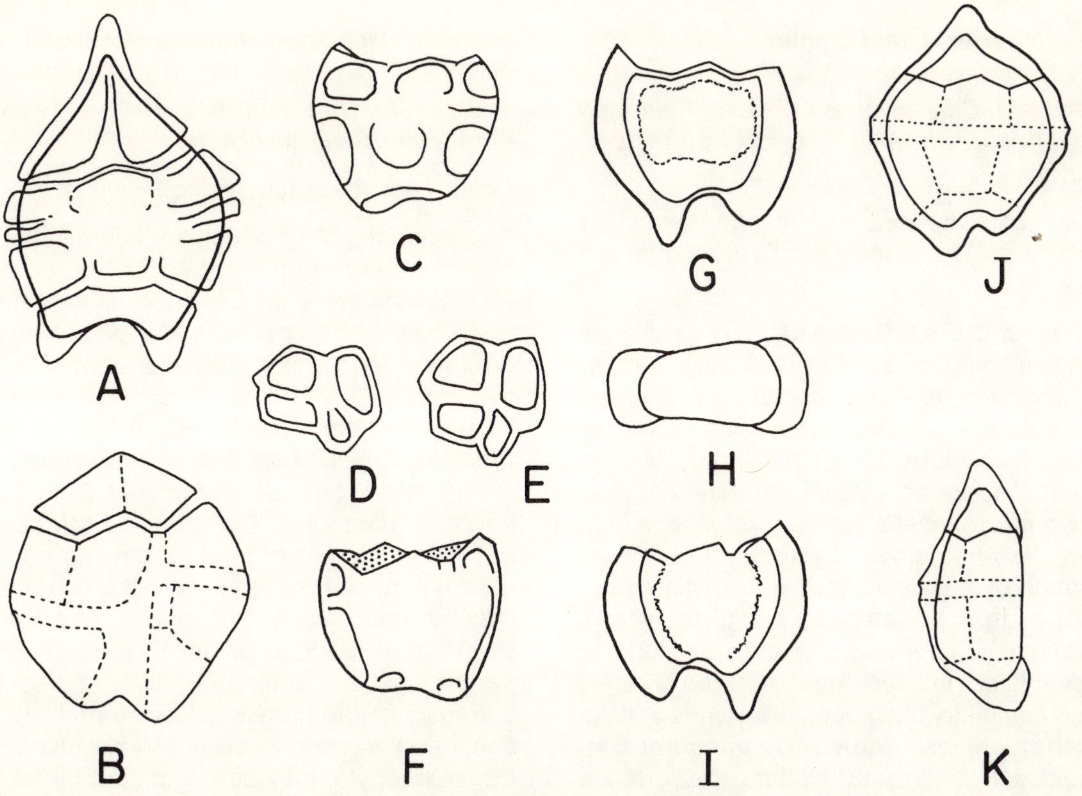

Figure 10.2 - REPRESENTATIVE Gv-CYSTS

A - *Areoligera* (?) sp. Complete specimen (dv) with prominent horns. Elaborately perforate membraneous projections are attached along lines shown and define paratabulation. Note middorsal reduction in surface detail and origin of both antapical horns from paraplate **Y**. Late Cretaceous; based on Wall and Evitt (1975).

B - *Canninginopsis denticulata* (vv). Parasutures (dashed) are clearly indicated by rows of denticles which outline major paraplates on ventral surface in an arrangement contrasting with Gc-cysts (compare Figure 10.1 and 10.6). Early Cretaceous; based on Morgan (1979).

C-D, F - *Areoligera senonensis*. C and F - Main part of marginate cyst (dv and vv); elaborate projections (bases indicated by arcuate lines) are ventrally restricted to lateral margins and reduced middorsally. D - Transversely elongate operculum of apical archeopyle (av) documents primary dorsoventral compression. Late Cretaceous.

E - Operculum (av) of *Systematophora* sp., a Gi-cyst from same assemblage as preceding, gives evidence of equidimensional archeopyle and uncompressed form. The paratabulation of this species and the projections which express it are virtually the same as in *A. senonensis*, except for distribution and form related to presence or absence of dorsoventral compression. Late Cretaceous.

G-I - *Membranophoridium aspinatum*. Circumcavate cyst lacking its operculum (dv, optical cross-section in av, and vv). Lines of junction between periphragm and endophragm are shown around central portions of dorsal and ventral surfaces. Not shown are slender funnel-like invaginations of periphragm along lateral margins which extend to endophragm. Paleogene.

J-K - *Canningia* sp. Entire cyst with operculum in place (dv and rlv). Wall, except on central ventral surface, is thick and complexly structured (but of low density); only inner and outer limits shown in figures. Localized variation in wall thickness or structure may define paratabulation (dotted lines) more or less completely. Depending upon the species, wall may be viewed as autophragm and ectophragm with interconnections, or as differentiated autophragm. In some species, blunt, lateral, hornlike prominences extend in line with paracingulum. Cretaceous.

Key to Complexes of Gv-Cysts

I. Wall 1- or 2-layered: an autophragm smooth or with features of low to high relief (with or without ectophragm); or distinct endophragm and periphragm, the latter forming prominent processes or septa, but periphragm neither extensive nor inflated .... *Areoligera* cpx

II. Wall 2-layered; periphragm (typically smooth or with features of low relief) separated from endophragm at least locally by distinct pericoel .... *Membranophoridium* cpx

III. Wall structure complex, spongy-fibrous layer of anastomosing filaments as external part of differentiated autophragm, or as filling of space between autophragm and ectophragm ....
.... *Canningia* cpx

Areoligera Complex

Taxa: *Areoligera*, *Canninginopsis*, *Cerbia*, *Chiropteridium*, *Circulodinium*, *Cyclonephelium*, *Glaphyrocysta*.

Occurrence: Late Jurassic to Miocene; especially common and varied in the Late Cretaceous and Paleogene.

See Figures 4.5 D-E and 10.2 A-D, F. The exteriors of cysts in this complex range from virtually featureless (*Circulodinium*) to highly modified by long and elaborate processes (*Areoligera* and *Glaphyrocysta*). The wall is frequently an autophragm, and an ectophragm may be developed as distal connections between processes (some *Cyclonephelium*, *Glaphyrocysta*). Alternatively, projecting features are clearly formed by a periphragm that is distinct from the endophragm at least in marginal areas (*Chiropteridium*, some *Areoligera*), but such pericoel as is present is not inflated. Paratabulation is rarely distinct and complete, but parasutural ridges or lines of fine ornament are known (*Canninginopsis*). Discontinuities in ornament distribution may define parasutures at least locally (some *Cyclonephelium*), penitabular arcs of ornament may be prominent in dorsal pre- and postcingular areas (*Areoligera* and *Glaphyrocysta*), and intratabular processes are occasionally developed (some *Chiropteridium*). The ornament tends to be coarser, longer, or more extensively interconnected in the marginal areas (marginate cycts), especially in one or two more or less meridional lines paralleling each lateral margin. The elements of ornament may have maximum length around the ventral surface, thereby creating (at least in the more complexly constructed cysts) a ventral concavity not evident if only the limits of the main cyst wall are considered.

Shape and ornament are highly variable intraspecifically, especially in those species with elaborate surface relief. This variability commonly exceeds in amount that which has been judged quite adequate to warrant specific distinctions in other groups of cysts. When it is not only extreme but also differs in amount from sample to sample of what appear to be the same species, this variation is very difficult to deal with satisfactorily. It plausibly reflects environmental conditions, and development of a practical way to assess it might lead to useful ecological interpretations. Unfortunately, at the present stage of our knowledge, despair by the microscopist and identification only to the generic level are more likely to be the result when large numbers of complex and varied cysts must be dealt with expeditiously.

Membranophoridium Complex

Taxa: *Membranophoridium*, *Palynodinium*, *Renidinium*, *Senoniasphaera*.

Occurrence: Late Jurassic to Miocene.

Characteristic of this complex is a distinct pericoel between periphragm and endophragm, its extent and location being the basis for the currently recognized genera. The periphragm typically lacks strong ornament (exception: *Palynodinium*). The pericoel may surround the cyst (circumcavate) as in *Senoniasphaera*, or be localized in broad or restricted lobes on the lateral ventral surfaces as in *Membranophoridium* (Figures 4.5 H, 10.2 G-I), *Palynodinium*, and *Renidinium*. Middorsal and midventral merger of periphragm and endophragm is common, and the line where the two layers meet may be conspicuous. The apex may be marked by a single rounded lobe or hornlike extension of the periphragm, the antapex by two, and each lateral extremity of the paracingulum by one. Paratabulation, rarely expressed except by angular margins of the archeopyle, may be suggested by folds of the periphragm. The pericoel may be crossed by occasional processes or funnelform structures joining periphragm and endophragm, especially laterally.

Canningia Complex

Taxa: *Canningia*.

Occurrence: Late Jurassic to Cretaceous.

Although forming a distinctive element in Late Cretaceous assemblages, cysts having the morphology of this complex have somehow largely escaped description, although many species have been referred to *Canningia* that do not have the wall structure characteristic of the type species. Most of these have features appropriate to the *Areoligera* complex as it has been distinguished here. Cysts in the *Canningia* complex are distinguished by a relatively thick wall of elaborate structure (Figure 10.2 J-K), but with a fairly fine "grain"; processes or septa extending from the surface are not present. Superficially two subparallel wall layers appear to be present, with a spongy-textured material filling the intervening space. Under high magnification it may or may not be possible to resolve a distinct ectophragm or to conclude that the autophragm is differentiated into an elaborately structured region covering a denser basal layer. Study of wall sections under the SEM will be informative.

One apical and two antapical lobes or hornlike prominences are typically present, and two similar lobes may appear at the lateral extremities of the paracingulum. Paratabulation may be recorded by surficial ridges (actually increases in overall wall thickness) along parasutures or by subtle changes in density within the spongy layer, but details are normally obscure. The contrast in wall modifications between the central and marginal areas of the dorsal and ventral surfaces is usually less marked in this complex than in the *Areoligera* and *Membranophoridium* complexes.

Cyst shape and details of wall structure are the principal criteria for distinguishing species. In some forms, accessory archeopyle sutures occur between precingular paraplates, causing the archeopyle to appear much larger than the operculum when the opening and its cover are seen separately.

Category of Gp-Cysts

Introduction

Diagnostic of cysts falling into this category is the topological peculiarity of the partiform hypocyst pattern: the anterior limit of **Z** (i.e., **Z/Im**) abuts against **Iu** (rather than against **X** as it does in cysts with quinqueform and sexiform gonyaulacoid patterns). However, features associated with this distinction may be more obvious than the position of the critical parasuture itself. Such features include: 1) large **Z** reaching to near the paracingulum where its width is constricted, 2) space between the ends of the paracingulum anterior to **Z** occupied by a cluster of small paraplates, 3) **Iu** relatively large, even approaching **VI** in size, and 4) tendency for translocation of **X/Y** and **X** onto the antapical surface in proportion to the size of **Iu**. This is the only group of gonyaulacoid cysts in which relatively large and prominent anterior intercalaries occur, but such paraplates are not present in all cysts within the category.

Shape, size, wall structure, and archeopyle type vary. Although size and ornament seem to be typical for certain groups, these apparent distinctions may disappear as more examples are discovered. The epicyst is characteristically smaller than the hypocyst. Conspicuous processes are not normally a part of the surface relief (exception: *Luehndea*) and intratabular processes have not been observed on the exterior.

Many of the cysts attributed to the two complexes discussed under the heading of this category are difficult to analyze owing to their small size, often complex wall structure, and near sphericity. In some cases, the relatively small epicyst (related to the unusual anterior position of the paracingulum) adds to the problem by making it difficult to confirm the presence of anterior intercalaries and, especially, to determine the precise composition of the archeopyle. To observe these critical features may require manipulation of individual specimens or diligent search for favorably oriented individuals. Conveniently, cysts of many of the small species having anterior intercalaries (*Microdinium* complex) are nearly spherical and can be found in nearly random orientations, while those small species whose preferred dorsoventral orientation results from primary compression (*Phanerodinium* complex) uniformly appear to lack anterior intercalaries, although details of their archeopyles may remain more or less obscure.

This is a diversified and widely represented category from which relatively few species have been described. The two complexes recognized here do not include all taxa with partiform hypocysts. Several genera of Gp-cysts that are distinct but as yet not well known are discussed among the short topics in the second portion of this chapter. These genera are *Carpathodinium*, *Eyachia*, "*Lacrymodinium*" and the similar genus *Pluriarvalium*, *Luehndea*, and *Paragonyaulacysta*.

Principal variables to be noted are:

- Shape
  - Length:width ratio, subspherical to ellipsoidal to pyriform
  - Degree of dorsoventral compression
- Walls
  - Number of layers and nature of interconnections, if any
  - Distribution of cavation
  - Details of relief, if present
- Paratabulation
  - How indicated
  - Precise pattern whenever discernible, with special attention to the number of anterior intercalaries, exsert or insert position of **1u**, relative size of **1i** and the number of other precingulars, details along midventral meridian, relative sizes of **Iu** and **X**, position of **X/Y**

Local suppression of paratabulation in dorsoventrally compressed forms

Archeopyle

All details observable

Key to Complexes of Gp-Cysts

I. Four to five anterior intercalaries; six to seven precingulars in addition to **1i** which may be much reduced; typically nearly spherical .... *Microdinium* cpx

II. Without anterior intercalaries; spherical to ellipsoidal or lenticular; parasutural ridges indicating paratabulation may be centrally reduced in dorsoventrally compressed forms .... *Phanerodinium* cpx

III. Not as in I or II.................Gp-Miscellaneous cpx

Microdinium Complex

Taxa: *Cladopyxidium*, *Fibradinium*, *Histiocysta*, *Microdinium* (in part, including *M. ornatum* = the type species, and *M. veligerum*), and *Subtilidinium*.

Occurrence: Late Cretaceous to Miocene.

Members of this complex are typically small (generally less than 40µm), spherical to ellipsoidal, without horns or prominently projecting surface features. The wall is especially varied in structure. It may be a simple autophragm or much more elaborate, with processes or septa supporting an ectophragm; in at least one species (*M. veligerum*) coarsely ornamented paraplates are separated by prominent parasutural pericoels beneath folds of periphragm. Strong convexity, intricate structure, and relatively great thickness make interpretation of the more complex walls difficult. Paratabulation (Figure 5.18) is frequently indicated more or less uniformly over the surface (but occasionally only on the operculum) by a variety of features, such as parasutural ridges or folds, discontinuities in surface texture or in structures between autophragm and ectophragm, or parasutural interruptions in the ectophragm. The archeopyle is apical in position so far as known, and includes the apicals and intercalaries (e.g., *Microdinium*) or only some of the apicals (e.g., *Cladopyxidium*).

Examination with the SEM will be essential to resolve many structural details, but cannot be expected to assist in interpreting the obviously varied structures within the walls in some species. Typical paratabulation patterns are shown in Figure 5.18. Paraplate **Iu** is typically large and **X** may be visible in ventral view or may lie nearly beside **Y** on the antapical surface.

Phanerodinium Complex

Taxa: *Druggidium*, *Gillinia*, *Glyphanodinium*, *Microdinium* (in part, including *M. opacum*, *M. spinosum*), *Phanerodinium*.

Occurrence: Cretaceous.

Cysts within this complex lack anterior intercalaries and have six or seven precingulars including **1i**, whose size is not severely reduced. Paratabulation is indicated by parasutural ridges or septa, which may be smooth or ornamented. Horns seem not to occur. In known examples, the hypocyst is considerably larger than the epicyst. Some have an approximately circular cross-section (*Druggidium*), but in others (Figure 10.3) dorsoventral compression occurs and may be accompanied by suppression of midventral parasutures. In extreme cases, the thin and featureless ventral surface may appear to be missing (Lejeune-Carpentier, 1951). The wall may be an autophragm, with or without internal structure, or periphragm and endophragm may be distinct with prominent pericoels developed over areas of quite local extent (*Gillinia*).

The archeopyle may be apical, precingular, or a combination of the two; it is often difficult to analyse with confidence owing to the small size of the epicyst and poor visibility of the opening in dorsoventrally oriented specimens. Paraplate **Iu** is large but varies in size; consequently the positions of **Iu/X** and of **X** also vary.

Known Cretaceous taxa are small (less than 40µm). Numerous species of *Gillinia* or *Phanerodinium* occur through the Late Cretaceous, but few have been described; *Druggidium* is a distinct form from the Early Cretaceous. A seventh precingular in some species apparently results from subdivision of the middorsal paraplate (Taylor **4**; Figure 5.18 G).

Gp-Miscellaneous Complex

Taxa: *Carpathodinium*, *Eyachia*, *"Lacrymodinium"*, *Luehndea*, *Paragonyaulacysta*, *Pluriarvalium*.

Occurrence: Early Jurassic to Early Cretaceous.

This category is a receptacle for a mix of species that have a partiform hypocyst but not necessarily other features in common. The taxa named above are imperfectly known to varying degrees. They are

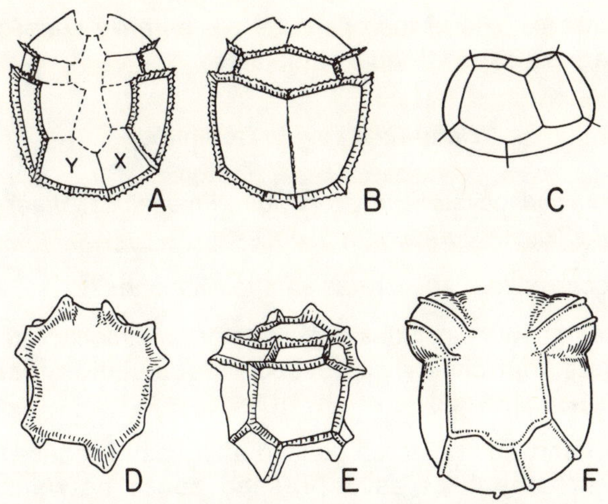

Figure 10.3 - Gp-CYSTS *PHANERODINIUM* AND *GILLINIA*

A-C - *Phanerodinium* sp. (vv, dv, and aav), showing suppression of ventral parasutures (dashed), prominent middorsal **III/IV** parasuture, and "companion" positioning of **X** and **Y** on ventral slope of antapical area.

D-E - *Phanerodinium* sp. (vv and dv). Note complete suppression of ventral paratabulation inside marginal septum, and absence of **III/IV** septum on dorsal hypocyst.

F - *Gillinia hymenophora* (vv). Compare with A to identify parasutural ridges on this cyst which has pericoel developed locally under lateral parts of paracingulum.

All are Late Cretaceous. Details of varied archeopyle form and location (apical, precingular, or apical-precingular combination) have not been established in these cysts which have a small epicyst that is difficult to examine in a direct view.

discussed as a short topic in the second part of this chapter (Figure 10.8).

Category of Gq-Cysts

Introduction

The key to major cyst categories in Chapter 8 makes clear that the limits of this category are drawn in an arbitrary manner to include certain skolochorate cysts, in addition to cysts in which the presence of a quinqueform gonyaulacoid hypocyst pattern is shown directly by parasutural lines. The distinctiveness of the elaborate Type A(3A)6P archeopyle in these skolochorate forms is the justification for this action, but skolochorate cysts with other kinds of archeopyles are excluded from the category, even if details visible in some of them strongly suggest or indicate quite conclusively that they, too, formed in thecae with a quinqueform pattern (see discussion of *Hystrichosphaeridium tubiferum* in Chapter 11). This course has seemed the most practical one because process formula and arrangement alone—even in the case of large intratabular processes—have proven not to be reliably diagnostic of the fine details of paratabulation needed to distinguish the quinqueform and sexiform patterns. With the exception of those skolochorate cysts having Type A(3a)6P archeopyles, all skolochorate cysts with intratabular processes are referred to the category of Gi-cysts, and those with nontabular processes to Gn-cysts.

Features of particular importance for defining genera and species, besides aspects of shape and type of archeopyle, include:

Wall

Process form, structure, and distribution

Number of layers and character of interconnections

Areal differentiation in surface ornament (beneath processes, other parts of intratabular area, pandasutural areas)

Paratabulation

Details relating to exsert, metasert, or insert character of **1u**

Relative sizes, positions, and contacts of **X**, **Y**, and **Z**

Presence and special form of parasulcal processes in skolochorate examples

Species with prominent pericoels are unknown, although complex walls, which can be interpreted as two-layered with interconnecting funnelform or cylindrical structures, occur.

Key to Complexes of Gq-Cysts

I. Skolochorate cysts with archeopyle Type A(3A)6P .... *Homotryblium* cpx

II. Others................................ *Heteraulacacysta* cpx

***Homotryblium* Complex**

Taxa: *Eocladopyxis*, *Homotryblium*, *Polysphaeridium* (as represented by the type species and by species originally described under the synonymous genus, *Hemicystodinium*); living *Pyrodinium bahamense*.

Occurrence: Paleocene to Recent.

The cysts are skolochorate with intratabular or nontabular processes. Some species suggest an evolutionary progression from originally dispersed nontabular processes, to clusters of processes, and

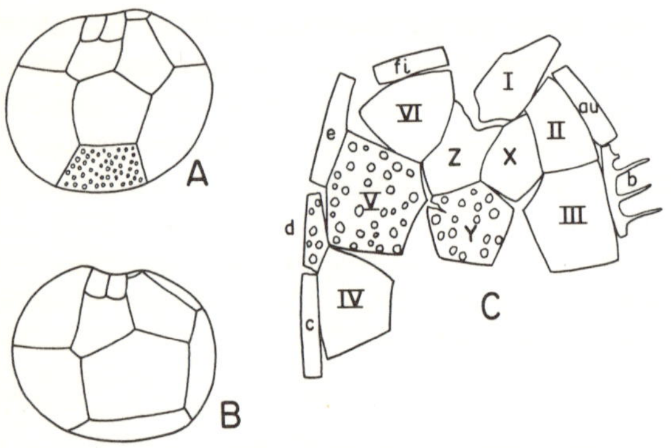

Figure 10.4 - REPRESENTATIVE Gq-CYSTS

A-B - *Dinopterygium* spp. Hypocysts (aav) of two species, showing only boundaries between paraplates and, for one paraplate in A, pattern of porelike markings in intratabular area. Note contrasting sizes of **Y**. Paleogene.

C - *Eocladopyxis* sp. cf. *E. peniculata*. Exterior view of "exploded" and flattened hypocyst whose paraplates have separated along parasutural lines of thinning in the wall. Form of processes visible in "profile" of cingular paraplate **b**; size and distribution of process bases shown by small circles on **d**, **V** and **Y**. Note paraplate **I** is undivided and **I/Z** parasuture is distinctively curved. Paleocene.

eventually to single intratabular units. Evidences of paratabulation on the hypocyst range from not apparent (*Polysphaeridium*), to suggested by intratabular processes (*Homotryblium*), and to direct delineation by parasutural grooves (*Eocladopyxis*; Figure 10.4 C).

Forms with intratabular processes show variable development of parasulcal processes. The process equivalent to **Z** is distinct, but **ai** is not always present, and a large and noticeably irregular, centrally located process (Figure 7.5 I, R) appears to represent several smaller sulcals (**li**, **lm**, **lu**, and possibly **fu**, of which **lu** occurs as a secondary spur).

Loss of the operculum and its fragmentation along the accessory archeopyle sutures commonly leaves hemispherical remnants as the largest objects representing these species. The distinctive shape of the sulcal tab and adjacent parts of the archeopyle margin on such a hemispherical specimen is a critical clue to its identity and orientation (Figure 6.10 R). With critical focusing this tab can be confirmed even in the polar view favored by the hemispherical shape. At times, authors (e.g., DeConinck, 1968; Morgenroth, 1968; and perhaps others) unaware of the distinctive archeopyle have confused *Homotryblium* with *Hystrichosphaeridium tubiferum*, but the latter is distinct in many ways, although process structure in the two genera can be exceedingly similar. Some of the opercular pieces have shapes diagnostic of individual paraplates.

Heteraulacacysta Complex

Taxa: *Alisocysta*, *Dinopterygium*, *Eisenackia*, *Heteraulacacysta*, *Tubidermodinium*.

Occurrence: Early Cretaceous to Miocene (?).

This category includes proximate or proximochorate cysts with clearly expressed paratabulation that may be shown by a variety of features:

1. Pandasutural areas defined as areas of reduced ornament, or as depressed zones between penitabular septa or between polygonal fields having distinctive surface texture or wall structure (including funnelform invaginations of the outer wall and barrel-shaped or cylindrical structures between wall layers).
2. Parasutural grooves or ridges without pandasutural areas (Figure 10.4 A-B).
3. Parasutural folds or septa which may have elaborate bladelike extensions that may complicate the outline and obscure the paratabulation.

The character imparted to the wall by the funnelform or tubercle-like structures is distinctive and unlikely to be confused with features in any other group except the *Hystrichodinium* complex of Gs-cysts, where other confusingly close parallels exist in the presence of bladelike or flattened spinelike extensions from the parasutures.

Archeopyles of Type (tA) and Type Ea predominate, although Type P also occurs, and Type 4A has been reported (Below, 1981).

Category of Gs-Cysts

Introduction

The ideal Gs-cyst has a complete sexiform gonyaulacoid paratabulation indicated by more or less continuous parasutural features, which may include ridges, septa, folds, or rows of various kinds of projections. Although many examples come close to the ideal, others fall short, especially through diminished clarity of the paratabulation, which leads, ultimately, toward intergradation with Gx-cysts. Normally the parasutural features are expressed in external relief, since most cysts having paratabulation indicated only by features within the wall (e.g., processes interconnecting ectophragm and autophragm) do not show the pattern clearly and, therefore, would be classed as Gx-cysts.

Five complexes are recognized in order to provide some subdivision of this large and varied category. One, however, the *Leptodinium* complex, remains large and diverse, including cysts with a great range in shape, wall structure, accessory ornament, and archeopyle type. The other four complexes are more narrowly defined in terms of a combination of less common features. A few forms included in some of these complexes are anomalous in that, while readily placed in the complex (and thereby in this category), they are not directly and objectively identifiable as Gs-cysts, because the paratabulation is faint or even absent. These cases require individual attention (see comments under each complex).

This category is the epitome of G-cysts, for it includes those forms of fossil gonyaulacoid dinoflagellates which bear the clearest and most complete expressions of paratabulation combined with the greatest variety of other features. It is also the focus of much debate and uncertainty. Criteria for recognizing genera within this category and, especially, within the *Leptodinium* complex, are currently in a remarkable state of flux, making it impossible to estimate how circumscriptions of even the most familiar genera (e.g., *Gonyaulacysta* and *Leptodinium*) will be revised in the near future. It is significant and, certainly, a mark of progress that recent works (e.g., Stover and Evitt, 1978; Below, 1981; Sarjeant, 1982b) have begun to focus on previously unattended details of wall structure (extent of cavation, complexities of internal wall structure) and paratabulation (presence of **P**, **Q**, and **K** paraplates; small topological differences on the ventral surface; general "style" of ventral and antapical patterns). Conflicting interpretations of these features—even on the same specimens by different specialists!—and contrasting opinions of their taxonomic significance (and of that of the archeopyle type and surface relief) dominate current discussions. Before these differences can be resolved, many further observations utilizing the best techniques of mechanical preparation (sectioning, selective orientation of individual specimens) and optical- and electron-microscopic study will be needed to establish the reality, frequency, and stratigraphic distribution of the features.

Many examples from this category of cysts already have been cited to illustrate features discussed in Chapters 4 through 6. In the remainder of this chapter, attention is drawn to a few other details and to some especially provocative questions. However, these pages must stand generally without such a comprehensive treatment of this varied and important category of cysts as may be possible in the relatively near future.

Regrettably, the increased flux of new morphological observations results in a temporary instability in taxonomy and nomenclature that, in this group more than most, has now reached crisis proportions. Readers must, therefore, either condition themselves to accept this instability—that is, to go without satisfactory answers to questions like: What really is the difference between genus "A" and genus "B"?—or make their own judgments on what taxa to recognize, what criteria to base them upon, and what terms to use for describing them. Regarding each of these points there is a plethora of available choices.

A further problem is that these features of paratabulation which, quite obviously, distinguish some taxa from others are details that are difficult to observe under the best of circumstances; all too often they are completely concealed, ill-preserved, or open only to ambiguous interpretation. Although distinctive combinations of such features can be shown to characterize groups of species and, therefore, are attractive generic criteria, there is a question of their applicability to the "average" specimen. What role should practical, but abiological, considerations (e.g., ease, frequency, and reliability of accurate observation) play in the choice of taxonomic criteria? It is easy to dismiss the question as irrelevant, since virtually all would agree that biological entities should be classified on biological properties. However, it is unrealistic not to recognize that the intimate historical and continuing association between the study of fossil dinoflagellates and the growth of "applied palynology" assures that the question will be debated in some quarters.

For purposes of the key that is presented here, the presence of distinctly elevated penitabular ornament, leaving only unornamented pandasutural areas along the parasutures, is considered to establish the ornament as intratabular, rather than parasutural. Accordingly, cysts with such features are treated as Gi-cysts, rather than Gs-cysts, although the likelihood of intergradation is obviously great. Cysts with paraplates separated by ornamented pandasutural areas alone, or in combination with linear folds or other features confined to the pandasutural areas (e.g., *Sirmiodinium*), are considered Gs-cysts.

Very few cysts in this category have a wall structure that would be described as comprising an autophragm and ectophragm with multiple interconnections. Cysts with such walls usually do not show paratabulation clearly or completely, although the distribution of connecting structures may suggest parasutures here and there. Therefore, most such cysts fall into the category of Gx-cysts. Problematic in this regard is *Stephanelytron*, a genus of few species distinguished by having an unusual antapical corona. Depending upon the species, processes between autophragm and ectophragm are clearly parasutural, are arranged in intratabular clusters, or are distributed without regard to paratabulation. On the grounds that parasutural alignment of processes is clear in the type species, the genus is arbitrarily placed among Gs-cysts as a rather atypical member of the *Leptodinium* complex.

The distinction between Gs-cysts and Gq-cysts depends upon the hypocyst pattern. To provide for consistent allocation of inconclusive cases, a sexiform pattern is here assumed and the cyst is referred to the Gs-cyst category unless there is positive evidence for a quinqueform pattern.

Details of ventral and apical paratabulation should be given careful attention whenever preservation or orientation permit. The distinction between L- and S-type ventral organization is frequently easy to make because many individual details of paratabulation associated with one or the other tend to be quite readily visible. However, the S- and L-type patterns are not used as criteria for recognizing complexes of Gs-cysts, for they occur in various combinations with other features. For the same reason, definite differences in details of paratabulation in the apical area are not used for subdividing the category, although they merit careful observation and description at the level of genus and species (see further discussion in the last part of this chapter).

Except in the cases of the *Ctenidodinium* and *Wanaea* complexes, archeopyle type has not been used as a factor in defining complexes. Consequently, a considerable variety exists within the large *Leptodinium* complex; however, Types (4A) and P are the most common and few genera as currently defined include more than one type. A well-documented example of a Type 4A archeopyle, in which an apical operculum is divided into its four constituent paraplates, occurs in this category (*Lithodinia jurassica*; Gocht, 1975a).

A lack of exact concordance between an archeopyle and the paraplate to which it chiefly corresponds (i.e., reduced and enlarged archeopyles) is easily demonstrated in species of Gs-cysts by reference to the prominent parasutures. Both reduced and enlarged archeopyles commonly involve rounding of the angles in the outline. A common type of enlargement of the Type P opening incorporates a portion of the paracingulum, with the result that a fraction of the anterior paracingular ridge lies on the operculum (Figure 6.1 E-F). An analogous type of enlargement extends along the entire paracingulum in the case of some epicystal openings; that is, the archeopyle suture runs through the paraplates of the paracingulum, rather than along the bounding parasutural ridge. Reduction, on the other hand, leaves a narrow band of the wall between the archeopyle suture and the bounding parasutural feature, whatever form that may have.

By analogy with Pp- and Pq-cysts, one would expect that presence of a definite pericoel in the region of the archeopyle would result in two separate openings with two opercula. However, Gs-cysts which have this type of wall relationship in combination with a Type P archeopyle show a peculiarity without known counterpart among other dinoflagellate cysts; namely, apparent primary absence of a perioperculum, which leaves the periarcheopyle always open, whereas the endoarcheopyle has a typical archeopyle suture and operculum (e.g., *Gonyaulacysta jurassica*, *Pentadinium laticinctum*, *Stephodinium coronatum*, *Tubotuberella apatela*).

Species with accessory ridges on major pre- and postcingular paraplates—features that apparently are related to the pattern of overlap and peripheral enlargement of thecal plates (Gocht, 1979a; see discussion in last part of this chapter)—mostly have been referred until now to the single genus, *Cribroperidinium*, although they include different combinations of other features that have been used elsewhere as the basis for distinctions between genera. Probably such accessory ridges should be expected to appear in occasional species, regardless of other characters present. The symmetrical pattern made by these ridges on the operculum of a Type P archeopyle can be especially striking (Figure 6.1 G-H), seeming to suggest that the operculum either encroaches severely on neighboring paraplates or represents an amalgamation of separate paraplates (Dörhöfer and Davies, 1980).

Characteristics critical for identification and description include:

Shape

Overall, dorsoventral compression

Apical horn: form and structure

Walls

Number and structure

Distribution of cavation

Openings in the periphragm

Parasutural features

External form and internal structure

Relation to wall layering and cavation

Other surface features

Form, structure, and distribution

Paratabulation

Careful notation of any features likely to be useful in comparing taxa, including: type of ventral organization; details of apical arrangement including preapicals and intercalaries; unusual or conspicuous aspects of individual paraplates or parasutures, especially paraplate size and shape as they relate to distinctive position or orientation of particular parasutures or the existence of unusual plate contacts; suppression or accentuation of particular parasutures; "extra" or "missing" paraplates

Archeopyle

Reduction or enlargement

Type, including presence and extent of accessory sutures; expression in different wall layers

Variation in all features

Key to Complexes of Gs-Cysts

I. Acavate or cavate cysts, with gonal, or gonal and intergonal, distally furcate projections of *Spiniferites* type (see discussion) .... *Spiniferites* cpx

II. Not as in I; archeopyle epicystal, the principal suture extends along paracingulum placing all or almost all precingulars in the operculum. Types E, Ea, 5P, 6P

A. Archeopyle Types E or Ea; operculum simple, parasutural features about equally strong on epicyst and hypocyst .... *Ctenidodinium* cpx

B. Archeopyle Types E, Ea, 5P, 6P; operculum simple or compound; parasutural features tend to be reduced or absent on epicyst, prominent to absent on hypocyst; one or both margins of paracingulum usually marked by prominent septum or fringe .... *Wanaea* cpx

III. Not as in I or II

A. Intratabular areas have porelike markings; parasutural features usually include ridges and prominent, usually non-furcate, gonal or intergonal spines or bladelike extensions (not of *Spiniferites* type) .... *Hystrichodinium* cpx

B. Cysts with other combinations of parasutural features, wall structure, surface relief, and archeopyles .... *Leptodinium* cpx

Spiniferites Complex

Taxa: *Achomosphaera*, *Avellodinium*(?), *Cannosphaeropsis*, *Hafniasphaera*, *Hystrichostrogylon*, *Nematosphaeropsis*, *Rottnestia*, *Spiniferites*; some living species of the *Gonyaulax spinifera* group.

Occurrence: (Late Jurassic ?) Early Cretaceous to Holocene.

The taxa grouped in this complex (Figures 4.3 D, L; 7.5 L-M; and 10.7) show such a range in morphology that superficial consideration certainly would not suggest placing them together. However, similar highly distinctive details of structure, which occur in all the forms named, not only provide a common basis for comparison, but also suggest close biological relationship.

The "common denominator" for the complex is what might be called the *Spiniferites*-type gonal process. This is a trifurcate process characteristically developed in many species of this widespread genus. Typical or recognizably modified expressions of this basic process design occur in other genera, in combination with varied wall structure and even archeopyle type, although exceptions to the Type P archeopyle are rare.

If the ventral paratabulation is visible, an S-type arrangement is usually pronounced. However, in exceptional species (e.g., *Spiniferites cornutus*, W.R.E., pers. obs.) whose paratabulation and relief are typical of the genus in every other respect, **1i** is not triangular as it is in the normal S-type pattern, but has parallel lateral margins as in the L-type pattern.

Details of the morphology and variations on the basic "theme" are discussed as a special topic later in this chapter. Variability within species may be great, and some of the many forms of *S. ramosus* which have been treated as subspecies may well be ecovariants. *Avellodinium*, with an epicystal archeopyle, is retained here tentatively, pending clarification of morphology in the type species (see comment under *Callaiosphaeridium* complex, Chapter 11).

Ctenidodinium Complex

Taxon: *Ctenidodinium*.

Occurrence: Middle Jurassic to Early Cretaceous.

The wall is typically acavate. The complex is distinguished by the prominent separation between epicyst and hypocyst that results from the principal archeopyle suture, which runs along the anterior limit of the paracingulum or between its bounding ridges. The operculum is simple, formed of the entire epicyst, and is typically adnate through a narrow and delicate link at the parasulcus that is frequently torn (Type Ea).

It is typical for paratabulation to be expressed clearly by parasutural relief. The two parasutures defining the paracingulum may be essentially equal in prominence (= separate genus, *Dichadogonyaulax*, of some authors), or the anterior one may be notably less prominent (even entirely undeveloped).

Wanaea Complex

Taxa: *Energlynia*, *Isthmocystis*, *Wanaea*.

Occurrence: Middle Jurassic to Early Cretaceous.

See Figure 6.10 N. The wall is typically acavate. The principal archeopyle suture extends along the paracingulum, and the operculum may be simple (e.g., *Wanaea*), or compound with free opercular pieces representing all or most of the individual precingular paraplates (*Isthmocystis*). Simple opercula are normally adnate at the parasulcus. In species whose operculum is divided by accessory archeopyle sutures, displacement of the free opercular pieces leaves a caplike apical region joined to the rest of the cyst by an isthmus-like connection. Precingular paraplate **1i** may be linked to the isthmus (rather than forming a separate opercular piece), effectively broadening it.

The two limits of the paracingulum may be expressed by equally or unequally developed structures (including broad and elaborately perforated flanges), or the paracingulum may not be indicated. Paratabulation is normally not expressed on the epicyst, unless by accessory archeopyle sutures. On the hypocyst, paratabulation may be expressed or not. If paratabulation is shown only by the archeopyle or cingular flange(s), reference to the category of Gs-cysts or to this complex is debatable and must be judged subjectively on the extent of similarity in other features.

Leptodinium Complex

Representative Taxa: *Acanthaulax*, *Carpodinium*, *Cassidium*, *Clathroctenocystis*, *Codoniella*, *Corrudinium*, *Cribroperidinium*, *Endoscrinium*, *Exiguisphaera*, *Gonyaulacysta*, *Hapsocysta*, *Heslertonia*, *Hystrichogonyaulax*, *Impagidinium*, *Lanterna*, *Leptodinium*, *Lithodinia*, *Meiourogonyaulax*, *Millioudodinium*, *Muratodinium*, *Occisucysta*, *Ochetodinium*, *Parvocavatus*, *Pentadinium*, *Prionodinium*, *Psaligonyaulax*, *Rhynchodiniopsis*, *Sirmiodiniopsis*, *Sirmiodinium*, *Stephanelytron*, *Stephodinium*, *Tubotuberella*; living *Gonyaulax apiculata* (the only living freshwater species of the genus) and several marine species of *Gonyaulax* (as yet unidentified) whose cysts are morphologically *Impagidinium*.

Occurrence: Middle Jurassic to Holocene.

The range of morphological features expressed in this complex is as broad as that in the *Deflandrea* complex among P-cysts, but here the limits of the complex can be drawn more rigidly and comparisons within the group made more comprehensively because, by definition, this complex comprises cysts in which a more or less complete paratabulation is determinable (Frontispiece D, G). Further, it is specified that the hypocyst pattern, whenever it can be identified, is sexiform, cysts with clearly quinqueform and partiform patterns being referred to Gq- and Gp-cysts, respectively.

Overall cyst shape; form, structure, and spacing of wall layers; distribution of cavation; surface ornament along parasutures and between them; type of archeopyle; the presence of a porichnion in the vicinity of **A/Q/1u** and of claustra (especially in the hypoperiphragm) are all variable. However, prominent intratabular processes are not present and large parasutural processes are rare. Suppression of parasutural clarity, whether as a species-specific character or individual variation, leads to morphological transitions toward Gx-cysts whose evaluation can only be subjective. Simple apical and precingular archeopyles, Types (4A) and P, are most common, but Types 4A, $2P_{3-4}$, $2P_{4-5}$, and (4A)Pa are also known. Species that might be referred to this complex except for having epicystal archeopyles fall in the *Ctenidodinium* complex. Both L-type and S-type ventral organizations occur, as well as other variations without special designations.

It is important to note the position of the antapical extremity as well as the position of openings in the periphragm (claustra) that may occur in this area. Most frequently, the antapical paraplate (**Y**) forms the extremity of the cyst, but its position actually varies from fully ventral to nearly fully dorsal, depending upon the species. Consequently, the antapical extremity may be formed in extreme cases by parasutures **Y/Z** or **Y/IV**. Persistent claustra may also occur in species-specific locations near the antapex—in **Z** in *Gonyaulacysta jurassica*, in the central part of **Y** in *Sirmiodinium grossii*, in the dorsal part of **Y** in *Tubotuberella apatela* (Figure 4.6 M-P).

Narrow pericoels, confined to the lines of parasutures over some or all of the cyst surface, are a distinctive feature of some members of this complex. The limits of these tunnel-like cavities are frequently visible as penitabular lines (in obvious pairs along the longer parastures) where the wall changes in color, texture, or thickness. Good examples occur in *Carpodinium*, *Gonyaulacysta*, *Impagidinium*, and *Pentadinium*. In some species of *Pentadinium*, a combination of fusion of closely spaced parasutural pericoels and suppression of certain parasutures results in a severely reduced paraplate count responsible for a suggested relationship to the modern genus *Ceratocorys* (Norris, 1978). Such a special explanation becomes unnecessary when the basic S-type ventral organization and sexiform gonyaulacoid characteristics of *Pentadinium* are recognized.

Pericoels, when present, are ordinarily not traversed by supporting structures. The special case of *Stephanelytron* (with ectocoel crossed by parasutural processes in some species, intratabular groups or nontabular processes in others) has already been mentioned in the introductory discussion to Gs-cysts.

Among the few species of dinoflagellate cysts found in open ocean sediments today are several having the morphology of *Impagidinium*. Their paratabulation strongly suggests that they represent the *Gonyaulax spinifera* group of species, but, so far, the relevant thecate stages have not been identified.

In many gonyaulacoid cysts with S-type ventral organization, parasutures in and around the parasulcus tend to be expressed weakly or not at all. This is also true in *Impagidinium*, but in this genus additional parasutures whose identity is species-specific may be affected as well. Especially likely to be reduced are the septa that separate **ai** and **lu** from their pre- or postcingular neighbors, and one or more of the septa that separate **au** and **fi** from the adcingulars **li**, **2**, **lu**, and **VI**. Such "missing" septa should be carefully noted in descriptions of this group. Although the S-type ventral organization is characteristic for *Impagidinium*, in one species (*I. pentahedrias*) **1i** is parallel-sided rather than triangular.

Muratodinium stands apart from the other genera included here. The external wall surface and the projections that rise from it (parasutural septa, intratabular spinelike processes, and a single paracingular ridge) are all conspicuously fibrous. More information on details of paratabulation may better establish its morphological affinities.

See also Figure 10.9 and the discussion on apical patterns at the end of this chapter.

Hystrichodinium Complex

Taxa: *Hystrichodinium*, *Raphidodinium*(?).

Occurrence: Late Jurassic to Late Cretaceous.

The genera placed together in this complex have prominent spinelike or bladelike projections extending from some or all parasutures (or, in a few species, rising from cysts without defined parasutures). In addition, a distinctive feature of wall structure, which is apparent in most individuals but whose exact nature and significance are not known at present, is used here as the basis for uniting these taxa in the one complex. Possibly what appear to be similar wall details in the two genera are not identical. The better known of the two genera is *Hystrichodinium* and it should be taken as typical.

The distinctive feature of wall structure appears as porelike (or, sometimes, tubercle-like) markings on the paraplates between parasutures. Only in one other group (*Heteraulacacysta* complex of Gq-cysts) have similar features been noted, and the microstructure may not be the same. In *Hystrichodinium*, this wall pattern presumably reflects an internal structure of the wall, surface relief visible under the SEM seeming insufficient to account for the appearance in the light microscope. Also characteristic of the complex are large, flattened or hollow spines, which may rise, individually or in linear groups with fused bases, from parasutural ridges (or possibly folds of periphragm), in contrast to the predominance of septa associated with one margin of prominent pandasutural areas in the *Heteraulacacysta* complex. In some species of *Hystrichodinium* type, the parasutural ridges are absent or limited to only certain parasutures, although the paratabular distribution of porelike markings may still be visible. Number of projections and evenness of their spacing vary with the species. The archeopyle in *Hystrichodinium* is Type P. An L-type gonyaulacoid arrangement can be seen in some species (W.R.E., pers. obs.) but the full paratabulation pattern is rarely traceable.

Raphidodinium is known from a single species. Projections and wall patterns make tentative association with *Hystrichodinium* reasonable, but its archeopyle type and some other details are uncertain.

ADDITIONAL SELECTED TOPICS

The Symmetrical Horns of Muderongia

Among fossil cysts of ceratioid design, *Muderongia* (and *Phoberocysta*, but to a lesser extent, chiefly because its outline is complicated by processes) stands out by virtue of having two closely similar equatorial projections which have been interpreted aspostcingular horns (Figure 10.5 A). However, in number, symmetry, and form, these projections differ from the postcingular horn as it is expressed in most living species of *Ceratium* and in the fossil cysts of the *Odontochitina* complex, which seem to represent close fossil counterparts (Figure 10.1 A-E).

In most species of *Muderongia*, these "postcingular" horns might be more appropriately described as equatorial, for their form is dominated by a more massive proximal portion which commonly extends laterally in direct line with the paracingulum. The distal part, extending posterolaterally from the proximal portion, may be so much smaller than the latter that it attracts little attention. In some species, the paracingulum clearly extends to the end of the basal section, where its concave cross-section contributes to a notch that is often conspicuous in the horn outline at the junction of proximal and distal portions.

In contrast, modern species of *Ceratium* living in the marine environment, which number around 100 and are highly varied in size and shape, have no more than one postcingular horn, and this is severely reduced or modified in a few species. It is constructed of extensions from two postcingular plates on the right side (**V** and **Vli** in the interpretation followed here), may be straight or arcuate, and directed more or less posteriorly or curved to point laterally, or even anteriorly. Differentiation of proximal and distal portions with an extension of the cingulum on the former, is not known, although the base of the horn may lie just posterior to the cingulum.

However, thecal structures known in modern *Ceratium* species living in fresh water provide a reasonable explanation of the distinctive number and form of the horns in *Muderongia*. Here, in two species (*C. hirundinella* and *C. furcoides*), we find frequently (but not always) two postcingular horns. The second extends from the left side and, like the first, is constructed from two postcingular plates, in this case **II** and **III**. Although the left postcingular horn in the living species seems never to achieve really symmetrical equivalence to the right, its occurrence is at least a step in the direction of *Muderongia*.

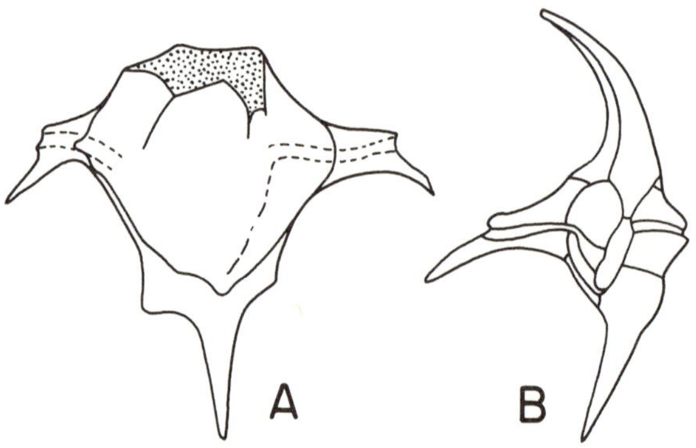

Figure 10.5 - HORNS IN *MUDERONGIA*

A - *Muderongia* sp. cf. *M. mcwhaei* (vv), showing typical form of postcingular horns projecting posteriorly from lateral extensions of cyst outline in line with paracingulum. Early Cretaceous.

B - *Ceratium carolinianum*. Theca of modern species (vv). Asymmetrical extension on right side in line with cingulum produces a structure such as may have enclosed the lateral projection of *Muderongia*. *C. carolinianum* does not produce sporopollenin cysts and no species of *Ceratium* today have symmetrical lateral projections of this type. Modern, freshwater; after Wall and Evitt (1975).

C. carolinianum (Figure 10.5 B), another of the five widely recognized freshwater species of the genus living today, is not known to have more than the left postcingular horn, but the form of that structure provides a clue to the probable construction of the theca within which the cyst we know as *Muderongia* formed. The right side of the theca in *C. carolinianum* is modified into what might now be described as a *Muderongia*-like prominence, in whose construction several plates participate: two postcingulars (**V** and **VIi**), two cingulars (**d** and **e**), and, to a somewhat lesser extent, two precingulars (**5** and **6**). The line of the cingulum extends laterally along this prominence, curving sharply around its extremity, and the postcingular horn proper originates just posterior to this point. As yet, the cyst of *C. carolinianum* has not been described.

The Distinctions between Gc-Cysts and Gv-Cysts

These two groups of cysts are readily distinguished from most others by a combination of conspicuous traits that include a pronounced dorsoventral compression, an exclusively apical archeopyle with sulcal notch offset to the left of the ventral surface, and an asymmetrical outline. Gc-cysts with long horns have an unmistakably ceratioid shape, but if the horns are short and rounded, the suggestion of *Ceratium* may not be at all clear. In fact, superficially, Gv-cysts may appear to be merely Gc-cysts whose postcingular horn has been reduced to insignificance. In many instances, the distinctions between the two types are certainly not obvious if parasutural markings are weakly expressed, which is most frequently the case. Possibly there are a few species which must be arbitrarily placed. However, examples of both types are now known which do show a clear paratabulation, and this exhibits consistent differences in hypocyst pattern which nicely match the differences between the corniform and sexiform gonyaulacoid configurations. When horns or lobes are developed around the margin of the hypocyst, these, too, are in consistent position. In Gc-cysts with two hypocystal horns, one prominence rises from the **X/Y** parasuture at the antapex, the other from the parasuture on the right side between the two postcingulars, **V** and **VIi**, just like the antapical and postcingular horns of the theca of *Ceratium*. In Gv-cysts, both prominences are associated with the antapical paraplate (**Y**) and are located in the vicinity of, or actually encompass, the **X/Y/Z** and **VI/Y/Z** triple junctions (as is clear in the fully paratabulate *Canninginopsis denticulata*). These differences are illustrated on Figure 10.6. Note that, although each type has six postcingulars, of which the paraplates visible on the dorsal surface appear very similar, different paraplates occupy the ventral surface. In Gc-cysts, there are two on the right (**VIi** and **VIu**) and one on the left (**II**), whereas in Gv-cysts, there is only one on the right (**VI**) but two on the left (**Iu** and **II**).

If, in a particular cyst under study, hypocyst paratabulation or ceratioid horns are clearly developed, it will be relatively easy both to place the cyst in the Gc- or Gv-cyst category and to estimate its biological affinity (say, relative to *Gonyaulax* vis-a-vis *Ceratium*). However, many cysts having the general attributes under consideration here show neither a distinct paratabulation nor ceratioid horns. Morphologically, they are Gv-cysts according to the criteria by which the category was defined, but their biological affinities will remain uncertain until additional criteria for making a judgment on this biological point are established. In other words, an isolated endocyst having the morphology of the endocyst that occurs within the ceratioid pericyst of *Endoceratium ludbrookiae* (Figure 10.1 I-J) is, morphologically, a Gv-cyst. Even though the biological affinities of such a fossil might be with *Ceratium*, this could not be judged without additional evidence.

Theme and Variations in the Spiniferites *Complex*

Long geologic range (earliest Cretaceous to Recent), distinctive and easily recognized morphology, numerous species, and frequent abundance of individuals have made *Spiniferites* (Figure 4.3 D) one of the most familiar genera of dinoflagellate cysts in the fossil record. However, distinctive though it may be, it does not stand alone, but is like the "theme" of a "theme and variations" when other, similar genera are compared to it. Analogous patterns of similarities and differences appear in other groups of genera and, indeed, are fundamental to the concept of cyst complexes. Only the clarity and ease with which the comparisons can be drawn in this case invite a more complete discussion than has been attempted elsewhere.

The feature of *Spiniferites* that seems the most strongly unifying character of the complex is the distinctive process structure, which will be described shortly. Other principal features of the genus, many of which—like the processes—find

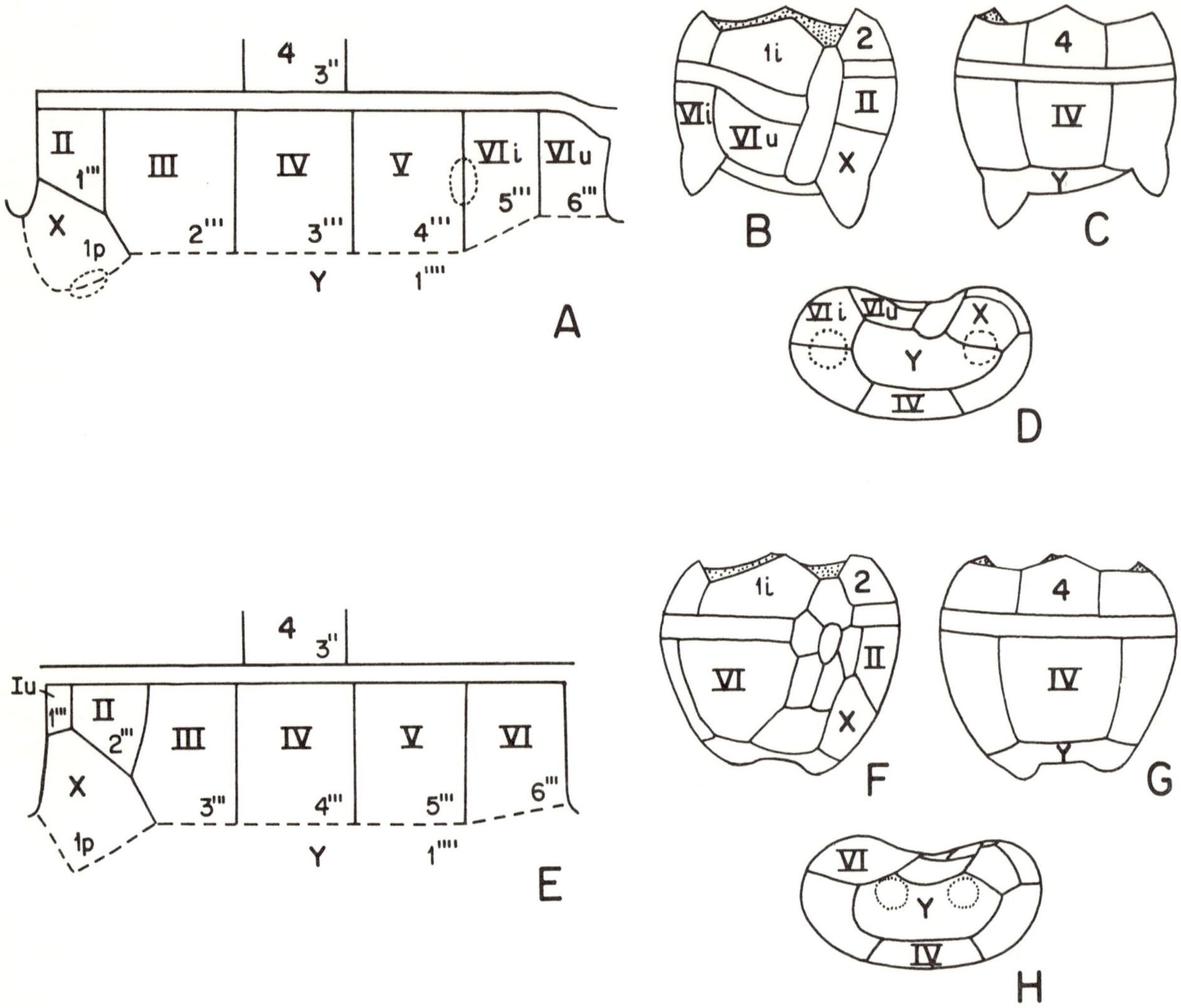

Figure 10.6 - COMPARISON OF PARATABULATION IN Gc- AND Gv-CYSTS

See text for discussion.
Corresponding diagrams to illustrate significant differences in paratabulation in Gc-cysts (A-D) and Gv-cysts (E-H). At left is a map-like projection of the paratabulation centered on the middorsal area and extending to the parasulcus in both directions. Both Taylor and Kofoid designations for paraplates are shown. At right are ventral, dorsal, and antapical views, with critical paraplates labeled. Dotted circles and ellipses show locations of horns and antapical lobes.

expression in similar or modified form in other genera of the complex, are the following:

1. A basically sexiform (and normally strongly S-type) gonyaulacoid paratabulation that is defined by parasutural ridges from which rise the distinctive processes.

2. Parasutures not expressed within the parasulcus, and certain others on the ventral surface (**1u/A**, **1i/ai**, **1i/fi**, **Im/lu**) suppressed selectively, so that the paratabulation, when recorded in Kofoidian terms, has the reduced, and rather un-*Gonyaulax*-like formula: *3′, 5″, 6c, 5‴, 1p, 1⁗, Xs.*

3. Narrow **Iu** almost always fully integrated into the parasulcus but its extent approximately indicated by two of its bounding parasutures, **Iu/II** and **Iu/X**.

4. Parasutures not expressed within the parasulcus.

5. Archeopyle Type P (with rare exceptions).

The arrangement and form of the processes are especially distinctive (Figures 10.7 A-H). Depending upon the species, gonal processes may occur alone, or may be supplemented by intergonal ones; intergonal processes never occur without gonals. The number of intergonals varies but is reasonably

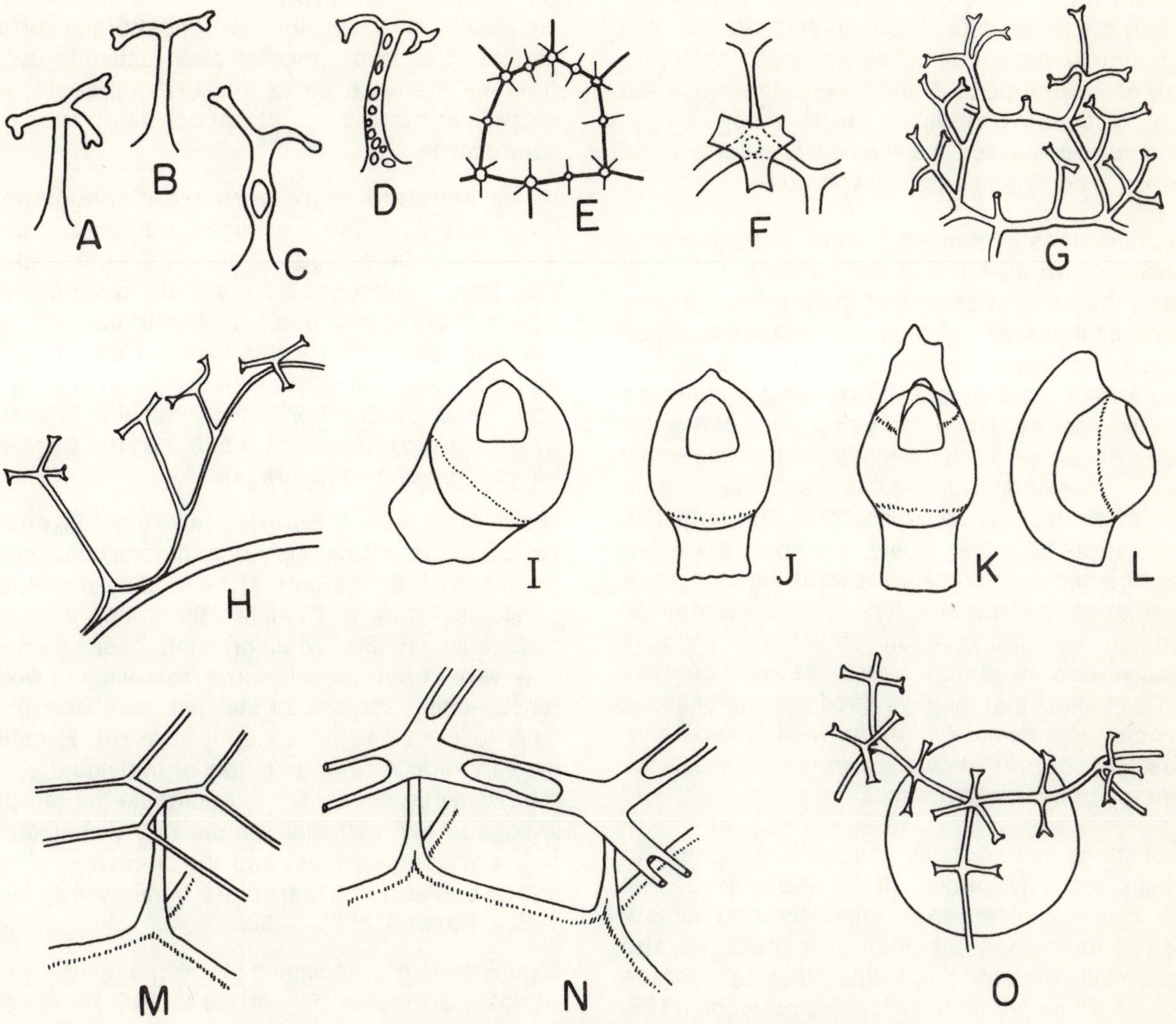

Figure 10.7 - REPRESENTATIVE CYST FEATURES IN *SPINIFERITES* COMPLEX

A-D - One trifurcate and three bifurcate processes with solid and vacuolated shafts. Secondary divisions for both types of processes are bifurcations.

E - Portion of paratabulation showing schematically the location of trifurcate processes at triple junctions (gonal processes) and bifurcate processes between them (intergonal processes). Note orientation of process tips relative to parasutures (heavy lines) and adjacent paraplates.

F - View directly down onto trifurcate tip of process in *Spiniferites pseudofurcatus*, showing tip, shaft (dotted circle), and ridges on cyst surface converging at process base. Miocene.

G - Parasutural ridges and associated projections from fragment of wall of *Spiniferites* sp., showing distribution of bifurcate and trifurcate processes.

H - Cross-section of portion of wall of cavate cyst of *Spiniferites* complex. At left, periphragm is in close contact with underlying endophragm and process has fully developed shaft and maximum length. As height of pericoel increases to the right, processes along parasuture become shorter, "losing" the lower portion of their shafts, until only trifurcate tip is left at the next triple junction.

I-L - Varied locations of pericoels shown by optical sections through polar axis. "Random" and antapical pericoels only (I-J) in some *Spiniferites*; at both poles (K) in *Hystrichosphaeropsis*, *Rottnestia*, and some *Spiniferites*; and consistently over other substantial portions of the surface (L) in *Hystrichostrogylon*. The archeopyle and the line of contact between periphragm and endophragm (finely hachured) are projected to the plane of section.

M-N - Perspective details showing two types of ectophragmal interconnections between process tips in *Nematosphaeropsis*. Pairs of trabeculae characterize *N. labyrinthea* (M), but perforate, ribbon-like connections (N) also occur.

O - In *Cannosphaeropsis utinensis*, the ectophragmal network (of which only a fragment is shown) is largely unsupported by processes but retains trifurcate and bifurcate structures at gonal and intergonal positions. These may extend directly from the single trabeculae, but the trifurcate ones especially tend to be elevated on short shafts. Circle suggests size of main body. Late Cretaceous.

constant within a species, especially if one considers only the parasutures between the large pre- and postcingular paraplates. The process shafts are rarely hollow throughout, and open to the exterior at the tips in one prominent Tertiary species (*S. pseudofurcatus*). Usually any cavity present within a process is confined to its basal portion.

The characteristic manner in which the processes branch toward their tips is the persistent element among the several genera of the complex. Those processes that are most regularly formed (i.e., those at widely spaced corners of the largest paraplates) show one or more orders of branching, depending upon the species. The primary branching of a gonal process is always a trifurcation; that of an intergonal one is a bifurcation. Subsequent branchings of both kinds of processes are bifurcations. The branches which radiate from the primary furcation are consistently directed approximately toward the centers of the adjacent paraplates; that is, they alternate in alignment with the parasutures between process bases when viewed from directly above a process tip. This means that the rays of bifurcate processes extend at approximately right angles to the parasutures. Of course, allowance must be made for distortion of flexible processes.

Although the more isolated gonal processes tend to be quite regularly formed, shafts of processes that stand close together and are joined by short parasutures tend to fuse, beginning at the base and occasionally extending to include their tips (Figure 7.5 L-M, Principle of Minority Suppression). This tendency accounts for the contrasting types of gonal processes frequently seen along the paracingulum; simple processes at the corners of pre- and postcingular paraplates alternate with larger, fused, and often irregularly elaborated processes that arise from the ends of short parasutures between cingular paraplates. It also explains partially fused processes in other locations, including especially elaborate processes that commonly rise from the apex.

Surface ornament and wall structure in *Spiniferites* are subject to such variation among species that it is difficult to generalize about them. The parasutural ridges or septa which run between processes are exceedingly varied in their expression, both in an absolute sense and in relation to the height of the processes. In some cases, relatively tall processes rise high above low ridges; in others, high septa, their crests characteristically "sagging" between processes, nearly reach the process tips.

Although two wall layers are often discernible, they are usually close together over most of the surface, and pericoel development is often limited to narrow "tunnels" beneath parasutural ridges/septa, with slight expansions at process bases (= suturocavate).

When separate wall layers are distinguishable (Figure 10.7 H), the processes, ridges, and septa are seen to be composed of periphragm, and this may have a microstructure visually quite different from that of the endophragm. Microtome sections of the wall reveal that a thin layer of periphragmal material may underlie these cavities (W.R.E., pers. obs.), raising a question in this case about the strict definition of pericoel as a cavity between periphragm and endophragm.

Most species of *Spiniferites* lack extensive pericoels, but a few species and occasional specimens of other species show locally developed pericoels (Figure 10.7 I-J), first thoroughly described by Lejeune (1936b). When present, these structures may vary in both position and frequency of occurrence within a species, or they may be characteristically located and persistently present. Excellent preservation or manipulation of individual specimens may be necessary to determine the morphological details, especially to distinguish between a fully enclosed pericoel and the presence of high septa surrounding paraplate **Y** (Harland, 1977, 1982a; Harland, et al., 1980).

Examination of a specimen in which a relatively high pericoel is present (Figure 4.3 L, 10.7 H) makes it immediately clear that in those species with two-layered walls the process shafts owe their existence to the periphragm alone. That is, each process projects outwardly only from the periphragm, and where a pericoel separates this layer from the endophragm, the length of the process shaft is reduced correspondingly. If the pericoel height approximates that of processes elsewhere, only the distally furcate process tips are present.

The external surface of the wall ranges from smooth to granulate, the ornament being evenly or only locally distributed. Intratabular verrucae distinguish a few species, and in others one wall layer or the other may be substantially thickened with a perceptibly organized microstructure, which is, however, unresolvable with light optics.

Other Genera. The other genera of the complex differ from *Spiniferites* in several different ways. Only minor differences in surface features or wall structures distinguish *Achomosphaera* and

Hafniasphaera from *Spiniferites*, leaving arguable the merit of their generic separation. In *Hystrichosphaeropsis*, *Rottnestia*, and *Hystrichostrogylon*, prominent pericoels are developed in two distinct arrangements. In *Nematosphaeropsis* and *Cannosphaeropsis*, pericoels are absent but distal interconnections of several types occur between processes which are, respectively with and without a full complement of process shafts. *Avellodinium* is distinguished by having a Type E archeopyle. The "variations on the theme" expressed by these genera should become evident from the following brief summaries of each:

Achomosphaera differs from *Spiniferites* only in the pronounced suppression of parasutural markings between process bases. In the holotype of the type species, *A. ramulifera*, no parasutural markings can be detected and their absence is reported in a few other species. However, variability in this feature appears to be intraspecific in *S. pseudofurcatus*—*Achomosphaera alcicornu* being the name for ridgeless specimens from the same samples when the generic value of this character is insisted upon—and some other species, suggesting that the character may not warrant distinction.

Hafniasphaera (Figure 10.7 D) differs from *Spiniferites* in having cavities within the wall. In some species (including the type, *H. hyalospinosa*), these appear as isolated small vesicles, but *H. septata* has hollow processes clearly divided by transverse partitions into a linear series of chamberlets. While such wall features are unquestionably distinctive, it seems debatable whether either the differences from *Spiniferites* or similarities among the species referred to *Hafniasphaera* justify a separate genus.

Rottnestia (Figure 4.3 L, 10.7 K) is bicavate, distinguished by development of pericoels at both poles and close appression of wall layers in the adcingular area. In accord with the nature of process shafts already discussed, adcingular processes are fully developed, whereas those rising above the pericoel have shaft lengths inversely proportional to the height of the pericoel. *Hystrichosphaeropsis ovum* (the type species) is similarly bicavate but has shorter processes, although species without processes (and, therefore, not clearly morphological allies of *Spiniferites* in the view taken here) have been referred to this genus (see Sarjeant, 1982b).

In *Hystrichostrogylon* (Figure 10.7 L), another genus in which the wall is extensively cavate, the pericoel is undivided, covering much of the ventral half of the cyst in the type species (*H. membraniphorum*) and in another species that might be referred here (Wilson, 1971, as yet undescribed).

In both *Rottnestia* and *Hystrichostrogylon*, parasutural ridges are well developed where the pericoel is absent. Where the pericoel is present, they are less pronounced, but also extend across the surface of the periphragm between the incompletely developed processes.

Nematosphaeropsis (Figure 10.7 M-N) differs from *Spiniferites* by having ectophragmal interconnections between the tips of its processes. These may be in the form of rodlike trabeculae (in the type species, *N. labyrinthea*) or more or less ribbon-like membranes in other species. In *N. labyrinthea*, the trabeculae extend between the process tips as continuations of the secondary furcations. That is, six trabeculae extend from each well-formed gonal process in a highly distinctive manner, suggesting electric power lines strung on poles. The pattern of trabeculae faithfully records a paratabulation, but in only the least distorted specimens is the pattern easily recognizable, even locally. Several species have ribbon-like interconnections (either continuous or with perforations intergonally) whose parasutural distribution is also recognizable only in the simpler types and when well preserved. Depending upon the species, parasutural ridges and intergonal processes may or may not be present. A thin and extensively folded wall, processes often overlapping like trees felled for timber, and a tangle of ectophragmal interconnections commonly obscure details of the archeopyle, wall layering, and surface pattern on the main body.

Cannosphaeropsis (Figure 10.7 O), a genus to which many species of contrasting morphology were formerly referred (Stover and Evitt, 1978), may be imagined as a kind of *Nematosphaeropsis* in which a variable number of processes have been abbreviated by deleting their shafts, while leaving their tips united by the trabecular network of ectophragm. In the type species (*C. utinensis*), the number of processes equipped with shafts is highly variable, and development of the trabecular network may also be quite irregular. However, occasional specimens show its character clearly. In contrast to *N. labyrinthea*, only single trabeculae interconnect the processes (i.e., three from each gonal process; two from each intergonal), and these commonly join the process shafts below the points of furcation. True to the basic *Spiniferites* pattern, the "remnants" of the furcate tips alternate with the parasutural interconnections between triple junctions expressed in the ectophragmal network. Parasutural ridges are not present between process bases, and the wall, if not an autophragm, is at least very thin and without trace of pericoel. As one consequence of its thinness, the wall is often extensively folded and the archeopyle correspondingly difficult to see. In fact, although the species is easy to recognize, many specimens are virtually impossible to orient and attempts to recognize particular paraplates as defined by the trabecular reticulum seems nearly hopeless; at most the general identity of large adcingular and elongate cingular paraplates may be readily apparent.

S-type ventral paratabulation and Type P_4 archeopyle are conspicuously uniform throughout the complex whenever they can be established unequivocally, making the few known exceptions that have been observed especially noteworthy. An undescribed Paleogene species which clearly possesses gonal and intergonal processes of typical *Spiniferites* type is distinguished by an L-type ventral organization, as indicated most clearly by the subparallel boundaries of the relatively large **1i**. Living cysts having the morphology of *Spiniferites* but a Type $2P_{3-4}$ archeopyle have been reported from the northeastern Pacific (Dobell and Taylor, 1981). The opercular pieces, which are free and essentially smooth, are separated only by the accessory archeopyle suture. There is no trace of the parasutural ridge that ordinarily would separate paraplates **3** and **4**, although parasutural features are normally developed elsewhere. A Type E archeopyle distinguishes *Avellodinium* (Duxbury, 1979); process form and parasutural pattern are reported to be as in *Spiniferites* but this may be a misinterpretation (see comment under *Callaiosphaeridium* complex, Chapter 11).

Table 10.1. DISTRIBUTION OF GENERA OF *SPINIFERITES* COMPLEX

	EK	LtK	P	N	H
Achomosphaera		X	X		
Avellodinium	X				
Cannosphaeropsis		X	X		
Hystrichosphaeropsis		X	X	X	
Hystrichostrogylon			X		
Hafniasphaera	X	X	X		
Nematosphaeropsis	X	X	X	X	X
Rottnestia		X	X		
Spiniferites	X	X	X	X	X

EK = Early Cretaceous, LtK = Late Cretaceous, P = Paleogene, N = Neogene, H = Holocene

Strong similarity (or, possibly, intergradation) exists between some members of the *Spiniferites* complex and some other taxa which have an S-type sexiform gonyaulacoid paratabulation but lack the process structure distinctive of *Spiniferites*. The boundary between *Spiniferites* and *Impagidinium* is especially difficult to draw. Typical *Impagidinium* has well-developed, even-crested septa and lacks processes, but it is closely approached by such species as *S.? cingulatus*, and even more so by *S.? pterotus*, in which septa are pronounced but the presence or absence of trifurcate projections at their intersections is difficult to establish.

Finally, it should be mentioned that several species formerly attributed to *Nematosphaeropsis* or *Cannosphaeropsis* on the basis of superficially similar ectophragmal interconnections between process extremities have now been referred elsewhere, since the processes are intratabular and the archeopyle is apical. The geologic distribution of genera of the *Spiniferites* complex is summarized in Table 10.1.

Checklist of Characters. In identifying or describing a species within the *Spiniferites* complex, the following features should be considered:

Overall shape and size.

General conformity of the paratabulation and archeopyle to the *Spiniferites* plan, including pseudotriangular **1i** and Type P_4 archeopyle.

Extent of suppression of parasutures in parasulcus and on neighboring parts of ventral surface (including **A/1u**).

Offset and separation of ends of paracingulum.

Shape of cingular paraplates (rectangular or hexagonal) and margins of paracingulum.

Width, shape, and inclination of parasulcus.

Structure, extent and prominence of parasutural ridges/septa.

Structure and types of processes; number of bifurcate processes (if present) along parasutures between major pre- and postcingular paraplates.

Details of process structure (hollow or solid, number of orders of branching and shapes of tips, nature and extent of fusions between closely spaced shafts, exact form of distal interconnections if present).

Location and extent of pericoel (regionally and under projecting structures).

Differentiation, character, and thickness of wall layers.

Nature and distribution of surface ornament.

Observed variability in all features.

Of course, not every species will have all these features, and many specimens that do will not be sufficiently well preserved or favorably oriented to show them. However, it is axiomatic that the confidence with which an identification can be made is proportional to the completeness with which these points can be observed.

Non-distinctiveness and high intraspecific variability —especially when combined with less than ideal preservation—seriously limit the usefulness of many species of *Spiniferites* with respect to any applications that depend upon accurate and consistent identification. In a local (intrabasinal) setting, some occurrences of the "more generalized" species may be significant, but from the point of view of their systematics, they are so difficult to deal with that the time required to obtain useful results often can be spent more profitably on other taxa. Therefore, the species of *Spiniferites* that warrant closest attention in stratigraphic practice are those immediately distinguishable by one or more outstanding features, whatever they may be.

As already noted, fewer species are known of the genera other than *Spiniferites*. Therefore, they pose less of a problem in discrimination.

General Taxonomic Considerations. Considering the genera reviewed here in a broad perspective, arguments could be advanced in favor of uniting them all in *Spiniferites*. For instance, their common characters certainly can be construed to suggest their close relationship as strongly as their differences suggest their separation. Also, with the exception of *Spiniferites* itself, each of the genera is represented by only a few known species. Furthermore, as in part already noted, the distinctions between genera are not always clear. The pericoels of *Hystrichostrogylon* have counterparts in some species of *Spiniferites*. Interconnections between process tips also occur occasionally in some species of *Spiniferites*, whereas not all the processes in *Nematosphaeropsis* and *Cannosphaeropsis* are always linked at their tips. The arguable importance of the distinguishing features of *Achomosphaera* and *Hafniasphaera* have already been pointed out. Finally, cysts with the morphologies of both *Spiniferites* and *Nematosphaeropsis* occur among living dinoflagellates, but are confined to members of the small and distinctive spinifera-group of *Gonyaulax*, whose thecae are closely similar. Nevertheless, "typical" representatives of most of the genera are distinctive,and formal steps to unite the genera would seem to serve little purpose for the present.

The *Spiniferites* complex provides an interesting example of the kind of intricate nomenclatural tangle that must be worked out if the theoretically undesirable and "legally" unacceptable, but historically inevitable and currently expedient, dual generic nomenclature of cysts and thecae is to be eliminated. In the literature on fossil dinoflagellates, *Spiniferites* was known as *Hystrichosphaera* until the priority of Mantell's name (1850) was discovered by Sarjeant, and a proposal to conserve the junior name was rejected (Sarjeant, 1970). A more troublesome conflict looms with the name *Gonyaulax*. Only current uncertainty about the exact morphology of the cyst of the type species, *G. spinifera*, effectively forestalls the next step. The cyst of this species is almost surely equivalent to either *Spiniferites* or *Nematosphaeropsis*. Should it prove to be like the latter, the seniority of *Gonyaulax* Diesing 1866 over *Nematosphaeropsis* Deflandre & Cookson 1955 would immediately settle the matter (unless *Nematosphaeropsis* by that time has been synonymized with *Spiniferites*!). On the other hand, should it prove to be like *Spiniferites*, formal conservation under the I.C.B.N. would be required to "save" *Gonyaulax*. At first glance, such action might seem virtually assured by the long and frequent use of *Gonyaulax* in the extensive literature on living dinoflagellates. However, another aspect of the situation may be critically important. The paratabulation pattern of *Spiniferites* is matched today by the tabulation of only a few species of living *Gonyaulax* —the spinifera-group—and the literature on dinoflagellate cysts provides a strong precedent for according separate generic status to cysts that differ as much as these do from others of generally gonyaulacoid design. If similar standards were applied to those living species that are now referred to *Gonyaulax* on the basis of thecal morphology, most of them would require a new name, leaving *Gonyaulax* appropriate for only a very few. With the prospect of such an alteration and reduction in the concept of *Gonyaulax*, the arguments for conserving the name against one now firmly established in the paleontological literature would not seem to be particularly potent.

Miscellaneous Genera of Gp-Cysts

Several interesting species from the Jurassic and Early Cretaceous have partiform hypocysts but clearly do not stand close to the younger and mostly small cysts of the *Microdinium* and *Phanerodinium* complexes. Although in some cases critical questions about details of paratabulation are unanswered, these taxa add considerably to our impression of the morphological scope and stratigraphic range of the partiform group of gonyaulacoid cysts. In the following comments, the principal characteristics and the problems or uncertainties associated with each example are mentioned. While not all the species involved are new, they have not been described in equal detail and the similarities in their paratabulation which are emphasized here have not been suggested. Unpublished observations (W.R.E.) are included in the information presented. The interpretation offered in the paratabulation diagrams presented here must, of course, remain tentative pending fully documented observations.

The genera to be discussed are *Carpathodinium* and *Paragonyaulacysta*, "*Lacrymodinium*" and *Pluriarvalium*, *Eyachia*, and *Luehndea*. Some comparative data are summarized in Table 10.2 and examples are illustrated on Figure 10.8.

Table 10.2 - COMPARISON OF MISCELLANEOUS Gp-CYSTS

	Carpatho.	*Eyachia*	*"Lacrymo."*	*Luehndea*
shape	pyrif.	polyg.	pyrif.	sph/ell.
spinose or fringed parasutures	no	no	some species	yes
number of apparent intercalaries	3-4	0	1	2-3
number of precingulars	6	5	6	6
state of 1u	exsert	metasert	exsert	insert?
archeopyle	xI or xA	4A	I	E
representative length (μm)	45-75	60-75	70-80	30-45
epicyst smaller than hypocyst?	no	yes	no	no
age	M.Jur.	E.-M.Jur.	L.Jur.-E.K.	E.Jur.

The hypocyst paratabulation of all examples is closely similar, including a large **Iu** and five other postcingulars (**II-VI**) in the usual partiform arrangement. The relative sizes of **Iu**, **X**, and **Y**—and, therefore, the positions of **X** and **X/Y** relative to the antapical surface—vary about as much as among the younger Gp-cysts. In those species which exhibit sulcal paratabulation (e.g., "*Lacrymodinium*" spp. and *Carpathodinium predae*), **Im/Z** abuts against **Iu** (W.R.E., unpublished observations).

Mutual similarities among *Carpathodinium*, "*Lacrymodinium*", *Paragonyaulacysta*, and *Pluriarvalium* include their pyriform shape and what appear to be intercalary archeopyles, supplemented (in the first two) by the exsert gonyaulacoid epicyst pattern (which also appears likely but is unproven in the others). The pyriform outlines and intercalary archeopyles also suggest the *Pareodinia* complex of Rr-cysts, but lack of information on paratabulation details in that complex prevents further comparison. *Luehndea* and *Eyachia* stand clearly in contrast to one another and to the others, with insert or metasert **1u** and individually distinctive properties.

The occurrence (although not in identical combinations) in these genera of particular and unusual elements that distinguish taxa of the *Microdinium* and *Phanerodinium* complexes hints strongly at common relationship.

Carpathodinium (Figure 10.8 A-C). References: Beju (1971), Drugg (1978). The epicyst paratabulation is exsert, while **A** and **B** (or most of it) are large and mostly ventral. The dorso-apical area includes a conspicuous archeopyle which appears to represent one or two out of three or four intercalaries. The absence of a conventional **C** and the topologic relationships of these "intercalaries" suggest they may have originated from primaries **B** and **C** as indicated by the labels applied to them in the figure (which shows a specimen interpreted to have three intercalaries). Except for the "intercalaries", the paratabulation appears to represent a combination of an exsert gonyaulacoid epicyst and a partiform hypocyst. However, additional information on paratabulation in the parasulcal and dorsoapical areas is needed.

This genus is probably close to (possibly synonymous with) *Paragonyaulacysta*. However, published documentation of the attributes of the latter genus as shown by the original material (Johnson and Hills, 1973) leaves the question unsettled.

"*Lacrymodinium*" (Figure 10.8 D-F). Reference: Albert et al. (in prep.). The epicyst paratabulation conforms to that of *Triadinium* (Figure 5.6 B) except in the presence of a single dorsal intercalary. However, unlike the "intercalaries" in *Carpathodinium*, this one seems to have resulted from a single transverse division of the usual middorsal gonyaulacoid paraplate **4**—hence, the labels **4v** and **4h** on the diagram. This intercalary forms the conspicuous archeopyle. A unique feature on the right side of the parasulcus is the presence of a paraplate between **VI** and **Z** that makes contact with **Y**. Possibly this represents a subdivision of **Z**. The SEM has shown that the parasutural ridges bear fringelike or finger-like projections which, collapsed against the surface in various directions, can

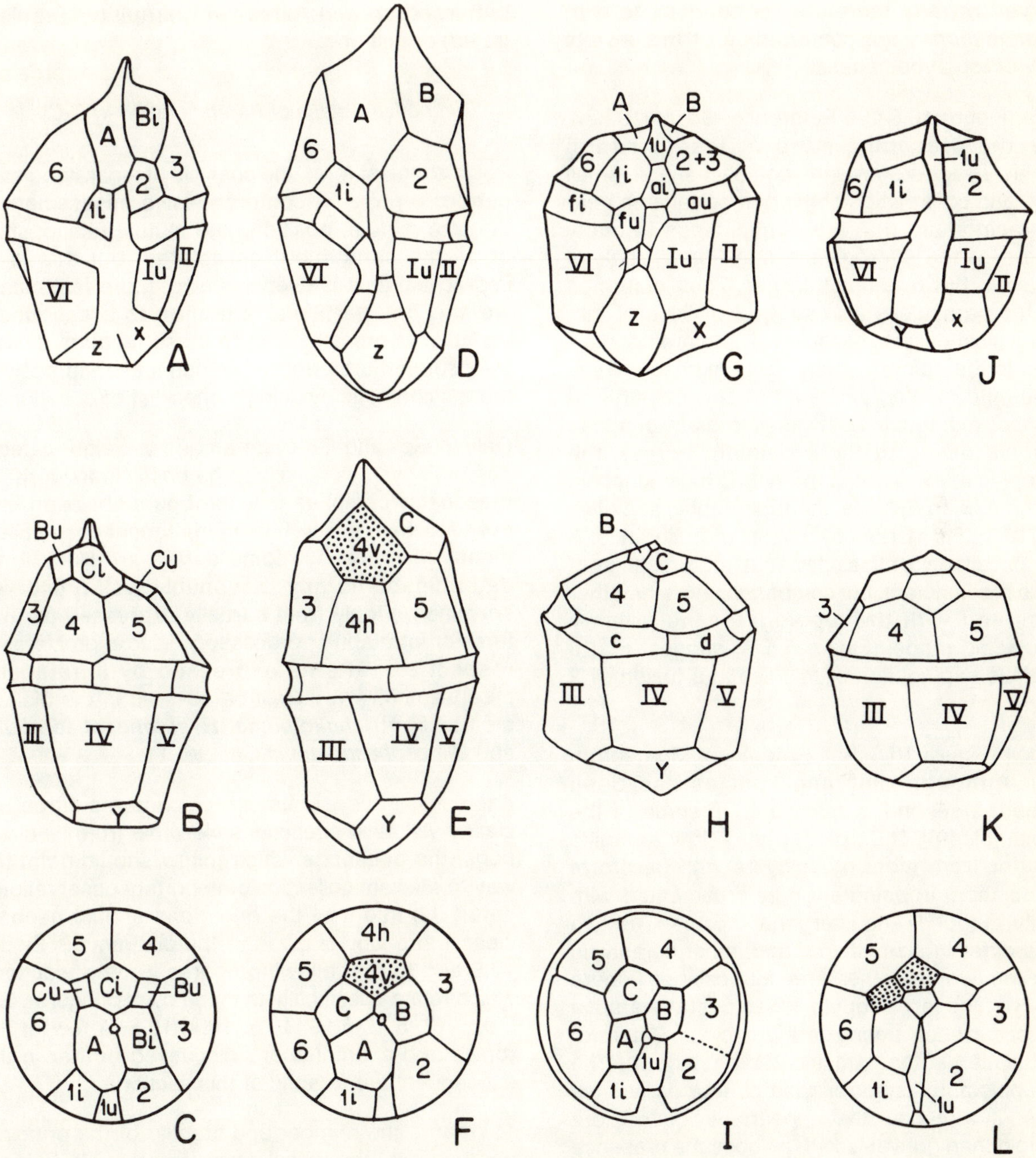

Figure 10.8 - FOUR EXAMPLES OF Gp-CYSTS

Approximate outline shape and paratabulation in vv (top row) and dv (middle row), and schematic av (bottom row). Small circle in C, F, and I shows location of apical horn relative to paratabulation. Taylor labeling of the apical and intercalary series is speculative to varying degrees. See text for discussion.

A-C - *Carpathodinium predae*, Middle Jurassic; in part after Drugg (1978).

D-F - "*Lacrymodinium*" sp. Late Jurassic. Archeopyle stippled.

G-I - *Eyachia prisca*. Middle Jurassic; based on Gocht (1979b). Hypothesized fusion of **2** and **3** along dotted line in I.

J-L - *Luehndea spinosa*. Early Jurassic. Processes omitted; stippled intercalaries form a single paraplate in undescribed species.

suggest additional parasutures under the light microscope. Herein, possibly, lies an explanation for the unusual paratabulation patterns attributed to *Pluriarvalium* (Sarjeant, 1962; Wiggins, 1975).

Pluriarvalium may therefore be congeneric with *"Lacrymodinium"*, but confirmation of this awaits restudy of topotype material.

Eyachia (Figure 10.8 G-I. Reference: Gocht (1979b). Unlike the preceding more or less pyriform examples, *Eyachia* shows the compact shape, small epicyst, and constriction of the parasulcal area near the paracingulum that are conspicuous in many members of the *Microdinium* and *Phanerodinium* complexes, but it is much larger. Faint markings within the parasulcus also suggest division of that area in typical partiform fashion. There are no intercalaries in the epicyst, where **1u** is metasert. As is also definite in "*Lacrymodinium*" and possible in *Carpathodinium* (but in both cases less obvious than in *Eyachia* owing to the prominent horns), the topographic apex is not at the **A/B/C** triple junction, but along **A/B**. In *Eyachia*, the topographic apex lies at **A/B/1u** and is marked by a small protuberance. A further peculiarity of this species is the presence of only five precingulars, but comparison with the other species and with the topology of the primary gonyaulacoid model (Figure 5.5 A) suggest that a reasonable explanation of this lies in a fusion of **2** and **3**.

Luehndea (Figure 10.8 J-L). Reference: Morgenroth (1970). Although only one species has been described, a second is associated in some of the same samples (W.R.E., pers. obs.). They are alike (and differ from most other cysts with partiform paratabulation) in being strongly ornamented with parasutural (gonal and intergonal) spines. The two species differ in size, shape, and, most notably, in one having two, the other three intercalaries (Figure 7.4 G-H). The archeopyle is epicystal—also a unique occurrence among partiform cysts. Details of paratabulation in the parasulcus are unknown. The topographic apex is surrounded by four paraplates arranged in a manner that appears decidedly more peridinioid than gonyaulacoid (as does the presence of three intercalaries). However, the six precingulars and apparently partiform hypocyst (with six postcingulars) present severe problems for a peridinioid interpretation. The topology of the apical view (Figure 10.8 L) is quite similar to that for *Carpathodinium* (Figure 10.8 C), except that in *Luehndea* **1u** appears to be fully insert. However, an attempt at this point to explain all the paraplates of the apical region by subdivision of primaries **A**,**B**, and **C** involves relatively unguided speculation. Therefore, some of these paraplates are not labeled in the diagram. The difference between the two species of *Luehndea* depends upon whether the patterned area includes two paraplates (as illustrated) or just one.

Distinctive Details of Paratabulation in G-Cysts

Put in simplest form, the analysis of a paratabulation pattern requires a comprehensive assessment of the size, shape, position, and mutual relationships of all the constituent paraplates; but this is as impractical as it is overwhelming. More realistically, we can pay particular attention to areas where variation is greatest, and to those features easiest to observe. Furthermore, we can make the analysis more systematic by using a checklist of questions.

Only in Gp- and Gs-cysts among the eight categories of G-cysts is it likely to be possible to work out reasonably complete details of paratabulation, and even for them this will be an infrequent accomplishment. Nevertheless, some details are likely to be determinable in any reasonably well-preserved specimen, if only from partially expressed parasutures or intratabular processes. Special problems in respect to patterns expressed by intratabular processes (Gi-cysts) will be covered in the discussion of *Hystrichokolpoma*, *Hystrichosphaeridium*, and *Oligosphaeridium* in Chapter 11.

Assuming an ideal case in which parasutures are clearly visible in specimens viewable from all directions, the questions which follow should point the way to efficient collection of important observational data. The intent of the questions should become clear if attention is given to the diagrams in Figure 10.9 and to the following figures in other sections which relate specifically to these points: Figures 5.8, 5.9, 5.12, 5.15-5.19, 10.3, 10.4, 10.8. A few of the topics and examples are discussed further in the subsequent paragraphs of this section.

1. What is the number and identity of the principal paraplates of the apical area—three (**A**, **B**, **C**), four (**A**, **B**, **C**, **1u**), or five (**A**, **B**, **Ci**, **Cu**, **1u**) for most G-cysts, but including conspicuous "extras" as intercalaries in some Gp-cysts?

2. Are preapicals present? If so, is there one (**P**) or two (**P** and **Q**), and what are their relationships to neighboring paraplates?

3. Where is the **A/B/C** triple junction (as actually expressed or as estimated by projecting the relevant parasutures) relative to the "topographic" apex as expressed by an apical horn or by the doming of the apical surface?

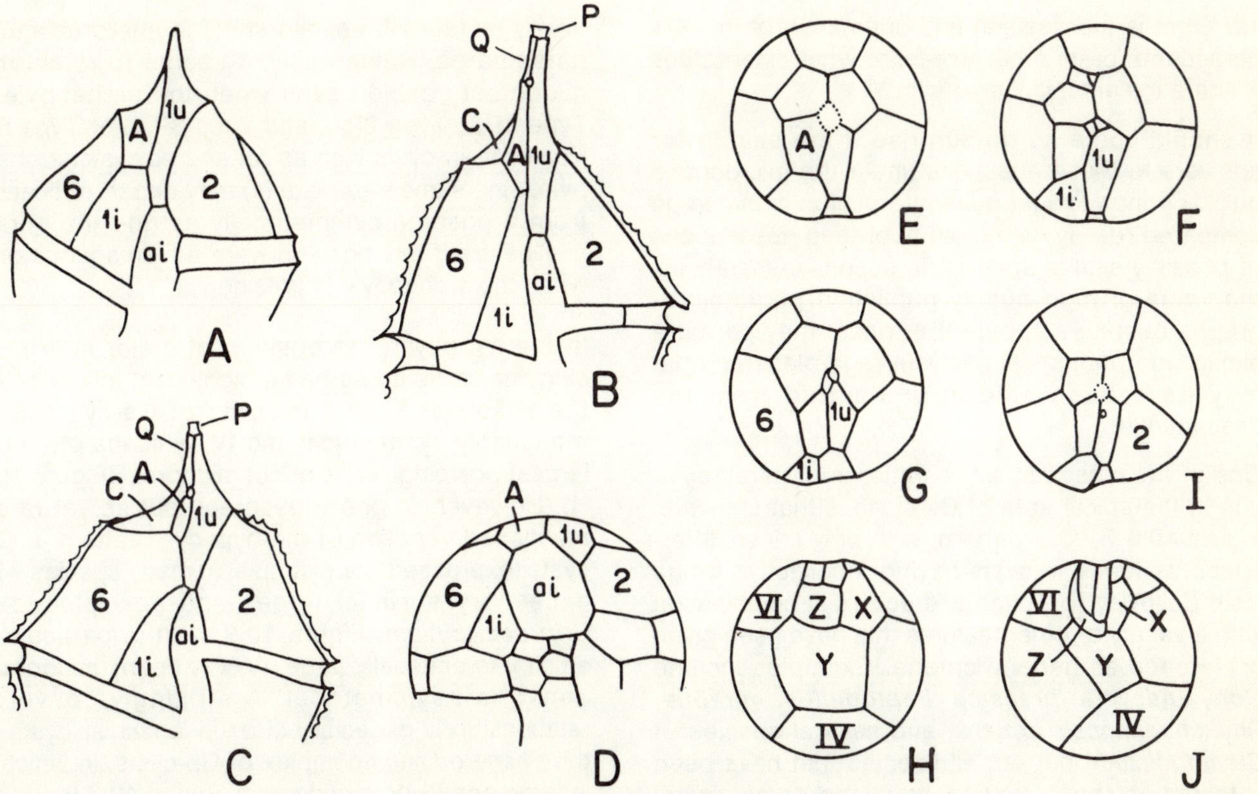

Figure 10.9 - SELECTED VARIATIONS IN PARATABULATION OF Gs-CYSTS

See text for discussion.

A - *Leptodinium* sp. (epicyst, vv). L-type ventral organization; **1i** about parallel-sided, relatively large; **A/1i** higher than **1u/ai**; **P** and **Q** present but not visible. Jurassic-Early Cretaceous.

B - *Gonyaulacysta jurassica* (epicyst, vv). S-type ventral organization; **A** is slender, ventral; **A/1i** higher than **1u/ai**; **1i** triangular; **P** and **Q** present. Late Jurassic.

C - *Gonyaulacysta cretacea* (epicyst, vv). S-type ventral organization; **A** is minute, out of contact with **1i**; **6** contacts **1u**; **P** and **Q** present. Early Cretaceous.

D - *Evittosphaerula paratabulata* (epicyst, vv). Unusually large **ai** contacts **6** and precludes contact between **A** and **1i**; "extra" paraplates in parasulcus; **P** and **Q** present but not visible. Miocene.

E - Epicyst (av) equivalent to A (horn shown by dotted circle without details of paratabulation). Jurassic-Cretaceous.

F - *Leptodinium mirabile*. Schematic diagram of epicyst (av). **P**, **Q**, **Ki**, **Ku** present; other features of paratabulation as in A. Late Jurassic.

G-H - *Impagidinium californiense*. Epicyst (av) and hypocyst (aav). S-type ventral organization; **1i** triangular; **A/1i** lower than or level with **1u/ai**; **A/1u** meets **ai** (not **1i**); **X** and **Z** not narrow; **Y** about symmetrical; **IV** about middorsal. Eocene.

I-J - *Cribroperidinium* spp. Epicyst (av) and hypocyst (aav). L-type ventral organization; **1i** about parallel-sided; **A/1i** higher than **1u/ai**; **A/1u** meets **ai** as in Figure A; **X** and **Z** narrow and arcuate; **Y** asymmetyrical; **IV** displaced to left from middorsal. Late Jurassic-Tertiary.

4. What is the shape, size, and degree of insertion of **1u**?

5. What are the relationships among **A**, **1i**, **1u**, **6**, and **ai**—their relative sizes, shapes, and the relative lengths and positions of their contacts?

6. Is a porichnion visible and where, precisely, is it located?

7. How do the positions of parasutures between the pre- and postcingular paraplates differ from those in the primary gonyaulacoid model, and how is this variance related to unusual shape, or size of paraplates?

8. Are there major anomalies in the number and identity of pre- and postcingular paraplates?

9. What are the relationships of **X**, **Y**, and **Z** to each other and to their neighbors?

10. What relationships among minor paraplates can be determined in the sulcal region?

11. How does **Iu** relate to its neighbors **au**, **II**, **X**, **Im**, and (in Gp-cysts only) **Z**?

12. What is the position and orientation of the **X/Y** parasuture, or, in other words, to what extent does **X** share the antapical area with **Y**?

It should come as no surprise if the search for answers to these questions under the microscope quickly brings to light more information than can be compared readily with most published descriptions of possibly similar species. In such circumstances, the value of high quality published photomicrographs becomes apparent at once: they permit a similar morphological analysis regardless of what may have been said in or omitted from the descriptions.

One of the most distinctive patterns of paratabulation in the apical area of Gs-cysts is that shown in Figure 10.9 F. This pattern, with only minor differences, is found in cysts having a collective range from Callovian to Aptian and occurs in combination with a variety of other features that have been given preference as generic criteria. Examples include *Gonyaulacysta jurassica*, *Leptodinium mirabile*, *Rhynchodiniopsis aptiana*, and several species of *Ctenidodinium*, but not all species that have been referred to these genera have the same apical patterns. One might think that such a particular combination of paraplate numbers, shapes, and sizes could serve as a useful taxonomic character, say, for a genus uniting the species in which it occurs. However, this detail of paratabulation has played no role in generic attribution, which has been based entirely on factors of horn development, wall structure, archeopyle type, and surface relief.

The ventral organizations that were distinguished as L- and S-type in Chapter 5 (Figure 5.8) are two particular combinations of features on the ventral surface that occur in a large number of species. That is justification for emphasizing them, but the variations illustrated in Figure 5.12 A-H show that they are not wholly discrete types, and other quite different arrangements occur.

The presence of relatively constant numbers of paraplates in both the pre- and postcingular series implies a considerable interdependency between paraplate size and parasuture position. Thus, the presence of one unusually large paraplate necessarily involves an unusual position for one or both of its lateral limits and a compensating reduction in size of one or more other paraplates in the series. However, variation in paraplate size and parasutural position is not unconstrained.

In species with Type P archeopyles, the midline of the opening (i.e., Taylor **4** = Kofoid *3″*) remains nearly middorsal, despite shifts in interprecingular parasutures. Remarkably, **4** seems to retain this middorsal position even when the archeopyle is Type $2P_{3\text{-}4}$, Type $2P_{4\text{-}5}$, and Type $3P_{3\text{-}5}$, or Type 6P. However, in cysts with apical and epicystal archeopyles, the same paraplate (**4**) may depart noticeably from a position symmetrically astride the dorsal midline, as if this position were advantageous only when the archeopyle is precingular.

In the hypocyst, the position of major interpostcingular parasutures has a significant influence on the shape of **Y**. In a majority of Gs-cysts, **Y** is reasonably symmetrical and **IV**, which is often the largest postcingular, is about middorsal (Figure 10.9 H). However, in one widespread but as yet rarely mentioned variation of the hypocyst pattern in Gs-cysts (expressed, for example, in many species with accessory surficial ridges and now placed in *Cribroperidinium*; Figure 10.9 J), **II** is particularly small,**V** is unusually large, **IV/V** is nearly middorsal, and **Y** is asymmetrical. A strikingly converse relationship is especially clear in lenticular cysts of the *Phanerodinium* complex of Gp-cysts, in which **II** is large and **III/IV** is middorsal (Figure 10.3 B-C).

Recent Observations Pertaining to Plate Overlap and Growth

The two phenomena of thecal plate overlap and growth of plates by marginal increments along bounding sutures have long been recognized in living Peridiniales (Chapter 1). Recent discussions of these matters (Gocht, 1979a, 1979b, 1981) underscore their relevance to the interpretation of dinoflagellate cysts.

The general pattern of plate overlap in peridinialean dinoflagellates (which is not conspicuous but can be demonstrated easily by appropriate preparation and inspection) is roughly from the middorsal area in all directions toward the midventral. Although detailed studies of overlap have not been reported for a large number of species, a remarkable degree of consistency appears to be the rule, albeit with exceptions. For example, in hexa style peridinioid species, *2a* overlaps *3″* and *5″* obliquely toward the cingulum, contrary to the general sense; and in *Ceratium cornutum*, Taylor **B** overlaps **2**, which is the reverse of the relationship in *Gonyaulax polyedra* (Happach-Kasan, 1982). It is tempting to think that plate overlap relations may one day be useful evidence in matters of plate homology and phylogenetic relationship, but first we must learn more about the number and identity of exceptions, and develop explanations for them.

The effect of incremental growth of thecal plates is especially obvious today in the thecae of certain peridinioid dinoflagellates and is also clear in many gonyaulacoids (including *Ceratium*). Since these are the two groups whose cysts are of principal interest to us, we need not be concerned here with the effects of growth in other groups. New plate material is added differently in some living gonyaulacoids than it is in others and in living peridinioids. In peridinioids, growth increments develop along all edges of the major plates (except contacts with the cingulum). Consequently, along any given suture between non-cingular plates, a two-sided growth band develops as new material is added to both plates. Viewed from the perspective of a plate center, each major plate appears to grow toward all its neighbors (except the cingulum). In species whose thecae enlarge in this manner, there is no visible correlation between patterns of growth and overlap.

Pandasutural bands on sporopollenin cysts of both peridinioid (e.g., living *Peridinium limbatum* and fossil *Palaeoperidinium pyrophorum*, Chapter 9) and gonyaulacoid origin (e.g., chorate cysts such as *Alisocysta*, *Areoligera*, *Schematophora*, *Systematophora*, and proximate cysts such as *Sirmiodinium*) either certainly or possibly represent similar two-sided growth increments along thecal sutures. However, for the same reasons that apply to the thecae, these bands are unlikely to tell us much about overlap relationships between thecal plates.

Living *Gonyaulax polyedra* serves as a model for a contrasting plan of thecal enlargement in which growth increments develop in a pattern closely correlated to plate overlap (Dürr, 1979a). In this and many other gonyaulacoid species, new material is added along any given suture only in a one-sided manner, to the overlapping plate, whereas the underlapping plate is not enlarged at all (or only to a trivial extent). If the growth band that results from this one-sided process shows a substantial contrast (e.g., in surface relief or pore distribution) to the portion of the plate present before growth—and this is the usual case—not only the pattern of growth, but the pattern of plate overlap as well is rendered visible.

Gocht (1979a) showed that a similar pattern of thecal growth is recorded in certain fossil gonyaulacoid cysts, but with added complications (Figure 10.10). Comparing the arrangement of "accessory ridges" on *Cribroperidinium* with the distribution of growth increments on the theca of *Gonyaulax polyedra*, he concluded that these ridges formed on the cyst in positions coordinated with the incremental bands of thecal material. However, he observed that two sets of accesory ridges in some Cretaceous species of *Cribroperidinium* evidently record two consecutive but differing stages of thecal growth. The ridges closer to the centers of the paraplates mark an initial stage during which material was added in a two-sided manner. These ridges bound the central portion of major plates (except on the side toward the paracingulum) without regard to the gonyaulacoid overlap pattern. A second set of more distal ridges records a subsequent stage of one-sided, overlap-coordinated growth like that in living *Gonyaulax polyedra*. In contrast to such complexly ornamented Cretaceous species of *Cribroperidinium*, certain Jurassic species of this genus, as well as some other Jurassic gonyaulacoid cysts (e.g., *Eodinia pachytheca*) record only the two-sided type of growth that proceeded independently from overlap patterns. Equivalent two-sided thecal growth is also known today in some living gonyaulacoid species (e.g., *G. diegensis* and *G. helensis*).

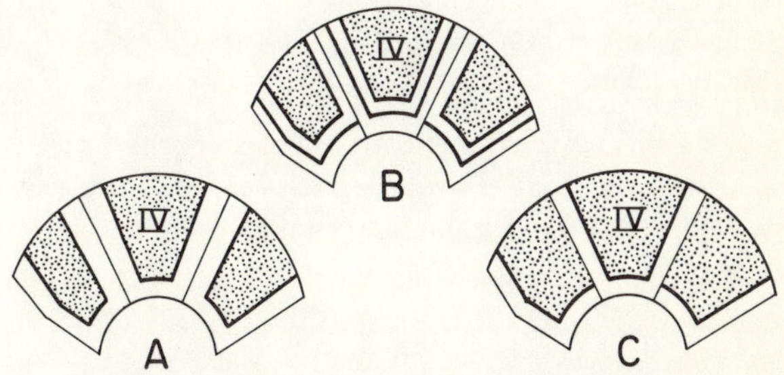

Figure 10.10 - GROWTH BANDS AND OVERLAP
See text for discussion.
Schematic exterior views of dorsal hypocyst showing **III**, **IV**, and **V** with margin of paracingulum at top. Patterned areas represent original portions of plates; unpatterned areas correspond to growth bands. Heavy lines are accessory ridges.
A - As in Jurassic species of *Cribroperidinium* and certain other gonyaulacoid cysts. Surficial pattern on cyst indicates growth along three sides of each major plate.
B - As in species of *Cribroperidinium* and certain other gonyaulacoid cysts and thecae with collective Cretaceous-Holocene distribution. Accessory ridges indicate two stages of growth, the first as in A; the second recording addition only on margin of overlapping plate.
C - As in theca of modern *Gonyaulax polyedra*. Major plates receive growth increments only on overlapping margins.
Based on Gocht (1979a).

Such observations, although few to date, lead one to speculate that particular types of thecal growth may have been characteristic of particular intervals of geological time. They also constitute interesting "grist for the mill" for those concerned with general biological affinities among major groups of dinoflagellates, as well as with the evolutionary origin of particular dinoflagellate features.

Gocht (1979b, 1981) also demonstrated that morphological details of sporopollenin cysts (other than those that quite obviously record growth bands) may signal the directions in which thecal plates overlapped. Specimens of Early Jurassic *Eyachia prisca*, for example, when seen under the SEM, suggest a pattern of plate overlap consistent with that observed in modern *Gonyauylax*. This is expressed by differences in relative height of adjacent paraplates and a certain appearance of "overhang" along parasutures. A more dramatic example is provided by Late Jurassic specimens attributed to an undescribed subspecies of *Hystrichogonyaulax cladophora*. The paratabulation in this species is delineated by comblike parasutural rows of many short processes. The unnamed subspecies has the special peculiarity that the comblike aggregates of processes along each parasuture extend at a low angle to the surface, shingle-like, over one of the paraplates that meet along the parasuture. The directions of inclination of these process aggregates match closely the directions of plate overlap observed between corresponding plates in living *Gonyaulax*. Here, then, is evidence of plate overlap in a fossil species that can be sensed immediately and observed directly.

An overlap-coordinated pattern of plate growth like that in living *Gonyaulax polyedra* can also be observed today in *Ceratium*, *Triadinium*, and probably other genera. It remains to be seen how extensively distributed among fossil cysts similar traces of growth and overlap will prove to be when due attention is paid to them.

Chapter 11. Morphology of G-cysts. II. Cysts Distinguished by Intratabular and Nontabular Features

The G-cysts to be considered in this chapter only rarely show traces of parasutural lines, aside from those directly associated with the principal and accessory archeopyle sutures, and, in some, paratabulation is not indicated at all. In many, however, paratabulation is shown by the arrangement of features within the areas of the paraplates, such as intratabular processes or groups of processes, penitabular rows of features, and polygonal fields of minor surface relief.

It is self-evident that the less clear and complete the paratabulation is, the less surely can it be used as the basis for suggesting biologic affinities. Many of the more nondescript cysts that will be encountered in the complexes to be reviewed here have been brought together on the basis of a common deficiency in features on which one would choose to establish an identification. Therefore, several of these categories are simply catchalls. Nevertheless, they include some easily recognized, morphologically interesting, and stratigraphically useful taxa.

CHARACTERISTICS OF CATEGORIES AND COMPLEXES

Category of Gi-Cysts

Introduction

This category includes skolochorate and proximochorate cysts whose surface relief is clearly organized into dominantly intratabular elements or groups. Parasutural features, if present at all, play a secondary role. This distinguishes them from Gs-cysts with relatively little ambiguity, but more serious intergrades occur with Gn-cysts, whose surface relief is basically nontabular. In that direction, the somewhat arbitrary distinction lies in the relative clarity of the paratabulation. When at least the major paraplates of the pre- and postcingular series can be identified through the arrangement of processes or process clusters, the cyst should be considered a Gi-cyst. If only the paracingulum and traces of paraplates in the pre- and postcingular areas are indicated, reference should be to Gn-cysts.

Equatorial views usually show the main cyst body to be spherical to somewhat ellipsoidal, but a small local convexity may mark the apex; cross-sections in polar view are normally circular. Accordingly, orientations are random or equatorial.

Paratabulation among Gi-cysts is most frequently indicated by 30 or fewer processes whose obviously consistent distribution is in accord with a one-to-one relationship between processes and thecal plates. Commonly, processes are differentiated in size and form, consistent with relative size and proportions of the respective paraplates. In agreement with the Principle of Minority Suppression, the smaller paraplates may be represented by quite diminutive processes or may bear no processes at all. Paratabulation may also be indicated quite directly; for example, by the outlines of expanded process bases which, in some instances, are even angularly penitabular. Similarly, outline shapes of expanded process tips are occasionally sufficiently clear to recognize equivalence to individual plates of a gonyaulacoid tabulation. Selective suppression of certain processes, as discussed in Chapter 4, may reduce the count to as few as 16 or 17. Rarely, two or more processes distinguish one or a few particular paraplates in a species in which most paraplates bear but a single process.

In some Gi-cysts, individual paraplates bear multiple processes, either as a cluster in which process bases may even fuse, or as arcuate to circular or penitabular arrays in which process tips may be joined by trabeculae, or process bases by ridges or septa.

Among cysts with surface relief in intratabular arrangement, clear distinction between quinqueform and sexiform patterns at the antapex is often impossible. Only in the penitabular arrangements of the *Systematophora* complex, or when special features are present in conjunction with intratabular processes (see discussion of *Hystrichosphaeridium tubiferum* later in this chapter), can this be determined with confidence. Accordingly, an arbitrary convention for assigning skolochorate cysts with intratabular processes to categories has been incorporated into the key and the definition of categories. It involves the following considerations: If the cyst has a Type A(3A)6P archeopyle and other features are compatible with a quinqueform pattern, then it is to be placed among the Gq-cysts (e.g., *Polysphaeridium*). This has the effect of keeping associated those cysts having in common this unique archeopyle type, even if the diagnostic shape and relations of paraplate **Y** are not demonstrable in individual cases. If, on the other hand, the chorate cyst with intratabular processes has another type of archeopyle, it should be referred to the category of Gi-cysts, regardless of evidence (e.g., process position or even distinct parasutures) that the paratabulation is quinqueform. This has the effect, for example, of avoiding separation of some species of *Oligosphaeridium* and *Hystrichosphaeridium* from others in the same genera, while providing for a uniform handling of cysts in which the hypocyst paratabulation is not clear.

When it comes to considering possible affinities based on inferred tabulation, however, it must be remembered that, as a result of the procedure just outlined, quinqueform patterns are also to be found among Gi-cysts, outside the category of Gq-cysts. Of course, quinqueform tabulation may have been present in dinoflagellate species represented by Gn- and Gx-cysts but is simply not reflected in their cyst morphology.

Process form and structure are highly diverse, but shapes commonly include some degree of flare toward base and tip, with a more or less cylindrical shaft in between. Internal structure usually falls into one of three general types: 1) hollow, with relatively thin wall and a large internal cavity that may be open or closed off at the tip, but is virtually never in communication with the cyst interior, 2) solid, with a dense and homogeneous texture and normally a smaller diameter than for the first type, or 3) either solid or hollow, but having a distinctly fibroid character, the density of fibril packing typically lowest toward the extremities. Species with hollow processes on the large paraplates often have solid ones on the small paraplates. The antapical process is often unique; it may be notably larger and/or longer than the rest, or have an otherwise distinctive form.

The wall is normally an autophragm, or endophragm and periphragm. Cavation, if it can be called that, is usually restricted to the hollow interior of processes. Inflated processes reach their extreme of development in some species of *Hystrichokolpoma* and *Achilleodinium*, in which their bases are separated by no more than narrow parasutural strips. A number of unusual features associated with *Hystrichokolpoma* and similar cysts are discussed later in this chapter.

Process number, design, and arrangement (including particular combinations of different process types on a single individual) are usually quite obviously species-specific and also have been used for some generic distinctions. Process length (relative, say, to main body diameter) also tends to be reasonably constant within a species (especially in specimens from one preparation), but some (ecological?) variation is apparent and a few individuals with dramatically atypical processes (e.g., exceptionally short or otherwise oddly shaped) can often be found when specimens are sufficiently numerous. Not surprisingly, such exceptional forms include holotypes of some rarely encountered "species".

A wide variety of archeopyles occur, but Types P and (4A) are decidedly the most common. In some instances, a difference in archeopyle type is the only significant difference between genera; for example, Type (4A) in *Hystrichosphaeridium* and Type P in *Kleithriasphaeridium*. In other cases, different archeopyle types have been retained in a single genus; for example, Types P and (4A)P in *Florentinia*. In still other cases, a particular archeopyle type occurs consistently in combination with other particular features; for example, Type P with fibroid process structure in most cysts of the *Cordosphaeridium* complex and Type E with the unique association of properties of the *Callaiosphaeridium* complex.

The *Callaiosphaeridium* and *Areosphaeridium* complexes illustrate special cases. The morphological integrity of each depends on a distinctive association of features that is sufficiently unusual to suggest that the members of each complex may also have close biological affinities. However, in each case, circumscription of the complex with respect to these distinctive features brings together cysts having other characters that, in other instances, have been used to make distinctions at levels of genera, complexes, or categories. Specifically, the *Areosphaeridium* complex, for whose definition details of ectophragm at process tips play a uniquely critical role, includes a few dorsoventrally compressed cysts with offset sulcal notch like any typical Gv-cyst. It also includes cysts with and without cingular processes—a distinction significant at the generic level in other examples of Gi-cysts. In the *Callaiosphaeridium* complex, the absence of immediately apparent intratabular features, combined with pronounced parasutural features, suggests placement among Gs-cysts; however, some of these features appear to result, in effect, from the abutting of hollow intratabular processes against one another, and they are here interpreted in that way.

Both these groups seem more coherent entities if the usual morphologic criteria are ranked somewhat differently than in other cases. Each group is dealt with further in the remarks on individual complexes.

Note that a few cysts having prominent processes whose intratabular arrangement is not clear are included for special reasons in the category of Gx-cysts (*Prolixosphaeridium* complex).

Differences among taxa within the category of Gi-cysts depend upon:

- Shape and size
  - Of main cyst body
  - Of imaginary surface determined by process tips
- Wall
  - Layers, their structure and relation
  - Features between process bases
- Processes
  - Shape and size
  - Particular features of base, shaft, and tip
  - Differentiation, if any, according to position on cyst, including selective reduction or absence
  - Linkages among elements from a single paraplate, and between elements from different paraplates
  - Relation to underlying wall
  - Details of paratabulation revealed by process form and distribution
- Archeopyle
  - Presence and extent of accessory sutures
- Variability in all features

Key to Complexes of Gi-Cysts

I. Major paraplates of precingular and postcingular areas represented by conspicuous intratabular processes, penitabular septa, or process complexes, rarely with but usually without distinct parasutural features.

A. Intratabular processes have characteristically perforate ectophragm expansions at tips .... *Areosphaeridium* cpx

B. Projections lack distal ectophragm expansions, or they are not as in A.

1. Intratabular projections (processes or arcuate to penitabular crests or septa) have distinctly fibrous appearance, surface of main body commonly also fibroid .... *Cordosphaeridium* cpx

2. Projections not notably fibroid.

a. Major plates bear individual processes or associations of two to several similar processes not linked by linear ridges or septa .... *Hystrichosphaeridium* cpx

b. Major paraplates bear arcuate to penitabular septa or ridges that may bear processes or not .... *Systematophora* cpx

II. Major paraplates of pre- and postcingular areas are not indicated, or indicated only by parasutural features; cingular paraplates bear free or coalesced hollow processes or penitabular features .... *Callaiosphaeridium* cpx

Areosphaeridium Complex

Taxa: *Areosphaeridium*, *Eatonicysta*, certain species of *Glaphyrocysta* (not including the type species).

Occurrence: Eocene to Miocene.

A morphological detail that would normally be considered trivial—the character of ectophragmal expansions from the tips of processes—is the feature that unifies the taxa comprising this complex (Eaton, 1971). These expansions are perforated by more or less closely spaced, approximately circular openings, and may form a continuous ectophragm (*Eatonicysta*), prominently polygonal process tips (e.g., *A. diktyoplokus* in Figure 4.2 O, 4.4 L; some *Glaphyrocysta*), or only distinctively denticulate edges to distal process branches (e.g., *A. arcuatum*).

To attach generic significance so emphatically to the feature utilized in this case requires deemphasizing other features that could be used as the basis for a quite different taxonomic allocation. Thus, we find associated here some species having cingular processes and others lacking them (e.g., *A. multicornutum* and *A. diktyoplokus*, respectively), yet this difference in another complex of Gi-cysts distinguishes two genera (*Hystrichosphaeridium* and *Oligosphaeridium*). However, certain variability in the development of cingular processes is apparent, for one or two have been observed in rare individuals of species that typically lack them. Similarly, although most species allocated to this complex are characterized by circular equatorial cross-sections (e.g., *A. diktyoplokus*), others exhibit the dorsoventral compression and offset sulcal notch typical for Gv-cysts (e.g., *A. multicornutum*, Figure 3.3 D, and *Glaphyrocysta texta*). Such dorsoventrally compressed forms also differ from those with uncompressed cross-section in having the processes on central ventral and central dorsal surfaces selectively suppressed. That this is concomitant with compression and not a significant indication of affinity has been suggested elsewhere (Chapter 10, under Gv-cysts).

The conflict of approaches that finds expression here can be viewed optimistically as an encouraging indication that, in some instances at least, we have reached the point of being able to distinguish subtle indications of real affinity from gross, but biologically less significant, morphological similarities and differences. The "catch" is that subjective judgment remains the sole determinant of which features point more reliably to natural affinities. In the case of the *Areosphaeridium* complex, the particular traits that argue for biological relationship (Eaton, 1971) are also relatively easy to recognize and use in the context of a basically morphological system.

Processes in species of this complex may appear solid and quite homogeneous, or definitely fibroid and more or less hollow toward their extremities. Complete polygonal outlines of process tips in exceptionally well-preserved and favorably oriented specimens of *A. diktyoplokus* document that the hypocyst pattern in this species was sexiform (Figure 11.1 A). In some species, a single distal polygon may be supported by several partly or completely separate shafts (e.g., *Glaphyrocysta texta*), while, in others, one paraplate (e.g., **Y** in *A. multicornutum*, Figure 11.1 B) may be represented by two completely separate processes.

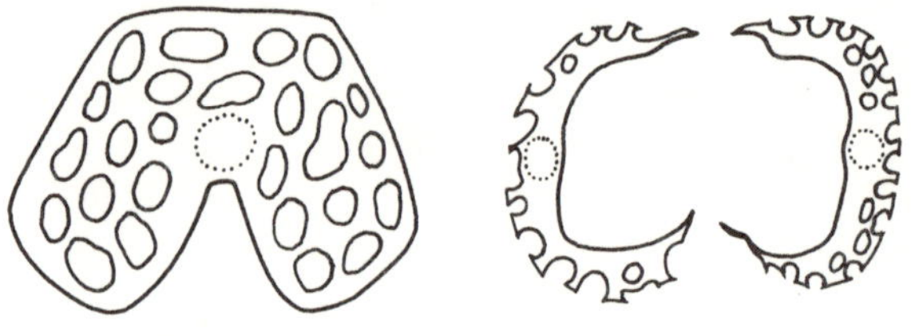

Figure 11.1 - PROCESS TIPS IN *AREOSPHAERIDIUM*

Shaft position is indicated by dotted circle.

A - Polygonal ectophragmal expansion at tip of antapical process (**Y**) in *Areosphaeridium diktyoplokus*. The deep indentation marks the ventral side.

B - Tips of pair of processes marking **Y** in *Areosphaeridium multicornutum*. Pre- and postcingular processes of this species have single shafts with similarly arcuate-denticulate tips, the open side directed toward the paracingulum.

Cordosphaeridium Complex

Taxa: *Aireiana*, *Araneosphaera*, *Cordosphaeridium*, *Cyclapophysis*, *Turbiosphaera*.

Occurrence: Late Cretaceous to Miocene; especially characteristic of the Paleogene.

Distinguishing features are the fibroid processes and consistently Type P archeopyle (Figure 4.3 F). Processes of *Aireiana* are of rather different form than those in the other genera and are especially affected by selective suppression. In some species of *Turbiosphaera*, the paracingular processes are modified into a continuous, slightly spiral, shelflike projection, but other species appear to show steps in the transition to this morphology from isolated paracingular processes. Distal interconnections over part or all of the cyst occur in *Araneosphaera* and *Cyclapophysis*. *Cordosphaeridium cantharellum* is unusual in having all processes of about equal size and the archeopyle outline often rounded, obscuring the equivalence of processes to paraplates, and even the general orientation of specimens.

Supernumerary postcingular processes may occur in some species, and have been interpreted by Gocht (1969) to indicate gonal positions of some processes. An alternative explanation, seemingly more compatible with the intratabular arrangement typical of most species, is that certain paraplates (**III-V**) are marked by pairs of processes. Selective fusion of processes (e.g., **au** with **lu**) also occurs in some species.

Hystrichosphaeridium Complex

Taxa: *Achilleodinium*, *Adnatosphaeridium*(?), *Compositosphaeridium*, *Conosphaeridium*, *Discorsia*, *Florentinia*, *Hystrichokolpoma*, *Hystrichosphaeridium*, *Kleithriasphaeridium*, *Litosphaeridium*, *Oligosphaeridium*, *Paucisphaeridium*, *Perisseiasphaeridium*, *Rigaudella*, *Surculosphaeridium*.

Occurrence: Late Jurassic to Miocene.

Processes in this complex are solid to hollow, but not of typically fibroid structure, although individual fibrils may stand out in the thin walls of some hollow processes. Selective suppression of processes may be important (Frontispiece E; Figure 7.5 G-H, K). In *Achilleodinium* and *Hystrichokolpoma*, process bases may be so expanded as to have only relatively narrow parasutural separations. Interconnections between tips of processes that represent different paraplates occur rarely but are diagnostic of *Rigaudella* and of species commonly referred to *Adnatosphaeridium* (although the morphology of the type species of the latter requires clarification).

Apical archeopyles of Type (4A) are dominant, and the parasulcal notch and angulations along the precingular-apical paraplate boundary are usually pronounced. Except in some *Hystrichokolpoma*, **1i**, although smaller than **6** and **2**, is larger than **ai** (in both process size and area of basal surface).

Explicit evidence of the hypocyst pattern in species of this complex is only rarely preserved, and then may be difficult to observe, although both quinqueform and sexiform patterns have been confirmed. The expanded bases of *Achilleodinium* and *Hystrichokolpoma* show a sexiform pattern, as do the parasutures visible in *Oligosphaeridium abaculum*; but process tips of a species close to *Hystrichosphaeridium tubiferum* show its hypotheca to have been quinqueform. The last two species named are discussed, along with many features relevant to this complex, in the second part of this chapter in a section on *Hystrichosphaeridium*.

Systematophora Complex

Taxa: *Amphorula*, *Emmetrocysta*, *Hemiplacophora*, *Hystrichosphaerina*, *Schematophora*, *Systematophora*.

Occurrence: Late Jurassic to Miocene.

This complex comprises cysts whose intratabular surface relief (for the larger pre- and postcingular paraplates) is organized in a ring, arc, or polygon of dimensions normally sufficient to enclose a large part of the respective paraplate area. The actual form of the relief is a variable within the complex and includes ridges, septa, a row of processes, or combinations of these features. Distal linkages between processes by ring trabeculae may occur. The smaller paraplates, notably those of the paracingulum and parasulcus, may lack relief or bear structures contrasting in form with those on the larger paraplates (e.g., isolated intratabular processes or rectilinear septa). The larger penitabular features adjoining the paracingulum often tend to be interrupted on the adcingular side. Many examples appear to be autocysts. The dominant archeopyle is Type (4A), although Type P also occurs.

The hypocyst paratabulation is exclusively sexiform if it can be determined. This is a matter of definition of the complex. Cysts in which the penitabular features clearly indicate a quinqueform pattern (i.e., pentagonal **Y** and contact between **V** and **Z**) are referable to Gq-cysts.

Callaiosphaeridium Complex

Taxa: *Actinotheca*, *Callaiosphaeridium*.

Occurrence: Cretaceous.

This complex brings together cysts which, like those assembled in the *Spiniferites* complex, may look very different but have a distinctive combination of traits in common. The diagrams of Figure 11.2 illustrate the principal characteristics of the complex. Generalized diagnostic features include: a) Cysts oblately sphaeroidal, their equatorial dimension further extended by isolated intratabular processes or intergrown membranes constructed on an intratabular plan, as explained below. These projections may suggest (misleadingly) a cavate equatorial structure like that present in the Gs-cyst, *Stephodinium* (cf. Figure 11.2 F). The structures representing **au** and **fi** tend to be smaller than those representing **b-e**, which are about equal. b) Hypocyst alone, or hypocyst and epicyst may bear parasutural features which define a sexiform gonyaulacoid pattern and may include gonal and intergonal processes. c) The ventral surface and parasulcus may be marked simply by an interruption in the equatorial complex or by several small processes or other structures. d) The suture of the epicystal archeopyle runs close to the anterior sides of the bases of the cingular structures. The nearly circular operculum is adnate in some (all?) species but the short ventral connection is frequently torn. Opercula thus isolated are quite inconspicuous and easily overlooked.

A seemingly trivial but distinctive feature of many species is the presence of hairlike to ribbonlike projections that at first appear to extend from the distal margin of the open cingular processes. On close inspection, however, it is clear that they consistently arise from the outer surface of the process wall a few micrometers from the margin.

What many authors have reported as *Callaiosphaeridium asymmetricum* from Neocomian to Maastrichtian strata at various geographic localities is actually a suite of distinct species, as a comparison of published illustrations will quickly confirm. Apparently, their most conspicuous feature—six equatorial "chimneys" which were described and illustrated admirably for the type species (Deflandre and Courteville, 1939)—desensitized most observers to the existence of many differences in detail, with the result that only relatively recently has a single additional species (Duxbury, 1977) been described. Combined with these cingular projections are parasutural ridges on the hypocyst, or on both hypocyst and epicyst. The parasutural ridges, which disappear toward the equator, may be unadorned, or may bear slender to relatively massive gonal processes, which may be distally furcate in a manner suggesting *Spiniferites*. If such ridges are otherwise present on the epicyst, those expressing the parasutures between apicals tend to be expressed weakly, with the consequence that a polar view of the epicyst shows an open apical area outlined by an irregularly angled line and having somewhat the appearance of the centrally located **Y** in a polar view of the hypocyst.

In a broad sense, one might characterize this genus as including cysts which are uniquely *Hystrichosphaeridium*-like in the equatorial region but *Spiniferites*-like at "higher latitudes". This is a highly unusual example of the restriction of prominent parasutural and intratabular features to different parts of the cyst surface. Differences between species of *Callaiosphaeridium* include the size, shape, spacing, and elaboration of the cingular processes; and the strength and intricacy of ridges and processes on the epicyst and hypocyst.

Avellodinium, a genus originally described as having the morphology of a *Spiniferites* with an epicystal archeopyle (Duxbury, 1977) may be a synonym of *Callaiosphaeridium* (as contended by Below, 1981), but here it has been included tentatively in the *Spiniferites* complex pending confirmation that *Callaiosphaeridium*-like cingular processes are present in the type species.

Actinotheca can be visualized as derived from *Callaiosphaeridium* (in a morphological sense, with no implication of evolutionary origin intended) by shortening of the cingular processes and extension of their cross-sectional dimension parallel to the equator until basal portions of adjacent processes meet. The result is an equatorial row of somewhat boxlike chambers (contiguous but apparently without open connections between them), with somewhat constricted distal openings. In polar view, this gives the impression of a segmented equatorial flange. In the two described species of the genus, parasutural (including gonal) features on epicyst and hypocyst are absent or only weakly developed.

In the form illustrated by Figure 11.2 G-H, the cingular "processes" are formed by low penitabular septa and have elongate cross-sections. Similarly low parasutural septa (or ridges) without gonal processes define the paratabulation on epicyst and hypocyst.

Although numerous species appear to represent this complex, individual specimens seem seldom to be abundant.

Categories of Gn-Cysts and Gx-Cysts

Introduction

Cysts in these two categories fail to show a paratabulation that is sufficiently clear and complete for it to be a useful feature, although parasutures associated with the archeopyle may be identifiable, and traces of other parasutures or intratabular groupings of projecting structures may be evident. Generally speaking, surfaces are either nearly smooth, or projecting structures are nontabular in distribution. Many of the types of wall structure and surface ornament present here are structurally analogous to those observed among Gs- and Gi-cysts, and transitional morphologies that span the gap between completely clear and quite undecipherable paratabulations are almost sure to be found. However, since surprisingly few currently recognized genera are defined in such a way that they actually span this gap, the categories can be utilized

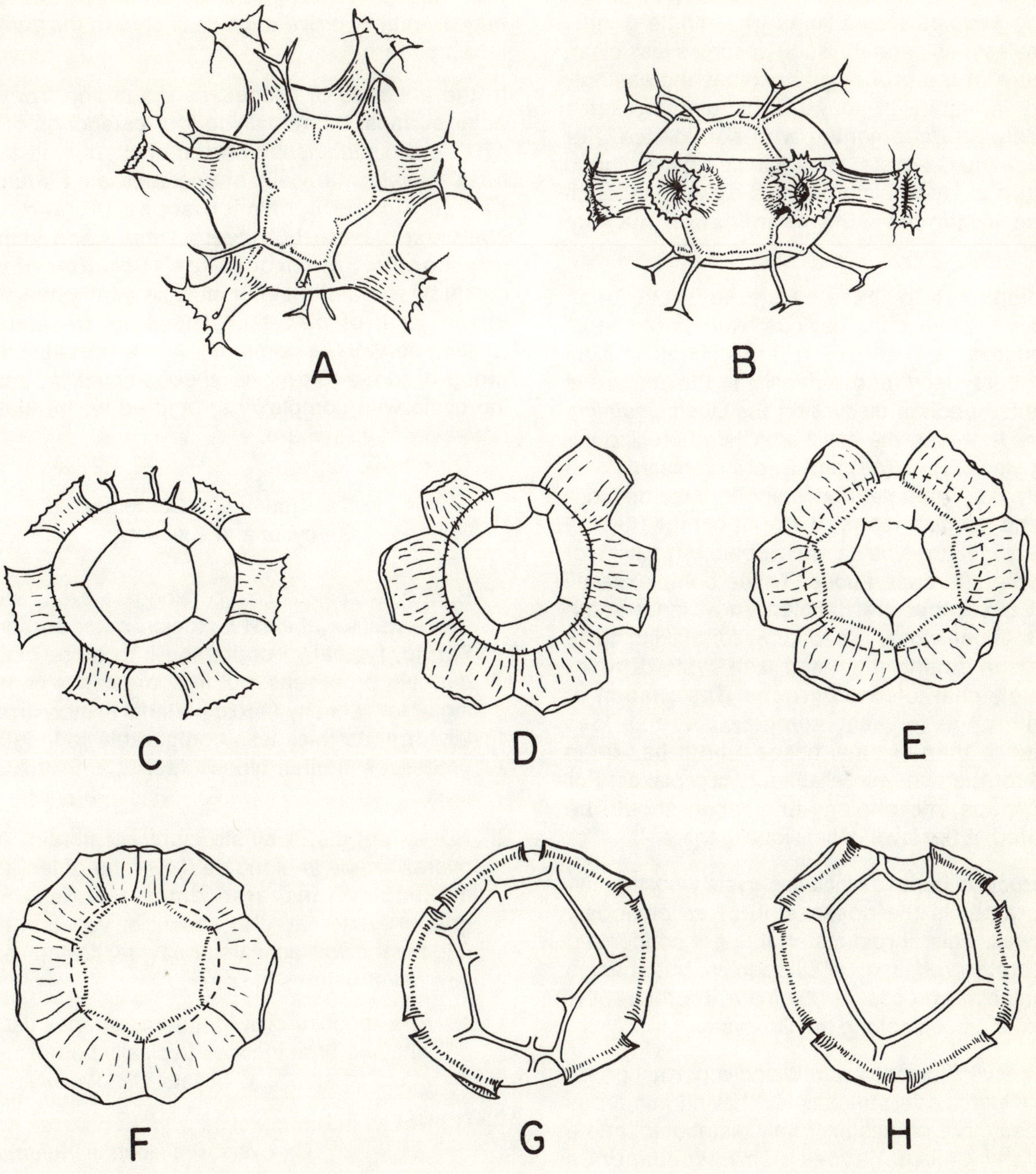

Figure 11.2 - CYSTS IN *CALLAIOSPHAERIDIUM* COMPLEX

An assortment of species showing basic characteristics of the complex and contrasting expressions of the six large, hollow, intratabular, paracingular projections. Parasulcus at top in antapical views, at bottom in apical views.

A-B - *Callaiosphaeridium* sp. cf. *C. trycherium* (av and dv); Early Cretaceous.

C-D - Two undescribed species of *Callaiosphaeridium* (aav); Early Cretaceous.

E - *Actinotheca ornata* (aav); Late Cretaceous.

F - *Actinotheca aphroditae* (aav); Late Cretaceous.

G-H - Undescribed genus and species (av and aav); Early Cretaceous.

E and F based on 35mm color transparencies of holotypes (Helby and Partridge, 1980).

without many awkward splits or arbitrary decisions affecting species within limits of a single genus. Nevertheless, as paratabulation becomes less clear, the nature of the groupings becomes increasingly arbitrary. Affinity with gonyaulacoid dinoflagellates can rarely be documented with confidence. For many Gn- and Gx-cysts, there seems little likelihood that future study will improve this picture, although in some instances SEM examination may be revealing.

The complexes in these categories have been defined with attention to obvious features of shape, wall structure, and surface relief. Archeopyle form has not been used, and variability in this feature is commented upon in discussing the *Operculodinium* complex; however, the basic apical and precingular archeopyles, Types (4A) and P, are dominant. With relatively few characters on which to base descriptions and comparisons, it is important that all observable features be carefully noted. An outline of these points is superfluous. These complexes all contain cysts that are deficient in at least some details that one would like to know for making close comparison among them and with cysts of more completely characterized groups. The citation of geologic range for each complex also has less significance than in other cases, and it has been omitted for the two "miscellaneous" complexes. For these groups, morphology and range should be considered at the level of individual genera.

The *Fibrocysta* complex includes cysts whose unifying character is the possession of an obviously fibroid wall. This fibroid material is not condensed into discrete processes in all species, but species with nontabular processes dominate. Therefore, the complex is placed among the Gn-cysts.

Other Gn-cysts include a varied and important group of taxa (*Operculodinium* complex) having numerous processes more or less uniformly distributed, and a miscellany of other types (Gn-miscellaneous complex) with diverse morphology.

Dinoflagellate cysts with a pronounced elongation are relatively infrequent and, therefore, quite distinctive. Those which have not been accomodated in any of the other cyst categories up to this point are placed here in two groups (*Batioladinium* and *Prolixosphaeridium* complexes), regardless of details of wall structure and surface relief. Placement of both these groups as categories of Gx-cysts is relatively arbitrary. Some constituents of the *Prolixosphaeridium* complex are exceptional in possessing well-developed processes, and it seems quite possible that paratabulation in the *Batioladinium* complex may eventually prove to be distinct from the gonyaulacoid pattern.

In the absence of processes extending from the outer surface, the remaining four categories of Gx-cysts are distinguished on the basis of wall structure: acavate with walls of uncomplicated structure as seen in the light microscope (*Apteodinium* complex), cavate but layers simple and without interconnecting processes (*Scriniodinimum* complex), cavate but with multiple interconnections in the form of discrete processes or lamellae (*Chlamydophorella* complex), and a miscellaneous group of forms (Gx-miscellaneous complex, including cysts with complexly structured walls) that do not easily fit elsewhere.

Key to Categories and Complexes of Gn-cysts and Gx-cysts

I. Spherical to ellipsoidal or elongate cysts whose entire wall (or at least its major, outer portion) is fibroid, typically condensed into more or less discrete processes, but in some cases consisting of low-density fibroid material in more irregular form, its thickness comparable to length of processes in other types .... Gn-cysts, *Fibrocysta* cpx

II. Not as in I (i.e., wall structure not fibroid) AND overall shape is elongate (i.e., ratio of length to maximum transverse dimension is 1.5 or greater); external relief absent or varied in form and height; wall acavate or cavate, having varied internal structures in some

   A. Apex angular or with a horn, antapex usually with two prominences (typically horns) .... Gx-cysts, *Batioladinium* cpx

   B. Not as in A.. .... Gx-cysts, *Prolixosphaeridium* cpx

III. Not as in I or II (i.e., wall structure not fibroid and overall shape not elongate)

   A. Chorate

      1. Cysts spherical to ellipsoidal, bearing numerous nontabular processes that are more or less evenly distributed and, if interconnected at all, are not joined extensively at their tips .... Gn-cysts, *Operculodinium* cpx

      2. Not as in 1 (i.e., having a shape other than spherical to ellipsoidal—e.g., lobate—OR

having projecting features that are not numerous processes more or less evenly distributed—e.g., concentrated locally—OR having processes extensively interconnected distally....................................
.... Gn-cysts, Gn-miscellaneous cpx

B. Not chorate

1. Acavate, wall without complex internal structure ....Gx-cysts, *Apteodinium* cpx

2. Not as in A (i.e., wall more complex: cavate, or internally structured if acavate, or exterior is elaborated into deep reticulum or less regular structures)

a. Cavate, wall includes two distinct layers and is not likely to be interpreted as a complexly structured or differentiated autophragm

1) Periphragm and endophragm; both of simple structure ....
...........Gx-cysts, *Scriniodinium* cpx

2) Distinct and more or less continuous ectophragm supported on autophragm by lamellae or discrete processes........................
....Gx-cysts, *Chlamydophorella* cpx

b. Not as in a (i.e., having other types of wall structure), with various types of modification within the wall or on its exterior surface) ....
.........Gx-cysts, Gx-miscellaneous cpx

Gn-Cysts, *Fibrocysta* Complex

Taxa: *Amphorosphaeridium*, *Exochosphaeridium*, *Fibrocysta*, *Lanternosphaeridium*.

Occurrence: Late Cretaceous to Miocene.

The cysts in this complex have a distinctly fibroid exterior, which may contrast with a denser and relatively thicker, inner portion of the wall. The fibroid portion may be organized into distinct processes or may appear relatively "uncondensed," as a general covering in which the fibers extend more or less radially from the main cyst wall. The outermost surface, in examples with diffuse fibrous material, may be irregular or have more or less definite form, even indicating paratabulation by its surface relief. However, such definite form is detectable only with exceptional material and is not a practical clue to paratabulation pattern, although transition is possible to forms with fibroid intratabular processes, such as *Turbiosphaera*, *Araneosphaera*, and *Cyclapophysis* in the *Cordosphaeridium* complex of Gi-cysts. *Fibrocysta* suggests *Cordosphaeridium* with nontabular processes; processes in *Exochosphaeridium* are more numerous and most of them are slender and simply terminated.

Cysts with a surface reticulum are not included here.

Gx-Cysts, *Batioladinium* Complex

Taxa: *Aprobolocysta*, *Batioladinium* (syn.=*Necrobroomea*).

Occurrence: Late Jurassic to Early Cretaceous.

Cysts in this group, which includes species that have also been referred at times to *Broomea*, *Imbatodinium*, and *Necrobroomea* (see Davey, 1982; Lentin and Williams, 1981), typically have a pyriform outline that is accented by one apical horn and two more or less equal antapical horns (Figure 11.3). The horns are strongly developed in *Batioladinium*, weakly so in *Aprobolocysta*. Dorsoventral compression is insignificant and there is no pronounced sulcal depression ventrally. Normally, paratabulation is indicated only by traces along the archeopyle suture. The paracingulum, if detectable, lies closer to the antapex than to the apex, and the wall is typically an autophragm without strong relief.

The outline of the apical archeopyle (or the course of the archeopyle suture that may be visible in an unruptured cyst) is marked by a deep notch, usually taken to be a sulcal notch on the ventral surface. The operculum is often observed in place or adherent ventrally, but the archeopyle suture is, in fact, complete, and isolated opercula occur. Angulations and short accessory sutures that might indicate the limits of paraplates along the archeopyle margin tend to be obscure to the right of the ventral notch, but to its left they define a small, planate paraplate followed by a much larger and prominently camerate one. A possible interpretation is that these are two precingular paraplates adjacent to a sulcal notch. However, increasing evidence now points to a very different interpretation (D. I. Wharton, pers. comm.); namely, that cysts of this type have a basically exsert gonyaulacoid epicyst (as in *Triadinium*, Figure 5.6) and probably should be compared to forms such as *Carpathodinium* and *"Lacrymodinium"* (Figure 10.8 A, D) — and, perhaps, *Pareodinia* — in which one or more anterior intercalaries occur in combination with a partiform gonyaulacoid hypocyst

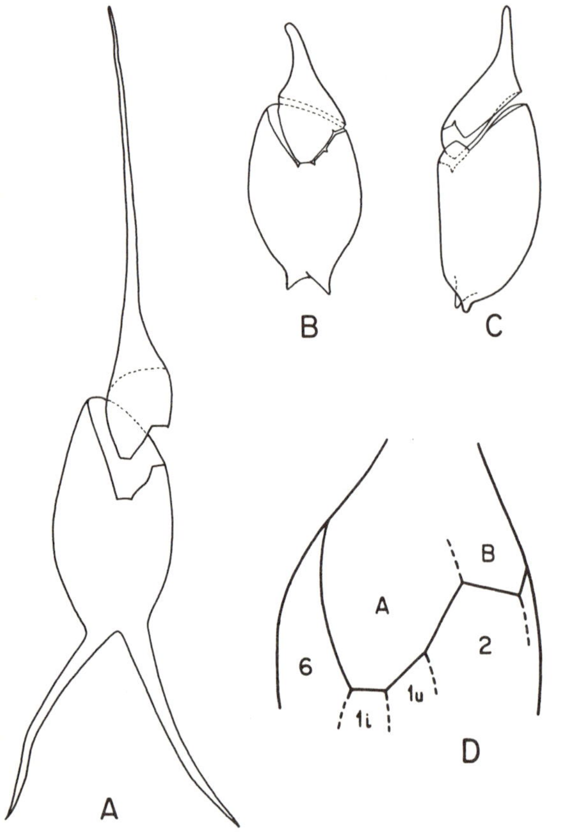

Figure 11.3 - Gx-CYSTS, *BATIOLADINIUM* COMPLEX

A - *Batioladinium longicornutum* (vv). Operculum and main part of cyst from two different individuals, drawn in proximity.

B-C - *B. micropodum*. Two individuals (vv and llatv) with opercula nearly in place (still adherent at sulcal notch in B).

D - Diagram of archeopyle suture and adjacent areas (vv), showing partial paratabulation (including planate **1u** and camerate **2**) indicated by accessory archeopyle sutures and by angulations in the course of the principal suture. Position of parasutures to the right of the sulcal notch is not known.

A-C represent actual specimens, Early Cretaceous, Netherlands.

and an exsert epicyst. By this interpretation, the extremity of the ventral notch represents the **A/1i** parasuture and the two paraplates to the left of the notch are **1u** and **2** (Figure 11.3 D).

Pending confirmation of the character of their hypocyst paratabulation, cysts of *Batioladinium*-type are retained here as a complex under the category of Gx-cysts. *Batioladinium* is used in preference to *Imbatodinium*, which Lentin and Williams (1981) regard as a senior synonym, because of lingering uncertainties about the morphology of the type of *Imbatodinium* (Davey, 1982). The author of this work considers the analysis of the morphology of cysts of this type offered by Dörhöfer and Davies (1980) to be a misinterpretation.

Gx-Cysts, *Prolixosphaeridium* Complex

Taxa: *Boreocysta*, *Bourkidinium*, *Distatodinium*, *Ellipsoidictyum*, *Omatia*, *Peridictyocysta*, *Prolixosphaeridium*, *Protoellipsodinium*, *Tanyosphaeridium*, *Wallodinium*; living *Polykrikos*.

Occurrence: Middle Jurassic to Holocene.

The cysts assembled here share the obvious and relatively uncommon feature of pronounced elongation—a length:width ratio of at least 1.5:1 exclusive of horns, in some cases much more. Other features of wall structure and archeopyle vary.

For comments relative to possible affinity of *Distatodinium*, see discussion of *Polykrikos* at the end of this chapter.

Gn-Cysts, *Operculodinium* Complex

Taxa: *Cleistosphaeridium* (of some authors), *Cometodinium*, *Coronifera*, *Dapsilidinium*, *Diphyes*, *Kiokansium*, *Lingulodinium*, *Operculodinium*; living *Gonyaulax grindleyi* (= *Protoceratium reticulatum*) and *G. polyedra*, whose fossil cysts are known, respectively, as *Operculodinium centrocarpum* and *Lingulodinium machaerophorum*.

Occurrence: Early Jurassic to Holocene.

Distinctive features for species in this complex are numerous nonfibroid processes in nontabular arrangement. The paracingulum may be suggested by alignment of a more or less equatorial row of processes. The apex may be marked by a group of clustered or fused processes otherwise like the rest, and the antapex may bear a uniquely large and distinctively structured process.

The archeopyle is notably variable in some genera and species. For example, a form of *Operculodinium* with Type 3P archeopyle was described as *Trioperculodinium* (now considered a synonym); archeopyles in *Lingulodinium* with three to five precingular opercular pieces have been observed. Archeopyle variation in *Diphyes* (Goodman and Witmer, 1980; Goodman, 1983) is especially clear and includes Types (tA), (tA)P, P, and 2P.

Gn-Miscellaneous Complex

Taxa: *Cauca*, *Dorocysta*, *Stenopyxinium*.

These rarely encountered forms are distinguished by a relatively small number of prominent spinelike projections, not evenly distributed, but concentrated in areas not clearly associated with paratabulation, although obviously coordinated with the general outline shape.

Gx-Cysts, *Apteodinium* Complex

Taxa: *Apteodinium*, *Batiacasphaera*, *Biorbifera*, *Breedoxella*, *Chytroeisphaeridia*, *Cobricosphaeridium*, *Ellipsodinium*, *Escharisphaeridia*, *Fromea*, *Fusiformacysta*, *Kallosphaeridium*, *Komewuia*, *Lagenorhytis*, *Lunatadinium*, *Mendicodinium*, *Muiradinium*, *Pyxidinopsis*, *Sentusidinium*, *Trichodinium*, *Xenicodinium*.

Occurrence: Early Jurassic to Pliocene.

This complex includes acavate cysts whose exterior is smooth, or in a few cases bears nontabular features of low relief relative to the size of the specimen; the individual elements may be coni, spines, or pila (e.g., *Batiacasphaera*, some *Kallosphaeridium*).

Traces of paratabulation may occur in the form of local alignment of surface relief. Rarely, faint parasutural ridges appear. Better definition of parasutures leads to intergradation with Gs-cysts. For example, *Apteodinium* is, in effect, a *Millioudodinium* whose parasutural markings have been reduced to the vanishing point. When concerned with such transitional forms, an arbitrary but probably workable principle is to place in the *Apteodinium* complex those species in which paratabulation is too imperfectly indicated to have any practical usefulness in identification. When the full range appears in a single species, reference to Gs-cysts should acknowledge the greater amount of useful detail revealed by the more informative specimens.

Gx-Cysts, *Scriniodinium* Complex

Taxa: *Balcattia*, *Craspedodinium*, *Dingodinium*, *Disphaeria*, *Duosphaeridium*, *Leberidocysta*, *Lophocysta*, *Membranilarnacia*, *Palaeotetradinium*, *Polygonifera*, *Rhombodella*, *Scriniodinium*, *Thalassiphora*, *Trigonopyxidia*.

Occurrence: Middle Jurassic to Miocene.

This complex includes cavate cysts without pronounced surface relief. Processes are not present and parasutural ridges, if indicated at all, are faint and incomplete. However, as in the similar group of acavate cysts just discussed, intergradation with otherwise morphologically similar Gs-cysts is to be expected.

Pericoel distribution varies. Some species are circumcavate (e.g., *Scriniodinium crystallinum*); some show separation between walls except along the archeopyle margin (e.g., *Leberidocysta*); in two cases (*Balcattia* and *Duosphaeridium*) the pericoel is restricted to a more or less simple body extending from the antapex and nearly as large as the main part of the cyst. Rarely (e.g., *Membranilarnacia*) the walls are interconnected by a few processes whose arrangement is uncoordinated with paratabulation.

The archeopyle in *Dingodinium* is usually developed on the anterodorsal surface in a deceptively intercalary position. Although its form is seldom distinct, occasional specimens show that the opening represents incomplete separation along the line of parasutures normal for a Type (A) archeopyle (Duxbury, 1980).

The similar forms of the endocyst and somewhat unusual apical archeopyles characteristic of *Palaeotetradinium*, *Rhombodella*, and *Trigonopyxidia* suggest that these forms are closely related biologically and distinct from the other genera mentioned here. In the anomalous genus *Lophocysta*, parasutural ridges are limited to the periphragm above the restricted, crestlike ventral cavation, although triple junctions are marked elsewhere by small elevations (Manum, 1979).

Gx-Cysts, *Chlamydophorella* Complex

Taxa: *Chlamydophorella*, *Gardodinium*, *Samlandia*, *Scriniocassis*.

Occurrence: Middle Jurassic to Oligocene.

In contrast to the *Scriniodinium* complex, the cavity between two wall layers in cysts of the *Chlamydophorella* complex is traversed by a large number of processes or other connecting structures. Accordingly the layers are considered to be autophragm and ectophragm. If the interconnections are made by solid or hollow processes, these are discretely formed and join the two walls in such a way that they do not give the impression of resulting simply by differentiation of a thick autophragm and dispersal of its elements (characters of one of the groups of the next complex discussed). If interconnections are lamellar or organized in a reticulum (*Samlandia*, *Scriniocassis*), the distinctions of individual elements may not be clear. Connections in some examples are funnelform invaginations of the ectophragm (*Gardodinium*). The ectophragm may be torn or incomplete between the tips of structures which link it to the underlying layer.

Gx-Miscellaneous Complex

Taxa: *Aldorfia*, *Balteocysta*, *Cassiculosphaeridia*, *Cyclodictyon*, *Eodinia*, *Epiplosphaera*, *Hapsidaulax*, *Kenleyia*, *Spongodinium*, *Tectatodinium*, *Valensiella*.

This is a catchall group and correspondingly varied. The common character, in addition to those that apply to the category as a whole, is a thick wall of complex structure not resolvable into two continuous layers with simple processlike connections and, therefore, different from the *Chlamydophorella* complex. At least three contrasting morphological groupings can be recognized:

1. Substantial processes or process-like structures support an incomplete ectophragm of varied form and extent. Examples: *Balteocysta*, *Cyclodictyon*, *Hapsidaulax*.

2. The wall has a complex internal structure of finer scale. Interpretation as either a differentiated autophragm or as two essentially complete layers internally modified or with connections between them is possible. Different interpretations may even be appropriate over different parts of an individual or for different individuals of obviously the same species. For example, the wall of *Spongodinium* in some areas is appropriately considered a thick autophragm enclosing isolated, empty, bubble-like vesicles, but this structure grades to one that appears rather differently beneath the apical horn or paracingulum. There the vesicles may be large and closely packed, leaving only membranes between them and giving the impression of autophragm and ectophragm joined by irregularly intersecting strands and lamellae. Gocht (1970, 1975b) has described the complex walls of *Aldorfia* and *Eodinia*. *Tectatodinium* is interpreted (Wall, 1967) as having appressed endophragm and periphragm, the latter with internal structure.

3. The external surface is elaborated into a prominent reticulum. Examples: *Cassiculosphaeridia*, *Epiplosphaera*, *Valensiella*.

The complex is named without using a generic name since no single genus can be in any way representative of the variety of form encompassed in this category of convenience.

ADDITIONAL SELECTED TOPICS

Hystrichosphaeridium *and Some Other Cysts with Intratabular Processes*

Gi-cysts—that is, skolochorate cysts with large intratabular processes—offer a special challenge to interpretation: the obvious orderliness of the processes suggests a decipherable pattern but customary absence of parasutural markings often leaves this far from obvious. In this section we will use species of three genera—*Hystrichokolpoma*, *Hystrichosphaeridium*, and *Oligosphaeridium*, all with prominent intratabular processes and apical archeopyles—as vehicles for exploring the manner and extent to which details of paratabulation can be indicated in such cysts.

With the work of Otto Wetzel (1933), the several species of fossil chorate cysts named by Ehrenberg were transferred from the misidentified genus, *Xanthidium* (a desmid), to *Hystrichosphaera*, a name later discovered (Sarjeant, 1967) to be a junior synonym of *Spiniferites*. Shortly after Wetzel's publication, Deflandre (1937) reassigned chorate cysts with intratabular processes (in contrast to *Spiniferites*-type parasutural features) to a new genus, *Hystrichosphaeridium*, of which he made Ehrenberg's species, *X. tubiferum*, the type. In *H. tubiferum*, a full gonyaulacoid paratabulation is indicated by one process per paraplate. During the following years many additional species of *Hystrichosphaeridium* were described, greatly expanding the morphological scope of the genus until revision by Davey et al. (1966) restricted it. Out of the older and broader *Hystrichosphaeridium*, these authors defined a suite of new genera distinguished by the number, form, position, and interconnections of the processes. *Oligosphaeridium* is one of these genera. Its principal difference from *Hystrichosphaeridium* is an absence of cingular processes. *Hystrichokolpoma*, the first of the three genera that we will consider here, was described (Klumpp, 1953) independently of the work of Deflandre and the revision of *Hystrichosphaeridium* by Davey et al. It is a characteristic Tertiary genus, whereas *Hystrichosphaeridium* and *Oligosphaeridium* are represented chiefly by Cretaceous species.

Hystrichokolpoma occupies a virtually unique position among Gi-cysts (shared with *Achilleodinium*, which is distinguished by having a precingular archeopyle) in that the inflated bases of its major processes occupy most of the area of their respective paraplates. Consequently the bases have large, subangular outlines, making it easy to recognize their paraplate equivalence and leaving the smaller, uninflated processes identifiable by a fairly simple process of elimination. Examples of *Hystrichokolpoma* are shown in Figures 4.3 G and 11.4 A-C.

Noteworthy in *Hystrichokolpoma* are the following:

1. The archeopyle, although apical, is often nearly circular. It lacks the strong angulations that alone suffice to suggest the boundaries between paraplates bordering the archeopyle in most other gonyaulacoid cysts having archeopyles of this type.

2. The epicyst and hypocyst are dominated by the large inflated processes. These are separated by a row of cingular processes which are slender and typically multiple, being clustered into transversely elongate groups, each group corresponding to a paraplate.

3. A longitudinally extended area occupied by small and mostly single processes along the ventral surface separates the large pre- and postcingular processes to the right and left (Figure 11.4 B-C). At the anterior end, this group includes **1i** and **ai**, which are side-by-side and longitudinally elongate but narrow. Behind these are the processes of the parasulcus (**fu**, **li**, **lm**, **lu**), followed by **X** and **Z**, again side-by-side. Even **II** may be conspicuously small and "crowded" toward the ventral midline.

4. Whereas **X** and **Z** are small and associated with the other small ventral processes, **Y** is a major structure, normally of smaller diameter but much higher than any postcingular and completely dominating the antapical area.

5. The bases of the apical processes on the operculum tend to merge into a single cavate structure of irregular form whose lobate contours do not clearly identify the contribution from each paraplate to the overall structure.

The strongly inflated process form exhibited by *Hystrichokolpoma* is seen in few other genera (*Achilleodinium*), but the size relationship and distribution of processes encountered in this genus are also to be found in a moderate number of other chorate G-cysts, some of them having quite different process form (e.g., *Turbiosphaera*). Although such details generally have not been described, the characteristics of paratabulation patterns they represent are certainly important.

Hystrichosphaeridium and *Oligosphaeridium* contrast with *Hystrichokolpoma* and resemble one another in process form and in most aspects of process position; they differ from one another in the number of paraplates that are represented by processes and (at least for certain species) in a subtle but significant detail of paraplate topology which can be recognized only under special circumstances. In these two genera the processes are typically tubular structures, flared at the tips. Their cross-sectional size is normally much less than that of the paraplates to which they correspond. Consequently each process rises from just the central portion of its paraplate and, contrary to the possibility in *Hystrichokolpoma*, one cannot identify a particular paraplate by direct inspection of its polygonal outline. Instead, the positional relation of one process to another must be assessed and, usually, several hypothetical identities tested before the most reasonable one can be singled out. *Hystrichosphaeridium tubiferum* and closely similar species are an ideal subject with which to pursue this procedure, and lessons learned here can be applied to more difficult cases.

In *H. tubiferum*, a full gonyaulacoid paratabulation is indicated by one process per paraplate. The individual processes are similar in form (cylindrical shafts flared toward tips and bases) and essentially equal in length, but differ in diameter in proportion to the sizes of the paraplates they represent. Specifically, that means that five processes are decidedly smaller than any others; namely, **P**, **fu**, **li**, **lm**, and **lu**. Their distribution, then, serves as a ready clue to orientation (even if the operculum of the apical archeopyle is completely in place) by marking the apex and the ventral surface. The cingular processes are also helpful for orientation, because their cross-section, especially close to the base, is distinctly elliptical, elongated in an equatorial direction, whereas cross-sections of the large pre- and postcingular processes are nearly equidimensional.

The schematic diagrams of Figure 3.3 B-C are based on *H. tubiferum* and relate the processes visible in ventral view to their respective paraplates. Figure 11.4 D shows an apical view of the cyst, with all processes (except for those of the operculum) indicated and labeled. Figure 11.4 E-G are based on an actual specimen, and can be taken as an exercise in process identification. They represent

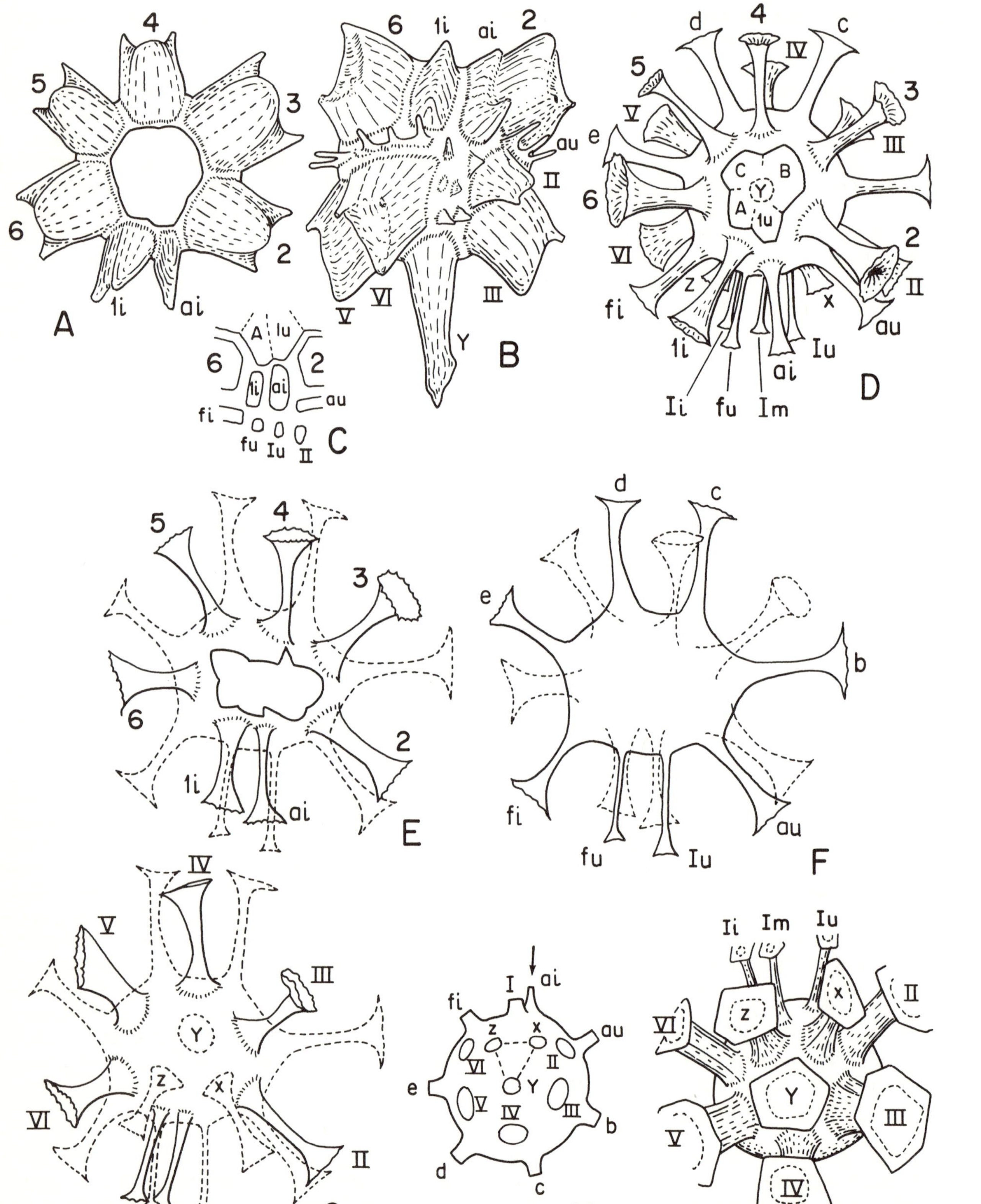
A
4
5
3
6
2
1i
ai
B
6
1i
ai
2
au
II
V
VI
III
Y
C
A
Iu
6
2
1i
ai
au
fi
fu
Iu
II
D
d
4
IV
c
5
3
V
III
e
C
B
Y
A
1u
6
VI
2
II
Z
X
fi
au
1i
ai
Iu
Ii
fu
Im
E
5
4
3
6
2
1i
ai
F
d
c
e
b
fi
au
fu
Iu
G
IV
V
III
Y
VI
Z
X
II
Ii
Im
H
ai
I
fi
au
z
x
VI
II
e
V
Y
III
IV
b
d
c
I
Ii
Im
Iu
VI
Z
X
II
Y
V
III
IV

Figure 11.4

three focus levels at successively lower positions. Consideration of these diagrams (in conjunction with Figure 3.3) will bring out the following points, listed in order of observation from high to low focus for the sequential views.

At high focus (Figure 11.4 E), the angular outline of the Type (4A) archeopyle is not ideally shown owing to slight distortion (compressed in a dorsoventral direction). Therefore, the succession of camerate and planate paraplates of the precingular series typical for the majority of fossil gonyaulacoid cysts (compare Figure 5.9 A) is not completely clear. The precingular and anterior sulcal processes are visible, and the latter (**ai**) makes, in effect, a seventh process in the precingular series but lies, actually, at a slightly lower focus level. Out of focus processes at the level of the paracingulum are shown by dashed outlines. In Figure 11.4 F, the six cingular processes are in focus, as are two very slender processes (**fu** and **lu**) which are in line with the previously identified **1i** and **ai**. The space between **fu** and **lu** corresponds to the flagellar scar. As the focus level is lowered to bring the postcingular processes into sharp focus (Figure 11.4 G), closely spaced **li** and **lm** also come clearly into view on the ventral surface. No flagellar scar separates these two and, in fact, their bases may touch and it is not uncommon to find their shafts partly coalesced (Figure 7.5 N-P). The bases of the three processes closest to the antapex (**X**, **Y**, and **Z**) are still out of focus and are shown by dashed outlines which make apparent their arrangement in an approximate triangle (hereafter referred to as the "**X-Y-Z** triangle"). In the overall apical view (Figure 11.4 D), the cluster of slender sulcal processes stands out as a marker of the ventral surface. Less obvious but also significant is that, while **Y** occupies the approximate center of the archeopyle outline in a directly apical view, **Z** and **ai** (i.e., the processes at opposite ends of the parasulcus) are not in line, suggesting a certain obliquity to the axis of the parasulcus. The reason for this will be explained shortly.

Figure 11.5 A-B show different focus levels of another specimen, this time in ventral view (upper end of specimen truncated by open archeopyle). The diagram at high focus (Figure 11.5 A) shows in plan view the spatial relationship of the sulcal processes just referred to. The ruled lines emphasize the different alignments of the parasulcal axis (s-s′) and the polar axis of the specimen (p-p′). Note also the spacing of **fu** and **lu** as compared to **li** and **lm**, and the positions of **X** and **Z** to the left and right of **Y**. Around the margin, the relative positions of precingular, cingular, postcingular, and antapical processes show up clearly. The cingulars appear particularly slender relative to the precingulars because the smaller diameter of their elliptical shafts is seen in this view. At deeper focus (Figure 11.5 B), the elongate bases of the cingular processes on the dorsal surface provide a useful guide to the approximate equator, above and below which lie the pre- and postcingulars.

H. tubiferum has been reported over a long stratigraphic range (Late Jurassic to Miocene). Some of these reports record misidentifications and, even near its type horizon in the Upper Cretaceous, similar but nevertheless distinct morphotypes occur. Clearly, the morphology varies considerably from sample to sample and locality to locality. Typically, the processes are trumpet-shaped and open at the distal ends, but fringes of ectophragm sometimes surround the flared openings and occasionally a thin and delicate membrane may cross it.

Examples of a *tubiferum*-like species from Montana (Bearpaw Formation) exhibit a cyst form, process number, and process distribution which are all as is typical for *H. tubiferum*, but each process merges distally into a clearly polygonal unit of irregularly perforate ectophragm that also extends without interruption over the process cavity. These ectophragm polygons show the outlines of paraplates just as clearly as would parasutural lines on the

Figure 11.4 - *HYSTRICHOKOLPOMA, HYSTRICHOSPHAERIDIUM,* AND *HOMOTRYBLIUM*

See text for discussion.

A-B - *Hystrichokolpoma* sp. cf. *H. cinctum*. A - Apical view, showing inflated processes surrounding open archeopyle. B - Slightly oblique vv, showing grouped cingular processes, diminutive processes of sulcal area (including **lu**, **X**, and **Z**), and uniquely long **Y**.

C - *Hystrichokolpoma* sp. Ventral detail showing reversal, in some species, of the usual size relationship between lengths of **A** and **1u** (and of **1i** and **ai**), which results in a "false sulcal notch" in the archeopyle margin. Paleogene.

D-G- *Hystrichosphaeridium tubiferum*. D - Apical view; archeopyle open with base of **Y** process visible through it; all other processes in view. E-G - Focus series through slightly distorted specimen (av); focus at levels of precingular, cingular, and postcingular process bases; out-of-focus processes in broken outline. Late Cretaceous.

H - *Homotryblium* sp. Antapical region (aav), showing cingular and hypocyst processes with **X-Y-Z** triangle emphasized; ventral side up; arrow points to position of anterior extremity of parasulcus. Compare with Figure 11.4 I. Eocene.

I - *Hystrichosphaeridium* sp. cf. *H. tubiferum* (aav). Polygonal outlines of process tips show shapes of paraplates characteristic of quinqueform gonyaulacoid design. Late Cretaceous.

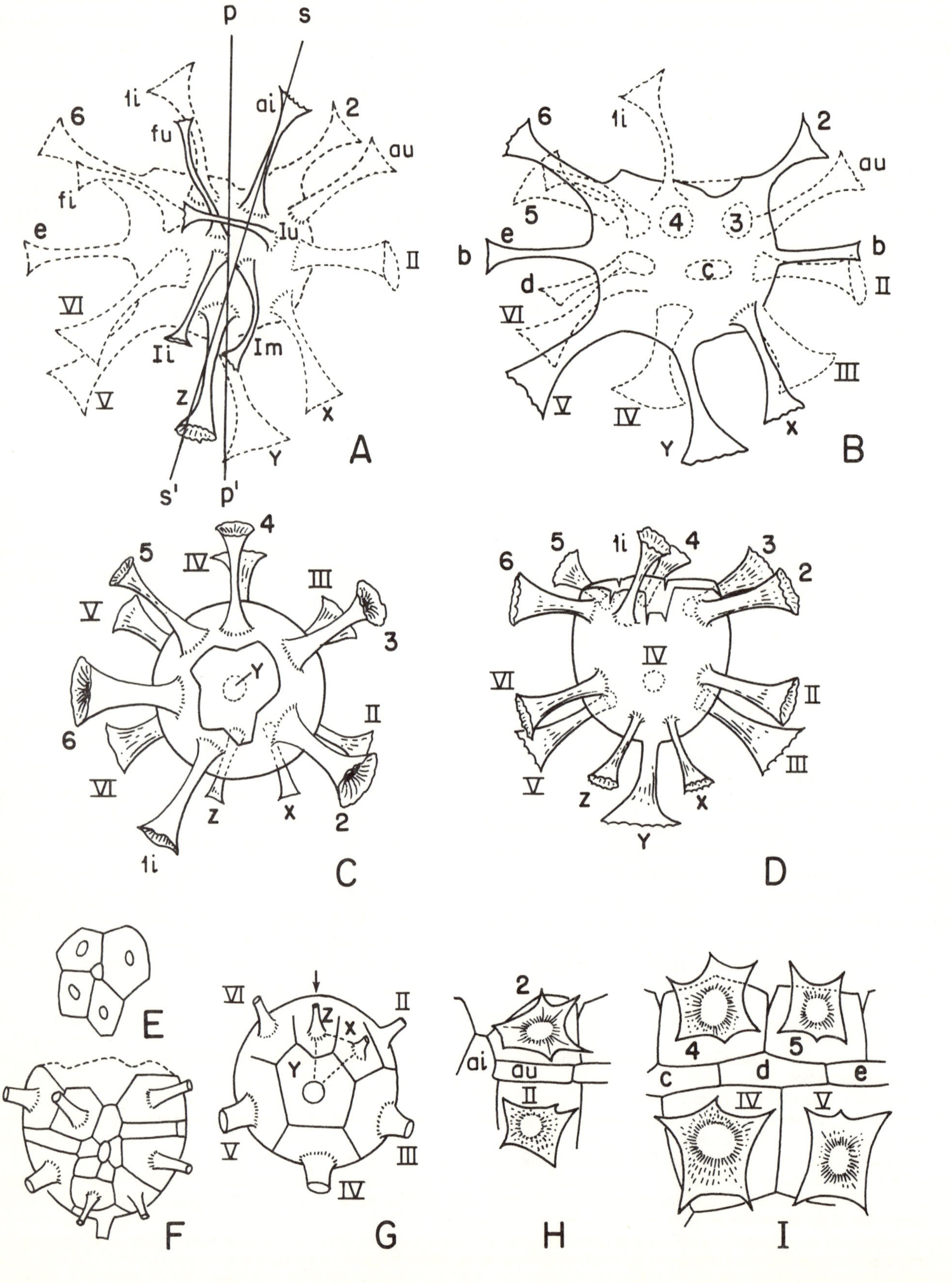
p
s
1i
ai
2
6
fu
au
fi
e
Iu
II
VI
Ii
Im
V
z
x
Y
A
s'
p'
6
1i
2
au
5
4
3
e
b
b
c
d
II
VI
III
V
IV
x
Y
B
4
5
IV
III
V
3
Y
6
II
VI
z
x
2
1i
C
6
5
1i
4
3
2
IV
VI
II
V
z
x
III
Y
D
E
VI
II
z
x
Y
V
III
IV
F
G
2
ai
au
II
H
4
5
c
d
e
IV
V
I

Figure 11.5

surface of the main cyst body, while the open strips between adjacent polygons can be compared with pandasutural areas. As shown in Figure 11.4 I, the antapical surface of such specimens reveals a distinct quinqueform pattern, quite like that in thecae of such a living genus as *Protogonyaulax* (Figure 5.7 D).

This, then, is the explanation for the obliquity (noted earlier) of the **Z-ai** sulcal axis relative to the longitudinal axis of the cyst as a whole: **Z** extends relatively far to the right owing to its interposition between **VI** and **Y**. The symmetrical "**X-Y-Z** triangle" formed by the centers of the processes for these paraplates is also a signature of this relationship, providing the form of the triangle can be accurately assessed in a polar view. Indeed, a similar triangle is obvious if one considers the process distribution in the genus of chorate Gq-cysts, *Homotryblium* (Figure 11.4 H).

Species of *Oligosphaeridium* are immediately distinguishable from *Hystrichosphaeridium* by the relative paucity of processes, especially obvious in a gap around the paracingulum (Frontispiece E, and compare Figure 11.5 C with 11.4 D, and 11.5 D with 11.5 B). Instead of the 30 processes of *H. tubiferum* (25 besides those on the operculum), there are usually only 18 (14 besides those on the operculum). The cingular processes and most of the sulcals are missing, leaving the following: **1u** and **A-C** on the operculum; **1i**, **2-6**, **II-VI**, **X**, **Y**, and **Z** on the main part of the cyst. Cyst size and details of process form, especially at their tips, are highly variable within the genus. Ordinarily there is not the slightest trace of parasutures, except around the archeopyle where accessory archeopyle sutures may partly define the boundaries between precingular paraplates. Nevertheless, the consistent correspondence of precingular processes to these indications of paraplates—and of other processes, in turn, to the precingulars—provide the basis for confidently identifying individual processes in species of this genus. In a few species, distinctive polygonal outlines of process tips may serve to confirm these paraplate identifications. Whatever the criteria, in sufficiently well-preserved specimens, the ventral surface can be told by the combination of the sulcal notch at the archeopyle margin and (usually) two processes—**X** and **Z**, which are also the smallest on the main part of the cyst—side-by-side close to the antapical process (**Y**). In occasional species, other processes (**Z** or **1i**) may also be consistently absent.

Oligosphaeridium abaculum is, so far, unique, not only among species of its genus but among all chorate cysts with widely spaced intratabular processes, in having distinct parasutural lines that define a complete paratabulation (Figures 1.7 F, 11.5 E-G). Many excellent specimens of this species (Davey, 1979) leave no doubt on four important points. First, the processes do, indeed, have a one-to-one relation to paraplates. Second, a "normal" pattern of paraplates exists in areas like the paracingulum and parasulcus where processes are absent. Third, the distribution of processes (both presence and absence) is as had been inferred for other species attributed to *Oligosphaeridium*. Fourth, the parasutural pattern on the hypocyst is typically sexiform in all respects. Thus, the possible inference (based on the known quinqueform character of *Eocladopyxis*, combined with arrangement of tubular processes in *Homotryblium*) that all members of the *Hystrichosphaeridium* group of cysts having intratabular processes are also quinqueform conflicts with direct evidence of the sexiform pattern expressed by parasutures in one of the members of this group, *O. abaculum*.

In other words, at least the hypocyst paratabulation in this species of *Oligosphaeridium* is quite unlike that of *H. tubiferum*. In uncompressed and ideally oriented specimens, the difference can even be

Figure 11.5 - *HYSTRICHOSPHAERIDIUM* AND *OLIGOSPHAERIDIUM*

See text for discussion.

A-B - *Hystrichosphaeridium tubiferum* (vv). Two focus levels of specimen without operculum (out-of-focus processes in broken outline).
A - Focused on sulcal processes. Straight lines show contrasting alignments of axis of parasulcus (s-s′) and polar axis of cyst (p-p′).
B - Focused midway through specimen. Late Cretaceous.

C-D - Process distribution typical for *Oligosphaeridium* shown in specimens (av and vv) without operculum. Absence of cingular and most sulcal processes is clearly apparent.

E-I - *Oligosphaeridium abaculum*. Specimens showing parasutural lines in addition to intratabular processes distributed as normal for the genus. E - Operculum (av), showing central preapical (without process) and bases of processes for four apicals. F - Paratabulation and processes on ventral surface (vv). G - Hypocyst (aav), showing bases of intratabular processes and parasutures delineating sexiform gonyaulacoid pattern; **X-Y-Z** triangle emphasized for comparison with Figure 11.4 H and I; ventral side up, arrow points to anterior end of parasulcus. H-I - Details of selected processes in cysts with unusually short process shafts, to show the correlation between the points around the margin of a process tip and the triple junctions of the paraplate which that process represents. Early Cretaceous; specimens courtesy R. J. Davey.

sensed from the configuration of the "**X-Y-Z** triangle" at the antapex (Figure 11.5 G). Its orientation and symmetry relative to the ventral direction are consistent with the sexiform design. However, this difference is subtle and can be judged with confidence only in polar views of exceptional specimens. Unanswered, of course, is the question of how fully these species are representative of other members of their respective genera!

The examples of *H. tubiferum* and *O. abaculum* establish that both quinqueform and sexiform hypocyst patterns are represented in chorate cysts of closely similar general appearance. The operculum of *O. abaculum* (Figure 11.5 E) shows that the apical organization in this species is fully insert and there is no reason to believe that it is otherwise in *H. tubiferum* (although confirmation in the shapes of the tips on apical processes remains to be shown). Archeopyle and operculum outlines among species of both genera are virtually identical. Relative to living thecate gonyaulacoid species, *H. tubiferum*, with its quinqueform hypocyst and insert **1u**, has the properties of *Protogonyaulax*, whereas the sexiform insert theca indicated by *O. abaculum* would match that of *Gonyaulax*. In view of the differences in tabulation that can be inferred from these chorate fossil cysts, it is interesting to speculate whether toxin production may also have been correlated with tabulation in the fossil species as it is in the living species of of *Protogonyaulax* (toxic) and *Gonyaulax* (nontoxic).

Variant Forms of Hystrichosphaeridium *and* Oligosphaeridium

Two probable variants of *Hystrichosphaeridium tubiferum* are the taxa originally named *H. bulbosum* and *H. tubiferum* subsp. *brevispinum*. They illustrate slight but quite striking differences in morphology that may occur occasionally in other skolochorate cysts. In the first variant, the process bases are unusually inflated; in the second, the process shafts are unusually short. Both conditions have a potential for yielding information on paratabulation independent of that usually available in cysts with tubiform intratabular processes of more "conventional" design.

H. bulbosum was named by Ehrenberg for specimens in the same matrix as the originals of *H. tubiferum* that show to an extreme a tendency subsequently observed many times in more typical examples of *H. tubiferum*. This involves the expansion of the hollow process bases in such a way that the lines of periphragm-endophragm contact enclose areas considerably larger than the process shafts. If these contact lines are displaced sufficiently toward those of neighboring processes, and the process bases are inflated as well, a configuration results that may strongly suggest the processes of *Hystrichokolpoma*. As in that genus, the paratabulation then may be indicated quite directly by the more or less polygonal outlines of the process bases. Such specimens of *H. tubiferum* were the basis for Morgenroth's (1968) identification of *H. bulbosum* and his transfer of that species to *Hystrichokolpoma*. However, other features which Morgenroth illustrated (e.g., open process tips, single paracingular processes, antapical process constituted like the rest, and paratabulation of the ventral and antapical areas) conform to *Hystrichosphaeridium*, not *Hystrichokolpoma*.

The lesson taught by this variant is that the outline shapes of expanded bases of intratabular processes may provide useful information on the shapes of paraplates and the contacts between them.

Specimens originally attributed to *H. tubiferum* subsp. *brevispinum* are like *H. tubiferum* subsp. *tubiferum* except in having drastically shorter processes. The meaning of this condition is unknown, but it seems much more likely to represent an anomaly in cyst formation than to justify special taxonomic status for the specimens which show it. At any rate, occasional specimens of other species of skolochorate cysts exhibit a similar abbreviation of process shafts. For example, in *Oligosphaeridium abaculum* brevispinose individuals occur as rarities among thousands of specimens having "normally" long processes. They can be identified confidently because they show the same parasutural markings that are such an unusual feature of this species, although they have been taken to represent a distinct species (Jux, 1980).

One effect of the short process shafts in brevispinose individuals is to hold the process tips in more rigid relation to their bases than when the extremities are separated by long and flexible shafts. In the case of brevispinose *O. abaculum*, this makes obvious a further feature of the process tips that is difficult to observe on specimens with long-shafted processes (Figure 11.5 H-I). Here we see that the attenuations of the process tips merit much closer attention than such features have generally been given. They are neither irregular in number nor random in orientation. Instead, the attenuations on each tip are directed approximately toward, and consistently match in number, the triple junctions along the parasutures that collectively enclose the paraplate from which the process rises. Consequently, in this species at least, identical details of the thecal tabulation (including, for example, the number of cingular paraplates, the positions of the sutures between them, and the sexiform aspect of the hypothecal tabulation) can be inferred from inspection of the points on the process tips and from the pattern of parasutures between process bases. Process tips in many other species of *Oligosphaeridium* are not so simple or precisely outlined as in *O. abaculum*. However, this species demonstrates that, under favorable circumstances, tip outlines have a hitherto unsuspected potential as clues to details of paratabulation and inferred tabulation.

Other Peculiarities of Hystrichokolpoma

Like the type species most species of the *Hystrichokolpoma* (Figure 11.4 B), are marked by conspicuously inflated processes, although a few (e.g., *H. rigaudiae*) are not. One interesting interspecific variable within the genus is the relative size of the two paraplates—**1i** and **ai**—that stand next to each other on the ventral margin of the apical archeopyle. As is customary in cysts with gonyaulacoid paratabulation, both these paraplates are smaller than the adjacent precingulars, **6** and **2**, but in *Hystrichokolpoma* **1i** is frequently smaller than **ai** —the reverse of the usual situation. Consequently, the deepest indentation in the margin of the archeopyle may not be, strictly speaking, the sulcal notch, where **1u** contacts **ai**, but to the right of the parasulcus, along **A/1i** (Figure 11.4 C). If, through ignorance of this relationship, the notch is assumed always to mark **1i/ai**, one of two misinterpretations is likely to follow. First, the next to smallest paraplate in the series along the archeopyle margin will be misidentified as a severely reduced **2**, with the consequence that the entire sequence of precingulars will be offset by one and "anomalous" alignments will appear to exist between the major pre- and postcingular processes. Second, the direction from which the archeopyle is being viewed—either in the form of the opening itself or the outline of its operculum—may be misconstrued, an exterior view being taken for an interior one.

Another peculiarity of many species of *Hystrichokolpoma* concerns the hollow processes themselves. These structures, when strongly inflated, take a variety of forms but generally appear quite different from the open tubiform processes of *Hystrichosphaeridium*. Normally, they would be interpreted as closed at the tip, but, actually, in most species there are one or more openings through the distal wall into the process cavity. The openings have circular to irregular outlines and may have noticeably thickened margins. They vary in size and may be flush with the distal surface or located at the tips of small elevations or discrete projections.

Two Contrasting Process Arrangements in Gonyaulacoid Hypocysts

Despite differences that may exist in the antapical area, the paratabulation of *Oligosphaeridium* and *Hystrichosphaeridium* as expressed by the processes in the pre- and postcingular series is essentially the same. This is also the arrangement most commonly encountered among other examples of quinqueform and sexiform gonyaulacoid thecae and cysts, and is illustrated in Figure 5.7. In these cases, each of the two series has three dorsal elements (**3**, **4**, and **5**; **III**, **IV**, and **V**), and the principal ventral elements are **2** and **II** on the left and **6**, **1i**, and **VI** on the right. About midventrally, **ai** lies between **1i** and **2** and the three derivatives of **I** span the distance between **II** and **VI** (but not quite at one level). This arrangement is shown schematically in Figure 11.6 A in a manner that emphasizes the relative positions of intratabular processes (such as those in *Oligosphaeridium* and *Hystrichosphaeridium*) as seen in apical view. Note the approximate alignment of **3**, **4**, and **5** with **III**, **IV**, and **V**, the middle element in each trio being about middorsal.

This arrangement holds for a large number of chorate cysts with intratabular processes, but it is not universal. We have seen among some Gp- and Gs-cysts that unusual proportions of certain paraplates may have the effect of "shifting" others out of their customary position (e.g., bringing **III/IV** and **IV/V** into middorsal positions in *Phanerodinium* and *Cribroperidinium*, respectively). A similar situation is expressed with special clarity in *Hystrichokolpoma*. Although it also occurs in some other taxa of chorate cysts, its frequency is not easily judged because the factors involved have received so little attention in descriptions.

This second arrangement is shown schematically in Figure 11.6 B. The contrast depends chiefly upon the relative positions of the postcingular paraplates, and this is shown in an alternative manner in Figure 11.6 C-E. In *Hystrichokolpoma* (Figure 11.6 B, E), the relatively small size of **II** leaves only four large postcingular processes, and these are positioned in marked contrast to the positions of corresponding

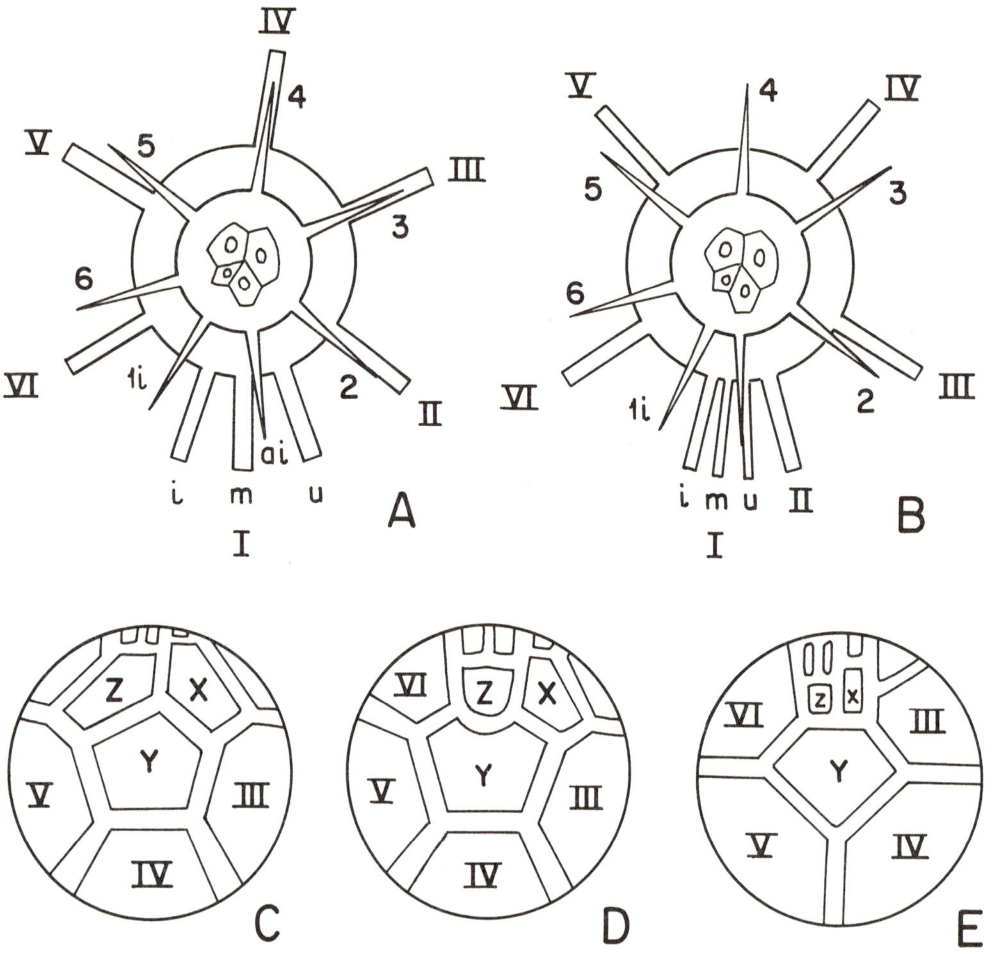

Figure 11.6 - CONTRASTING HYPOCYST PATTERNS

See text for discussion.

A-B - Schematic diagrams (av) of chorate cysts with intratabular processes and apical archeopyle. Pointed processes extending from inner circle represent precingulars and **ai**. Parallel-sided processes from outer circle represent postcingulars and sulcal derivatives of **I**. Figure A shows processes aligned as in *Hystrichosphaeridium* and *Oligosphaeridium*; B shows alignments as in *Hystrichokolpoma*.

C-D - Two hypocyst patterns of paraplates (aav) in which the distribution of postcingular processes shown in A is combined, in C, with a quinqueform arrangement at the antapex (as in *Homotryblium* and *Hystrichosphaeridium* sp. cf. *H. tubiferum*) and, in D, with a sexiform arrangement (as in *Oligosphaeridium abaculum*).

E - Hypocyst pattern of paraplates (aav) in *Hystrichokolpoma*, with the postcingular distribution of processes shown in B.

processes in *Hystrichosphaeridium* and *Oligosphaeridium* (Figure 11.6 A, C-D). Visual key to the difference is the position of middorsal **4** relative to the postcingulars. In *Hystrichokolpoma*, process **4** anterior to the paracingulum is aligned with the gap between **IV** and **V** posterior to it. In addition, **2** and **3** are far from being aligned with **II** and **III**, respectively. The concentration of small processes toward the ventral midline and the reduction in size of **X** and **Z** relative to their state in, say, *Hystrichosphaeridium*, are associated differences that are also obvious in *Hystrichokolpoma*.

It should be noted that, in some cases, this "variation on the theme" of sexiform patterns results in the antapical paraplate (e.g., as represented by the antapical process in *Hystrichokolpoma*) being effectively five sided, but this is not to say that there is at all close resemblance to the quinqueform condition, as the diagrams in Figure 11.6 C-E should make clear.

These characteristics define a style of process arrangement which will probably be observed widely once it is sought. It occurs in taxa having processes constructed as differently as those of *Hystrichokolpoma* (*Hystrichosphaeridium* complex) and *Turbiosphaera* (*Cordosphaeridium* complex). The possible affinities of these taxa to Gs-cysts of *Cribroperidinium* type, whose hypocyst paraplates

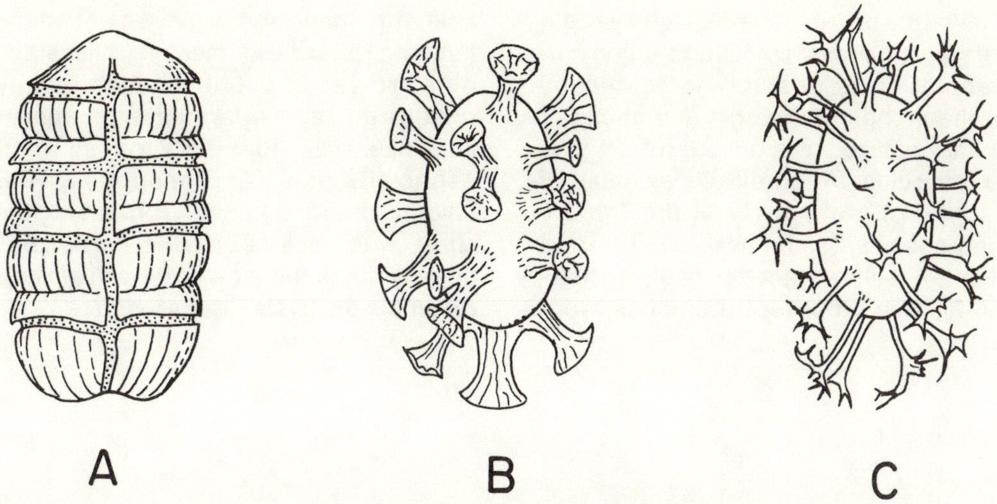

Figure 11.7 - Gx-CYSTS *POLYKRIKOS* AND *DISTATODINIUM*

A-B - Modern *Polykrikos kofoidii*. A - Motile pseudocolony (vv) consisting of four cells, each with its flagellar furrows. After Dodge (1982). B - Cyst from Pacific coastal plankton.

C - Paleogene *Distatodinium* sp.

have somewhat similar properties, remains to be explored.

Cysts of Living Polykrikos *and Possible Fossil Relatives*

The living athecate dinoflagellate, *Polykrikos*, which includes non-photosynthetic predatory species that feed on other dinoflagellates, has been referred to the Gymnodiniales, but it is really unique and its closest affinities are unknown (Morey-Gaines and Ruse, 1980). Initially, the motile cell has the form normally associated with the gymnodinioid genus, *Gyrodinium*, but, by a process of vegetative division without complete separation of daughter cells, this soon elaborates into a pseudocolonial entity which is multinucleate and has multiple flagellar furrows and flagella (Figure 11.7 A). Although the genus has been recognized for over a century, only relatively recently was it established through culturing (Dale, 1976) that certain morphologically distinctive cysts which had been known almost equally long actually represent the same genus. For references, a brief review, and a discussion of cysts of one living species, see Harland (1981).

The sporopollenin cysts of *Polykrikos* are ellipsoidal (L:W ranging to well over 1.5) and relatively thick-walled. In *P. schwartzii*, the wall is externally reticulate, with or without an enclosing ectophragm spanning the crests of the muri (Harland, 1981), while in *P. kofoidii*, there are numerous discrete processes in place of a reticulum (Figure 11.7 B). To a degree that varies greatly from specimen to specimen, the meshes of the reticulum, as well as the processes, tend to be aligned in rows about parallel to the equator; the processes may also show a rough longitudinal alignment. The form and quasi-regular distribution of processes in *P. kofoidii* may even suggest its dinoflagellate affinity by a general similarity to the intratabular processes of such familiar fossil cysts as *Hystrichosphaeridium* and *Oligosphaeridium*. However, variability (in number of processes and number of rows of processes or meshes) is too great for a regular or consistent paratabulation pattern to be inferred from the surface relief; indeed, it is difficult even to decide if paracingula or parasulci are indicated by the alignment of features on the cyst wall. A typical archeopyle is not developed, excystment being through an opening that apparently forms where the wall is especially thin at one end of the cyst. The resultant opening is non-operculate and lacks the sharp definition of a sutured margin, although an irregularity in its outline may give the impression of a sulcal notch in occasional specimens.

Taken individually, the features exhibited by modern cysts of *Polykrikos* are not highly distinctive, but their combination, especially in cysts of definitely ellipsoidal form, is unusual and suggests the possibility of recognizing fossil analogs. Two that have been suggested (Harland, 1981) are the Jurassic genera, *Ellipsoidictyum* and *Valensiella*. These both include slightly ellipsoidal cysts with a rather irregular external reticulum. Although the details of their morphology have not been studied carefully, both have operculate apical archeopyles, and a gonyaulacoid paratabulation is possibly indicated by angulations in the archeopyle margin and traces of paraplates partially outlined by the more robust muri in the surface recticulum. A more likely example of the *Polykrikos* type of cyst may lie in the Tertiary process-bearing species which have been referred to *Distatodinium*. These are rather strongly ellipsoidal (L:W = 1.5-2); their processes are notably variable in number and arrangement within a species, but tend toward alignment in transverse and longitudinal rows; and one end of the cyst has an unsutured, non-operculate opening with rather indefinite limits, apparently formed where the wall is especially thin. (Another fossil, which is quite enigmatic and is known from a single specimen in chert, was once referred to *Polykrikos* and subsequently made the type of *Wetzelodinium*; it does not resemble the cysts discussed here.)

Chapter 12. Morphology of Other Cyst Categories

The cysts to be treated in this chapter comprise five very distinct groups, contrasting in their morphology and stratigraphic distribution, and ranging from exceedingly poorly known to quite exhaustively studied.

CHARACTERISTICS OF CATEGORIES AND COMPLEXES

Category of M-Cysts

Representative Taxa: *Caligodinium*, *Cryptarchaeodinium*, *Tuberculodinium*; living *Diplopsalis*, *Zygabikodinium*.

This is a catchall category and, accordingly, comprises taxa that may have little in common and should be considered individually in regard to both morphology and stratigraphic occurrence. A cyst included here gives sufficient evidence of paratabulation to show that its particular pattern, however incompletely it may be expressed, is distinct from any that would place it in one of the other categories. Actual traces of paratabulation may be limited to small areas (even to the archeopyle suture(s) only), or the pattern may be essentially complete. Biological affinity may be quite definitely known or entirely uncertain. Misinterpretation of paratabulation may account for the presence of several genera in this category; however, their referral to another category cannot be justified on the basis of available information (e.g., *Cryptarchaeodinium* is probably a Gs-cyst).

Caligodinium - an archeopyle with three to five opercular pieces of uncertain paraplate equivalence. Polars **A-C** are a probability (or possibly antipolars **X-Z**) but supporting evidence is inconclusive.

Cryptarchaeodinium - Parasutural ridges and general configuration suggest the *Ctenidodinium* or *Leptodinium* complexes of Gs-cysts, but Type E archeopyle is uncertain and reported paratabulation (including seven postcingulars and three paraplates at the antapex) is anomalous.

Tuberculodinium - An antapical archeopyle with two or three opercular pieces (Figure 6.10 W-X) accords with the unusual tabulation of living *Pyrophacus* (Figure 1.6 P) and cysts of this type have been cultured to produce motile cells of *Pyrophacus*.

Categories of R-Cysts

Introduction

It has been important throughout Chapters 8-12 to remember that we have been dealing with morphological categories and complexes which do not necessarily represent biological relationships. The distinction is especially important in the case of R-cysts. The diversity that is assembled here almost certainly represents several major biological groupings, some perhaps restricted to early Mesozoic strata, others possibly holding clues to the basic interrelationships of such modern biological entities as the gonyaulacoids and peridinioids.

Clearly, the paratabulation patterns of R-cysts combine certain distinctive properties of their own with others that are also represented among P-cysts and G-cysts. However, too many of the patterns are incompletely known to permit drawing boundaries between the groups with great confidence. Consequently, their limits are defined somewhat arbitrarily, although guided by a practical rationale. As new observations fill in our knowledge of paratabulation patterns and provide sounder grounds for evaluating similarities and differences, it will surely become advisable to shift some of these taxa from the places now allocated to them in this morphological scheme.

Two categories of R-cysts are recognized here—Rp-cysts and Rr-cysts. They share with P-cysts and with some Gp-cysts the presence of one or more relatively large anterior intercalaries. It will be easier to specify the limits among these groups if we first restate the relevant characteristics of P-cysts and Gp-cysts.

P-cysts have a paratabulation pattern that is equivalent (except in the matter of cingulars and sulcals, which are not considered) to one of the thecal patterns typical for living *Protoperidinium* subgenus *Protoperidinium*, *Protoperidinium* subgenus *Archaeperidinium*, or *Peridinium*. That means they have a paratabulation that can be represented by the Kofoidian formula: *0-2pr, 4′, 2-3a, 7″, 5‴, 0p, 2⁗*. It also happens that most fossil P-cysts have a paratabulation like the symmetrical varieties of this arrangement that occur in *Protoperidinium* subgenus *Protoperidinium* and *Peridinium* today; that is, with three intercalaries in a hexa or quadra dorsal arrangement. In other words, *2a* is normally middorsal and linteloid, separated from both *3′* and *4″* by rectilinear, rather than geniculate, boundaries.

Gp-cysts are distinguished by having the partiform version of gonyaulacoid hypocyst pattern, whose distinctive feature vis-a-vis the other cysts immediately concerning us is the presence of six postcingulars in combination with two principal paraplates (**X** and **Y**) at the antapex. Anterior intercalaries may occur in varied number and arrangement.

Now, the distinction of Rp- and Rr-cysts from Gp- and P-cysts can be phrased as follows: Although relatively large anterior intercalaries are present in all four groups, the epicyst pattern in both categories of R-cysts fails to meet the specifications for P-cysts. In addition, the hypocyst pattern in Rp-cysts is like that of P-cysts (*5‴, 0p, 2⁗*) and is not like that of Gp-cysts with its six postcingulars. On the other hand, the hypocyst pattern in Rr-cysts (to whatever extent it can be determined) contrasts with that of both P-cysts and Gp-cysts. Until these distinguishing properties become familiar, perhaps the following key will be helpful:

Key to Cysts with Large Anterior Intercalaries

I. Hypocyst pattern determinable
   - A. Hypocyst paratabulation: *5‴, 0p, 2⁗*
     1. Epicyst as specified for P-cysts....... P-cyst
     2. Epicyst not as in 1 .... Rp-cyst
   - B. Hypocyst paratabulation not *5‴, 0p, 2⁗*
     1. Hypocyst partiform .... Gp-cyst
     2. Hypocyst not partiform .... Rr-cyst

II. Hypocyst pattern not determinable
   - A. Epicyst as specified for P-cysts............ P-cyst
   - B. Epicyst not as in A .... Rr-cyst

Examples of R-cysts that have been known for many years from Late Jurassic and Early Cretaceous strata include the genera *Broomea*, *Pareodinia* (Wiggins, 1975), *Gochteodinia*, and a few others. These are not, however, either the most interesting or the most informative examples, for their paratabulation patterns are relatively obscure. Of much greater potential significance are substantially older forms whose paratabulation is displayed with relative clarity over half or more of the cyst (e.g., *Dapcodinium*, *Rhaetogonyaulax*). Until relatively recently, such examples have been so few in number, so different from one another and from younger forms with better known patterns, and so infrequently encountered in an excellent state of preservation that they have been more attractive as topics for isolated special studies (Evitt, 1961; Sarjeant, 1962; Wiggins, 1972; Harland et al., 1975; Fisher and Van Helden, 1979) than as the basis for defining the scope of one or more major taxonomic groups. Fortunately, new vistas are being opened onto this relatively unknown terrain as assemblages of well-preserved cysts from the Late Triassic and Early Jurassic come under study (Wille and Gocht, 1979; Bjaerke, 1980; Wille, 1982). Early results indicate great diversity and novelty, but, as already noted, it will be some time before a stable context is developed in which these discoveries can be viewed in satisfactory perspective. Therefore, the approach in this section will be rather different from that followed in earlier sections. Here we will not attempt to define complexes in number appropriate to the known diversity. Instead, selected examples from both categories will be discussed under general headings. In addition, the *Pareodinia* complex of Rr-cysts, including a rather close morphological grouping of chiefly Late Jurassic-Early Cretaceous cysts, will be defined.

In many R-cysts, one or more of the anterior intercalaries is involved in forming the archeopyle. Conveniently, this highlights aspects of the epicyst paratabulation that would be much less easily observed if one had to rely on parasutural features unrelated to the archeopyle. Normally, the anterior intercalaries (especially the one or two that are situated middorsally) contrast in shape with those most frequently encountered in P-cysts; they are commonly also six-sided, but it is their lateral edges rather than the adcingular and adpolar edges that tend to be parallel. This difference reflects their position, alternating between precingulars and apicals, and consequently the geniculate (in contrast to rectilinear) contacts both toward the cingulum and toward the pole. Typical as this feature is, variability at intraspecific and higher levels assures that no such simple criterion will apply without exception throughout a group as diverse as R-cysts. However, it occurs frequently enough to be highly useful. This is the character that rules out a referral to P-cysts even if other traces of paratabulation are lacking.

Category of Rr-Cysts

The general characteristics of the category and criteria for distinguishing it from Rp-cysts and others with certain similar properties have been set out in the preceding introductory discussion. The paratabulation of some Rr-cysts (*Pareodinia* complex) is indicated only to the extent that a dorsal intercalary archeopyle gives information about the immediately adjacent paraplates. In other cases (e.g., *Dapcodinium*), a complete paratabulation is indicated by parasutural features. Intermediate between these extremes are examples in which a relatively large and complex intercalary archeopyle, supplemented by some parasutural markings, illuminates something less than a complete pattern. The variety encompassed in this category suggests the practical need for several morphological groups. However, it seems premature to try to establish them now on the incomplete evidence currently available. Accordingly, only the *Pareodinia* complex is defined, and other genera are placed under a miscellaneous heading.

***Pareodinia* Complex**

Taxa: *Broomea*, *Cantulodinium*, *Crussolia*, *Gochteodinia*, *Kalyptea*, *Pareodinia*.

Occurrence: Middle Jurassic to Early Cretaceous (Aptian).

Various aspects of some of the genera assembled here were reviewed by Wiggins (1975). The common features are a conspicuous archeopyle in the intercalary area and a strong tendency toward a basically pyriform shape. Other details vary. The archeopyle involves one to three paraplates and may be the only indication of paratabulation. Its shape, even when corresponding to only a single paraplate, implies the presence of a minimum of three intercalaries. The paracingulum, when indicated at all, may lie at only about a third the length from the posterior extremity.

In *Broomea* (Late Jurassic), what is taken to be the median intercalary forms the Type I archeopyle. Traces of parasutural lines are visible in the original material, whose quality suggests that the entire paratabulation could be analysed if a sufficient number of specimens were available. A unique feature of the type species is the strange construction of the antapical horns, which appear not to contain extensions of the autocoel and which break into elongate "splinters."

Cantulodinium (Early Cretaceous) is apparently a rare form, outstanding for the presence of five hornlike extensions from the hypocyst; the archeopyle is Type 2I. *Gochteodinia* is a relatively long-ranging (Tithonian-Valanginian) type whose strongly developed surface ornament distinguishes it in a complex where smooth or nearly smooth exteriors are the norm.

Pareodinia is perhaps the least elaborate form in the complex, without paratabulation except for the archeopyle, which occurs as both Types 2I and 3I in a generally pyriform cyst with little surface relief other than the apical horn. *Hebecysta* is similar. *Kalyptea* is also similar but departs from the pyriform shape by having an antapical horn as well as an apical one. *Crussolia* is cavate, but the existence and extent of cavation in cysts attributed to some of the other genera have been debated.

Miscellaneous Genera of Rr-Cysts

Taxa: *Arpylorus* (?), *Comparodinium*, *Dapcodinium*, *Hebecysta*, *Mancodinium*, *Maturodinium*, *Moesiodinium*, *Parvocysta*, *Phallocysta* (?), *Rhaetogonyaulax*, *Sverdrupiella*, *Valvaeodinium*.

Occurrence: Silurian?; Late Triassic to Middle Jurassic.

Suggestions of both peridinioid and gonyaulacoid properties can be recognized in the paratabulations of these genera, along with distinctive properties of their own. Large anterior intercalaries are typical, but vary in number; other aspects of morphology are also diverse. For discussions, see Evitt (1961, and later in this chapter) for *Dapcodinium*, Morgenroth (1970) for *Mancodinium* and *Maturodinium*, Harland et al. (1975) and Fisher and Van Helden (1979) for *Rhaetogonyaulax*, and Wille and Gocht (1979) for *Comparodinium*. A major study including many new forms representing this group is in progress by Wille (pers. comm.).

Hebecysta and *Sverdrupiella* (Bujak and Fisher, 1976) and *Valvaeodinium* (Morgenroth, 1970) are very imperfectly known. Although the archeopyles in the first two appear to be Type I, 2I, or 3I, the tendency toward biconical shape sets them apart from most other members of this complex and suggests *Rhaetogonyaulax*. *Sverdrupiella* is cavate and *Valvaeodinium* has a definite Type 2I archeopyle. All need more study before a useful analysis of their morphology is possible.

Phallocysta and *Parvocysta* show traces of parasutures and may be close to the Rp-cyst, *Susadinium*. Dörhöfer and Davies (1979) considered the archeopyle structure in *Phallocysta* to be significant for their concept of the evolution of tabulation patterns.

Moesiodinium, with a Type I archeopyle, has a cavate construction and gives an impression that can be shockingly suggestive of a *Deflandrea*-like dinoflagellate for so low a stratigraphic position. It certainly would not be surprising if it should turn out to have a hypocyst pattern that would place it among the Rp-cysts.

The paratabulation of Silurian *Arpylorus*, the only known pre-Mesozoic microfossil whose general affinities with dinoflagellates have been widely accepted, was interpreted by Sarjeant (1978) in a way that places it reasonably in this category, although some of its details are either somewhat uncertain or atypical of this group. The large intercalaries with the middorsal one functioning as the archeopyle, the short precingulars, and the relatively low position of the paracingulum are all features encountered among R-cysts. However, the detailed organization of *Arpylorus* will not be established finally until better-preserved material becomes available and can be studied with the SEM. Renewed questions have been expressed recently about its dinoflagellate affinities (Bujak and Davies, 1983).

Suggested tabulation formulae and archeopyle types are given in Table 12.1 and a few examples are diagrammed in Figure 12.1. Almost certainly, several lines of "experimentation" in cyst design are represented here. One of them, whose implications might be especially interesting if we could read them clearly, is represented by *Dapcodinium* (Figure 12.1 D-G, which is the topic of a short discussion at the end of this chapter.

Table 12.1 - PARATABULATION AND ARCHEOPYLES IN SOME R-CYSTS

Genus	Paratabulation Formula	Archeopyle Type
Comparodinium	*5′, 5-6a, X″, Xc, X‴, 5p, 1⁗*	2A
Dapcodinium	*4′, 4a, 7″, 6c, 7‴, 2p, 1⁗*	uncertain
Mancodinium	*4′, 4a, 7″*	(4A4I)7P
Maturodinium	*4′, ?4a, 7″, 7c, 6‴, 1p, 1⁗*	(4A)a4I
Rhaetogonyaulax	*6′, 5a, 7″*	(tAtI) or A5I
Valvaeodinium	*4′, 2a, 5″, Xc, 5‴, 1⁗*	2I or (tA)2I?
Arpylorus	?	I
Heibergella	?	I, 2I, or 3I
Moesiodinium	?	I
Phallocysta	?	3I
Sverdrupiella	?	I, 2I, or 3I

The paratabulation patterns of both *Dapcodinium* and *Rhaetogonyaulax* have figured prominently in the scheme of tabulation analysis and hypothesized evolution proposed by Eaton (1980) and discussed briefly in Chapter 13. Although a detailed paratabulation for both epicyst and hypocyst of *Rhaetogonyaulax* has been described (Harland, et al., 1975), the almost universally poor to mediocre preservation of specimens that have been illustrated to date leaves some uncertainty about the pattern attributed to the hypocyst.

Category of Rp-Cysts

Taxa: *Angustidinium*, *Susadinium*.

Occurrence: Early Jurassic to mid-Cretaceous.

This category is quite narrowly defined to include cysts whose hypocyst paratabulation is definitely like that of P-cysts (i.e., *5‴,0p, 2⁗*), but whose epicyst paratabulation is distinct from P-cysts and, therefore, unlike that seen today in most familiar *Protoperidinium* and *Peridinium*. In view of this identity in hypocyst patterns, one might argue that it would be better to place Rp-cysts as a separate category of P-cysts. That option was rejected for three reasons. Firstly, the few epicyst patterns among these cysts that are as yet fully known do not occur today in *Protoperidinium* or *Peridinium*. Secondly, many of these fossils appear to be separated stratigraphically from the abundant and diversified P-cysts in mid-Cretaceous and younger strata. Thirdly, associated with them in the same assemblages are other cysts having similar outlines and wall structures but lacking (or having only ambiguously expressed) hypocyst paratabulation. Retaining them all as R-cysts, at least until their similarities and differences (among themselves and relative to the typical Cretaceous and later P-cysts) are much better known, seemed to have a practical advantage over other approaches. However, it must not be forgotten that these are morphological

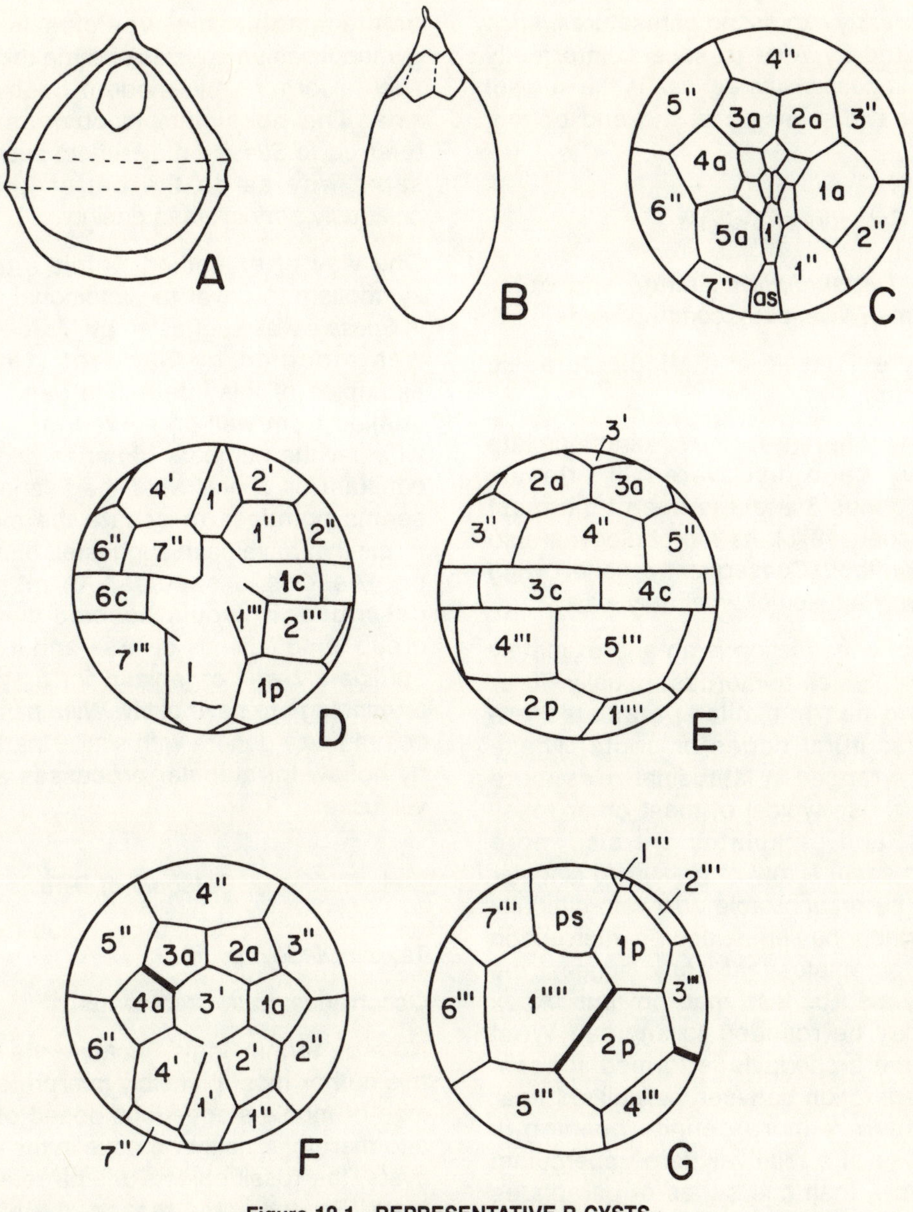

Figure 12.1 - REPRESENTATIVE R-CYSTS

A - *Moesiodinium* (Middle Jurassic), a circumcavate cyst with middorsal six-sided archeopyle (dv).

B - *Pareodinia* (Middle Jurassic-Early Cretaceous), an autocyst with Type 3I archeopyle (dv).

C - Schematic epicyst paratabulation (av) in *Rhaetogonyaulax* (Late Triassic). Paratabulation as interpreted by Fisher and Van Helden (1979); two preapicals and apicals *2'-6'* unlabeled.

D-G - Schematic paratabulation of *Dapcodinium priscum* (Late Triassic-Early Jurassic) in vv, dv, av, and aav. Removal of two heavy parasutures in G would produce a hypocyst pattern topologically matching the "standard" sexiform gonyaulacoid model. The epicyst pattern matches the tabulation in living freshwater *Glenodinium polonicum*. Removal of the heavy parasuture in F would make it conform also to modern freshwater *Peridiniella sphaeroidea* (See discussion at end of this chapter).

assignments, not intended either to restrain or to direct ideas about the possible biological relationships of the fossils to peridinioid dinoflagellates or to any others.

This category is exemplified by *Angustidinium acribes*, a mid-Cretaceous species whose complete paratabulation is expressed by fine parasutural ridges (Goodman and Evitt, 1981), and by Middle Jurassic fossils having the same or similar paratabulation patterns combined with a variety of cyst shapes and wall structure (Bjaerke, 1980; Wille, 1982). These older representatives include *Susadinium* (Dörhöfer and Davies, 1980 = *Facetodinium* Bjaerke, 1980) whose paratabulation is expressed by intratabular convexities, as well as species having clearly expressed parasutures, such as those allocated by Wille to several informally named genera. *Angustidinium acribes* is the subject of a short topical discussion at the end of this chapter.

Category of S-Cysts

Taxa: *Suessia*, *Dapsilidinium*? *langii* and other undescribed forms (Wille, pers. comm.).

Occurrence: Late Triassic and Middle Jurassic (Wille, pers. comm.).

Among the many different forms of dinoflagellate cysts that have been discovered in Triassic sediments, the genus *Suessia* is one of the most remarkable (Morbey, 1975). Its morphology is also very poorly understood. Consequently the following comments are only a token of what may exist.

The basic plan of construction is an approximately spherical cyst on which numerous (usually 60 or more) small and nearly similar paraplates are defined by parasutural ridges or septa. These paraplates are arranged in latitudinal rows more numerous than those typical of most other fossil dinoflagellates, and paraplates are also more numerous within each. A row of presumed cingular paraplates may be recognizable only with difficulty from similar neighboring series, and an interruption in the bands of paraplates that would suggest the parasulcus may also be less than obvious. Apex and antapex may be rounded to angular. What appears to be the archeopyle is formed in some species by a separation between paraplates near the apex; in others a more definite opening is formed by release of a relatively large operculum that includes more than one series of paraplates surrounding the apex.

The wall appears to be a single layer, and the observed surface ornament ranges from low ridges to low septa that are smooth or have denticulate crests. The variability of characters has not been adequately determined. Although species with different numbers of paraplates and paraplate series, as well as different types of archeopyle, have been observed, only one generic name has been used and one species described.

Cysts of *Suessia* type have been recovered from Late Triassic strata (Carnian-Rhaetian) in widely separated areas of central and northwestern Europe and northern Canada. No cysts of living species show similar features. However, the multiplicity of small polygonal paraplates is reminiscent of the amphiesmal vesicles in some modern gymnodinialean species and the thin thecal plates in such a form as the modern freshwater *Woloszynskia*. The possibility is obvious that the cysts referred to *Suessia* may reflect such patterns in the same way as do the better known fossils of obviously peridinialean design.

One way of extending Kofoid's terminology and symbolism to cover the additional paraplate series in *Suessia* was suggested by Morbey (1975) and this was modified by Sarjeant (1982a). Too few examples of this interesting genus have yet been studied from well-preserved material to establish whether the numbers of series and paraplates are constant or, if not, what their variability may be. It seems pointless to debate the merits of special terms that have been suggested on the basis of only one described species. As more species are described, it should become clear whether the group is sufficiently diverse and its paratabulation sufficiently well organized for a special paraplate terminology to have utility. Wille has reported (pers. comm.) new genera with similar patterns expressed by hollow intratabular processes and intratabular verrucae.

Category of D-Tests

Taxon: *Dinogymnium*.

Occurrence: Late Cretaceous.

Note on terminology. Because the fossils seem to this author most plausibly interpreted as the coverings of motile stages, composed of sporopollenin, and therefore, neither thecae in the usual sense nor cysts, the fossil objects will be referred to here as tests. For the same reason, the para-terminology applied to features on the cyst is not used. However, with admitted inconsistency but in the interest of simplicity, the opening at the apex will be called the "archeopyle", using quotation marks to indicate that this may be an extension of the customary connotation.

See Frontispiece F and Figure 5.3 G. The test is essentially biconical, circular in cross-section and constructed of a single wall layer, usually deeply indented by a cingulum at or posterior to midlength and by a sulcus which extends from the cingulum posteriorly only (or also somewhat anteriorly as a lesser depression). At the apex an "archeopyle" is formed by the opening of two arcuate slits, one to the right and one left of a lozenge-shaped operculum. Other openings in the wall include what are taken to be flagellar pores at the intersection of cingulum and sulcus, and minute trichocyst pores distributed in large numbers either evenly or selectively over the test. Epitest and hypotest may be equally lacking in major relief. More often, the wall is folded into similar sets of more or less meridional folds or ribs anterior and posterior to the cingulum; or one portion may be smooth and the other ribbed. The external surface is smooth to granulate.

Features useful in discrimination of species are:

1. Overall shape, including rounding of apices, position and indentation of furrows. A useful indication of cingulum position and relative sizes of epi- and hypotest is the **cingulum index**, which is the decimal quotient of the distance from apex to midwidth of cingulum divided by the total length of the polar axis.

2. Magnitude, distribution, and extent of ribs. Patterns with ribs of alternating length and with ribs convergent along the ventral meridian occur, as well as meridional ribs that reach from cingulum to pole.

3. Distribution of trichocyst pores. These may be uniformly distributed or show concentration in or restriction to certain areas, especially at summits of surface relief and at the apical and antapical extremities.

4. Nature and distribution of surface granulation.

5. Size and shape of "archeopyle," its relationship to immediately adjacent features, and details of the "archeopyle" sutures, which may be confluent or separated dorsally and ventrally so that the operculum is normally adnate at one or both points.

Establishing the limits of inter- and intraspecific variability is the greatest problem in species identification. What appear to be monospecific associations suggest a wide range in size (up to a linear factor of three) and some aspects of outline shape within the same species. Cingulum index, distribution of trichocyst pores and surface granulation, and general pattern of ribbing appear to be less variable.

Although specimens have been observed in post-Cretaceous sediments, they seem most likely to be reworked, and no examples have been described from the Albian or older strata. Therefore, occurrence of these fossils, easily recognized as to genus, are excellent and quick indicators of Late Cretaceous age.

Although many species have been described, few can be considered to be circumscribed precisely enough to permit recognition with confidence and detailed stratigraphic ranges are, therefore, also uncertain. Only occasionally does a single form occur in large numbers, but geographically broad distribution suggests that their ultimate stratigraphic utility may be great when ranges and species limits are firmly established.

Motile or Encysted Stage?

Current uncertainty on this point illustrates the persistent challenge presented by taxa—modern or fossil—that are "different". It also underscores the opportunity for progress in the study of dinoflagellates that lies in focusing research by neontologists and paleontologists on matters of mutual interest, in this case an unusual fossil, *Dinogymnium*, and an unusual modern form, *Balechina*.

Like fossil dinoflagellates in general, the forms we now ascribe to *Dinogymnium* were originally taken to represent the remains of motile cells and were referred to the modern genus *Gymnodinium* (Deflandre, 1936). Subsequently, as the cyst nature of fossil dinoflagellates was recognized, these fossils, which are also composed of sporopollenin, were considered to be cysts (Evitt et al., 1967; Evitt, 1968). Two more recent contributions to discussions on these fossils have raised again the question of the nature of fossil *Dinogymnium* and its affinities. In one, May (1976) has suggested that they represent fossilized motile dinoflagellates. In the other, Tappan and Loeblich (1977) have restated arguments that they are cysts, adding that their affinities are with peridinialean *Lophodinium* rather than with gymnodinialean *Gymnodinium*.

The following points support the argument that the fossils were the walls of motile dinoflagellates:

1. Strong resemblance of the fossils to the motile cells of living *Gymnodinium* and its close allies.

2. Presence of openings through the wall whose size, form, and distribution are compatible with their being trichocyst pores.

3. Openings appropriate for flagellar pores located at the junction of cingulum and sulcus (however, the detailed morphology of these pores remains yet to be documented).

4. Complete compatibility between the form of the depressions here considered to be flagellar furrows and actual occupation of the depressions by flagella of design and function typical for dinoflagellates.

5. Presence in some living gymnodinioids (e.g., *Balechina coerulea*) of a relatively thick and resistant

wall (probably a thickened pellicle) which contrasts with the normally thin-walled *Gymnodinium* (Taylor, 1980).

Critical to further testing of this idea will be SEM examination of the interior surface of the *Dinogymnium* wall in the region of the flagellar pores, and study of the wall in *Balechina* or similar thick-walled living species, to compare the wall in composition and morphology with the test of *Dinogymnium*.

Less compelling seem the principal arguments advanced by Loeblich and Tappan that 1) the resemblance of *Dinogymnium* is closer to *Lophodinium* than to *Gymnodinium* and that *Dinogymnium* is, therefore, a peridinialean, and 2) that the presence of an archeopyle proves the fossils are cysts. Regarding the first contention, it is a matter of opinion whether the general form of *Dinogymnium* shows greater resemblance to *Gymnodinium* or *Lophodinium*. The latter is a poorly known genus represented by two species, whereas the former is a major taxon represented widely in fresh, brackish, and marine waters by many species. Both cell form and diversity in fossil *Dinogymnium* are strikingly parallel to those in *Gymnodinium*, whereas no species of *Dinogymnium* is a close match for one of *Lophodinium*. Furthermore, the status of *Lophodinium* as a typical peridinialean is uncertain. The type species has been said to be based on a cyst, and the second species to have a theca of many thin hexagonal plates, allying it with *Woloszynskia*, a genus whose species were removed from *Gymnodinium* only when their possession of tenuous thecal plates was discovered (Loeblich III, 1970). In other words, the known species of *Lophodinium* are either imperfectly understood or stand very close to whatever boundary line one might choose to recognize between the Peridiniales and Gymnodiniales. The test of *Dinogymnium* shows no traces of a paratabulation pattern, just as *Gymnodinium* lacks thecal plates.

The second argument, that the existence of an archeopyle establishes *Dinogymnium* as a cyst, is also specious. The presence of a pellicle modified by substantial addition of sporopollenin would almost certainly require a special exit mechanism not needed by thin-walled *Gymnodinium* for release of cell contents in conjunction with cell division. This would be a plausible function of the "archeopyle" in *Dinogymnium*. Although such a function would be analogous to that of the archeopyle in a dinoflagellate cyst, the argument that *Dinogymnium* must be a cyst because it has an opening that has been called an archeopyle is clearly circular. In this connection, it may be enlightening to determine how the protoplast of a living form such as *Balechina coerulea* escapes the confines of its wall.

If *Dinogymnium* represents a motile dinoflagellate, the sporopollenin composition of its wall is certainly unusual, but this may become less impressive as more is learned about the occurrence of sporopollenin in the normally thin pellicle of thecate dinoflagellate cells and about the peculiarly thick wall of *Balechina* and its allies. As pointed out early in Chapter 1 and documented by much that has been said on subsequent pages, diversity is a hallmark of dinoflagellates. The ability of these organisms to produce a variety of non-living building materials—cellulose, sporopollenin, calcite, and silica—is patently clear. With this capability established in the greater dinoflagellate genome, the general proposition that some dinoflagellates may utilize sporopollenin as the principal constituent of the covering for the motile cell does not seem absurd. Specifically, that *Dinogymnium* may have produced sporopollenin in its motile stage seems no more exceptional than the morphological features of these fossils when compared with other fossil dinoflagellates.

Category of N-cysts

Taxon: *Nannoceratopsis*.

Occurrence: Jurassic.

See Frontispiece H, Figures 5.3 F and 12.2. This easily recognized genus has been reported from many geographic areas, always from Jurassic strata (except where plausibly reworked). Relatively few species (under 10) have yet been described, but probably there are many more, some of them distinguishable on the basis of criteria not yet applied and perhaps determinable only with the aid of the SEM.

The outline shape is unique, with a minute epicyst anterior to a distinct and indented paracingulum and an expansive, laterally compressed hypocyst, which may have one to three (usually two) posteriorly directed angles or horns. The lateral surfaces of the hypocyst, which account for most of the surface area, are joined by a narrow sagittal band. Species have been differentiated on overall shape and the character of the ridges margining this band. Differences in archeopyle and surface sculpturing of the wall probably are also significant. The archeopyle is formed by release of one or more cingular paraplates. The wall is apparently composed of a single layer, but the outer portion at least may be porous to fibrous.

Studies under the SEM (Gocht, 1964 and 1972; Piel and Evitt, 1980) have revealed details of surface sculpture and paratabulation not normally visible in the light microscope. They show that this Jurassic genus can no longer be considered a fossil representative of the Dinophysiales, the order to which it had been referred. Cyst formation is unknown in dinophysialean dinoflagellates today and no fossils except *Nannoceratopsis* have been attributed to the order. The paratabulation of *Nannoceratopsis* combines an epicystal arrangement suggestive of modern Peridiniales, with a hypocystal arrangement almost identical to that found today in the Dinophysiales. The union of these otherwise disjunct characteristics in *Nannoceratopsis* is the basis for recognizing the order Nannoceratopsiales. It also raises an as yet unanswerable question about the evolutionary relationship of the Dinophysiales and the Peridiniales, which have been considered in the past to have little in common.

Figure 12.2 shows the paratabulation in *Nannoceratopsis* compared to the tabulation in a modern dinophysialean. Designation of the epicystal and cingular paraplates follows the familiar scheme of Kofoid. Although the pattern here is not exactly like that known from any fossil or living peridinialean, the organization into apicals, precingulars, and cingulars is similar. There is no trace of a sagittal parasuture across the cingulum middorsally and over the apex of the cyst that might correspond to the prominent sagittal suture over corresponding regions of the theca in modern dinophysialeans. By contrast, all essential dinophysialean features are present in the hypocyst. There are two large (lateral) and two small (ventral) paraplates, and a sagittal band (transversely striate with a narrow median groove) is prominent in some specimens. The latter is apparently equivalent to the megacytic band and suture in a modern dinophysialean. There is one different element in the *Nannoceratopsis* hypocyst, however. This is a striate band on each large hypocystal paraplate (H_2 and H_3) adjoining the paracingulum between the dorsal and ventral terminations of the sagittal band. These striate bands suggest that the theca of *Nannoceratopsis* may have had a lateral megacytic zone on each side, just posterior to the cingulum.

The organization of the parasulcus in *Nannoceratopsis* is also similar to that in modern dinophysialeans, but as that is also similar to the arrangement in modern Peridiniaceae, its implications in terms of relationships are ambiguous.

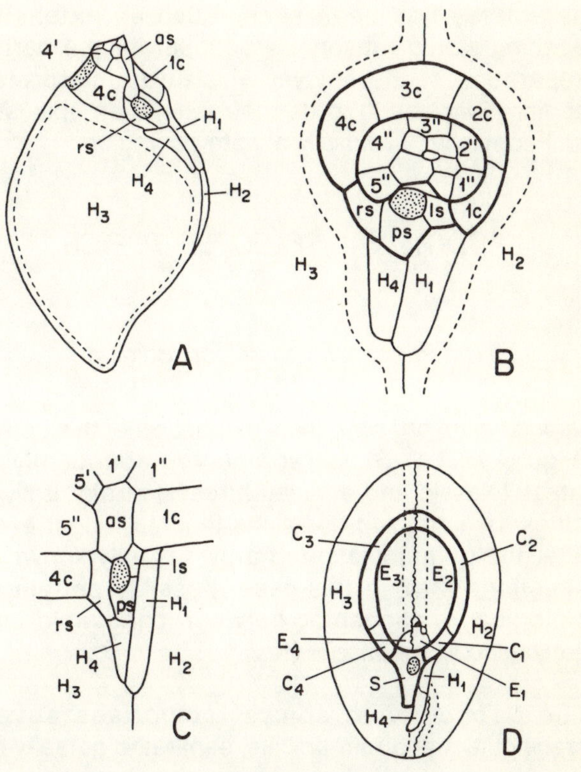

Figure 12.2 - PARATABULATION IN *NANNOCERATOPSIS*

A-C - *Nannoceratopsis senex* (Middle Jurassic). A - Rlv, with selected paraplates labeled (H = hypocystal paraplates) and showing Type C_{3c} archeopyle (patterned). B - Paratabulation detail in av. For clarity, epicyst, paracingulum, and parasulcus are enclosed by heavy lines of parasutures; only fraction of hypocyst near paracingulum is shown. Five apicals (unlabeled) cluster around depression apparently reflecting apical pore and are enclosed by precingular series of five paraplates. Parasulcus incompletely visible owing to oblique perspective. C - Diagrammatic detail of parasulcus and adjacent paraplates in direct view.

D - Tabulation detail of dinophysialean theca in av, for comparison with B. Epitheca, consisting of four principal plates (E_1-E_4), contrasts sharply with the peridinialean aspect of the *Nannoceratopsis* epicyst but hypotheca and hypocyst are nearly identical.

Dashed lines mark limits of sagittal band of intercalary thecal growth in D and presumed similar zone differentiated around margins of hypocyst paraplates in A and B.

The archeopyle in species of *Nannoceratopsis* as reported on the basis of observations under the SEM is formed by release of the middorsal cingular paraplate (*3c*). This feature can also be recognized in well-preserved specimens under the light microscope. In some specimens a larger opening, corresponding to two or, possibly, three cingular paraplates has been seen. Such an extensive opening would reasonably account for the partial separation of the epicyst which was interpreted earlier (Evitt, 1967) to be the result of an epicystal archeopyle with ventrally adnate operculum.

ADDITIONAL SELECTED TOPICS

The Paratabulation of Dapcodinium

Dapcodinium priscum, the type species of this genus (Figure 12.1 D-G), is represented abundantly in latest Triassic and earliest Jurassic strata. It illustrates the intriguing questions that may linger even after the paratabulation of a cyst is fully known—questions that, in this case, possibly pertain to biological relationships between peridinioid and gonyaulacoid dinoflagellates.

The *Dapcodinium* epicyst comprises seven precingulars and four apicals, separated dorsally by four similar, geniculate intercalaries that alternate regularly between precingulars and apicals in a distinctly symmetrical arrangement. This epicyst pattern, in which it is not difficult to perceive a certain peridinioid aspect, is shared with only one other group of fossil dinoflagellates, typified by the Early Jurassic genus *Mancodinium* (which is treated by some authors as a junior synonym of *Dapcodinium*).

The hypocyst, without known equivalent in the fossil record, is also distinctive, having seven postcingulars and three other major paraplates which can be construed as three antapicals or as two posterior intercalaries and one antapical. Although the hypocyst pattern is quite unique, a distinctly gonyaulacoid "flavor" is imparted to it by the configuration of the paraplates surrounding the parasulcus and by traces of parasutures within the parasulcus itself. This gonyaulacoid aspect becomes strikingly evident when one realizes that removal of just two parasutures (heavy lines in Figure 12.1 G) would convert the pattern into a "standard" sexiform gonyaulacoid design. The examples of *Helgolandinium* and *Fragilidium* (Figure 5.17) show that such "extra" sutures and plates occur in the thecae of living dinoflagellates, although an exact equivalent of the *Dapcodinium* pattern is not known.

If we look among living species for epithecal patterns that suggest *Dapcodinium*, we find a virtually identical arrangement in freshwater *Glenodinium polonicum* (Woloszynska, 1916, as *Sphaerodinium*). Another and only slightly less similar epithecal pattern occurs in *Peridiniella sphaeroidea* (Balech, 1980), but in this case one parasuture (equivalent to the heavy line in Figure 12.1 F) would need to be removed for the *Dapcodinium* diagram to match the modern species. If we now consider the hypothecae of these two modern species which are so similar to *Dapcodinium* in their epithecae, we find puzzling differences. The hypotheca of *G. polonicum* comprises five postcingulars and two antapicals in a typically peridinioid pattern, thus contrasting sharply with *Dapcodinium*, but the hypotheca of *P. sphaeroidea* is sexiform gonyaulacoid. Taking epicyst and hypocysts together, the pattern of *D. priscum* differs from that of *P. sphaeroidea* chiefly in having two parasutures in the hypocyst and one in the epicyst that are not represented in the theca of the modern species. It is not known what other differences may occur in the cingulum and sulcus.

The implications of this comparison are not altogether clear, either for *D. priscum* or *P. sphaeroidea*. The uniqueness of *Dapcodinium* in the fossil record is nearly matched by that of *P. sphaeroidea* today. The latter species owes its generic name to the perceived epithecal similarity to *Peridinium*, but the gonyaulacoid properties of its hypotheca preclude classification near that genus. *Dapcodinium* presents equal problems.

Perhaps the best way to close this discussion is to articulate a few of the questions it raises:

1. Is the hypocystal paraplate pattern of *Dapcodinium* evidence of a basic biological affinity to gonyaulacoid dinoflagellates?

2. If so, how does one explain the somewhat peridinioid aspect of the epicyst?

3. What does it mean that epicyst patterns closely similar to that in *Dapcodinium* occur today in association with a gonyaulacoid hypotheca in *P. sphaeroidea* and with a peridinioid hypotheca in *G. polonicum*?

4. Does the combination of a peridinioid epitheca and a gonyaulacoid hypotheca suggested by *P. sphaeroidea* and *D. priscum* tell us something significant about relationships between peridinioids and gonyaulacoids? (For further comment, see discussion of the YY-type of thecal design in Chapter 13.)

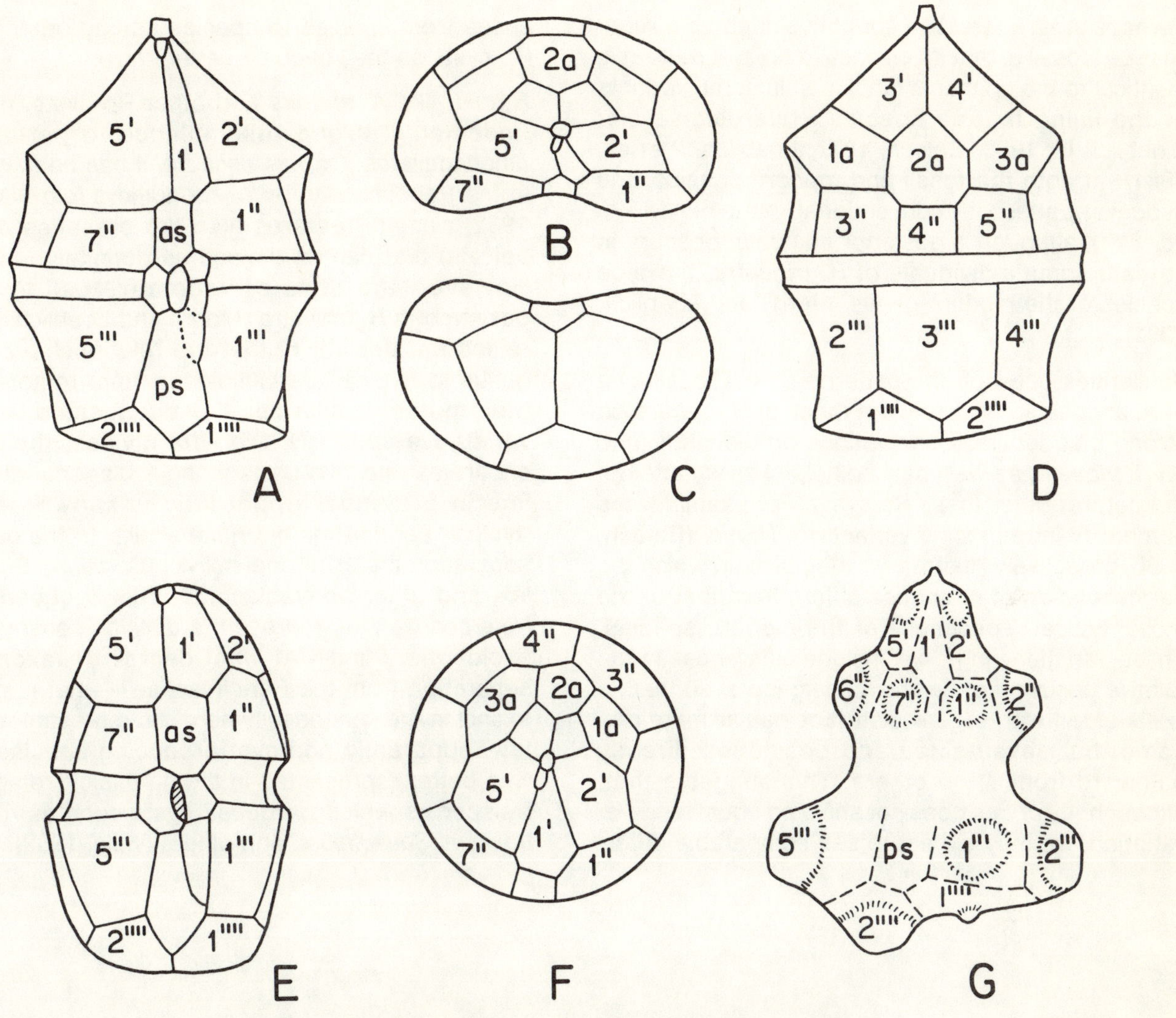

Figure 12.3 - MODERN *HETEROCAPSA* AND FOSSIL Rp-CYSTS

A-D - *Angustidinium acribes*, a mid-Cretaceous Rp-cyst (vv, av, aav, dv). After Goodman and Evitt (1981).

E-F - *Heterocapsa niei*. E - Tabulation of a thecate cell (vv). After Loeblich III (1968). F - Tabulation of epitheca (av), combining the typical relationship of *2a* to *3''* and *4''* (in contrast to the relationship shown in B, which occurs only rarely in this species) with a variation on the ventral surface (contrast E), which brings *1'* into contact with *7''* as reported by Balech (1977).

G - *Susadinium scrofoides* (vv), an Early Jurassic Rp-cyst, showing intratabular convexities and inferred position of parasutures (broken lines). After Bjaerke (1980, as *Facetodinium inflatum*).

Fossil Angustidinium *and Living* Heterocapsa

Specimens of mid-Cretaceous *Angustidinium acribes* (Figure 12.3 A-D) are small, bicavate, and of superficially unremarkable peridinioid appearance. They were first attributed to *Deflandrea* (Davey and Verdier, 1971), but transferred to a new genus as its type on the basis of peculiarities of the epicyst paratabulation brought to light by the SEM (Goodman and Evitt, 1981). The epicyst pattern is unusual in having five apicals, and a middorsal intercalary with an "extra" side because it contacts two of them. As a result, *2a* is camerate instead of planate. This can be recognized clearly in the fossils by the shape of the archeopyle, which indicates the presence of a middorsal parasuture between apicals even if that parasuture itself can be seen only in the SEM. Further unusual relations between paraplates visible in the SEM on the ventral surface include: a narrow *1'*, contact between the anterior sulcal and the last apical (*5'*), and presence of a second preapical which is peculiarly out of alignment with *1'*. The combination of an epicyst paratabulation that is unlike that of P-cysts with a hypocyst paratabulation that is typically peridinioid (*5''', 0p, 2''''*) is the justification for placing *Angustidinium* in the morphological category of Rp-cysts.

Perhaps most interesting about this hundred million-year-old fossil is that its distinctive epicyst pattern is identical to the epithecal tabulation in some variants of the living marine species, *Heterocapsa niei* (Loeblich III, 1968; Balech, 1977; Loeblich III et al., 1981). In both the fossil and modern species, the hypocyst pattern is "conventional" peridinioid (*5‴, 0p, 2⁗*), although a posterior intercalary occurs in at least some individuals of *H. triquetra*, the type species of the modern genus (Morrill and Loeblich, 1981).

Similarities between the patterns on the fossil cyst of *A. acribes* and the modern theca of *H. niei* are so strong that detailed observations on the latter (and on *H. triquetra* as well) may be helpful as we attempt to interpret other fossil Rp-cysts. For example, the numerous intraspecific variants of *H. niei* (Balech, 1980) collectively illustrate all the points in which the paratabulation of *A. acribes* differs from that of the most "typical" specimens of the modern species. These variations in *H. niei* include differences in the relative positions of *2a* and its neighbors, so that *2a* is six-sided with a geniculate adcingular margin in some, but seven-sided and positioned directly poleward from *4″* in others. This suggests that, although *A. acribes* does not show an equal range of variation, we may expect to see comparable differences from species to species among other fossil Rp-cysts as they become better known.

H. niei, like *A. acribes* and other Rp-cysts, has an undeniably strong resemblance to peridinioid dinoflagellates. On this account, it has been placed in the Peridiniaceae by some authors (e.g., Dodge, 1982) despite features (like the presence of five apicals) that can be viewed as definitely unusual. However, the style of vegetative cell division observed in *H. triquetra* (Morrill and Loeblich, 1981) raises the possibility that the relationship of this genus to "typical" peridinioids is more remote than their mutual similarities have suggested. As the vegetative cell of *H. triquetra* divides, the theca separates into two parts along a diagonal division line in a manner apparently unknown among "typical" peridinioids but quite similar to the oblique separation that is normal in living *Ceratium*, *Gonyaulax*, and other gonyaulacoids. This suggests that *Heterocapsa* may represent a distinct, conservative evolutionary line—At what degree of taxonomic separation from the Peridiniaceae?—that is nearly extinct in the sea today (where it is represented by a few, apparently non-cyst-producing species) but was better represented in the Jurassic-Cretaceous by species which produced cysts such as *Angustidinium*, *Susadinium*, and others (Wille, 1982).

Chapter 13. Thoughts on Classification and Evolution

The treatment in this chapter will be quite selective. First we will review some of the considerations that bear on the classification of peridinialean dinoflagellates and the interpretation of their phylogeny. Then, after brief attention to the topic of cyst taxonomy at the level of genera and species, we will examine several approaches to the suprageneric taxonomy of dinoflagellates (chiefly peridinialean) and some recently published ideas on phylogeny and evolution. Finally, we will return to the ideas of Taylor, some of which have already received our attention in this volume, but now we will pursue them in several speculative directions.

No new classification of dinoflagellate cysts is presented here, nor is the intent to survey in any comprehensive way the broader topic of the classification and evolution of dinoflagellates in general. The reader wishing to pursue these matters should consult such works as those by Sarjeant and Downie (1964), Wall and Dale (1968b), Loeblich III (1970, 1980), Sarjeant (1974), Taylor (1976, 1980), Williams (1977), Norris (1978a, 1978b), Eaton (1980), Tappan (1980), Bujak and Williams (1981), and Taylor (1983). Especially the work of Tappan (1980) should be consulted for a concise yet comprehensive review of ideas relating to the place of dinoflagellates among the protists, as well as separate synoptic classifications of living and fossil dinoflagellates to the family level, with listings of included genera. The classification of fossil dinoflagellates adopted by Tappan is not discussed separately here since it generally follows the Downie-Sarjeant-Norris approach which is commented upon later in this chapter.

GENERAL CONSIDERATIONS

To better appreciate some of the ideas on classification and evolution of dinoflagellates that are currently abroad, it may be useful to reflect briefly on the nature of classifications in general.

Classifications of organisms come in two guises, phenetic and phylogenetic. A purely phenetic classification provides a method for organizing and cataloguing information by expressing various degrees of similarity and difference among the items being classified. It can be designed at any stage in the study of a group and is advantageously based on objective and arbitrary criteria. Multiple phenetic classifications of the same group may be equally useful despite marked contrasts in the criteria used and divergent views of serious students regarding biological relationships among its taxa. The success of a phenetic classification depends largely on orderliness, consistency and completeness. As it matures, there is a strong tendency for categories with a large number of items to be subdivided, essentially for the convenience of moderate size, with the consequence that the number of taxa maintains a rough proportionality to the number of observations. Accordingly, the complexity of the classification generally reflects the level of detail to which the group concerned has been studied and described.

A purely phylogenetic classification, in contrast, is an expression of concepts of evolution and biological relationship. Its successful design cannot precede the accumulation of a substantial minimum of observational data and it is based on subjective evaluation of phenetic characters believed to indicate biological affinity. Multiple phylogenetic classifications of the same group represent fundamentally different opinions on relationships and evolution and cannot be equally acceptable to a single investigator. Serious students having substantially different ideas about the biological relationships and evolution of the taxa within a group could not conceivably all find equally acceptable a single phylogenetic classification of those taxa. The success of a phylogenetic classification is in the eyes of the beholder. To the convinced, it has reality and predictive ability; to the skeptic, it is wrong. The practicality of having only a moderate number of constituents in a category is irrelevant, since the number of taxa and the complexity of their arrangement theoretically have more to do with the biological success of the group than with the skill of the classifier.

The ideal classification would be comprehensively both phenetic and phylogenetic, and would, of course, encompass modern and fossil representatives of the group in question. In practice, most classifications are compromises of the two approaches, but there is normally a progressive increase in the phylogenetic component as a group becomes better known and as conflicts with phenetic aspects of the classification are eliminated. It is as if the classification of a group had an ontogeny of its own. In the juvenile stage of discovery and early study, the framework of a classification is sketchy and categories are broadly defined on gross, immediately apparent features. Early impressions about biological affinities are vague and tentative, and a classification at this stage is apt to serve chiefly as an aid in cataloguing information. A mature classification, on the other hand, approaches the ideal to the extent of being comprehensive in scope, incorporating weighted evaluations of selected phenetic criteria, and presenting an integrated concept of phylogeny, however incomplete and subject to further refinement may be its details. Between such juvenile and mature stages lies a long period of gradual transition during which phylogenetic concepts slowly develop as those phenetic characters that are significant indicators of the general evolutionary history of the group are distinguished from those that are not. With regard to dinoflagellates, we are perhaps somewhere late in the long adolescent period, confident about biological relationships on the broad scale and able to speculate about them here and there in detail, but as yet unable to resolve obvious incompatabilities among the phylogenetic schemes that are being advanced currently by serious and experienced scientists.

The extent to which the classification of a particular group of organisms may approach the ideal depends upon many factors. Six factors that have important bearing on the classification of peridinialean dinoflagellates are discussed in the following paragraphs.

1. The cyst and theca will continue to play different roles in the taxonomy of fossil and living dinoflagellates. Only in a relatively few instances is there the possibility of examining cyst and theca of a single species, for many modern species do not produce cysts and no fossil species are known from fossilized thecae. While both neo- and paleobiologists are now aware of the importance of integrating information from theca and cyst where it is available, up until now the classification of living dinoflagellates has experienced virtually no revision in light of cyst characters. Similarly, although inferred thecal patterns are becoming an increasingly important factor in the taxonomy of fossil cysts, much opportunity remains for extending this trend. Current knowledge supports the following assertions:

a. The *cyst* has an important but definitely limited potential as a factor in the classification of *modern* dinoflagellates, a field in which thecal morphology will continue to be a prime factor.

b. The *theca* has an important but definitely limited potential as a factor in the classification of *fossil* dinoflagellates, a field in which many features not suggested by thecal morphology will continue to play a large role.

c. In neither case have the potentials of these factors, however limited they may be, yet been realized. Consequently, the taxonomy of modern dinoflagellates will surely experience some substantial changes based on a consideration of the cyst, just as will the taxonomy of fossil dinoflagellates based on further consideration of the theca.

d. In the long run, the currently existing disparate taxonomies that have resulted from the differences between fossil cysts and modern thecae must be compromised in building any prospectively successful and comprehensive classification.

2. External morphology has been and will continue to be critical in the systematics of both cysts and thecae. The principal criteria for determining genera and species, the fundamental building blocks of any classification, are aspects of external morphology, especially the counterpart features of tabulation and paratabulation. Although biochemical, genetic, ultrastructural, and other factors determined though sophisticated modern techniques are potentially applicable to the study of living forms, they do not yet play a significant role in the recognition of genera and species or the higher categories based upon them. Tabulation and paratabulation are especially important because they are, at once, the established basis for the morphological comparison of modern taxa and the most promising basis for extensive comparisons between living dinoflagellates and those in the fossil record.

3. Cyst and theca are companion expressions of the species in which they occur. Although cyst and theca are distinguished by compositional, structural and preservational differences, and the cells they enclose differ in ploidy, motility, and intracellular organization, each is critical to the life cycle of species in which both occur. Neither can be considered, in principle, superior to or inferior to the other in its ability to convey useful taxonomic information about its genus and species; and neither, taken alone, establishes fully the morphotype of its species. Considered in this light, no close parallel exists between the relationship of cyst to theca and the relationship of dispersed spores and pollen to parent plant that would favor comparable treatment of cysts as form taxa in contradistinction to "real" taxa. Nevertheless, in practice, cysts and thecae present the taxonomist with quite different problems and possibilities. For instance, the theca for any particular peridinialean species provides no information from which the morphology of its cyst can be inferred, nor even whether or not a sporopollenin cyst is produced. The cyst, on the other hand, may be a key to considerable thecal detail even though the theca itself is never seen. Balancing this "advantage" of the cyst, the theca necessarily includes complete information about tabulation, whereas the effectiveness of the cyst as a conveyor of that information about the theca varies between great extremes, depending upon the species. A cyst that has virtually no traces of paratabulation may be distinctive and easily identified to genus and species by comparison with other cysts on the basis of criteria other than paratabulation, but, without paratabulation, it cannot be compared to clearly paratabulate cysts or to thecal patterns. This is an important consideration in regard to cyst classification, for it severely affects the confidence with which many cysts can be placed in any scheme that relies on tabulation.

Examples at one extreme are provided by species we have already discussed: living *Peridinium limbatum* and fossil *Palaeoperidinium pyrophorum*. Cysts of both reflect thecal features (shape, complete tabulation, intercalary bands) quite faithfully—in the fossil example to the point of indicating minute details of the theca, including the pattern of trichocyst pores relative to the ridges of the intratabular reticulum. At the other extreme are cysts such as that of living *Scrippsiella faeroensis*, which is a sphere bearing spines in wholly nontabular arrangement and lacking any other trace of paratabulation; in fact, there is no visible clue that it is even a dinoflagellate (Dale, 1978). Only slightly more informative is the cyst of modern *Gonyaulax grindleyi*, which is also a spinose sphere (known from sediments under the name of *Operculodinium centrocarpum*). Although the spine arrangement is nontabular, the wall is locally modified by a typical Type P archeopyle (verifiable as Type P_4 by reference to the theca) that at least signals its dinoflagellate affinity and gives a trifle of information about thecal tabulation.

4. The discovery of cysts for species of living thecate dinoflagellates has illuminated differences between the taxonomic concepts of neobiologists and paleobiologists, showing that many taxa based on cysts in the fossil record are much more narrowly defined than those based on modern thecae. Factors contributing to this situation are the greater diversity of morphological characters provided by cysts, the premium placed on fine degrees of morphological discrimination by the application of dinoflagellates to stratigraphy, and the failure of neobiologists yet to deal successfully with taxa that produce closely similar thecae but highly distinctive cysts. Examples of the last point are the cysts formed today by species placed by neobiologists in the single genus *Gonyaulax*, which fall into at least seven paleontological genera: *Apteodinium* (or *Trichodinium*), *Impagidinium*, *Lingulodinium*, *Nematosphaeropsis*, *Operculodinium*, *Spiniferites*, and *Tectatodinium*. Similarly, Wall and Dale (1968b) reported culturing motile cells identifiable as *Gonyaulax spinifera* from several different cyst morphotypes.

Comparably narrow definition of fossil genera is apparent among ceratioid and peridinioid cysts, although comparative scopes of the taxa are less easily judged because no living species of the former produce sporopollenin cysts and most living species of the latter produce cysts with a simpler wall structure than is common among fossil representatives. Although some students of fossil dinoflagellates (for example, Norris and Dörhöfer) would disagree, many others would consider that fossils placed under at least seven genera (*Aptea*, *Endoceratium*, *Muderongia*, *Odontochitina*, *Phoberocysta*, *Pseudoceratium*, and *Xenascus*) probably had thecal characters that would justify placing them in the single genus *Ceratium* if their thecae could be observed directly. In the generic taxonomy of modern peridinioid dinoflagellates, details of sulcal and cingular tabulation are becoming increasingly important, but these features are frequently impossible to observe in cysts (modern or fossil). With regard to the fossils, we can only say that a host of cyst genera—Stover and Evitt (1978) list over 30—give evidence of having been associated with thecae whose tabulation patterns (aside from sulcus and cingulum) are today restricted to *Peridinium*, *Protoperidinium*, and a few closely similar genera.

In the long run, we can expect that lessons learned from further intensive study of living species for which both cysts and thecae are known will lead to some reduction in the taxonomic "over-splitting" of many fossil dinoflagellates and to recognition of polyphyletely among some modern thecate taxa.

5. New taxa based on cysts are being described more rapidly than new taxa based on thecae. That is to say, the flux of fundamentally new information on morphology is much greater for the fossil forms. The taxonomy of modern dinoflagellates is nearly static at the level of genera insofar as newly discovered forms are concerned (i.e., in contrast to genera that result from taxonomic revision of known forms). At the species level, a moderate number of new living forms continue to be described, but these are distinguished generally by minute differences from described species to which they are reasonably similar, rather than by gross contrasts. On the other hand, additions to the list of fossil genera result more from new discoveries than from taxonomic revisions, and even many of the new species contrast more or less sharply with any previously described. This means that, in contemplating the design of a comprehensive classification, the paleontologist cannot hope to survey his material with the thoroughness possible for the neobiologist. Any classification of the fossils made at this time, no matter how perceptively it recognizes the "true" phylogeny of the described taxa, is certain to require substantial expansion and change just to accommodate unanticipated new forms.

6. Finally, the distribution of fossil cysts through the stratigraphic column must be regarded as an ambiguous record of dinoflagellate evolution. As was emphasized in Chapter 2, the selectivity of the fossil record of dinoflagellates severely reduces the advantage which the paleontologist usually enjoys in discussions about evolution. Whereas the neobiologist's inferences about evolution depend entirely on comparative studies of an array of contemporary organisms, the paleontologist normally has the potential to "see" the comings and goings of species on the world scene through time, even though his senses are largely blind to details of living processes. The paleontologist commonly makes with confidence assertions, such as: "The fossil record shows that group X had a long and varied history before any organisms representing group Y appeared on the scene," or: "Group Z underwent a profound diversification in the Mesozoic." However, the student of fossil dinoflagellates is less privileged. His evidence as to the timing of evolutionary events may be nearly as meager as the neobiologist's. The failure to recover fossil cysts representing a particular group of dinoflagellates from strata of a particular age may record the real absence of the group at that time or only the absence of cyst-producing species among its members. We have been slow to recognize the limitations that this situation places on the interpretation of the dinoflagellate record and it will be some time before our perceptions of dinoflagellate history accommodate to these restrictions.

CYST TAXONOMY AT THE LEVEL OF GENUS AND SPECIES

Dinoflagellate cyst morphology is the focal subject of this book and even the reader who knew little about dinoflagellates before opening it has been exposed in the preceding chapters to a considerable introduction to the subject. Genera and species have been cited in great numbers in both text and figures as examples of this or that feature. However, relatively little has been said about the *ranking* of morphological features that have been, or may be, used in the discrimination of species and genera. Virtually nothing of a lastingly definitive nature can be said on this topic, because we are dealing here with especially fluid aspects of taxonomy—and especially volatile aspects of taxonomists!—where criteria change almost with the season and where each specialist's fancy becomes embodied, at least temporarily, in the circumscriptions of new or emended taxa. Actually, contemporary systematic literature is our principal guide, and it has both contestable authority and temporary impact.

Taxonomic practices applied today to fossil dinoflagellates at the level of species and genera have developed over a period of 50 years and have been strongly influenced by the works and methods of Deflandre, Eisenack, and Cookson. To be sure, other investigators in the last quarter century, using improved and expanded techniques of preparation and study to explore vast amounts of new material, have added new criteria and many new taxa. However, neither the basic way of characterizing species and genera of fossil dinoflagellates nor of assessing their affinities with living forms has changed in principle. It seems to this author that the taxonomy of fossil dinoflagellates at the level of genus and species is near a turning point that is likely to be marked in the retrospective view of a decade or two from now by several developments:

1. A heightened appreciation that wall structure, surface ornament, and the archeopyle are features whose variability and significance are themselves highly variable; and, accordingly, a severe reduction in their taxonomic importance (especially that of the archeopyle).

2. A greatly increased emphasis on all aspects of paratabulation wherever it is observable, including details in the parasulcus and the shape and contact relations of individual paraplates.

3. A much clearer perception that taxa based on cysts are of unlike character and significance: some afford a basis for inferring many aspects of thecal tabulation and "outrank" those that provide relatively little information on this point. Consequently, the more informative cysts will be clustered into well-defined groups, some of them clearly allied to living forms, while the more nondescript genera will be accepted for what they are, effectively catchall categories.

The taxonomy that evolves from these developments may contrast markedly with that familiar to us today, and the recent synthesis by Stover and Evitt (1978), or any comparable summary published in the next few years, may appear grossly anachronistic, retaining interest only as a convenient summary of outdated concepts.

Another contrast with current nomenclature is likely to develop as progress is made toward resolving the question of how to handle the names of those cysts, modern and fossil, whose equivalence with living thecate taxa is beyond doubt. Here is an important and contentious issue. Strict adherence to the International Code of Botanical Nomenclature as it stands now, insertion of a new provision for use of alternative names in works dealing chiefly with either cysts or thecae, and continued *de facto* use of alternative names are all possibilities likely to have their vigorous proponents and opponents. Whichever is adopted in the long run will have a distinctive effect on the vocabulary of generic names that appears in future publications, quite apart from the taxonomic views they represent.

A first step along these lines is the recent paper by Harland (1982b) which makes a substantial attempt to integrate within one closely circumscribed group of taxa the disparate nomenclatures created by neontologists and paleontologists who have been concerned heretofore with different stages of the life cycle.

Harland confined his attention to Quaternary and Recent species which he believed should be referred to the single genus, *Protoperidinium*. Using as criteria tabulation/paratabulation, cyst wall structure, surface ornament, shape, and archeopyle type, he allocated 15 species that produce sporopollenin cysts to three sections of *Protoperidinium* subgenus *Archaeperidinium* and seven sections of *Protoperidinium* subgenus *Protoperidinium*, reducing to sections seven genera that had been based on cysts. He assigned both living and fossil species to sections according to the information available on their thecal tabulation and/or cyst morphology. Homeomorphic development of cyst features was called upon in several instances to solve problems in assignment of taxa. The resultant subdivision of *Protoperidinium* is summarized in Table 13.1.

Table 13.1 - SUBGENERA AND SECTIONS OF QUATERNARY-HOLOCENE *PROTOPERIDINIUM* ACCORDING TO HARLAND (1982b)

Genus	*Protoperidinium*
Subgenus	*Minusculum*
Subgenus	*Archaeperidinium*
Sectio	*Archaeperidinium*
Sectio	*Stelladinium*
Sectio	*Fuscusasphaeridium*
Subgenus	*Protoperidinium*
Sectio	*Votadinium* (ortho quadra)
Sectio	*Asymmetropedinium* (ortho penta)
Sectio	*Brigantedinium* (ortho hexa)
Sectio	*Selenopemphix* (ortho hexa)
Sectio	*Quinquecuspis* (ortho hexa)
Sectio	*Trinovantedinium* (ortho hexa)
Sectio	*Protoperidinium* (para hexa)

(Species having meta quadra, meta penta, meta hexa, para quadra, and para penta thecae are not known to produce sporopollenin cysts today.)

SUPRAGENERIC TAXONOMY AND PHYLOGENY

Classifications Proposed Through 1974

Sarjeant (1974) and Norris (1978a) have reviewed the history of development of suprageneric classifications applied to dinoflagellate cysts. The first classification of consequence was that proposed by Eisenack (1961), in which all genera of fossils then recognized as dinoflagellates were referred to 11 families: five new, and three others of long standing based on modern thecate genera. One further family of fossil chorate cysts, now but not then accepted as dinoflagellates, had also been named at that time. Although Eisenack's families of fossil dinoflagellates were based on morphology, there is no doubt that he conceived of this as a phylogenetic arrangement in which the fossils were placed relative to major groups of living taxa. At that time Eisenack, like many of his predecessors, treated dinoflagellates under the International Code of Zoological Nomenclature.

Subsequent commentaries on classification and contributions to the developing picture are to be found in works by Eisenack (1964), Norris and Sarjeant (1965), and Vozzhennikova (1967). Although the last mentioned author made a comprehensive attempt to merge taxa of fossil dinoflagellates with the classification of modern dinoflagellates, she drew largely on earlier morphological concepts and her proposals have not been widely followed. In a more recent survey (Vozzhennikova, 1979), she classified fossil dinoflagellates into 25 families, including nine new ones. A scheme that strongly influenced subsequent workers was that proposed by Sarjeant and Downie (1966). These authors devised a phenetic classification of cyst-families based on genera of fossil cysts, and presented it as a parallel construction to the classification of families based on living dinoflagellates.

In a subsequent revision and enlargement, the same authors (Sarjeant and Downie, 1974), modified the concept of cyst-families, and, in its most recent version (Sarjeant, 1974), the scheme was presented, alongside a classification of modern dinoflagellates (taken from Loeblich III, 1970), as a comprehensive suprageneric classification of fossil dinoflagellates arranged in 39 families. However, there was no analysis of the taxonomic overlap between the two classifications and no intimation of the inherent nomenclatural conflicts that would ensue from their joint adoption. For instance, the four families of peridinioid genera (Palaeoperidiniaceae, Deflandreaceae, Nelsoniellaceae, and Phthanoperidiniaceae) stand in the classification of fossil forms while the unique Peridiniaceae stands in the classification of modern forms. Similarly, Spiniferitaceae appears as a family of fossils, while Gonyaulacaceae appears in the modern list.

Families of fossil genera in Sarjeant and Downie's classification are not grouped into higher infraordinal categories but are listed in a sequence that proceeds generally from proximate to cavate to chorate. Each family is an assemblage of genera with morphological similarities that include general features of tabulation, archeopyle type, overall shape, wall structure (including cavation and degree of contraction), and surface ornament. The classification has the effect of carrying to the family level the concept of formally and distinctively named cyst-taxa as contrasted with theca-taxa. Tappan (1980) adopted a similar approach and generally followed Sarjeant and Downie's classification of dinoflagellate cysts, but she incorporated some new information and numerous technical corrections.

This Author's Personal Comment: That many authors have adopted the classification which has evolved from Sarjeant and Downie's original proposal (1966), despite the substantial problems just alluded to and others of a more mechanical and editorial nature, probably reflects the perceived advantage of the orderly taxonomic framework it provides for any work dealing with numerous taxa. However, the "families" of Sarjeant and Downie must be regarded as analogues of the subturmae encompassing form-genera of dispersed spores and pollen (although few would argue that all the included genera of dinoflagellate cysts should be treated as form genera).

Since phylogenetic concepts played no role in formulating the classification, one should be cautious in criticizing its authors for seeming to suggest relationships at suprageneric levels when that was not intended. However, one can also question whether, under such circumstances, a detailed family-like classification of "non-families" is any more useful than an alphabetical listing or some other systematic but frankly arbitrary ordering.

The Classification of Norris (1978)

A comprehensive classification of dinoflagellate cysts was presented by Norris (1978a, 1978b—hereafter cited as Norris, 1978). Its design rests on the foundations laid by Sarjeant and Downie, but it purports "to remove chronistic and cladistic absurdities" from its phenetic predecessor and, thereby, to approach a phylogenetic classification. Its scope is significantly greater than that of the earlier work, as it takes into account subsequent major contributions on particular groups (by Wiggins, 1975; Lentin and Williams, 1976; and others) and incorporates more recently described genera.

In all, 34 families (numbered without implied significance to their sequence) are recognized and these are assembled in three orders and four suborders of the Peridiniales, as outlined in Table 13.2.

Table 13.2 - SUPRAGENERIC CLASSIFICATION OF NORRIS (1978)

Note: Family names incorrectly applied by Norris (1978) appear in parentheses and are followed by an older legitimate family name typified by the name of a genus which Norris assigned to the family. Numerals in parentheses give the number of genera of cysts referred to each family; additional genera not listed (including some that are possibly synonymous) are discussed in Norris' text.

Order PERIDINIALES

A. Suborder RHAETOGONYAULACINEAE

1. (Rhaetogonyaulacaceae) Nelchinopsiaceae (12)
2. (Sverdrupiellaceae) Broomeaceae (5)
3. Pareodiniaceae (12)

B. Suborder GONYAULACYSTINEAE

4. Gonyaulacystaceae (17)
5. Apteodiniaceae (13)
6. (Spiniferitaceae) Hystrichodiniaceae (11)
7. (Endoscriniaceae) Stephodiniaceae (13)
8. Scriniocassiaceae (5)
9. Cordosphaeridiaceae (14)
10. (Lingulodiniaceae) Exochosphaeridiaceae (9)
11. (Ctenidodiniaceae) Eodiniaceae (9)

C. Suborder HYSTRICHOSPHAERIDIINEAE

12. Hystrichosphaeridiaceae (24)
13. Cleistosphaeridiaceae (7)
14. (Florentiniaceae) Homotrybliaceae (6)
15. Areoligeraceae (4)
16. (Lithodiniaceae) Belodiniaceae (16)
17. Canningiaceae (8)
18. Xiphophoridiaceae (2)
19. Senoniasphaeraceae (11)
20. Membranilarnaciaceae (6)
21. Muderongiaceae (2)
22. Pseudoceratiaceae (4)
23. Odontochitinaceae (4)

D. Suborder DEFLANDREINEAE

24. Deflandreaceae (21)
25. Palaeoperidiniaceae (10)
26. Wetzeliellaceae (6)
27. Phthanoperidiniaceae (14)
28. Ovoidiniaceae (4)

E. Suborders incertae

29. Ceratocorythaceae (given as Ceratocoryaceae)
    Subfamily Thalassiphoroideae (3)

30. Pyrophacaceae (1)
31. Heteraulacacystaceae (3)
32. Microdiniaceae (2)

Genera incertae familiae (7)

Order DINOPHYSIALES

33. Nannoceratopsiaceae (1)

Order GYMNODINIALES

34. Dinogymniaceae (2)

*Dates and authorship of corrected names are as follows (correct name cited first in each pair): Belodiniaceae Eisenack 1961 for Lithodiniaceae Norris 1978, Broomeaceae Eisenack 1969 for Sverdrupiellaceae Norris 1978, Eodiniaceae Eisenack 1961 for Ctenidodiniaceae Sarjeant & Downie 1966, Exochosphaeridiaceae Sarjeant & Downie 1966 for Lingulodiniaceae Sarjeant & Downie 1974, Homotrybliaceae Sarjeant & Downie 1966 for Florentiniaceae Harker & Sarjeant 1975, Hystrichodiniaceae Deflandre 1936 for Spiniferitaceae Sarjeant 1970, Nelchinopsiaceae Wiggins 1972 for Rhaetogonyaulacaceae Norris 1978, Stephodiniaceae Eisenack 1964 for Endoscriniaceae Vozzhennikova 1967. If *Nelchinopsis* is treated as a junior synonym of *Gonyaulacysta* (following Duxbury, 1977, rather than Norris), then "*Rhaetogonyaulax*" should provide the root for both the family and suborder names, replacing Nelchinopsiaceae and Rhaetogonyaulacineae, respectively.

Norris firmly rejects the concept of a dual nomenclature for cysts and thecae and provides, in principle, for integrating the nomenclature at suprageneric levels by recognizing as the proper name of a taxon the first legitimate name available from either cysts or thecae. In practice, names based on cysts that were proposed for categories at the family level and above are maintained, pending clarification of cyst-theca relationships; only the Pyrophacaceae is based on the name of a living genus. At the generic level, he accepts the names applied by Bradford (1975) to cysts of living peridinioid genera.

The three modern families Gonyaulacaceae, Peridiniaceae, and Ceratiaceae, and their namesake genera, to which many fossil cysts show more or less resemblance, are accounted for explicitly or implicitly. *Gonyaulax*, which is considered "clearly polyphyletic", is represented in three families, but none of these is called Gonyaulacaceae. Instead, all three bear names based on cysts (Apteodiniaceae, Hystrichodiniaceae, Exochosphaeridiaceae), awaiting determination of the precise cyst morphology of the type species of *Gonyaulax* (*G. spinifera*). Similarly, pending confirmation of the cyst morphology of the type species of *Peridinium* (freshwater *P. cinctum*), Peridiniaceae does not appear in the classification, although Phthanoperidiniaceae is recognized as a possible synonym. Since the classification is not concerned with any dinoflagellates that do not produce sporopollenin cysts, no mention is made of the possible relation of modern *Ceratium* or Ceratiaceae to any fossil taxa.

The phylogeny of Dinophyceae since the Triassic, which Norris believes his classification to reflect, is expressed in the arrangement of families into suborders, and in the substantially revised makeup of many families as compared to Sarjeant and Downie's scheme. However, there is no change in the kinds of morphological features employed as taxonomic criteria, only in the phylogenetic significance attributed to them.

Norris bases the four major suborders of the Peridiniales on differences in tabulation pattern and archeopyle type, and he holds these two characters to be of fundamental and probably equal importance. Contraction (i.e., formation of proximate, chorate, and cavate cysts) is considered to be in part related to water temperature and latitude; however, owing to the effects of land-sea distribution and ocean circulation on evolutionary activity, different degrees of contraction are construed to characterize genetically distinct families. General affinities of the families are still indicated by similarities in paratabulation and archeopyle. Additional features in the diagnosis of families include the presence, number, and distribution of horns; the distribution of ornament; and the presence and character of special ornamental features.

Evolutionary and stratigraphic relationships are treated in an extensive discussion and illustrated by one figure showing stratigraphic ranges of families and another suggesting their phylogenetic affinities. A "profound differentiation of the Peridiniales during the Mesozoic" is inferred.

This Author's Personal Comment: The classification, which is comprehensive in scope and presentation, eliminates some of the inconsistencies of Sarjeant and Downie's scheme without greatly changing their way of ordering cysts on morphological grounds. In both schemes, morphological similarities among some of the groups of genera almost certainly reflect biological affinity. However, in this author's opinion, Norris fails to achieve his principal purpose and his classification is not, as he claims it to be, a substantially closer approach to a phylogenetic arrangement. Many of the criticisms of Sarjeant and Downie's classification voiced by Wall and Dale (1968b) apply equally to Norris' scheme.

While no extensive critique of Norris' arguments is intended here, two matters will be mentioned, one relating to the archeopyle, the other to the stratigraphic record. Observed variations in archeopyle type (see Chapter 6 and examples in Chapters 9-12) suggest that this feature has been highly overrated as a taxonomic criterion in the past, even at the levels of species and genera. Therefore, attributing a fundamental importance to relatively minor differences in archeopyle type at the still higher levels of family and suborder appears inappropriate. If the significance of this criterion were diminished, major changes in Norris' classification would be required.

Regarding selectivity of the fossil record (see Chapter 2), Norris acknowledges that dinoflagellates may have had an extensive but unrecorded pre-Triassic history. However, his phraseology, especially in the sections on evolutionary and stratigraphic relationships, frequently illuminates his failure to make an important distinction: the appearance, diversification, and extinction *of dinoflagellates* is not the same as the appearance, diversification, and extinction *of dinoflagellates that produced particular sporopollenin cysts*. The fossil record tells us a great deal about the latter but only an uncertain fraction of what we would like to know about the former. Our familiarity with many abundant and well-preserved fossil dinoflagellates does not reduce the possibility that the major patterns of dinoflagellate evolution—even the principal participants in terms of evolving taxa—may be grossly different from the patterns (and taxa) recorded by fossil cysts. Such a possibility, which in this author's view should be emphasized as a major qualification on any theory of dinoflagellate phylogeny based on the fossil record, appears not to have been considered. Under the interpretation that is suggested here, those of Norris' conclusions about dinoflagellate phylogeny that rest solely on the observed stratigraphic distribution of cysts must be viewed as speculative and not directly supportable by fossil evidence (beyond the fact that all fossil dinoflagellate cysts document that the organisms they represent were in existence by certain minimum dates in the geologic past).

With regard to the suborders of Peridiniales which Norris recognizes (Table 13.2) and the phylogeny that he suggests for them, the Deflandreineae and Rhaetogonyaulacineae present minimal grounds for dissent. These divisions acknowledge roughly the same features used in this volume to distinguish the most typical peridinioid and rhaetogonyaulacoid cysts—P-cysts and R-cysts—with the minor difference that *Suessia* and similar forms (in Broomeaceae, which was given as Sverdrupiellaceae by Norris) are here set apart (S-cysts). However, Norris' view that the remaining peridinialean cysts comprise (in addition to several groups with uncertain affinities) two major and distinct groups (Gonyaulacystineae and Hystrichosphaeridiineae) with independent origins and wholly separate histories appears to this author to be contrary to the weight of the evidence. Essentially the same taxa have been interpreted in this volume as comprising diverse and morphologically intergrading cysts with a single (gonyaulacoid) affinity. These contrasting viewpoints are illustrated well by Norris' placement of six families (Gonyaulacystaceae, Hystrichodiniaceae, Stephodiniaceae, Hystrichosphaeridiaceae, Homotrybliaceae, and Belodiniaceae) that include genera characterized by well-expressed paratabulation. To Norris, morphological similarities between the first three families (referred to suborder Gonyaulacystineae) and the second three (referred to suborder Hystrichosphaeridiniineae) reflect convergence, not close affinity, since differences in archeopyle type and contraction are held to outweigh similarities in paratabulation. To the author of this volume, all the taxa appear closely related, the significance of similar paratabulation far outweighing that of differences in the manner in which it is expressed (parasutural or intratabular features in proximate, chorate, or cavate cysts) and in the archeopyle.

The Informal "Lineage Concept"

Wall and Dale (1968a) were the first to compare a large number of modern peridinealean cysts with fossil cysts on a morphological basis. Since the identities of the modern cysts they studied had been established by culturing excysted cells to the thecate stage, it was possible for them to suggest affinities for many groups of fossil cysts having morphologies closely similar to the modern ones. Their paper includes an astute discussion of many aspects of dinoflagellate classification and nomenclature but, in communicating their conclusions about modern and fossil cyst morphology they eschewed formal nomenclature. Instead, they used the informal term "lineage" for an assemblage of morphologically similar genera within a single family. As labels for these lineages, they coined adjectival terms from names of typical taxa by adding the ending "-oid." By this means they could outline and discuss a variety of ideas about evolution within the Peridiniales without commitment to a formal hierarchy of categories that would have required new names as well as new and emended circumscriptions.

Wall and Dale listed 110 genera in 24 lineages, representing two families (Gonyaulacaceae and Peridiniaceae), and took strong exception to the principles by which Sarjeant and Downie (1966) had classified approximately the same genera into nearly 40 families.

Williams (1977), in a work dealing with only sporopollenin cysts and including many more recently described genera, adopted an approach similar to that of Wall and Dale. However, he reassembled some of the earlier genera, used a two-step organization in which lineages are divided into sublineages or groups, and deleted reference to Gonyaulacaceae and Peridiniaceae, or to any other families.

This Author's Personal Comment: When the scopes of formal suprageneric taxa are unsettled, or when attribution of genera to suprageneric categories is quite peripheral to an author's purpose, something like these informal procedures allows great flexibility without enormous nomenclatural complexity. As explained in the introduction to Part III, Chapters 9-12 are organized in rough accord with these ideas.

Views of Dörhöfer and Davies (1980) on Early Dinoflagellate Evolution

Focusing on the suborder Rhaetogonyaulacineae, which Norris (1978) considered the most primitive of the peridinialean suborders, Dörhöfer and Davies (1980) interpreted details of paratabulation and archeopyle formation in a large number of fossil cysts. Only an abbreviated list of their conclusions can be presented here, but these should suffice to illustrate the broad scope and conceivable significance of their ideas about dinoflagellate phylogeny, were they to prove correct. Among their conclusions are the following:

The morphologically similar genera of families Odontochitinaceae, Muderongiaceae, and Pseudoceratiaceae are combined in a single family (Pseudoceratiaceae) and this is removed from the Hystrichosphaeridiineae to the most primitive of Norris' suborders, the Rhaetogonyaulacineae, where it is viewed essentially as ancestral to the Gonyaulacystineae, Hystrichosphaeridiineae, and possibly the Peridiniineae. The Pseudoceratiaceae is said to be distinguished by an apical archeopyle whose operculum includes two anterior intercalary paraplates. These paraplates, which have not been recognized previously, are held to be an important indicator of the critical position of the family in early evolution of dinoflagellate cysts. New genera of a new family (Phallocystaceae) are ancestral to the Pareodiniaceae (suborder Rhaetogonyaulacineae) and are derived from suessioid forms of the same suborder. "Sequential lineages are documented explaining all major archeopyle types (I, AI, A, AIP, P) within the Rhaetogonyaulacineae from a common Triassic stock" (p. 42). Postulated evolutionary steps include: "possible evolution of proximate dinoflagellate cysts from *Micrhystridium*-type acritarchs with disintegration archeopyles and suessioid paratabulation within the Triassic" (p. 42); fusion of neighboring plates as the principal mechanism for reducing the number of paraplates and paraplate series in the primitive suessioid design to the numbers characteristic of gonyaulacoid and possibly peridinioid derivatives; formation of a Type (4A2I) archeopyle in Pseudoceratiaceae by fusion of paraplate 2a to the apical series; formation of the Type (4A) archeopyle of such genera as *Cyclonephelium* by fusion of apical and intercalary paraplates into one apical series; possible formation of Type P archeopyles of the Gonyaulacystineae through fusion of a cluster of adjacent dorsal intercalary and precingular paraplates; and possible evolution of peridinioid cysts through such pseudoceratiaceans as *Muderongia simplex*.

This Author's Personal Comment: The value of this paper as a provocative inspiration to further research probably cannot be overestimated. Many of the issues Dörhöfer and Davies raise will not be settled before we know a great deal more than we do now. On the way, unanticipated discoveries will help define other potentially fruitful studies. However, much time and effort will also be consumed in rewriting what appears to this author to be some very poor paleontology.

The dinoflagellates that are less well known—whether owing to deficiency in number of available specimens, inferior quality of preservation, or limited informativeness of morphological features—offer special challenge and opportunity. Especially is this true of dinoflagellates from the Late Triassic and Jurassic within the categories encompassed by the Rhaetogonyaulacineae (sensu Norris; i.e., without Pseudoceratiaceae) and the Pseudoceratiaceae (sensu Dörhöfer and Davies), for, in both temporal and morphological respects, they occupy positions of intense interest. However, for precisely that reason, they should be investigated with meticulous care and interpreted with special caution.

Some of Dörhöfer and Davies' observations undoubtedly are correct and some of their interpretations also may be, although the evidence offered often seems meager to justify the confidence expressed. Other interpretations (e.g., that patterns on the operculum of forms like *Cribroperidinium* indicate origin of *3″* from fusion of three intercalaries and three precingulars) are contraindicated by substantial evidence (relative to the example cited, see Gocht, 1979). Some of their SEM's and light micrographs show clearly what is claimed; others do not.

This author can strongly support their conclusions regarding the close relationship of genera within the three families they unite in the Pseudoceratiaceae, the probable existence of a significant relation between the imbrication pattern of thecal plates and the location and composition of the archeopyle, the contrast between the "suessioid" (in their sense) paratabulation pattern of the Rhaetogonyaulacineae and that represented by typical gonyaulacoids of the Middle Jurassic and later, and others. On the other hand, he cannot escape a very strong feeling that their principal ideas on phylogeny have the substance of a "house of cards." It appears to him that, in a conspicuous number of instances, critical interpretations of paratabulation are based on suggestive but probably fortuitous alignments of folds, tears, and elements of ornament on poorly preserved and relatively featureless specimens. On such foundations stand sweeping generalizations and imaginative constructions where logic is set aside and fancy is cited as proof.

A point of internal conflict is of more than incidental interest. The thrust of their arguments rests heavily on the assumption, whose validity is contested vigorously in this volume, that the time and sequence in which new features appeared in the evolution of dinoflagellates are faithfully recorded by the stratigraphic succession of fossil cysts accessible to us. Nevertheless, they nowhere mention Silurian *Arpylorus*, with its few and highly organized paraplates and paraplate series, and, to use one of their terms, a "stabilized" archeopyle involving a single large middorsal paraplate. Judged by their own morphological standards, this vexingly precocious harbinger of things to come would have to be accepted as considerably "more advanced" than their "primitive" rhaetogonyaulacineans from the Late Triassic.

Ideas of Eaton (1980) on Peridinialean Evolution

A thorough and truly novel reevaluation of tabulation patterns lies behind the comprehensive views of Eaton (1980) regarding relationships and evolution within the Peridiniales. These views illustrate the flood of new ideas that can be the reward of a perceptive and critical "new look at old evidence." With an approach that resembles to some degree Bütschli's (1885), Eaton replaces the established Kofoidian (1909a) analysis of peridinialean tabulation with one in which the cingulum forms an absolute line of reference for recognizing several latitudinal series of plates toward each pole. The result requires redefinition of plate series near the poles (especially anterior and posterior intercalaries) and leads to an interpretation of plate homologies quite in contrast to any suggested heretofore.

After careful exposition of his basic thesis and development of a terminology and symbolism to express it (with critical differences from Kofoid's, which it superficially resembles), Eaton analyses a large number of tabulation patterns from both modern thecae and fossil cysts. In broad terms, Eaton's ideas include:

1. The primitive tabulation of peridinialean dinoflagellates consisted of many series of plates with many plates in each series (cf. *Suessia*).

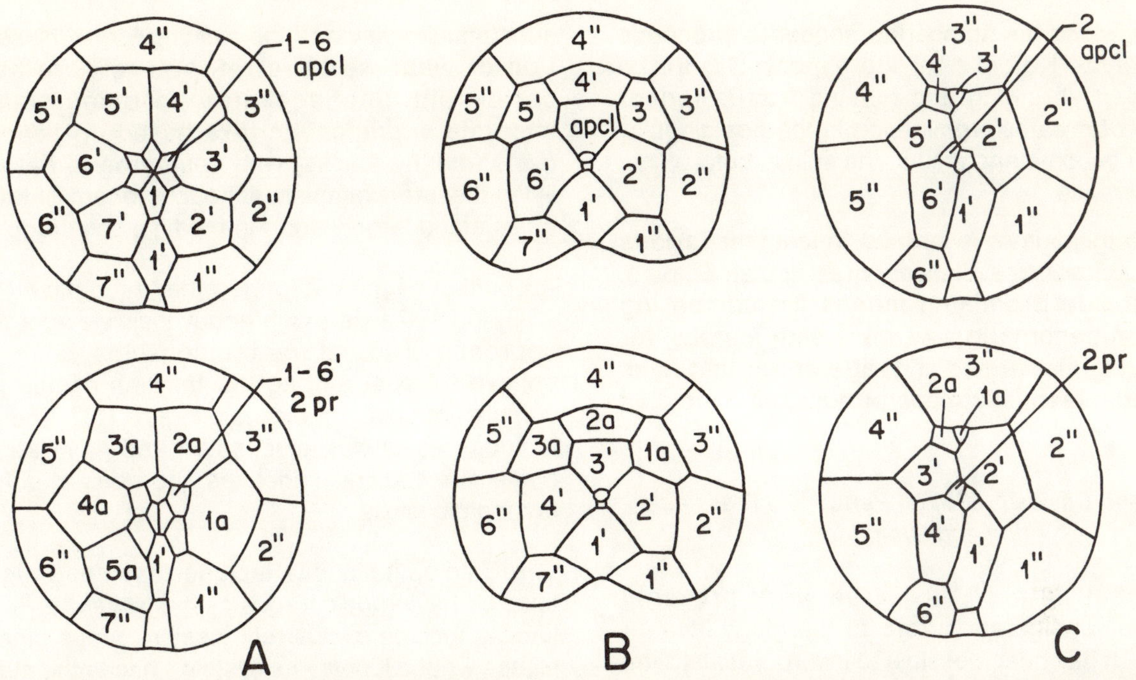

Figure 13.1 - THREE EXAMPLES OF THE APPLICATION OF EATON AND KOFOID SYMBOLS FOR THE EPITHECA

Plates in the lower diagrams are labeled according to Kofoid's system, essentially without regard to whether plates bearing the same labels in the three diagrams are considered homologs. The Kofoid-like symbols in the upper diagrams are those employed by Eaton to indicate his concept of plate homologies among the three patterns. apcl = apical closing series

A - *Rhaetogonyaulax rhaetica* (Late Triassic). Paratabulation in lower diagram as reinterpreted by Fisher and Van Helden (1979). B - *Protoperidinium depressum* (living species). C - *Ctenidodinium* sp. (Jurassic).

2. A more advanced ancestor of modern peridinialeans had plates in five latitudinal series in a 4-7-7-7-4 sequence; such forms as *Rhaetogonyaulax* and *Dapcodinium* are rather close derivatives from this. (Although this enumeration resembles the 3-6-6-6-3 primary count of Taylor, the arrangements within the series, especially toward the poles, are fundamentally different, and the two primitive patterns have no conceptual relationship.)

3. With evolution, the number of plates was reduced by one or a combination of: fusion of adjacent plates, outright deletion of a plate anywhere in a series, or selective loss (through progressive shrinkage and eventual disappearance) of those plates at the right end of plate series in the epicyst and at the left end of series in the hypocyst.

Figure 13.1 will convey some of his ideas, but careful study of the original paper will be necessary to sense their full impact.

This Author's Personal Comment: The scope and potential significance of the study, the novelty of the ideas expressed, and the volume of material considered render premature any attempt at a detailed summary or critical evaluation. Certain it is that both controversy and new insights will ensue. Eaton's notions bear little resemblance to those of Taylor, which have received close attention in this volume, and the two sets of ideas will not easily accomodate one another. Significantly, both systems are substantial departures from recent practice and purport to identify homologous plates, labeling them in a consistent way that facilitates discussion of affinity and evolution. However, many intriguing conflicts between the two systems become apparent when they are compared. For example, the small anterior intercalaries (in Kofoidian sense—**Ki** and **Ku** in the Taylor terminology used in this volume) in a basically gonyaulacoid pattern such as *Ctenidodinium* (Figure 5.8 H) are accomodated in Taylor's system as incidental acquisitions without great significance, but, for Eaton, they are significant relicts from an ancestral pattern in which they were larger plates equal to others of the apical series. Another example relates to the first Kofoidian postcingular in the typical gonyaulacoid pattern. For Taylor, this is a subdivision of a midventral primary that has, so to speak, moved out onto the flank of the hypocyst from the sulcal region. For Eaton, the same plate (which stands as the last numbered element of the postcingular series in his notation, since his numbering sequence proceeds oppositely to Kofoid's for the hypocyst) is in the process of shrinking and moving from a former position of greater prominence in the postcingular series to become engulfed in the sulcus before final disappearance.

Although this author is inclined to feel that Taylor's ideas are closer to evolutionary reality than Eaton's, both provide promising means for comparing tabulation patterns in new ways and, thereby, for discovering similarities and differences that have remained essentially hidden under the Kofoidian system.

Bujak and Davies (1983) on Peridinioid Classification and Evolution

The recent date of this large work prevents adequate consideration here of the many ideas it contains. It has been possible to incorporate into the earlier part of this volume only a few of these authors' contributions, and the discussion of their views on classification and evolution must be similarly limited. Bujak and Davies present a

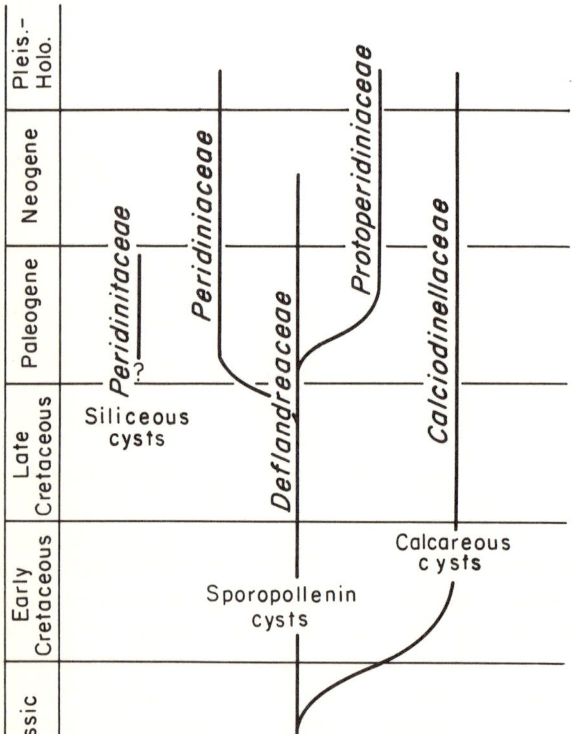

Figure 13.2 - INTERRELATIONS AMONG FAMILIES OF SUBORDER PERIDINIINEAE ACCORDING TO BUJAK AND DAVIES (1983)

substantial revision of the existing formal classification of what we have referred to here as the peridinioid dinoflagellates. In doing so, they integrate suprageneric taxa based on genera of cysts (mostly fossils) with counterparts based on living thecate examples, although genera of the two types stand largely apart in the final construction.

An outline of the categories they recognize (Table 13.3) in itself tells much about their viewpoint and approach. The natural relationships which they believe to exist among the families of this great complex of taxa are shown in Figure 13.2. Their text includes extensive discussion and illustration of numerous additional lineages and trees at different taxonomic levels.

The Calciodinellaceae include fossil and modern species having calcareous cysts, while the Peridinitaceae include exclusively fossil siliceous dinoflagellates whose identity as cysts or thecae the authors consider undecided. Species producing sporopollenin cysts are confined to the remaining three families—Deflandreaceae, Peridiniaceae, and Protoperidiniaceae. The Deflandreaceae, the major group of fossil peridinioids with sporopollenin cysts, is divided into six subfamilies (all extinct), one of which (Palaeoperidinioideae) was the stock which has yielded modern freshwater Peridiniaceae. The modern marine Protoperidiniaceae was also derived from the Deflandreaceae (through its subfamily Senegalinioideae, comprising three fossil genera; *Senegalinium*, *Phelodinium*, and *Uvatodinium*).

Bujak and Davies maintain as separate genera in the subfamily Protoperidinioideae most of the cyst genera based on Pleistocene and Holocene sporopollenin cysts of peridinioid type which had been proposed largely by Reid, Bradford, and Harland. These are the genera which Harland (1982b) allocated to subgenera and sections of *Protoperidinium*. Bujak and Davies do not assess the taxonomic infrastructure of *Protoperidinium* but discount the role of homeomorphy claimed by Harland to explain certain similarities between cysts in different sections of this genus.

This Author's Personal Comment: Bujak and Davies' thorough treatment of thecal and cyst morphology in the first part of this work includes much that is new, as well as a welcome synthesis of many previously published observations. Undoubtedly it will serve as an important and useful guide for future descriptive work. The uniqueness and significance of the taxonomic part of their work lie in the careful attempt to build a comprehensive classification of a major group of dinoflagellates on foundations of both cyst and thecal morphology, while giving due attention also to stratigraphic occurrence. It is, perhaps, still uncertain whether the morphological criteria chosen (archeopyle, cyst composition and wall structure, details of tabulation and paratabulation) will prove effective in the long term at the levels at which they have been applied. Nevertheless, the proposed lineages and linkages provide a focus for future investigation and discussion about evolution within the group. In a broad view, progress on many fronts should be stimulated by the proposals set forth in this work.

Regarding selectivity of the dinoflagellate fossil record and the reliability of stratigraphic occurrence as evidence in support of particular phylogenies of dinoflagellates, some of the criticisms mentioned in the commentary on Norris' views also apply here. Bujak and Davies, however, take pains in their introduction to explain their reasons for discounting the significance of the points raised by Evitt (1980, and elaborated in Chapter 2 of this work). Readers interested in this debate will wish to assess for themselves the logic and substance of the argument.

Table 13.3 - SUPRAGENERIC CLASSIFICATION OF PERIDINIOID DINOFLAGELLATES BY BUJAK AND DAVIES (1983)

Order	Peridiniales Haeckel 1894
Suborder	Peridiniineae Fott 1959 emend.
Family	Calciodinellaceae Deflandre 1947 emend.
Family	Deflandreaceae Eisenack 1954 emend.
Subfamily	Deflandreoideae nov.
Tribe	Ceratiopsieae nov. (as Ceratiopseae)
Tribe	Chatangielleae Vozzhennikova 1967 stat. nov. em.
Tribe	Deflandreae nov.
Tribe	Lentinieae nov.
Tribe	Manumielleae nov.
Tribe	Nelsonielleae Eisenack 1964 stat. nov. emend.
Tribe	Spinidinieae nov.
Subfamily	Ovoidinioideae Norris 1978 stat. nov. emend.
Subfamily	Palaeocystodinioideae nov.
Subfamily	Palaeoperidinioideae Vozzhennikova 1961 stat. nov. em.
Tribe	Chichaouadinieae nov.
Tribe	Luxadinieae nov.
Tribe	Palaeoperidinieae nov.
Tribe	Phthanoperidinieae Drugg & Loeblich 1967 stat. nov. em.
Subfamily	Subtilisphaeroideae nov.
Subfamily	Wetzelielloideae Vozzhennikova 1961 stat. nov. em.
Family	Peridiniaceae Ehrenberg 1832 emend.
Family	Peridinitaceae nom. subst. pro Lithoperidiniaceae Deflandre 1947 emend.
Family	Protoperidiniaceae nov.
Subfamily	Diplopsaloideae nov.
Subfamily	Protoperidinioideae nov.
Subfamily	Senegalinioideae nov.

TAYLOR'S IDEAS REVISITED

Up to this point, we have considered the Taylorian system only in the context of gonyaulacoid patterns. However, Taylor (1980) also applied his reasoning to the peridinioid forms, and it is to them and their possible relationship to gonyaulacoid types that we will now turn our attention.

One objective of the discussion relative to gonyaulacoid dinoflagellates in Chapter 5 was to advocate the greater utility of Taylor's system relative to Kofoid's, particularly for comparing different patterns. There are three reasons why a comparable replacement of Kofoid's system by Taylor's in reference to peridinioid patterns is not advocated with equal vigor here. First, the Kofoidian system is satisfactory for describing and comparing the limited variety of highly symmetrical peridinioid patterns normally encountered by the paleontologist. Second, the evidence for hypothesizing a primary peridinioid model and the steps from it to the patterns we see today are less clear than in the gonyaulacoid case. Third, although some striking similarities between gonyaulacoid and peridinioid patterns can be pointed out, the exact homologies that may exist between them remain speculative.

Our objective in returning to Taylor's ideas in a broader perspective now is expressly to introduce some of these speculations and to suggest some of the avenues they open up for looking at familiar tabulation patterns in a new light. Questions and ideas that may be evoked in the curious reader-observer —whether generally in support of or contesting the notions presented here—may lead to discovering what others have not seen: some bit of evidence that confirms unproven, or even unsuspected, relationships between these major groups of dinoflagellates.

To superficial inspection, a generalized ortho hexa peridinioid tabulation (Figure 5.2 D-I) may appear to have little in common with a generalized gonyaulacoid pattern (Figure 5.10 A-L). However, in Taylor's view, both these examples are derivable from simpler primary models by calling upon the same mechanisms already discussed in conjunction with gonyaulacoid patterns in Chapter 5, namely: 1) moderate migration of certain plates along sutures between adjacent plates, and 2) subdivision of a small number of primaries. Furthermore, the primary models envisaged as starting points for these disparate products show remarkable similarities in general and intriguing differences in detail.

Although the discussion which follows adheres to the "spirit" of Taylor's approach, some of the interpretations and speculations mentioned differ in details from those already suggested by him (Taylor, 1980). The reader should consult his work for these other possibilities.

A comparison of the hypothesized paths from primary models to existing gonyaulacoid and peridinioid patterns is best begun by considering the hypotheca (Figure 13.3). Figure 13.3 A-C recapitulate early steps in the gonyaulacoid sequence presented more fully in Chapter 5. These diagrams are labeled throughout with the now-familiar Taylor symbols. Figure 13.3 F shows a representative peridinioid hypothecal pattern labeled with Kofoid symbols. The hypothesized sequence of changes leading up to this completed peridinioid product is suggested by Figure 13.3 D-E, which depict stages roughly equivalent to those in the gonyaulacoid series portrayed above them in Figure 13.3 A-B.

Note that the modification of the primary model to produce the terminal peridinioid tabulation involves only rather moderate changes in plate proportions. No topologic contrasts are introduced except in relation to subdividing the midventral postequatorial primary into secondary sulcals. Note also how similar to one another the two primary models are. The six postequatorials occupy identical positions; the three antipolars have identical shapes and relate to the postequatorials in identical ways. In fact, the figures are topologically identical throughout, differing only in orientation relative the ventral point of flagellar insertion. This difference involves an apparent rotation of one pattern by 60 degrees with respect to the other. Significant effects of the rotation are, first, that the antipolar series includes a midventral *plate* in the peridinioid pattern but a midventral *suture* in the gonyaulacoid, and second, that the middorsal postequatorial is camerate in the peridinioid pattern but planate in the gonyaulacoid, and other plates are similarly affected.

The letters "A" and "Y" superposed on the antipolar triads (Figure 13.9), with their bases directed ventrally (i.e., toward the point of flagellar insertion), were proposed as graphically suggestive designates for these two distinctive orientations. Thus, one may speak of the primary gonyaulacoid hypothecal pattern as an "A" pattern and the primary peridinioid hypothecal pattern as a "Y" pattern.

In both primary patterns, one of the antipolars can be imagined as the source of the posterior sulcal plate of more advanced types. In the gonyaulacoid case, as already discussed (Chapter 5), modification beyond the quinqueform stage shown here (Figure 13.3 C) is hypothesized to have involved an oblique displacement (to the left and toward the equator) of primary plate **Z** as contact was established between **VI** and **Y** to produce the sexiform hypotheca of *Gonyaulax*. In the peridinioid case, a slight migration of the midventral antipolar directly toward the equator would have sufficed to achieve the fully "advanced" condition in modern *Protoperidinium* (Figure 13.3 F).

Comparable gonyaulacoid-peridinioid similarities can be found in the epitheca (Figure 13.4), but they are less readily apparent, "disguised" by features (more plates and additional series) not present in the gonyaulacoid epitheca. Such contrast between hypothecal and epithecal similarities should not be surprising in view of the generally more highly differentiated character of the epitheca.

Figure 13.4 A-C recall the progression from the hypothetical gonyaulacoid epithecal pattern, with full radial symmetry, through the *Pyrodinium*-stage with metasert **1u**, to the *Gonyaulax*-stage in which insert **1u** makes the ventral apical. Figure 13.4 F shows a representative peridinioid epithecal pattern, incorporating the ortho hexa features most common in the fossil record and labeled with Kofoid symbols. The hypothesized sequence of changes leading up to the completed peridinioid product is suggested by Figure 13.4 D-E. In part, these are similar to and, in part, different from the changes in the gonyaulacoid series. The ventral apical (*1′*) is similarly derived from the ventral preequatorial—but from all of it, and by a direct poleward migration along the ventral suture between primary polars. *Protoperidinium steidingerae*, a modern species whose "first apical" is an "intercalary" in that it fails by far to reach the apex (Balech, 1979), expresses what might be imagined as an intermediate stage in this migration.

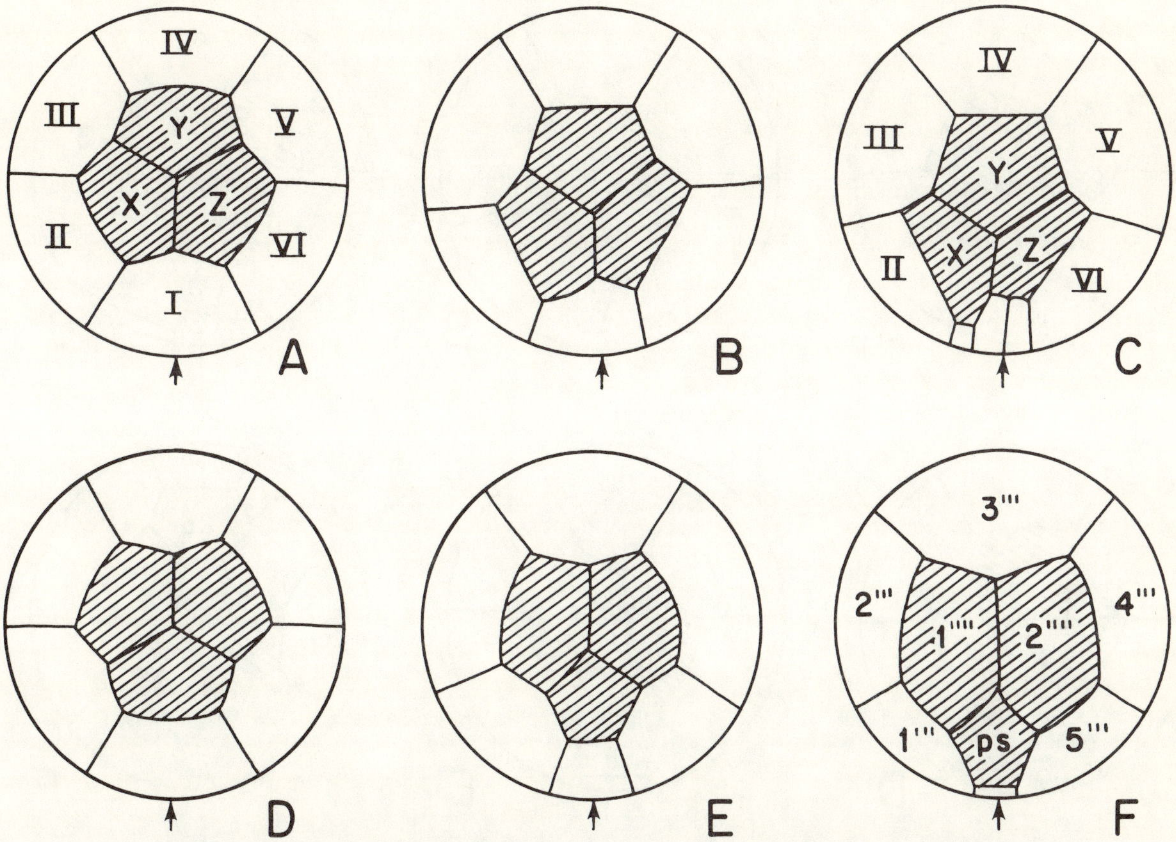

Figure 13.3 - COMPARISON OF PERIDINIOID AND GONYAULACOID HYPOTHECAL PATTERNS

Patterns in antapical/antipolar view. Shaded = primary antipolars and their derivatives. Arrows point to locus of flagellar insertion.

A-C - Gonyaulacoid patterns. A - Hypothetical primary arrangement. B - Intermediate stage showing distortion of antipolar and ventral areas. C - Schematic, representative of modern *Triadinium*, showing three ventral secondaries derived from I.

D-F - Peridinioid patterns. D - Hypothetical primary arrangement. E - Intermediate stage showing distortion of antipolar and ventral areas. F - Schematic, representative of modern *Protoperidinium*, *Peridinium*, and similar genera.

The absence of a similarly positioned suture extending poleward from gonyaulacoid primary **1** can be viewed as the reason for an initial oblique migration of **1u** along **A/2** before direct poleward migration along **A/B** could establish **1u** as a full apical.

Although seven precingulars (instead of six) and three anterior intercalaries distinguish the final peridinioid epitheca from the gonyaulacoid, the "additional" plates can be accounted for in a relatively simple fashion by dividing primaries along the dotted lines in Figure 13.4 E (commented upon further below).

In comparing the peridinioid and gonyaulacoid epithecal patterns, note that, as for the hypotheca, the hypothesized primary models differ not at all in topology, but only in orientation relative to the ventral point of flagellar insertion. Again, the primary epithecal patterns can be distinguished as an A-pattern and a Y-pattern, but it is now the peridinioid which is A and the gonyaulacoid which is Y. Combining the letter designations for the two hemispheres and specifying first the epithecal pattern, followed by the hypothecal pattern, we can distinguish the hypothetical YA primary model of the gonyaulacoids from the AY primary model of the peridinioids.

A further interesting contrast between gonyaulacoids and peridinioids can be correlated with this "reciprocal" relationship between the AY and YA primary tabulations. If one considers the sequence of sutures and plates along a middorsal line extending from pole to antipole, the differences between the two primary models are exactly those which actually distinguish the symmetrical dorsal tabulations of gonyaulacoid and peridinioid species living today (Figure 13.5).

Diagrams to suggest how the known assortment of peridinioid epithecal patterns might have arisen from the primary model have been presented by

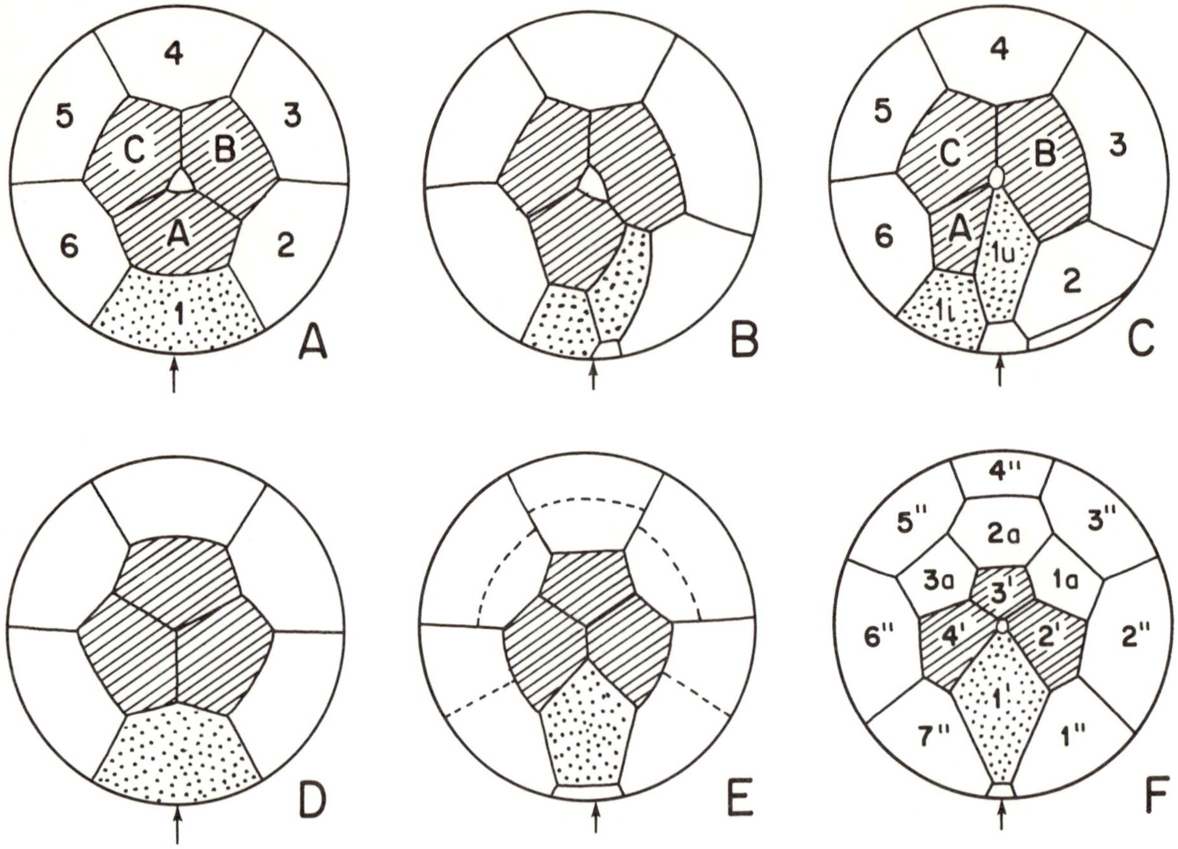

Figure 13.4 - COMPARISON OF PERIDINIOID AND GONYAULACOID EPITHECAL PATTERNS

Patterns in apical/polar view. Lined = primary polars and their derivatives. Stippled = primary midventral preequatorial and its derivatives. Arrows point to locus of flagellar insertion.

A-C - Gonyaulacoid patterns. A - Hypothetical primary arrangement. B -Intermediate stage, schematic, representative of modern *Pyrodinium*. Radial symmetry has been distorted and metasert **1u** results from oblique poleward migration of a portion of the midventral preequatorial. C -Schematic, representative of modern *Gonyaulax*, with insert **1u** forming a nearly symmetrical ventral apical.

D-F - Peridinioid patterns. D - Hypothetical primary arrangement. E -Intermediate stage in which radial symmetry has been distorted (although bilateral symmetry has been retained) by moderate changes in plate shapes and by poleward migration of the midventral preequatorial. Dotted lines show incipient longitudinal division of two ventral preequatorials and latitudinal division of three dorsal ones. F - Schematic, representative of a modern ortho hexa species of *Protoperidinium*.

Taylor (1980) and Bint (1983). Their illustrations involve detail beyond the scope of this discussion, but Figure 13.6 shows that only slightly different positions for the secondary sutures by which the preequatorial primaries are divided would account for the distinctive ortho-meta-para and quadra-penta-hexa arrangements, the differences depend merely upon whether the secondary suture lies to one side or the other of a particular triple junction. It is interesting that, in living *Protoperidinium*, thecal plate *2a* overlaps both *3''* and *5''* in a direction obliquely toward the cingulum. Although contrary to the general tendency for dorsal epithecal plates to overlap toward the apex, this is what would be expected if the intercalaries had originated by division of primary preequatorials as suggested by Figure 13.6.

According to the Taylor hypothesis, the plates of the gonyaulacoid sulcus and immediately adjacent area involve division of three primaries—the first and last equatorials (**a** and **f**) and the first postequatorial (**I**)—as well as minor shifts in their spatial relationships. These changes are reviewed in Figure 13.7 A-D. The subjacent Figure 13.7 E-H hypothesize a sequence of comparable steps by which the plates of the peridinioid sulcus and adjacent cingular region may have originated from the peridinioid primary model. Note that the ventral primary patterns bear a "reciprocal" relationship to one another, just like the dorsal ones shown in Figure 13.5.

By this hypothesis, the first and last equatorials of the peridinioid model are divided to yield terminal cingulars and plates occupying positions analogous

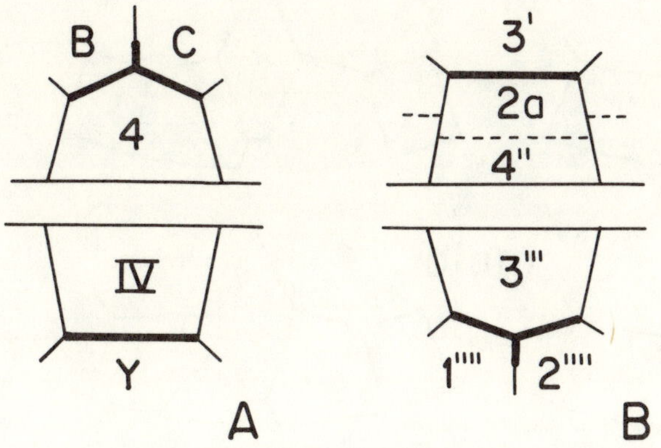

Figure 13.5 - MIDDORSAL PLATES AND SUTURES IN PERIDINIOID AND GONYAULACOID PATTERNS

The diagrams in each case apply equally well to bilaterally symmetrical thecae of many living species and to the hypothetical primary models. Dotted lines show location of divisions by which peridinioid anterior intercalaries are hypothesized to have arisen from preequatorial primaries. Heavy lines emphasize the reciprocal orientations of the two patterns.

A - Gonyaulacoid middorsal sequence: interapical suture (**B/C**) - camerate preequatorial plate (**4**) - planate postequatorial plate (**IV**) - antapical plate (**Y**).

B - Peridinioid middorsal sequence: apical plate (*3'*) - intercalary + preequatorial plates (*2a* + *4''*) one above the other separated by rectilinear *3'/2a* - camerate postequatorial plate (*3'''*) - interantapical suture (*1''''/2''''*).

to **fu** and **ai** in the gonyaulacoid sulcus. The midventral postequatorial is divided—but into two parts rather than three like the gonyaulacoid **I**—to provide two of the principal sulcals. Fusion of two of the secondary products of these divisions along the primary suture that originally separated them is called upon to construct the peridinioid right sulcal. Although such fusion is without counterpart in the hypothesized development of the gonyaulacoid sulcus, union of plates that are otherwise separate is almost certainly recorded elsewhere in the thecae of living peridinioids; for example, in the reduced number and greater size of dorsal cingulars in *Protoperidinium* (Figure 1.6 H), and in the single antapical in living *P. unipes* (Figure 1.6 T).

Eventual contact (Figure 13.7 G-H) between the seventh precingular (a derivative from the preequatorial lying to the right of the midventral one) and the anterior sulcal (derived from the first equatorial) is a topologic novelty in the peridinioid sequence that may be associated with the poleward "withdrawal" of the midventral preequatorial (eventual *1'*) from its primary preequatorial position. The arrow in Figure 13.7 F-G explains this "new" contact between *7''* and the anterior sulcal as resulting from a leftward shift in the adcingular end of the *1'/7''* suture relative to the midventral suture extending from the flagellar pore.

Variations in the sulcal tabulations of both gonyaulacoids and peridinioids include the presence of one or more additional small plates, some of which may be internal. It is not intended that the hypothetical lines of development suggested here should account for such minor and variable features. Although unlikely to be observed in fossil cysts, "extra" sulcal elements occasionally may be conspicuous (Figure 10.9 D).

The thrust of the foregoing discussion has been to emphasize perceived similarities between peridinioid and gonyaulacoid tabulations, and to suggest ways in which they may have developed by similar mechanisms from similar but distinct starting points. Now we are ready to consider the kind of a case that can be made for the existence and recognition of homologies between the plates of peridinioid and gonyaulacoid thecae.

Both the AY (peridinioid) and YA (gonyaulacoid) primary models have 24 plates in only four different shapes, arranged regularly and symmetrically in series of 3, 6, 6, 6, and 3 plates. If one considers the relative simplicity, order, and symmetry of these two models, several questions come readily to mind. Do exact homologies exist between plates in the peridinioid and gonyaulacoid patterns? If so, to what extent can they be recognized? Are the AY and YA combinations the only expressions of this 3-6-6-6-3 design among dinoflagellates, or do some living or fossil dinoflagellates represent derivatives of the other two combinations that can be imagined—AA and YY? Pursuit of speculative answers to these questions will bring our discussion of the Taylor approach to a close.

If the AY and YA patterns are taken as idealized geometrical figures on a spherical surface, there are two ways to transform one directly into the other. The first is by a 180-degree rotation about an equatorial axis passing either through the center of an equatorial plate or the midpoint of an intra-equatorial suture. The second is by a rotation of 60 or 180 degrees about the polar axis. A third but less direct conversion is possible if one imagines the two patterns, not as static geometric figures, but as constructs of 24 plates that became associated in two stages. One stage would be represented by an encircling band of 18 plates organized into three

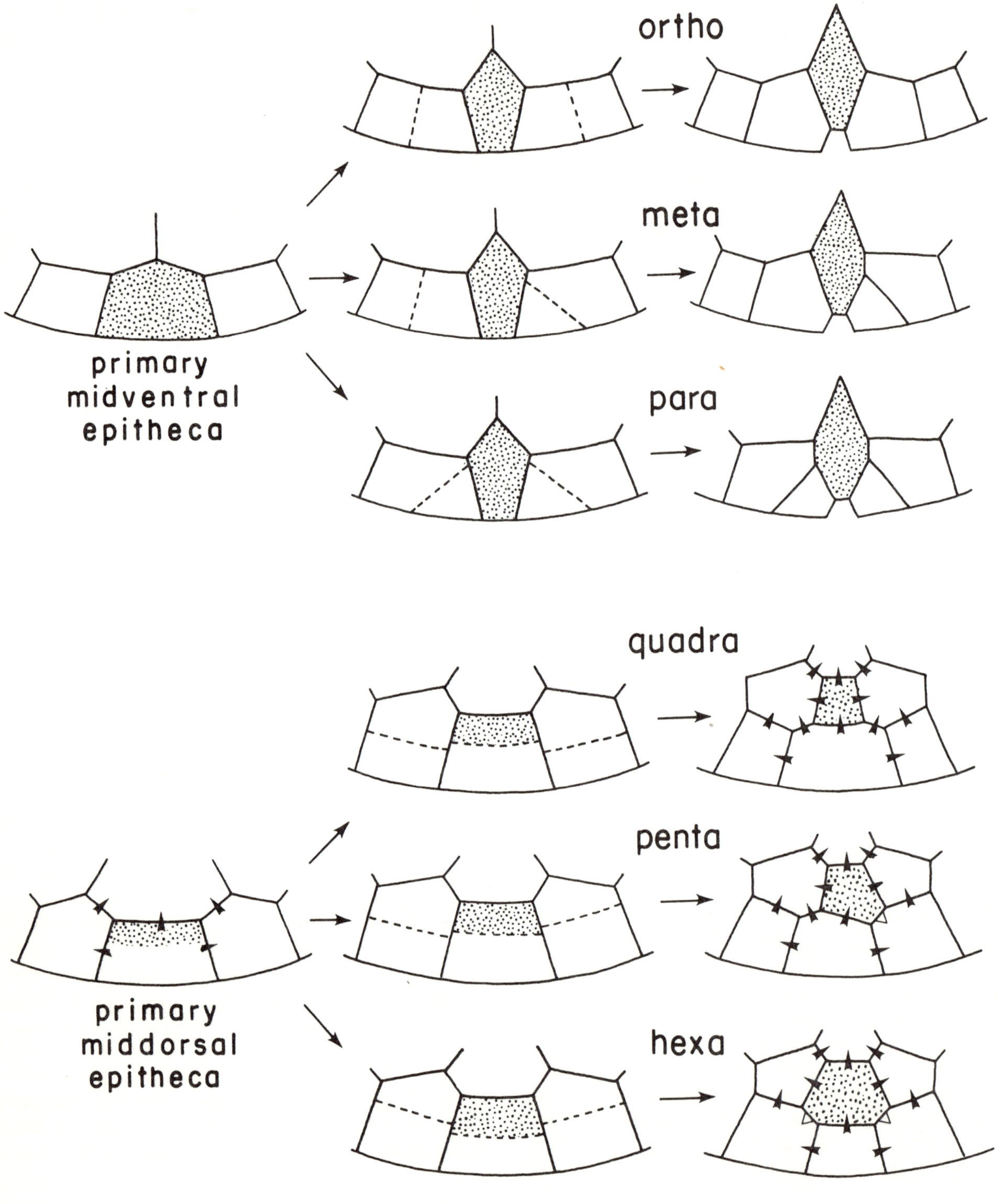

Figure 13.6 - ORIGIN OF ORTHO-META-PARA AND QUADRA-PENTA-HEXA PATTERNS

Hypothesized subdivision of primary preequatorial plates that would explain the origin of seven precingulars and three intercalaries characteristic of the most common peridinioid primary patterns while, at the same time, accounting for the observed differences between ortho, meta, and para ventral arrangements (upper diagrams) and between quadra, penta and hexa dorsal arrangements (lower diagrams).

Dashed lines = incipient sutures subdividing primary plates. Stippling identifies areas that form the first apical and second intercalary paraplates. Arrowheads on sutures point in direction of overlap—"anomalous" overlap directions indicated by points of open triangles (see text for discussion).

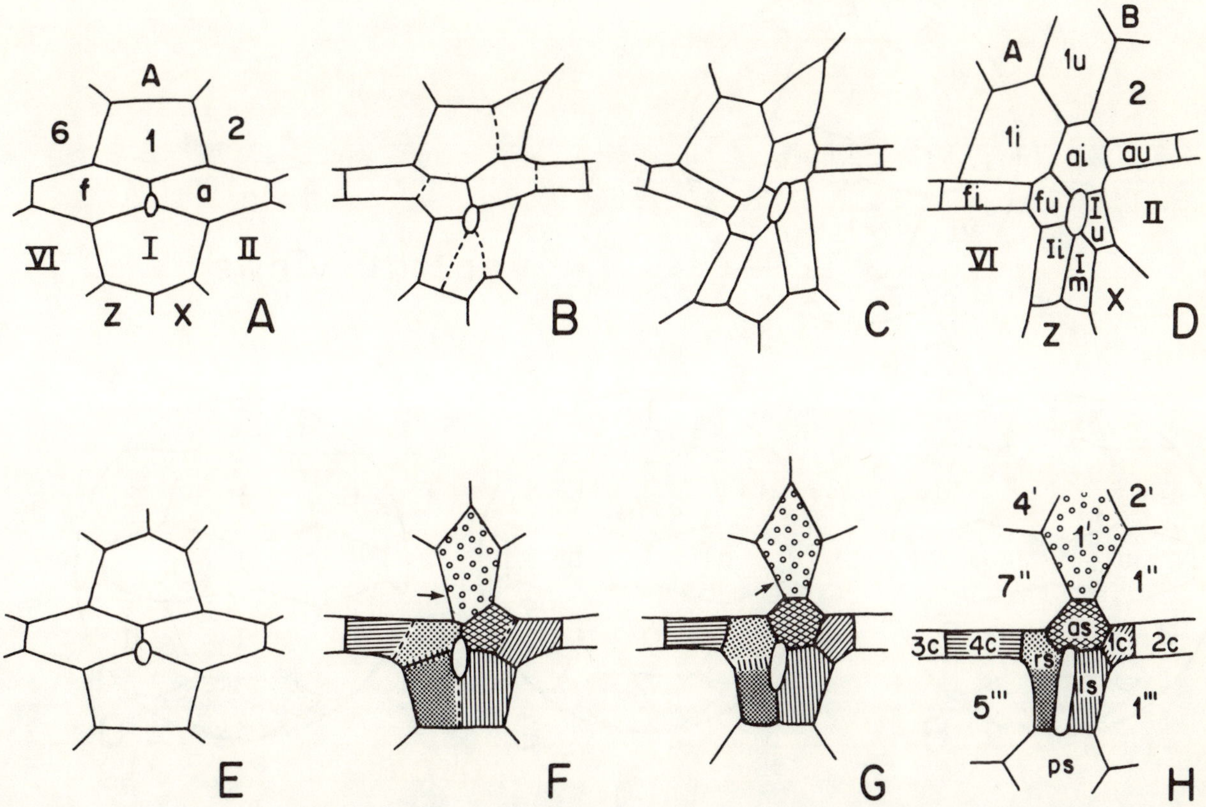

Figure 13.7 - HYPOTHETICAL DERIVATION OF SULCAL PLATES

A-D - Gonyaulacoid Pattern. These diagrams are repeated here from Figure 5.7 A-D to facilitate comparison with the following.

E-H - Peridinioid Pattern. Hypothesized changes are similar to those in gonyaulacoids but differ in part owing to different topology of the midventral pre- and postequatorial primaries (**1** and **I**). Areal patterns identify regions thought to be homologous. Dotted lines indicate incipient division of primaries. Hachures mark primary suture along which secondary fusion is hypothesized. Arrows indicate shift in *1′/7″* suture relative to midventral suture extending from the flagellar pore. Labels on cingulars in H are appropriate for *Protoperidinium* (*1c* = *t* of some authors).

series of six each. The other stage would be represented by the organization of six plates into a triad at each pole. It would not be critical which stage occurred first, although early formation of the equatorial complex might seem the more plausible alternative.

These three alternative ways in which the peridinioid and gonyaulacoid primary patterns might relate to one another are illustrated in Figure 13.8, with the gonyaulacoid (YA) model seen in ventral and apical views and labeled with Taylor symbols (Figure 13.8 A, E). Homologies that would exist following each of the alternative "conversions" from YA to AY are indicated by Taylor symbols on the peridinioid (AY) equivalents of each view.

The existence of one unique locus on the dinoflagellate surface—the point of flagellar insertion—is assumed to predate the origin of the AY and YA arrangements. The position assumed for this locus, along an intraequatorial suture that remains ventral in both patterns, also serves to determine the one equatorial axis of rotation (out of 12 possibilities) about which the first method of "conversion" (Figure 13.8 B, F) might operate. For the second method of "conversion," only one out of three possible positions following rotation about the polar axis is shown (Figure 13.8 C, G); the other two are identified only by the Taylor symbols in parentheses on the midventral preequatorial plates.

Note that, in the left and middle peridinioid diagrams which show the results of direct rotation, each peridinioid primary is necessarily topologically and geometrically identical to its counterpart in the gonyaulacoid model. But this is not so for the third possibility represented by the righthand diagrams. In them, a rotation (about a polar axis) of the polar and antipolar triads by 60 degrees in relation to their pre- and postequatorial neighbors is proposed to

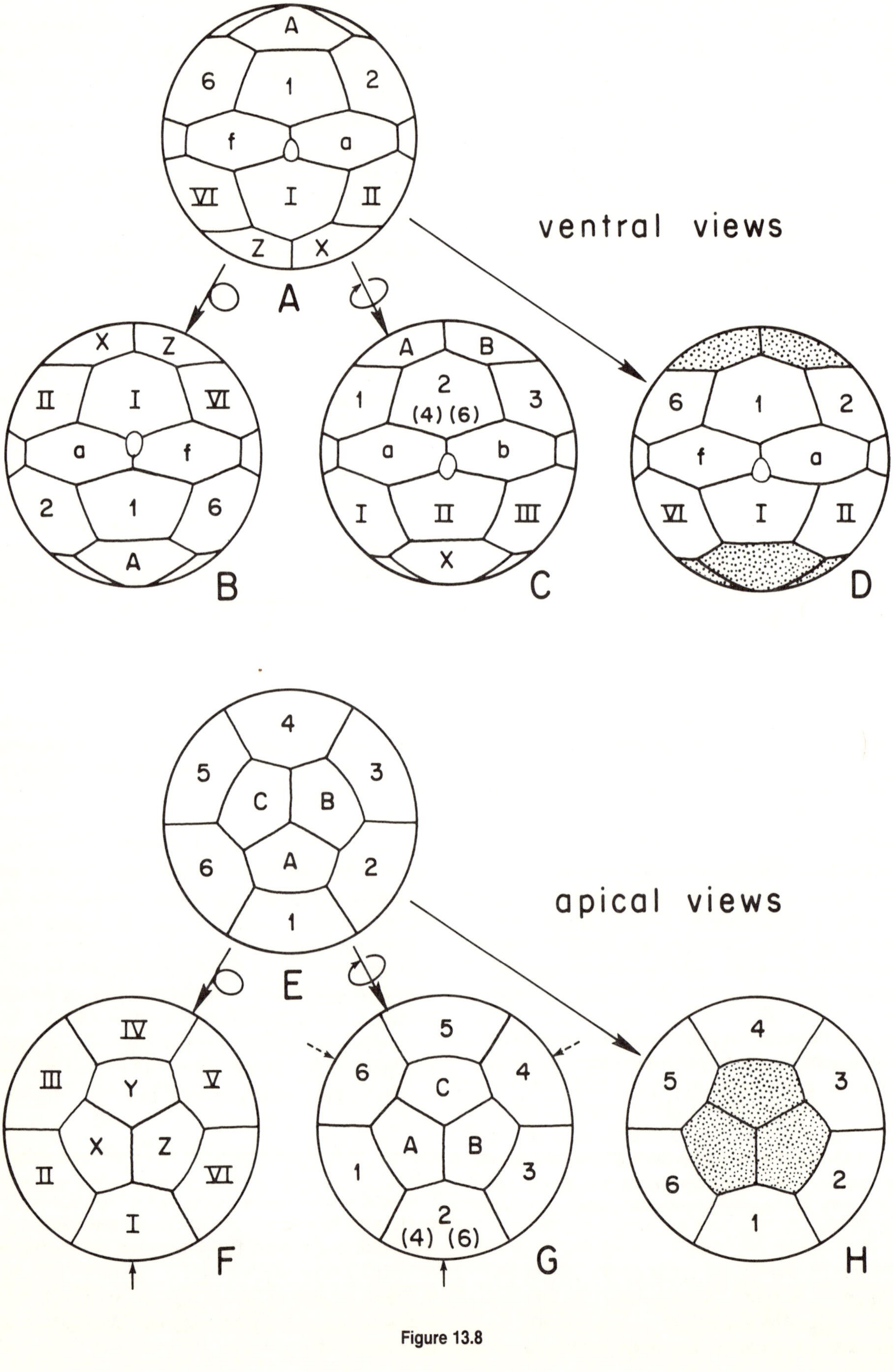
ventral views
A
6
1
2
f
a
VI
I
II
Z
X
B
X
Z
II
I
VI
a
f
2
1
6
A
C
A
B
1
2
(4) (6)
3
a
b
I
II
III
X
D
6
1
2
f
a
VI
I
II
apical views
E
4
5
C
B
3
6
A
2
1
F
IV
III
Y
V
X
Z
II
VI
I
G
5
6
C
4
A
B
1
3
2
(4) (6)
H
4
5
3
6
2
1

Figure 13.8

establish the topologic relationships that are diagnostic of the peridinioid pattern. However, the positions of and relationships among the preequatorials, equatorials, and postequatorials are the same as in the gonyaulacoid model.

Taylor (1980) pointed out that if peridinioids and gonyaulacoids were related to one another according to the first method of "conversion," they would, in effect, swim in opposite directions relative to the primary tabulation (i.e., one with **A-B-C** foremost, the other with **X-Y-Z**). Significant evidence in support of such a relationship might be claimed if the transverse flagellum extended in opposite directions from the flagellar pore in the two groups. That, in fact, it does not do so makes the option less appealing, although one might further heighten the level of speculation by supposing that the transverse flagellum became associated with the cell surface only after the direction of movement had been determined.

The second method of "conversion"—rotation about the polar axis—implies that the flagella in the two groups emerge at different points along the equatorial series of plates. The topologic requirements of the peridinioid pattern would be met equally well by this method if the point of flagellar insertion (expressed in terms of the gonyaulacoid primary labeling) were at any one of three loci: along the **a/b**, **c/d**, or **e/f** sutures (i.e., lying adjacent to primary preequatorials **2**, **4**, or **6**, respectively). Although topologically equivalent, these three options differ in other respects. Flagellar insertion along **c/d** would retain the same plane of approximately bilateral symmetry as in the gonyaulacoid model but the shift of flagellar insertion to the "dorsal" surface would seem extreme. The other alternatives would retain the flagellar insertion closer to its gonyaulacoid position but would shift the sagittal plane 60 degrees to the right or left, implying that the approximation of bilateral symmetry is different in gonyaulacoids and peridinioids.

The third method of "conversion" involves no alterations in swimming direction and no shift in the point of flagellar insertion. Instead, 18 out of the 24 plates would be exactly homologous between the peridinioid and gonyaulacoid primary models, and would occupy identical positions relative to one another and to the flagellar pore. Portions of homologous primary plates would be involved in the formation of the first apical, the posterior sulcal, and the other principal plates of the sulcus close to the flagellar pore. However, this option requires a mechanism to orient differently the two triads of primary plates at the ends of the polar axis. While a biologically suitable mechanism to accomplish this can only be guessed (e.g., division of one amphiesmal vesicle into three parts in two different ways, or expansion of three amphiesmal vesicles from three differently situated primordia), the geometrical requirements and possibilities inherent in the pattern are limited and can be specified as follows: if the low-latitude surface area of a sphere is divided into 18 portions arranged in three latitudinal rows of six equal polygons each, alternating in position, and, further, if the design is radially symmetrical, a pattern approximating that of the primary model is the only possible result. However, if a triad of plates closing over the polar area is to be combined with this arrangement without diminishing the symmetry or regularity of the pattern over one hemisphere, there are two possibilities—in effect, an A solution and a Y solution. With two possibilities at each pole, that means four possible combinations: AA, AY, YA, and YY (Figure 13.9). The hypothesized differences in shapes (camerate versus planate) of the pre- and postequatorials in the two primary models would follow from these alternative orientations of the polar triads.

The existence of a dinoflagellate tabulation that could be related with confidence to an AA or YY primary version of the 3-6-6-6-3 design would contraindicate the first two methods of "conversion" but support the plausibility of the third. This follows from the fact that the AA and YY patterns could not

Figure 13.8 - THREE ALTERNATIVE RELATIONSHIPS BETWEEN GONYAULACOID AND PERIDINIOID PRIMARY TABULATIONS

In each of the two groups of four diagrams, the transverse row shows three alternative ways in which the peridinioid primary topology (AY-arrangement) may relate to the gonyaulacoid pattern (YA-arrangement) hypothesized by Taylor and illustrated by the upper figure. The diagram at the left of the row shows the gonyaulacoid pattern turned pole-for-pole. The middle diagram depicts the gonyaulacoid pattern with the point of flagellar insertion shifted to primary suture **a/b** from **f/a**. The topological effect of this is identical to that which would result if the shift were to either **c/d** or **e/f** (dashed arrows in apical view). In the righthand diagram, positions of the preequatorial, equatorial, and postequatorial plates, and the point of flagellar insertion are exactly as in the gonyaulcoid pattern but the polar and antipolar triads (patterned) are placed as if they had been rotated about the polar axis 60 degrees relative to the other group of 18 plates.
Taylor symbols identify all homologous plates, except that only one of the three equivalent alternatives of the middle diagram is fully labeled, the others being indicated only for one plate by the symbols in parentheses.

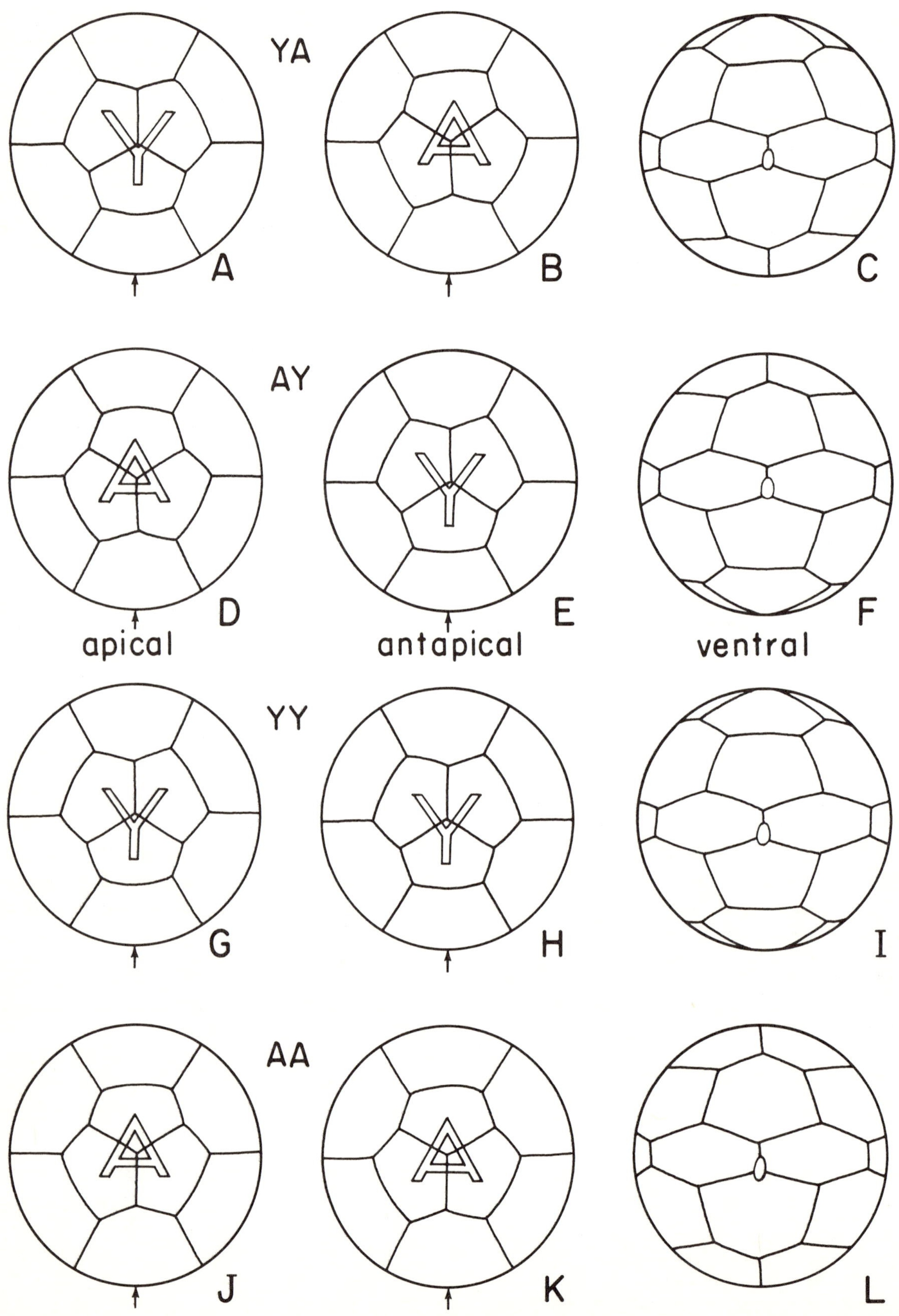

Figure 13.9

be converted by simple rotation into either the AY or YA combination, but they would be the two possible additional ways in which the polar triads might be combined with the low-latitude group of 18 plates. Although future study may bring to light patterns suggesting that AA or YY primary models exist, none has yet been identified clearly among either living or fossil dinoflagellates. However, as pointed out in Chapter 12, the paratabulation of fossil *Dapcodinium priscum* and the tabulation of living *Peridiniella sphaeroidea* can be viewed as suggesting a YY origin. Perhaps, more convincing AA or YY primary patterns will be suggested by other Triassic-Early Jurassic species of R-cysts when their paratabulations become better known.

Two aspects of the third possibility make it attractive as a working hypothesis in comparison to the other two. First, it seems intuitively more stable and conservative, for the living cell need not exchange anterior and posterior or shift its thecal covering relative to the point of flagellar insertion. Second, it is consistent with a high degree of equivalence between plates of the peridinioid and gonyaulacoid lines. For example, the same primaries and their derivatives occupy similar areas or, if they change position, the same plates tend to be involved and in similar ways.

If one were to accept this third method of "conversion" as an hypothesis, then the Taylor symbols in Figure 13.8 D,H would indicate the homologies in the equatorial and adjacent series of primary plates, but homologies in the polar and antipolar triads are not so readily determined. Are there any "signals" that might identify gonyaulacoid-peridinioid homologs among these plates? Perhaps. In thecae of some peridinioids and some gonyaulacoids, a structure at the apical pole relates to a particular suture between what we have identified as triads of primary plates. In gonyaulacoids with a quinqueform hypotheca (such as *Triadinium* and *Pyrodinium*, also *Helgolandinium* and *Fragilidium*), the conspicuous prepolar, **P**, is consistently elongate along the **A/B** suture, which is also the route by which **1u** becomes inserted in the apical series. In many peridinioids, the apical pore plate is also associated with a special structure—in this case a separate plate, the so-called canal plate—that extends along a particular radial suture, which also happens to be the one along which the midventral preequatorial primary migrates into the apical series. Unfortunately, the function of these apical structures is not known in either peridinioids or gonyaulacoids. At the antipole, no such distinctive structures as pores and canal plates—whatever their function—are associated with the primary triad, but it is possibly significant that in each case one of the primary antipolars tends to shift position toward the equator and to become incorporated into the sulcus as the Kofoidian posterior sulcal.

If these developments—**1** inserts between **A** and **B**, and **Z** becomes the posterior sulcal—are used as the basis for judging homologies between elements of the apical triads, then the Taylor symbols for the gonyaulacoid primary model could be applied to the peridinioid model and extended to secondary patterns as in Figure 13.10.

This discussion has not been intended to argue for replacing Kofoid's system by Taylor's relative to peridinioids, but rather to present a specific hypothesis of homologies between peridinioids and gonyaulacoids that may provoke debate and focus future inquiry. Meanwhile, Kofoid's system remains broadly satisfactory for description and comparison within the peridinioid group, especially for the limited spectrum of patterns encountered in paratabulate cysts.

Taylor considered the thecal organizations of gonyaulacoids and peridinioids to be sufficiently distinct to warrant recognizing a new order, Gonyaulacales, for those forms formerly included in the Peridiniales that have a thecal tabulation derivable from the YA pattern. However, he did not redefine the Peridiniales in similar terms, thus leaving this order a composite of some taxa whose thecal patterns are derivable from an AY primary arrangement (i.e., peridinioids of this discussion)

Figure 13.9 - FOUR MODELS OF PRIMARY TABULATION

Diagrams A-C represent polar, antipolar, and ventral views of the primary arrangement of plates hypothesized for the Taylor gonyaulacoid model. The two contrasting polar and antipolar patterns (diagrams A and B) are distinguished by the superposed letters "Y" and "A" (which are oriented relative to the ventral locus of flagellar insertion indicated by the arrow). The properties of this YA-arrangement are such that there could exist three other similarly organized and equally symmetrical, but, in part, topologically different, models having the same number of plates arranged in the same 3-6-6-6-3 sequence. These are shown in the other diagrams. The AY-model (diagrams D-F) is a prospective starting point for the peridinioid patterns, as explained in the text. Tabulation patterns among living or fossil dinoflagellates for which the AA- and YY-models (diagrams G-I and J-L) might seem especially appropriate starting points have not been recognized (—yet?).

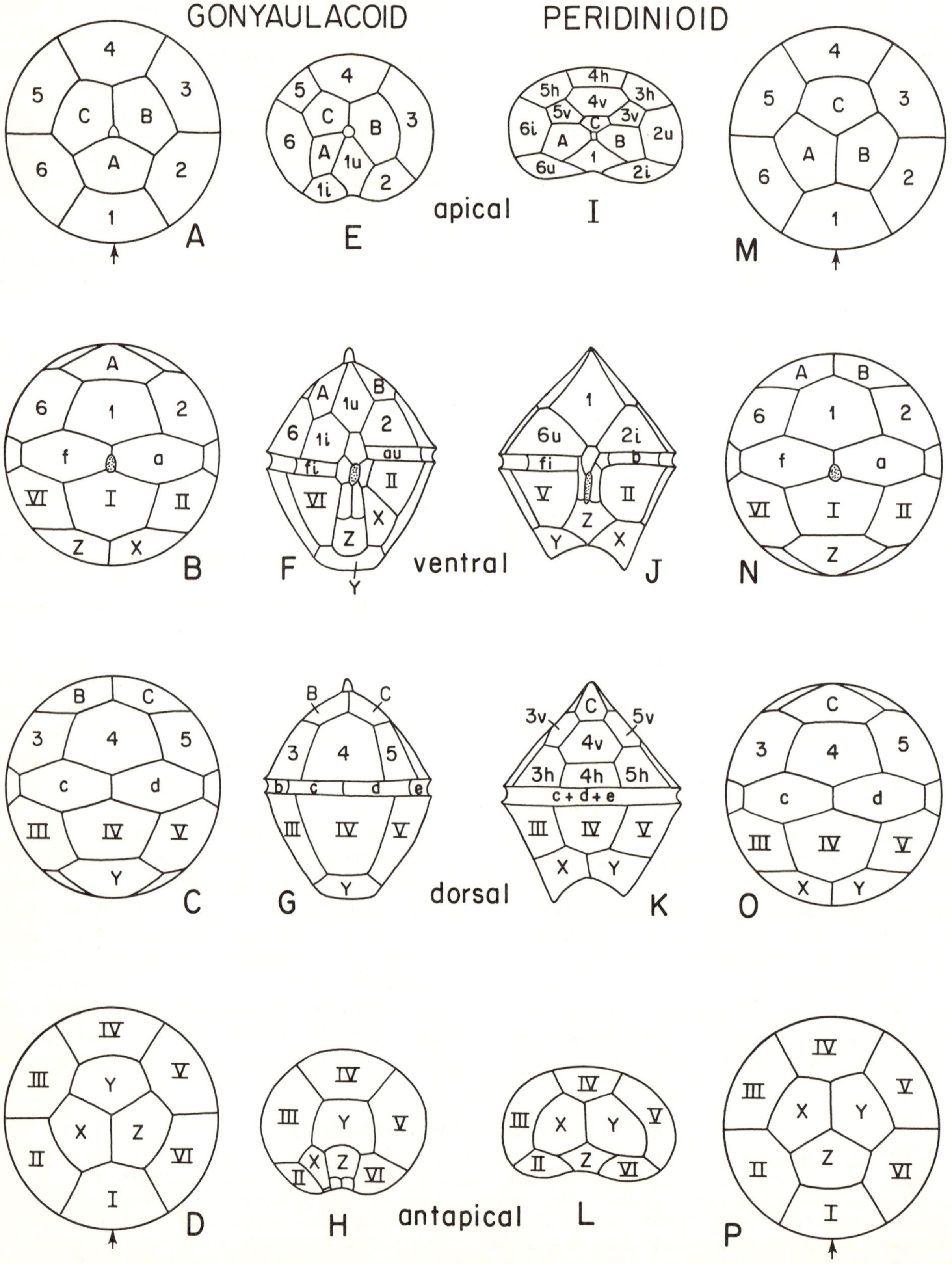
GONYAULACOID
PERIDINIOID
apical
ventral
dorsal
antapical
A
B
C
D
E
F
G
H
I
J
K
L
M
N
O
P

Figure 13.10

and many others whose thecal patterns cannot be analyzed currently in a comparable way.

The course adopted here has been to retain Peridiniales as formerly defined. This is not meant to question that a significant distinction exists between the peridinioids and gonyaulacoids (on the basis of thecal organization and other features, such as nutrition and the mechanism of thecal division during vegetative cycle; see Taylor, 1983), but to resist other implications of setting off a separate Order Gonyaulacales while leaving the remainder of the Peridiniales unaltered. The possible derivation of peridinioids (AY) and gonyaulacoids (YA) from similar starting points suggests to this author a relationship between these two groups that is closer than either may have with other taxa traditionally associated in the Peridiniales, and also closer than exists elsewhere at the ordinal level (e.g., between Peridiniales and Dinophysiales). Although formal taxonomic changes are not within the scope of this work, a better expression of the perception supported here would be to segregate peridinioids and gonyaulacoids (characterized, respectively, by their AY and YA patterns and other features) at subordinal rank from each other and from other peridinialeans which cannot yet be analyzed in a similar way.

Figure 13.10 - EXAMPLES OF GONYAULACOID AND PERIDINIOID PATTERNS COMPARED WITH HYPOTHETICAL TAYLOR MODELS

Taylor symbols have been applied to the larger plates in the peridinioid examples according to the hypothesized gonyaulacoid-peridinioid homologies discussed in the text.

A-D - Primary gonyaulacoid model.

E-H - Secondary gonyaulacoid pattern as represented by a theca with insert sexiform organization (e.g., many species of *Gonyaulax*).

I-L - Secondary peridinioid pattern as represented by a theca with ortho hexa organization and four cingular plates (e.g., many species of *Protoperidinium*).

M-P - Primary peridinioid model. The primary patterns shown here are equivalent to the righthand diagrams (D, H) in Figure 13.8.

Taylor labels for the plates of the gonyaulacoid sulcal area (Figure 13.10 F) are shown in Figure 13.7 D, and the shadings in Figure 13.7 E-H imply the Taylor labels appropriate for the four peridinioid sulcal plates (unlabeled in Figure 13.10 J) according to the hypothesis presented here. These Taylor labels would be as follows: **ai** for *as* (= anterior sulcal), **au** for *1c* (= first cingular, or *t*), **lu** for *ls* (= left sulcal), and fused **fu** + **li** for *rs* (= right sulcal).

Summary and Perspective

In Part I of this volume, dinoflagellates were introduced with a review of aspects of their biology and their fossil record. The roles of the resting zygotic cyst in the life cycle of living dinoflagellates and the morphological characteristics of cysts in representative modern species were reviewed. Special emphasis was laid on the selective nature of the dinoflagellate fossil record and the severe constraints which this places on the interpretation of dinoflagellate evolution and the ecological roles of these organisms during the geologic past.

In Part II, attention was directed to the morphology of the sporopollenin cyst—its general configuration, wall structure and surface patterns, formative processes, excystment aperture—and a descriptive terminology. The well-established Kofoid system for analyzing tabulation was applied to peridinioid cysts, but a newer system proposed by Taylor was endorsed strongly as preferable for the gonyaulacoids. Whereas the former has clear merit as an objective approach, the latter greatly facilitates comparisons of patterns based on apparent plate homologies within the gonyaulacoid group. Since Taylor's system is unfamiliar to most specialists, it was explained in detail, and its application to gonyaulacoids was explored in conjunction with discussions of four types of gonyaulacoid hypothecal patterns termed quinqueform, sexiform, partiform, and corniform.

Part III concerned the interpretation of dinoflagellate cysts. Introductory chapters covered principles of interpretation and an informal system of 17 morphological cyst categories to serve as a framework for ensuing discussion, in the absence of a satisfactory overall classification of cysts above the generic level. In the following four chapters, the morphological characteristics of many fossil cysts were reviewed, mostly under the headings of generic complexes within the 17 categories. In each chapter, selected important taxa, or examples of particular problems or principles of interpretation, were considered in more detail. In the volume's last chapter, various views on classification and evolution, especially as they relate to sporopollenin cysts, were surveyed, and ways in which the Taylor system of primary gonyaulacoid tabulation might be applied to peridinioid patterns were the subject of a final speculative foray.

In bringing this work to a close, it seems appropriate to paraphrase and amplify conclusions already reported by Taylor (1980) which place sporopollenin-producing dinoflagellates in perspective with the dinoflagellates as a whole:

1. The dinoflagellates occupy a position close to the base of the eukaryotic radiation; probably they were present and may already have been highly diverse in the Precambrian.

2. An early differentiation separated forms with flagella laterally inserted and lying in furrows (dinokonts) from more primitive forms with terminal flagella and no surface furrows (desmokonts).

3. The theca of dinoflagellates living today consists of cellulosic plates organized in four major ways which express different arrangements of vesicles within which the plates formed.

a. In the most primitive living thecate dinoflagellates, the Prorocentrales, the theca has paired lateral valves and a cluster of small plates around the two flagellar pores. There is only slight diversity in cell shape and thecal tabulation.

b. In the Dinophysiales, paired lateral valves are also present but they are more subdivided than in the Prorocentrales and have distinct epitheca, cingulum (in far anterior position), and hypotheca. The extensive modern diversity in this group represents an adaptive radiation accompanied by great modification of cell shape but only minor changes in thecal tabulation.

c. In the Peridiniales (of this volume), the thecal plates are organized in an essentially radially symmetrical plan. Two particular variations living today (and possibly more as yet unrecognized) differ by insertion of flagella or orientation (i.e., peridinioids and gonyaulacoids of this volume, which are distinguished at the ordinal level by Taylor as Peridiniales and Gonyaulacales).

d. In the gymnodinioid dinoflagellates (considered a polyphyletic group by Taylor, but treated for convenience in this volume under the familiar ordinal name Gymnodiniales), there is a proliferation of vesicles and diminution of plate thickness (or even complete loss of plates). Cell form is moderately diverse, vesicle patterns are inconspicuous and largely unknown, and there are many adaptations in nutrition and motility. Taylor, like Bergh (1881) and Bütschli (1885) considers the most highly specialized thecate or vesiculate dinoflagellates to occur among the gymnodinioids.

4. A fifth type of thecal organization, apparently unrepresented today but characteristic of an important group of Jurassic dinoflagellates, is the basis for the Nannoceratopsiales. In this order, an epicyst paratabulation strongly suggestive of the peridinialean design is combined with a hypocyst organization essentially identical to the dinophysialean pattern.

5. In several groups of dinoflagellates, living and fossil, nonmotile endogenous cysts develop(ed) continuous walls probably based on the pellicular layer beneath the vesicular layer. These cysts serve(d) various purposes. They include(d) relatively delicate, asexual ecdysial cysts and, in a minority of species, more robust resting cysts (hypnozygotes) with resistant, thick walls (commonly of sporopollenin but also rarely of calcite or silica). Geologically preservable sporopollenin cysts are unknown today in the Prorocentrales and Dinophysiales, and these groups have no certain fossil record. Preservable cysts are rare in the Gymnodiniales today; possible fossils of the group include rare sporopollenin cysts (of *Polykrikos* type, Jurassic-Holocene) and sporopollenin tests that may have enclosed motile cells (Late Cretaceous). Sporopollenin cysts are formed today by around ten percent of species within the scope of Peridiniales as used in this volume. Clearly, it is cysts representing this order which dominate the fossil record. The Nannoceratopsiales also yielded sporopollenin cysts (Jurassic), and many sporopollenin fossils of unestablished affinity (mostly Paleozoic) may include cysts of desmokont dinoflagellates. Despite an abundance of recognizable fossil dinoflagellates from Mesozoic and younger sediments, the fossil record of this group appears to be extremely incomplete relative to the variety and numbers of living species and to the great geologic age of the group suggested by its apparent primitiveness.

REFERENCES

Abé, T. H., 1981, Studies on the family Peridinidae, an unfinished monograph of the armored Dinoflagellata: Kyoto Univ., Publ. Seto Marine Biol. Lab., Spec. Publ. Ser., v. 6, viii + 412 p.

Albert, N. R., Evitt, W. R., and Stein, J. A., (in prep.), *Lacrymodinium*, a new genus of Late Jurassic to Early Cretaceous dinoflagellates.

Alberti, Gerhard, 1961, Zur Kenntnis Mesozoischer und Alttertiärer Dinoflagellaten und Hystrichosphaerideen von Nord- und Mitteldeutschland sowie einigen anderen europäischen Gebieten: Palaeontographica, Abt. A, v. 116, p. 1-58.

Allman, G. J., 1855, Observations on Aphanizomenon Flos-aquae, and a species of Peridinea: Quart. Jour. Micro. Sci., v. 3, p. 21-25.

Anderson, D. M., and Wall, David, 1978, Potential importance of benthic cysts of *Gonyaulax tamarensis* and *G. excavata* in initiating toxic dinoflagellate blooms: Jour. Phycology, v. 14, p. 224-234.

Artzner, Darrah; Davies, E. H.; Dörhöfer, Gunter; Fasola, Armando; Norris, Geoffrey; and Poplawski, Silvana, 1979, A systematic illustrated guide to fossil organic-walled dinoflagellate genera: Royal Ontario Mus., Life Science Misc. Publ., 119 p. Toronto.

Balech, Enrique, 1964, El género *Cladopyxis*: Mus. Argentino Cien. Nat. Bernardino Rivadavia, Rev., Hidrobiologia, v. 1, p. 27-39.

______ 1967a, *Palaeophalacroma* Schiller, otro miembro de la familia Cladopyxidae (Dinoflagellata): Neotropica, v. 13, p. 105-112.

______ 1967b, Observaciones sobre dinoflagelados fosiles: Rev. Mus. Argentino Cien. Nat. Bernardino Rivadavia, Rev., Paleontologia, v. 2, p. 5-50.

______ 1974, El género "*Protoperidinium*" Bergh 1881 ("*Peridinium*" Ehrenberg, 1831, partim): Museo Argentino Cien. Nat. Bernardino Rivadavia, Rev., Hidrobiologia, v. 4, no. 1, p. 1-79.

______ 1977a, Cuatro especies de *Gonyaulax* sensu lato y consideraciones sobre el género (Dinoflagellata): Museo Argentino Cien. Nat. Bernardino Rivadavia, Rev., Hidrobiologia, v. 5, p. 115-135.

______ 1977b, *Cachonina niei* Loeblich (Dinoflagellata) y sus variaciones: Physis, Sec. A, v. 36, p. 59-64.

______ 1978, El género *Pyrophacus* Stein (Dinoflagellata): Physis, sec. A., v. 38, p. 27-38.

______ 1979, Tres dinoflagelados nuevos o interesantes de aguas brasilenas: Bolm Inst. Oceanogr., Sao Paulo, v. 28, p. 55-64.

______ 1980, On thecal morphology of dinoflagellates with special emphasis on cingular and sulcal plates: Univ. Nat. Auton. Mexico, Centro Cienc. Mar. Limnol., Ann., v. 7, p. 57-68.

Baltes, Nicolae, 1967, Albian microplankton from the Moesic Platform, Rumania: Micropaleontology, v. 13, p. 327-336.

Barrows, A. L., 1918, The significance of skeletal variation in the genus *Peridinium*: Univ. California Publ. Zoology, v. 18, p. 397-478.

Beam, C. A., and Himes, M., 1974, Evidence for sexual fusion and recombination in the dinoflagellate *Crypthecodinium cohnii*: Nature, v. 250, p. 435-436.

______ and ______ 1980, Sexuality and meiosis in dinoflagellates: p. 171-206 *in* Levandowsky, M., and Hutner, S. H. (eds.), Biochemistry and physiology of the Protozoa: v. 3. New York: Academic Press.

______ 1976, Clave illustra de dinoflagelados antarticos: Inst. Antartico Argentino, Pulb. No. 11, 99p. Buenos Aires.

Beam, C. A., Himes, M., Himelfarb, J., Link, C., and Shaw, K., 1977, Genetic evidence of unusual meiosis in the dinoflagellate *Crypthecodinium cohnii*: Genetics, v. 87, p. 19-32.

Below, Raimond, 1981, Dinoflagellaten-Zysten aus dem oberen Hauterive bis unteren Cenoman

Süd-West-Marokkos: Palaeontographica, Abt. B, v. 176, p. 1-145.

Benedek, P. N. von, and Gocht, Hans, 1981a, *Thalassiphora pelagica* (Dinoflagellata, Tertiär): Elektronenmikroskopische Untersuchung und Gedanken zur Paläobiologie: Palaeontographica, Abt. B, v. 180, p. 39-64.

______ and ______ 1981b, Zur Entstehung des Netzwerks von *Thalassiphora reticulata* Morgenroth (Dinoflagellata, Oligözan): Neues Jahrb. Geol. Paläontologie, Mh. 1981, p. 385-394.

Benedek, P. N. von, Gocht, Hans, and Sarjeant, W. A. S., 1982, The dinoflagellate cyst genus *Pentadinium* Gerlach: a re-examination: Neues Jahrb. Geol. Paläontologie Abh., v. 162, p. 265-285.

Bergh, R. S., 1881, Der Organismus der Cilioflagellaten. Eine phylogenetische Studie: Morphologisches Jahrb., v. 7, p. 177-288.

Bibby, B. T., and Dodge, J. D., 1972, The encystment of a freshwater dinoflagellate: a light and electron microscopical study: British Phycological Jour., v. 7, p. 85-100.

Bint, A. N., 1983, *Umbodinium crustov* gen. et sp. nov., a peridinioid dinoflagellate with two intercalaries from the Albian of Kansas: Palynology, v. 7, p. 171-182.

Bjaerke, Tor, 1980, Mesozoic palynology of Svalbard IV. Toarcian dinoflagellates from Spitzbergen: Palynology, v. 4, p. 57-77.

Bold, H. C., and Wynne, M. J., 1977, Introduction to the algae. Structure and reproduction: 706 p. New York: Prentice Hall.

Boltovskoy, Andrés, 1973, Formación del arqueopilo en tecas de dinoflagelados: Rev. Española Micropaleontología, v. 5, p. 81-98.

______ 1976, Estructura y estereoultraestructura tecal de dinoflagelados. II. *Peridinium cinctum* (Müller) Ehrenberg: Physis, Sec. B, v. 34, p. 73-84.

______ 1979, Estudio comparativo de las bandas intercalares y zonas pandasuturales en los géneros de dinoflagelados *Peridinium s.s.*, *Protoperidinium*, y *Palaeoperidinium*: Limnobios, v. 1, p. 325-332.

Bouck, G. B., and Sweeny, B. M., 1966, The fine structure and ontogeny of trichocysts in marine dinoflagellates: Protoplasma, v. 61, p. 205-253.

Bourrelly, Pierre, 1968, Notes sur les péridiniens d'eau douce: Protistologica,v. 4, p. 5-13.

______ 1970, Les algues d'eau douce. Initiation à la systématique. Tome III: Les algues bleues et rouges, les Eugléniens, Péridiniens et Cryptomonadines: 512 p. Paris: E. N. Boubée et Cie.

Braarud, Trygve, 1945, Morphological observations on marine dinoflagellate cultures (*Porella perforata*, *Goniaulax tamarensis*, *Protoceratium reticulatum*): Avh. Norsk Vidensk.-Akad., Oslo, Mat.-Naturv. Klasse, 1944, no. 11, p. 1-18.

Bradford, Martin, 1975, New dinoflagellate cyst genera from the Recent sediments of the Persian Gulf: Canadian Jour. Botany, v. 53, p. 3064-3074.

______ 1978a, An annotated bibliographic and geographic review of Pleistocene and Quaternary dinoflagellate cysts and acritarchs: Amer. Assoc. Stratigraphic Palynologists, Contrib. Ser., no. 6, 191 p.

______ 1978b, Acritarchous cysts of *Peridinium faeroense* Paulsen: implications for dinoflagellate systematics. A discussion: Palynology, v. 2, p. 195-197.

Brideaux, W. W., 1977, Taxonomy of Upper Jurassic-Lower Cretaceous microplankton from the Richardson Mountains, District of Mackenzie, Canada: Bull. Geol. Survey Canada, v. 281, p. 1-89.

Brooks, James, and Shaw, G., 1972, Geochemistry of sporopollenin: Chemical Geology, v. 10, p. 69-87.

Brunel, Jules, 1970, Le phytoplancton de la Baie de Chaleurs: 365 p. Montreal: Presses Univ. Montreal.

Bujak, J. P., 1980, Dinoflagellate cysts and acritarchs from the Eocene beds of southern England: p. 36-91 *in* Bujak, J. P., Downie, Charles, Eaton, G. L., and Williams, G. L., Dinoflagellate cysts and acritarchs from the Eocene of southern England: Special Papers Palaeontology, no. 24.

Bujak, J. P., and Davies, E. H., 1983, Modern and fossil Peridiniineae: Amer. Assoc. Stratigraphic Palynologists, Contrib. Ser., no. 13, 216 p.

Bujak, J. P., and Fisher, M. J., 1976, Dinoflagellate cysts from the Upper Triassic of arctic Canada; Micropaleontology, v. 22, p. 44-70.

Bujak, J. P., and Williams, G. L., 1979, Dinoflagellate diversity through time: Marine Micropaleontology, v. 4, p. 1-l2.

______ and ______ 1981, The evolution of dinoflagellates: Canadian Jour. Botany, v. 59, p. 2077-2087.

Bütschli, Otto, 1885, Unterabteilung (Ordnung) Dinoflagellata: p. 865-1088 *in* Dr. H. G. Bronn's Klassen und Ordnungen des Thier-Reichs, Wissenschaftlich Dargestellt in Wort und Bild, v. 1, Protozoa, Abt. 2, Mastigophora, p. 865-1088. Leipzig: C. F. Winter'sche Verlagshandlung.

Calandra, Francois, 1964, Sur un presumé dinoflagellé *Arpylorus* nov. gen. du gothlandien de Tunisie: Acad. Sci. (Paris), Comptes Rendus, v. 258, p. 4112-4114.

Carter, H. H., 1858, Note on the red colouring matter of the sea around the shores of the island of Bombay: Ann. Mag. Nat. History, ser. 3, v. I, p. 258-262.

Chapman, D. V., Dodge, J. D., and Heaney, S. I., 1982, Cyst formation in the freshwater dinoflagellate *Ceratium hirundinella* (Dinophyceae): Jour. Phycology, v. 18, p. 121-129.

Chapman, D. V., Livingston, David, and Dodge, J. D., 1981, An electron microscope study of the excystment and early development of the dinoflagellate *Ceratium hirundinella*: British Phycol. Jour., v. 16, p. 183-194.

Châteauneuf, J.-J., and Gruas-Cavagnetto, Carla, 1978, Les zones de Wetzeliellaceae (Dinophyceae) du bassin de Paris: Bull. Bur. Rech. Geol. Minières, Sér. 2, p. 59-93.

Chatton, Edouard, 1952, Classe des Dinoflagellés ou Péridiniens: p. 309-406 *in* Grasse, P.-P., Traité de Zoologie, Anatomie, Systématique, Biologie, v. 1, fasc. 1, Protozoaires, Généralités, Flagellés. Paris: Masson et Cie.

Churchill, D. M., and Sarjeant, W. A. S., 1963, Freshwater microplankton from Flandrian (Holocene) peats of south Western Australia: Grana Palynologica, v. 3, p. 29-53.

Claparède, Edouard, and Lachmann, Johannes, 1859, Études sur les Infusoires et les Rhizopodes: Mém. Inst. Nat. Génevois, v. 6, p. 392-484 (1858).

Clarke, R. F. A., and Verdier, J.-P., 1967, An investigation of microplankton assemblages from the chalk of the Isle of Wight, England: Verh., Konink. Nederlandse Akad. Wetensch., Afd. Natuurkunde, ser. 1, v. 24, p. 1-26.

Cookson, I. C., and Eisenack, Alfred, 1958, Microplankton from Australian and New Guinea Upper Mesozoic sediments: Proc. Royal Soc. Victoria, n.ser., v. 70, p. 19-79.

______ and ______ 1982, Microplankton aus australischen mesozoischen und tertiären Sedimenten, Zweiter Teil: Palaeontographica, Abt. B, v. 184, p. 23-63.

Corliss, J. O., 1975, Nuclear characteristics and phylogeny in the protistan phylum Ciliophora: Biosystems, v. 7, p. 338-349.

Costa, L. I., and Downie, Charles, 1976, The distribution of the dinoflagellate *Wetzeliella* in the Paleogene of northwestern Europe: Palaeontology, v. 19, p. 591-614.

______ and ______ 1979, The Wetzeliellaceae; Palaeogene dinoflagellates: Proc. IV Internat. Palynology Conference, Lucknow (1976-77), v. 2, p. 34-43.

Cox, E. R., 1978, Dinoflagellates: p. 489-501 *in* Laskin, A. I., and Lechevalier, H. A., CRC Handbook of microbiology (2nd ed.), v. 2. West Palm Beach: CRC Press.

Cox, E. R., and Arnott, H. J., 1971, The ultrastructure of the theca of the marine dinoflagellate *Ensiculifera loeblichii* sp. nov.: p. 121-136 *in* Parker, B. C., and Brown, R. M., Jr. (eds.), Contributions in Phycology. Lawrence: Allen Press.

Cox, R. L., 1971, Dinoflagellate cyst structures: walls, cavities, and bodies: Palaeontology, v. 14, p. 22-23.

Dale, Barrie, 1976, Cyst formation, sedimentation, and preservation: factors affecting dinoflagellate assemblages in Recent sediments from Trondheimsfjord, Norway: Rev. Palaeobotany Palynology, v. 22, p. 39-60.

______ 1977, Cysts of the toxic red-tide dinoflagellate *Gonyaulax excavata* (Braarud) Balech from Oslofjorden, Norway: Sarsia, v. 63, p. 29-34.

______ 1978, Acritarchous cysts of *Peridinium faeroense* Paulsen: implications for dinoflagellate systematics: Palynology, v. 2, p. 187-193.

______ 1983, Dinoflagellate resting cysts: "benthic plankton": p. 69-136 *in* Fryxell, G. A. (ed.), Survival strategies of the algae. Cambridge: Univ. Press.

Davey, Roger, 1969, Non-calcareous microplankton from the Cenomanian of England, northern France, and North America. Part I: Bull. British Mus. (Nat. Hist.), Geology, v. 17, p. 105-180.

_______ 1971, Palynology and palaeo-environmental studies, with special reference to the continental shelf sediments of South Africa: p. 331-347 *in* Farinacci, A. (ed.), Proc. Second Planktonic Conference, Roma 1970, v. 1. Rome: Edizioni Tecnoscienza.

_______ 1979, Two new early Cretaceous dinocyst species from the northern North Sea: Palaeontology, v. 22, p. 417-437.

_______ 1982, Dinocyst stratigraphy of the latest Jurassic to Early Cretaceous of the Haldager No. 1 borehole, Denmark: Danmarks Geol. Undersogelse, Ser. B., Nr. 6, p. 1-36.

Davey, R. J.; Downie, Charles; Sarjeant, W.A.S.; and Williams, G.L., 1966, Studies on Mesozoic and Cainozoic dinoflagellate cysts: British Mus. (Nat. History) Bull Geology, Suppl. 3, p.1-248

Davey, R. J., and Verdier, J.-P., 1971, An investigation of microplankton assemblages from the Albian of the Paris Basin: Verh., Koninkl. Nederlandse Akad. Wetensch., Afd. Natuurkunde, ser. 1, v. 26, p. 1-58.

Davies, E. H., 1981, Archeopyle patterns in Late Cretaceous peridiniaceans: Palynology, v. 5, p. 234 (abstract).

DeConinck, Jan, 1968, Dinophyceae et Acritarcha de l'ypresien du Sondage de Kallo: Mém. Inst. Roy. Sci. Nat. Belgique, no. 161, 67 p.

Deflandre, Georges, 1933, Note préliminaire sur un péridinien fossile *Lithoperidinium oamaruense* n.g. n.sp.: Bull. Soc. Zool. France, v. 58, p. 265-273.

_______ 1935, Considérations biologiques sur les microorganismes d'origine planctonique conservés dans les silex de la craie: Bull. Biologique France Belgique, v. 69, p. 213-244.

_______ 1936, Microfossiles des silex crétacés. Première Partie: Généralités. Flagellés: Annales Paléontologie, v. 25, p. 151-191.

_______ 1937, Microfossiles des silex crétacés. Deuxième Partie: Hystrichosphaeridés. Sarcodines. Organismes divers: Annales Paléontologie, v. 26, p. 51-103.

_______ 1938, Microplancton des mers jurassiques conservé dans les marnes de Villers-Sur-Mer (Calvados). Étude liminaire et considérations générales: Trav. Station Zool. Wimereux, v. 13, p. 147-200.

_______ 1941, Le microplancton kiméridgien d'Orbagnoux et l'origine des huiles sulfurées naturelles: Mém. Acad. Science Inst. France, v. 65, p. 1-32.

_______ 1943-1966 (in part with Deflandre-Rigaud, Marthe): Fichier micropaléontologique générale: Arch. Orig. Centre Docum., C.N.R.S (Paris): 1943, ser. 1 (Arch. 155); 1944, ser. 3 (Arch. 159); 1945, ser. 5 (Arch. 207); 1946, ser. 6 (Arch. 235); 1946, ser. 8 (Arch. 257); 1958, ser. 10 (Arch. 366); 1962, ser. 11 (fiches 1751-1947); 1965, ser. 14 (fiches 2522-2875); 1966, ser. 15 (fiches 2876-3175); 1966, ser. 16 (fiches 3176-3422).

_______ 1945, Microfossiles des calcaires siluriens de la Montagne Noire: Annales Paléontologie, v. 31, p. 41-76.

_______ 1947, Le problème des hystrichosphères: Bull. Inst. Océanographique (Monaco), no. 918, p. 1-23.

_______ 1948, Les calciodinellidés: dinoflagellés fossiles à thèque calcaire: Le Botaniste, v. 34, p. 191-219.

_______ 1952a, Dinoflagellés fossiles: p. 391-406 *in* Grassé, P.-P. (ed.), Traité de Zoologie. Anatomie, Systématique, Biologie, v. 1, fasc. 1. Paris: Masson et Cie.

_______ 1952b, Classe des Dinoflagellés: p. 116-124 *in* Piveteau, J. (ed.), Traité de Paléontologie, v. 1, Les stades inférieurs d'organisation du Règne Animal. Paris: Masson et Cie.

Deflandre, Georges, and Cookson, I. C., 1955, Fossil microplankton from Australian late Mesozoic and Tertiary sediments: Australian Jour. Marine Freshwater Research, v. 6, p. 242-313.

Deflandre, Georges, and Courteville, Henri, 1939, Note préliminaire sur les microfossiles des silex cretacés du Cambrésis: Bull. Soc. française Microscopie, v. 8, p. 95-106.

Dettmann, Mary, 1965, Techniques used in the study of megaspores: p. 699-706, *in* Kummel, Bernhard, and Raup, David (eds.), Handbook of paleontological techniques. San Francisco: W. H. Freeman.

Deunff, Jean, 1961, Un microplancton à hystrichosphères dans le Tremadoc du Sahara: Rev. Micropaléontologie, v. 4, p. 37-52.

Diesing, C. M., 1866, Revisionen der Prothelminthen, Abteilung: Mastigophoren: Akad. Wiss. Vienna, Sb., v. 52, p. 287-401.

Diwald, Karl, 1938, Die ungeschlechtliche und geschlechtliche Fortpflanzung von

Glenodinium lubiniensiforme spec. nov.: Flora, v. 132, p. 174-192.

______ 1939, Ein Beitrag zur Variabilität und Systematik der Gattung *Peridinium* (nach dem auf den österreichischitalienischen Terminfahrten der "Najade" in den Jahren 1911-1914 gesammelten Material): Archiv Protistenkunde, v. 93, p. 121-184.

Dobell, P. E. R., and Taylor, F. J. R., 1981, Viable *Spiniferites* cysts with 2P archeopyles from Recent marine sediments, British Columbia, Canada: Palynology, v. 5, p. 99-106.

Dodge, J. D., 1971, Fine structure of the Pyrrhophyta: Bot. Rev., v. 37, p. 481-508.

______ 1981, Three new generic names in the Dinophyceae: *Herdmania*, *Sclerodinium*, and *Triadinium* to replace *Heteraulacus* and *Goniodoma*: British Phycol. Jour., v. 16, p. 273-280.

______ 1982, Marine dinoflagellates of the British Isles: 303 pages. London: H. M. Stationery Office.

Dodge, J. D., and Crawford, R. M., 1968, Fine structure of the dinoflagellate *Amphidinium carteri* Hulburt: Protistologia, v. 4, p. 231-242.

______ and ______ 1970a, The morphology and fine structure of *Ceratium hirundinella* (Dinophyceae): Jour. Phycology, v. 6, p. 137-149.

______ and ______ 1970b, A survey of thecal fine structure in the Dinophyceae: Jour. Linnaean Soc., v. 63, p. 53-67.

Dörhöfer, Gunter, and Davies, E. H., 1980, Evolution of archeopyle and tabulation in rhaetogonyaulacinean dinoflagellate cysts: Royal Ontario Mus., Life Sciences, Misc. Publs., p. 1-91.

Downie, Charles, Evitt, W. R., and Sarjeant, W. A. S., 1964, Dinoflagellates, hystrichospheres and the classification of the acritarchs: Stanford Univ. Publs., Geol. Sci., v. 7, p. 1-16.

Downie, Charles, and Sarjeant, W. A. S., 1966, The morphology, terminology, and classification of fossil dinoflagellates: p. 10-17, *in* Davey, R. J., Downie, Charles, Sarjeant, W. A. S., and Williams, G. L., Studies on Mesozoic and Cainozoic dinoflagellate cysts: Bull. British Mus. (Nat. Hist.), Supplement 3.

Drugg, W. S., 1978, Some Jurassic dinoflagellate cysts from England, France and Germany: Palaeontographica, Abt. B, v. 168, p. 61-79.

Dumitrică, Paulian, 1973, Cenozoic endoskeletal dinoflagellates in southwestern Pacific sediments cored during Leg 21 of the DSDP: p. 819-884, *in* Burn, R. E., et al., Initial reports of the Deep Sea Drilling Project v. 21.

Dürr, Georg, 1979a, Elektronenmikroskopische Untersuchungen am Panzer von Dinoflagellaten. I. *Gonyaulax polyedra*: Archiv Protistenkunde, v. 122, p. 55-87.

______ 1979b, Elektronenmikroskopische Untersuchungen am Panzer von Dinoflagellaten. II. *Peridinium cinctum*: Archiv Protistenkunde, v. 122, p. 88-120.

______ 1979c, Elektronenmikroskopische Untersuchungen am Panzer von Dinoflagellaten. III. Die Zyste von *Peridinium cinctum*: Archiv Protistenkunde, v. 122, p. 121-139.

Dürr, Georg, and Netzel, Harald, 1974, The fine structure of the cell surface in *Gonyaulax polyedra* (Dinoflagellata): Cell Tissue Research, v. 150, p. 21-41, 6 pls., 3 text-figs.

Duxbury, Stanley, 1977, A palynostratigraphy of the Berriasian to Barremian of the Speeton Clay of Speeton, England: Palaeontographica, Abt. B, v. 160, p. 17-67.

______ 1979, Three new genera of dinoflagellate cysts from the Speeton Clay (Early Cretaceous) of Speeton, England: Micropaleontology, v. 25, p. 198-205.

Eaton, G. L., 1971, A morphogenetic series of dinoflagellate cysts from the Bracklesham Beds of the Isle of Wight, Hampshire, England: p. 355-379 *in* Farinacci, A. (ed.), Proc. Second Planktonic Conference, Roma 1970, v. 1. Rome: Edizioni Technoscienza.

______ 1976, Dinoflagellate cysts from the Bracklesham Beds (Eocene) of the Isle of Wight, southern England: Bull. British Mus. (Nat. Hist.), v. 26, p. 230-332.

______ 1980, Nomenclature and homology in peridinialean dinoflagellate plate patterns: Palaeontology, v. 23, p. 667-688.

Edwards, L. E., 1982, Biostratigraphically important species of *Pentadinium* Gerlach 1961 and a likely ancestor, *Hafniasphaera goodmanii* n. sp., from the Eocene of the Atlantic Coastal Plain: Palynology, v. 6, p. 105-117.

Edwards, L. E., and Bebout, J. W., 1981, Emendation of *Phthanoperidinium* Drugg and Loeblich 1967, and a description of *P. brooksii* sp. nov. from the Eocene of the mid-

Atlantic outer continental shelf: Palynology, v. 5, p. 29-41.

Ehrenberg, C. G., 1832 (1830 as separate), Beiträge zur Kenntniss der Organisation der Infusorien und ihrer geographischen Verbreitung, besonders in Sibirien: Preuss. Akad. Wiss., Berlin, Abh. 1830, p. 1-88.

______ 1836, Mittheilungen über die in den Feuersteinen bei Delitzsch vorkommenden mikroskopischen Algen und Bryozoen als Begleiter der fossilen Infusorien: Preuss. Akad. Wiss., Berlin, Bericht Verh. 1836, p. 114-115.

______ 1837, Über die fossilien Infusorien-Gattungen *Xanthidium* und *Peridinium*: Archiv Naturgeschichte, v. 3, p. 273-278.

______ 1854, Mikrogeologie. Das Erden und Felsen schaffende Wirken des unsichtbaren kleinen selbständigen Lebens auf der Erde: 374+31+88 p. Leipzig: Leopold Voss.

Eisenack, Alfred, 1931, Neue Mikrofossilien des baltischen Silurs I.: Palaeont. Zeitschr., v. 13, p. 74-118.

______ 1935, Mikrofossilien aus Doggergeschieben Ostpreussens: Zeitschr. Geschiebeforschung, v. 11, p. 167-184.

______ 1936a, *Eodinia pachytheca* n.g. n. sp., ein primitiver Dinoflagellat aus einem Kelloway-Geschiebe Ostpreussens: Zeitschr. Geschiebeforschung, v. 12, p. 72-75.

______ 1936b, Dinoflagellaten aus dem Jura: Annales Protistologie, v. 4, p. 59-63.

______ 1938, Die Phosphoritknollen der Bersteinformation als Überlieferer tertiären Planktons: Schriften Phys.-ökon. Gesellschaft Königsberg, v. 70, p. 181-188.

______ 1939, Die Wandung fossiler Dinoflagellaten: Archiv Protistenkunde, v. 93, p. 81-86.

______ 1954, Mikrofossilien aus Phosphoriten des samländischen Unteroligozäns und über die Einheitlichkeit der Hystrichosphaerideen: Palaeontographica, Abt. A, v. 105, p. 49-95.

______ 1958, Mikroplankton aus dem norddeutschen Apt nebst einigen Bemerkungen über fossile Dinoflagellaten: Neues Jahrb. Geol. Paläontologie, v. 106, p. 383-422.

______ 1959, Fossile Dinoflagellaten: Archiv Protistenkunde, v. 104, p. 43-50.

______ 1961, Einige Erörterungen über fossile Dinoflagellaten nebst Übersicht über die zur Zeit bekannten Gattungen: Neues Jahrb. Geol. Paläontologie, Abh., v. 112, p. 281-324.

______ 1963, Hystrichosphären: Biol. Reviews, v. 38, p. 107-139.

______ 1964 and 1967, Katalog der fossilen Dinoflagellaten, Hystrichosphären und verwandten Mikrofossilien: v. 1, 888 p. (1964), v. 1, suppl. 1, 241 p.(1967).

______ 1966, Symbionten in fossilen Protisten: Paläontol. Zeitschrift, v. 40, p. 103-107.

Eisenack, Alfred, and Fries, Magnus, 1965, *Peridinium limbatum* (Stokes) verglichen mit der tertiären *Deflandrea phosphoritica* Eisenack: Geol. För. Stockholm, Forh., v. 87, p. 239-248.

Eisenack, Alfred, and Kjellström, Göran, 1971-1981, Katalog der fossilen Dinoflagellaten, Hystrichosphären und verwandten Mikrofossilien: v. 1, suppl 2, 215 p. (1971); v. 1, suppl. 3, 254 p. (1975); v. 1, suppl. 4, 124 p. (1981); v. 2, 1132 p. (1972); v. 2, suppl. 1, 518 p. (1975); v. 2, suppl. 2, 412 p. (1981).

Engelhardt, D. W., 1976, *Geiselodinium tyonekensis* sp. nov., a dinoflagellate cyst from the nonmarine Tertiary (Miocene) of Alaska: Geoscience and Man, v. 15, p. 121-124.

Entz, G., 1925, Über Cysten und Encystierung der Süsswasser-Ceratien: Archiv Protistenkunde, v. 51, p. 130-183)

Erdtman, Gunnar, 1950, Fynd av *Hystrichosphaera furcata* i Gullmarn: Geol. För. Stockholm, Förh., v. 72, p. 221.

______ 1954, On pollen grains and dinoflagellate cysts in the Firth of Gullmarn, SW Sweden: Bot. Notiser, v. 2, p. 103-111.

Eren, Jacob, 1969a, Cyst formation in *Peridinium cinctum* fa. *westii*: Jour. Protozoology, v. 16 (suppl.), p. 35 (abstract).

______ 1969b, Studies on development cycle of *Peridinium cinctum* fa. *westii*: Verh. Internat. Vereinigung Limnologie, v. 17, p. 1013-1016.

Evitt, W. R., 1961, Observations on the morphology of fossil dinoflagellates: Micropaleontology, v. 7, p. 385-420.

______ 1963, A discussion and proposals concerning fossil dinoflagellates, hystrichospheres, and acritarchs, I and II: Nat. Acad. Sci. (Washington), Proc., v. 49, p. 158-164.

______ 1965, A method for making serial sections of pollen and other microfossils: p. 696-699 *in* Kummel, Bernhard, and Raup, David (eds)., Handbook of paleontological techniques. San Francisco: W. H. Freeman.

_______ 1967, Dinoflagellate studies II. The archeopyle: Stanford Univ. Publs., Geol. Sci., v. 10, no. 3, 83 p.

_______ 1968, Progress in the study of fossil *Gymnodinium* (Dinophyceae): Rev. Palaeobotany Palynology, v. 2, p. 355-363.

_______ 1970, Dinoflagellates—a selective review: Geoscience and Man, v. 1, p. 29-45.

_______ 1973, Maestrichtian *Aquilapollenites* in Texas, Maryland and New Jersey: Geoscience and Man, v. 7, p. 31-38.

_______ 1974, Restudy of an Oligocene freshwater dinoflagellate from Vermont: Geoscience and Man, v. 9, p. 1-6.

_______ 1975a, (ed.) Proceedings of a forum on dinoflagellates held at Anaheim, California, October 6, 1973 as part of the sixth annual meeting, A. A. S. P.: Amer. Assoc. Stratigraphic Palynologists, Contrib. Ser., no. 4, p. 1-76.

_______ 1975b, Introduction to the forum and remarks on classification and morphology: p. 7-13 *in* Evitt (1975a): Amer. Assoc. Stratigraphic Palynologists, Contrib. Ser., no. 4.

_______ 1975c, Georges Deflandre 1897-1973: Geol. Soc. America, Memorials 1975, 11 p.

_______ 1975d, The archeopyle in Cretaceous *Palaeoperidinium eurypylum* (Manum and Cookson) comb. nov., and similar dinoflagellates: Geoscience and Man, v. 11, p. 77-86.

_______ 1978, Special connotations of "quadra," "penta," and "hexa" in descriptive terminology of dinoflagellates: Palynology, v. 2, p. 199-204.

_______ 1981, The difference it makes that dinoflagellates do it differently: Amer. Assoc. Stratigraphic Palynologists, Newsletter, v. 4, no. 1, p. 6-7; also Palynology, v. 5, p. 236 (abstract).

Evitt, W. R., Clarke, R. F. A., and Verdier, J.-P., 1967, Dinoflagellate studies III. *Dinogymnium acuminatum* n. gen., n. sp. (Maastrichtian) and other fossils formerly referable to *Gymnodinium* Stein: Stanford Univ. Publs., Geol. Sci., v. 10, no. 4, p. 1-27.

Evitt, W. R., and Davidson, S. E., 1964, Dinoflagellate studies I. Dinoflagellate cysts and thecae: Stanford Univ. Publs., Geol. Sci., v. 10, no. 1, p. 1-12.

Evitt, W. R., Lentin, J. K., Millioud, M. E., Stover, L. E., and Williams, G. L., 1977, Dinoflagellate cyst terminology: Geol. Survey Canada, Paper 76-24, p. 1-11.

Evitt, W. R., Steidinger, K. A., and Stover, L. E., 1979, Modern and fossil dinoflagellates: Geology, Feb. 1979, p. 110-112 (Penrose Conference Report).

Evitt, W. R., and Wall, David, 1968, Dinoflagellate studies IV. Theca and cyst of Recent freshwater *Peridinium limbatum* (Stokes) Lemmermann: Stanford Univ. Publs., Geol. Sci., v. 12, no. 2, p. 1-15.

Fauré-Fremiet, G., 1908, Étude déscriptive des péridiniens et les infusoires ciliés du plankton de la Baie de la Hougue: Ann. Sci. Nat. (Zool.) Paris, sér. 9, v. 7, p. 209-242.

Fensome, R. A., 1981, The Jurassic dinoflagellate genera *Wanaea* and *Energlynia*: their morphology and evolution: Neues Jahrb. Geol. Paläontologie, Abh., v. 161, p. 47-61.

Fisher, M. J., and Van Helden, B. G. T., 1979, Some observations on the fossil dinocyst genus *Rhaetogonyaulax* Sarjeant, 1966: Palynology, v. 3, p. 265-276.

Foucher, J.-C., 1979, Distribution stratigraphique des kystes des dinoflagellés et des acritarches dans le Crétacé supérieur du Bassin de Paris et de l'Europe septentrionale: Palaeontographica, Abt. B, v. 169, p. 78-105.

Francis, D., 1967, On the eyespot of the dinoflagellate, *Nematodinium*: Jour. Experimental Biology, v. 47, p. 495-502.

Fütterer, Dieter, 1976, Kalkige Dinoflagellaten ("Calciodinelloideae") und die systematische Stellung der Thoracosphaeroideae: Neues Jahrb. Geol. Paläontologie, v. 151, p. 119-141.

Gerlach, Ellen, 1961, Mikrofossilien aus dem Oligozän und Miozän Nordwest-deutschlands, unter besonderer Berücksichtigen der Hystrichosphären und Dinoflagellaten: Neues Jahrb. Geol. Paläontologie, Abh., v. 112, p. 143-228.

Gocht, Hans, 1964, Planktonische Kleinformen aus dem Lias-Dogger Grenzbereich Nord- und Süddeutschlands: Neues Jahrb. Geol. Paläontologie, Abh., v. 119, p. 113-133.

_______ 1968, Zur Morphologie und Ontogenie von *Thalassiphora* (Dinoflagellata): Palaeontographica, Abt. A, v. 129, p. 149-156.

_______ 1969, Formengemeinschaften altertiären Mikroplanktons aus Bohrproben des Erdölfeldes Meckelfeld bei Hamburg: Palaeontographica, Abt. B., v. 126, p. 1-100.

——— 1970, Dinoflagellaten-Zysten aus dem Bathonium des Erdölfeldes Aldorf (NW-Deutschland): Palaeontograpahica, Abt. B, v. 129, p. 125-165.

——— 1972, Zur Morphologie der Dinoflagellaten-Gattung *Nannoceratopsis* Deflandre: Lethaia, v. 5, p. 15-29.

——— 1975a, Morphologie und Wandstruktur von *Lithodinia jurassica* Eisenack 1935 (Dinoflagellata, Oberjura): Neues Jahrb. Geol. Paläontologie, Mh. 1975, p. 343-359.

——— 1975b, Neuuntersuchung von *Eodinia pachytheca* Eisenack 1936 (Dinoflagellate, Oberjura): Neues Jahrb. Geol. Paläontologie, Abh., v. 148, p. 12-32.

——— 1979a, Korrelation von Überlappungssystem und Wachstum bei fossilen Dinoflagellaten (*Gonyaulax*-Gruppe): Neues Jahrb. Geol. Paläontologie, Abh., v. 157, p. 344-365.

——— 1979b, *Eyachia prisca* n. g. n. sp. (Dinoflagellata) aus dem Lias/Dogger Grenzbereich: Neues Jahrb. Geol. Paläontologie, Mh. 1979, v. 10, p. 305-319.

——— 1981, "Direkter" Nachweis der Platten-Überlappung bei *Hystrichogonyaulax cladophora* (Dinoflagellata, Oberjura): Neues Jahrb. Geol. Paläontologie, Mh. 1981, p. 149-156.

——— 1983, Morphogenetische Deutung und Bezeichnung ausgewählter Merkmale bei Dinoflagellaten-Zysten: Neues Jahrb. Geol. Paläontologie, Mh. 1983, p. 257-276.

Gocht, Hans, and Netzel, Harald, 1974, Rasterelektronenmikroskopische Untersuchungen am Panzer von *Peridinium* (Dinoflagellata): Archiv Protistenkunde, v. 116, p. 381-410.

——— and ——— 1976, Reliefstrukturen des Kreide-Dinoflagellaten *Palaeoperidinium pyrophorum* (Ehr.) im Vergleich mit Panzer-Merkmalen rezenter *Peridinium*-Arten: Neues Jahrb. Geol. Paläontologie, Abh., v. 152, p. 380-413.

Goodman, D. K., 1983, Morphology, taxonomy, and paleoecology of Cretaceous and Tertiary organic-walled dinoflagellate cysts: Stanford University, Ph. D. dissertation, 478 p.

Goodman, D. K., and Evitt, W. R., 1981, The dinoflagellate *Angustidinium acribes* (Davey & Verdier) gen. et comb. nov. from the mid-Cretaceous of the northern California Coast Ranges: Grana, v. 20, p. 43-54.

Goodman, D. K., and Stover, L. E., 1981, Comparison of *Microdinium* and allied dinoflagellate genera: Palynology, v. 5, p.215-216 (abstract).

Goodman, D. K., and Witmer, R. J., 1980, Archeopyle variation in the *Diphyes*-complex: Palynology, v. 4, p. 24l (abstract).

Górka, Hanna, 1969, Microorganismes de l'ordovicien de Pologne: Palaeontologia Polonica, No. 22, p. 1-102.

Graham, H. W., 1942, Studies in the morphology, taxonomy, and ecology of the Peridiniales: Carnegie Institution Washington, Publ. 542, v + 129 p.

Gray, Jane, 1965, Extraction techniques: p. 530-587 *in* Kummel, Berhnard, and Raup, David, Handbook of palynological techniques. San Francisco: W. H. Freeman.

Greuet, Claude, 1970, Ultrastructures de l'ocelle du dinoflagellé *Nematodinium* comparées à celles d'autres représentants de la famille des Warnowiidae: Sept. Congr. Internat. Microscopie Électronique, Grenoble, 1970, p. 385-386.

Habib, Daniel, 1973, Taxonomy, morphology and suggested phylogeny of the dinoflagellate genus *Druggidium*: Geoscience and Man, v. 7, p. 47-55.

Hall, R. P., 1925, Mitosis in *Ceratium hirundinella* O.F.M., with notes on nuclear phenomena in encysted forms and the question of sexual reproduction: Univ. California Publ. Zoology, v. 28, p. 29-63.

Hand, W. G., Collard, P. A., and Davenport, Demorest, 1965, The effects of temperature and salinity change on swimming rate in the dinoflagellates, *Gonyaulax* and *Gyrodinium*: Biol. Bull. Mar. Biol. Lab. Woods Hole, v. 128, p. 90-104.

Happach-Kasan, Christel, 1982, Beobachtungen zum Bau der Theka von *Ceratium cornutum* (Ehrenb.) Clap. et Lachm. (Dinophyta): Archiv Protistenkunde, v. 125, p. 181-207.

Harker, Stuart, 1979, Archeopyle formation in *Palaeohystrichophora infusorioides* Deflandre, 1935: Neues Jahrb. Geol. Paläontologie, Mh. 1979, p. 369-377.

Harland, Rex, 1977, Recent and late Quaternary (Flandrian and Devensian) dinoflagellate cysts from marine continental shelf sediments around the British Isles: Palaeontographica, v. 164, Abt. B, p. 87-126.

______ 1979, The *Wetzeliella* (*Apectodinium*) *homomorphum* plexus from the Paleogene/earliest Eocene of north-west Europe: Proc. IV Internat. Palynology Conference, Lucknow (1976-1977), v. 2, p. 59-70.

______ 1982a, Recent dinoflagellate cyst assemblages from the southern Barents Sea: Palynology, v. 6, p. 9-18.

______ 1982b, A review of Recent and Quaternary organic-walled dinoflagellate cysts of the genus *Protoperidinium*: Palaeontology, v. 25, p. 369-397.

Harland, Rex, Morbey, S. J., and Sarjeant, W. A. S., 1975, A revision of the Triassic to lowest Jurassic dinoflagellate *Rhaetogonyaulax*: Palaeontology, v. 18, p. 847-864.

Harland, Rex, Reid, P. C., Dobell, Patricia, and Norris, Geoffrey, 1980, Recent and sub-Recent dinoflagellate cysts from the Beaufort Sea, Canadian Arctic: Grana, v. 19, p. 211-215.

Harland, Rex, and Sarjeant, W. A. S., 1970, Fossil freshwater microplankton (dinoflagellates and acritarchs) from Flandrian (Holocene) sediments of Victoria and Western Australia: Proc. Roy. Soc. Victoria, v. 83, p. 211-234.

Harland, Rex, and Sharp, Jane, 1980, *Phthanoperidinium obscurum* sp. nov., a non-marine dinoflagellate cyst from the late Eocene of England: Rev. Palaeobotany Palynology, v. 30, p. 287-296.

Harland, W. B., Cox, A.V., Llewellyn, P. G., Pickton, C. A. G., Smith, A. G., and Walters, R., 1982, A geologic time scale: 131 p. Cambridge: Univ. Press.

Harris, W. K., 1974, Tertiary non-marine dinoflagellate cyst assemblages from Australia: Geol. Soc. Australia, Spec. Publ., no. 4, p. 159-166, pl. 1-2.

Hasle, G. R., and Nordli, Erling, 1951, Form variation in *Ceratium fusus* and *tripos* populations in cultures and from the sea: Avh. Norske Vidensk. Akad. Oslo, Math.-Naturv. Klasse, v. 4, p. 1-25.

Helby, R. J., in press, *Muderongia* and related genera in the Late Jurassic-Early Cretaceous of Australia: Alcheringa.

Helby, R. J., and Partridge, A. D., 1980, A catalogue of 35mm transparencies of Australian Mesozoic dinoflagellate and acritarch species described by Isabel C. Cookson and coworkers 1955-1974, Roger Morgan, 1975-1980, John Filatoff 1975: Earth Resources Foundation, Univ. Sidney, Sidney. 15 p., 287 color transparencies.

Helby, R. J., and Stover, L. E., in press, *Ternia balmei* gen. et sp. nov., a Jurassic dinoflagellate with possible dinophysialian affinity: Alcheringa.

Helenes, Javier, 1983, Evaluation of Jurassic-Cretaceous dinoflagellates in the *Ascodinium-Ovoidinium* complex: Micropaleontology, v. 29, p. 255-266.

Himes, M., and Beam, C. A., 1975, Genetic analysis in the dinoflagellate *Crypthecodinium* (*Gyrodinium*) *cohnii*: evidence for unusual meiosis: Proc. Nat. Acad. Sci. (Washington), v. 72, p. 4546-4549.

Huber, Gottfried, and Nipkow, Fr., 1922, Experimentelle Untersuchungen über die Entwicklung von *Ceratium hirundinella* O. F. M.: Zeitschr. Botanik, v. 14, p. 337-37l.

______ and ______ 1923, Experimentelle Untersuchungen über die Entwicklung und Formbildung von *Ceratium hirundinella* O. F. M.: Flora, v. 116, p. 114-215.

Huber-Pestalozzi, Gottfried, 1950, Klasse: Peridineae: p. 94-310 *in* Das Phytoplankton des Süsswassers. Systematik und Biologie. Cryptophyceen, Chloromonodineen, Peridineen: v. 16, pt. 3, of Thienemann, August (ed.), Die Binnengewässer. Stuttgart: Schweitzerbart'sche Verlags.

Islam, M. A., 1982, Archeopyle structure in the fossil dinoflagellate *Phthanoperidinium*: Rev. Palaeobotany Palynology, v. 36, p. 305-316.

Jahn, T. L., Harmon, W. M., and Landman, M., 1961, Mechanisms of locomotion in flagellates. I. *Ceratium*: Jour. Protozoology, v. 10, p. 358-363.

Jansonius, Jan, 1962, Palynology of Permian and Triassic sediments, Peace River area, western Canada: Palaeotographica, Abt. B, v. 110, p. 35-98.

Jollos, Victor, 1910, Dinoflagellatenstudien: Archiv Protistenkunde, v. 19, p. 178-206.

Jux, Ulrich, 1968-1979; 1968a, Uber den Feinbau der Wandung bei *Cordosphaeridium inodes* (Klumpp 1953): Palaeontographica, Abt. B, v. 122, p. 48-54. 1968b, Uber den Feinbau der Wandung bei *Hystrichosphaera bentori* Rossignol: *ibid*. v. 123, p. 147-152. 1971, Uber den Feinbau der Wandungen einiger tertiarer Dinophyceen-Zysten und Acritarcha – *Hystrichosphaeridium*, *Impletosphaeridium*, *Lingulodinium*: *ibid*., v.

132, p. 165-174. 1976, Über den Feinbau der Wandungen bei *Operculodinium centrocarpum* (Deflandre & Cookson) Wall 1967 und *Bitectatodinium tepikiense* Wilson 1973: *ibid*., v. 155, p. 149-156. 1980, Uber den Feinbau der Wandung bei *Oligosphaeridium abaculum* Davey, 1979: Palaeontographica, Abt. B, v. 17, p. 1-6.

Keupp, Helmut, 1979a, Calciodinelloidea aus der Blätterton-Fazies des nord-deutschen Unterbarremian: Naturhist. Gesellschaft Hannover, Berichte, v. 122, p. 7-69.

______ 1979b, Lower Cretaceous Calcisphaerulidae and their relationship to calcareous dinoflagellate cysts: Centre Recherche, Explor.-Prod. Elf-Aquitaine, Bull., v. 9, p. 651-663. Pau.

______ 1980, *Pithonella patriciagreeleyae* Bolli 1974, eine kalkige Dinoflagellaten-Zyste mit interner Paratabulation (Unter-Kreide, Speeton/SE-England): Neues Jahrb. Geol. Paläontologie, Mh. 1980, p. 513-524.

______ 1981, Die kalkigen Dinoflagellaten-Zysten der borealen Unter-Kreide (Unter-Hauterivium bis Unter-Albium): Facies, v. 5, p. 1-190.

Klebs, G., 1883, Ueber die Organisation einiger Flagellatengruppen und ihre Beziehungen zu Algen und Infusoriengruppen: Bot. Inst. Tübingen, Untersuchungen, v. 1, p. 233-262.

Klement, Karl, 1960, Dinoflagellaten und Hystrichosphaerideen aus dem unteren und mittleren Malm Südwestdeutschlands: Palaeontographica, Abt. A, v. 114, p. 1-104.

Klumpp, Barbara, 1953, Beitrag zur Kenntnis der Mikrofossilien des mittleren und oberen Eozän: Palaeontographica, Abt. A, v. 103, p. 377-406.

Kofoid, C. A., 1907, The plates of *Ceratium* with a note on the unity of the genus: Zoolog. Anzeiger, v. 32, p. 177-183.

______ 1909a, On *Peridinium steini* Jörgensen, with a note on the nomenclature of the skeleton of the Peridinidae: Archiv Protistenkunde, v. 16, p. 25-47.

______ 1909b, Mutations in *Ceratium*: Bull. Mus. Comp. Zoology (Harvard), v. 52 p. 213-257.

Kofoid, C. A., and Swezy, Olive, 1921, The free-living unarmored Dinoflagellata: Univ. California, Mem. no. 5, viii + 538 p.

Krutzsch, Wilfried, 1962, Die Mikroflora der Geiseltalbraunkohle, Teil III. Süsswasserdinoflagellaten aus subaquatisch gebildeten Blätterkohlenlagen des mittleren Geiseltales: Hallesches Jahrb. Mitteldeutsche Erdgeschichte, v. 4, p. 40-45.

LeBlond, P. H., and Taylor, F. J. R., 1976, The propulsive mechanism of the dinoflagellate transverse flagellum reconsidered: Biosystems, v. 8, p. 33-39.

Lebour, Marie, 1925, The dinoflagellates of northern seas: vi + 250 p., Plymouth (U.K.): Marine Biol. Assoc.

Lefèvre, Marcel, 1933, Les péridinites des Barbades: Ann. Cryptogamie Exotique, v. 6, p. 215-229.

Leffingwell, H. A., and Hodgkin, Norman, 1971, Techniques for preparing fossil palynomorphs for study with the scanning and transmission electron microscopes: Rev. Palaeobotany Palynology, v. 11, p. 177-199.

Lejeune, Maria, 1936a, L'étude microscopique des silex. Son enseignement dans les domains paléontologiques et géologiques (Prémier note): Soc. Géol. Belgique, Annales, v. 59, p. 190-197.

______ 1936b, L'étude microscopique des silex. Encore *Hystrichosphaera ramosa* Ehrbg.: les coques "dedoublées," le "flagelle" (Troisième note): Soc. Géol. Belgique, Annales, v. 60, p. 321-333.

Lejeune-Carpentier, Maria, 1938, *Peridinium pyrophorum* Ehrenberg: Mus. Roy. Hist. Nat. Belgique, Bull., v. 14, p. 1-13.

______ 1940, L'étude microscopique des silex. Systématique et morphologie des "Tubifères" (Huitième note): Soc. Géol. Belgique, Annales, v. 63, p. 216-236.

______ 1951, L'étude microscopique des silex. *Gymnodinium* et *Phanerodinium* (Dinoflagellates) de Belgique (Treizième note): Soc. Géol. Belgique, Annales, v. 74, p. 307-313.

Lentin, J. K., and Williams, G. L., 1973-1981, Fossil dinoflagellates: index to genera and species: Geol. Survey Canada, Paper 73-42, 176 p. (1973); Supplement: Canadian Jour. Botany, v. 53, p. 2147-2157 (1975); 1977 edition: Bedford Inst. Oceanography, Report BI-R-77-8, v + 209 p.; 1981 edition: Bedford Inst. Oceanography, Report BI-R-81-12, vii + 345 p.

______ and ______ 1976, A monograph of fossil peridinioid dinoflagellate cysts: Bedford Inst. Oceanography, Report Ser. BI-R-75-16, p. 1-237.

Loeblich, A. R., Jr., and Loeblich, A. R., III, 1966-1974, Index to the genera, subgenera and sections of the Pyrrhophyta. (I), 1966: Univ. Miami, Studies Tropical Oceanography, no. 3, p. i-x + 1-94. II, 1968: Jour. Paleontology, v. 42, p. 210-213. III, 1969: Jour. Paleontology, v. 43, p. 193-198. IV, 1970a: Jour. Paleontology, v. 44, p. 536-543. (V), 1970b: Phycologia, v. 9, p. 199-203. VI, 1971: Phycologia, v. 10, p. 309-314. VII, 1974: Phycologia, v. 13, p. 57-61.

Loeblich, A. R., III, 1967, Aspects of the physiology and biochemistry of the Pyrrhophyta: Phykos, v. 5 (1966), p. 216-255.

_______ 1968, A new marine dinoflagellate genus *Cachonina* in axenic culture from the Salton Sea with remarks on the genus *Peridinium*: Proc. Biol. Soc. Washington, v. 81, p. 91-96.

_______ 1970, The amphiesma or dinoflagellate cell covering: Proc. North Amer. Paleont. Convention Sept. 1969, pt. G, p. 867-929.

_______ 1980, Dinoflagellate nomenclature: Taxon, v. 29, p. 321-324.

Loeblich, A. R., III, and Loeblich, L. A., 1979, The systematics of *Gonyaulax* with special reference to the toxic species: p. 41-46 *in* Taylor, D. L., and Seliger, H. H. (eds.), Toxic dinoflagellate blooms, Proc. 2nd Internat. Conf. on Toxic Dinoflagellate Blooms, v. 1.

Loeblich, A. R., III, Schmidt, R. J., and Sherley, J. L., 1981, Scanning electron microscopy of *Heterocapsa pygmaea* sp. nov., and evidence for polyploidy as a speciation mechanism in dinoflagellates: Jour. Plankton Res., v. 3, p. 67-80.

Loeblich, A. R., III, and Volcani, B. E., 1969, Morphology and chemical composition of the cell wall of the marine dinoflagellate *Cachonina niei*: Abstracts, XI Internat. Botanical Congress, p. 131.

Lohmann, Hans, 1904, Eier und sogenannte Cysten der Plankton-Expedition. Anhang: Cyphonautes: Ergebnisse Plankton-Expedition. Humboldt-Stiftung, n. ser., v. 4, 61 p. Kiel.

López, J., 1955, Variación alometrica en *Ceratium tripos*: Investigación Pesquera, v. 2, p. 131-159.

_______ 1966, Variación y regulación de la forma en el género *Ceratium*: Investigación Pesquera, v. 30, p. 325-427.

Mantell, G. A. 1850, A pictorial atlas of fossil remains, consisting of colored illustrations selected from Parkinson's "Organic remains of a former World" and Artis's "Antediluvian Phytology": xii + 208 p. London: H. G. Bohn.

Manum, S. B., 1979, Two new Tertiary dinocyst genera from the Norwegian Sea: *Lophocysta* and *Evittosphaerula*: Rev. Palaeobotany Palynology, v. 28, p. 237-248.

Matsuoka, Kazumi; Kobayashi, Satoru; and Iizuka, Shoji, 1982, Cysts of *Protoperidinium divaricatum* (Meu.) P. & D. 1976 from surface sediments of Omura Bay, West Japan: Rev. Palaeobotany Palynology, v. 38, p. 109-118.

May, F. E., 1976, Dinoflagellates: Fossil motile stage tests from the Upper Cretaceous of the northern New Jersey Coastal Plain: Science, v. 193, p. 1128-1130.

_______ 1980, Dinoflagellate cysts of the Gymnodiniaceae, Peridiniaceae, and Gonyaulacaceae from the Upper Cretaceous Monmouth Group, Atlantic Highlands, New Jersey: Palaeontographica, Abt. B, v. 172, p. 10-116.

McIntyre, D. J., 1975, Morphologic changes in *Deflandrea* from a Campanian section, District of Mackenzie, N.W.T., Canada: Geoscience and Man, v. 11, p. 61-76.

McKee, E. D., Chronic, John, and Leopold, E. B., 1959, Sedimentary belts in lagoon of Kapingamarangi Atoll: Bull. Amer. Assoc. Petroleum Geologists, v. 43, p. 501-562.

McLean, D. M., 1972, *Cladopyxidium septatum*, n. gen., n. sp., possible Tertiary ancestor of the modern dinoflagellate *Cladopyxis hemibrachiata* Balech 1964: Jour. Paleontology, v. 46, p. 861-863, pl. 1.

Michaelis, G. A., 1830, Uber das Leuchten der Ostsee nach eigenen Beobachtungen: 52 p. Hamburg: Perthes und Besser.

Morbey, S. J., 1975, The palynostratigraphy of the Rhaetian Stage, Upper Triassic in the Kendelbachgraben, Austria: Palaeontographica, Abt. B, v. 152, p. 1-75, pl. 1-19.

Morey-Gaines, Gregory, and Ruse, R. H., 1980, Encystment and reproduction of the predatory dinoflagellate, *Polykrikos kofoidii* Chatton (Gymnodiniales): Phycologia, v. 19, p. 230-236.

Morgan, Roger, 1977, Elucidation of the Cretaceous dinoflagellate *Diconodinium* Eisenack and Cookson, 1960, and related peridinioid species from Australia: Palynology, v. 1, p. 123-127.

Morgenroth, Peter, 1968, Zur Kenntnis der Dinoflagellaten und Hystrichosphaerideen des Danien: Geol. Jahrb., v. 86, p. 533-578.

_______ 1970, Dinoflagellate cysts from the Lias Delta of Lühnde/Germany: Neues Jahrb. Geol. Paläontologie, Abh., v. 136, p. 345-359.

Morrill, L. C., and Loeblich, A. R., III, 1981, A survey for body scales in dinoflagellates and a revision of *Cachonina* and *Heterocapsa* (Pyrrhophyta): Jour. Plankton Research, v. 3, p. 53-66.

_______ and _______ 1981, The dinoflagellate pellicular wall layer and its occurrence in the Division Pyrrhophyta: Jour. Phycology, v. 17, p. 315-323.

Müller, O. F., 1773, Vermium terrestrium et fluviatilium . . . : Haineck et Faber typis Martini Hallager, Havniae et Lipsiae v. 1, 32 + 135 p.

_______ 1777, Zoologiae Danicae prodromus . . . : Hallageriis Hauniae 1777, 274 p.

_______ 1786, Animalcula infusoria fluviatilia et marina . . . : Hauniae, Nicolae Mölleri, 8 + lvi + 367 p.

Nevo, Zvi, and Sharon, Nathan, 1969, The cell wall of *Peridinium westii*, a noncellulosic glucan: Biochim. Biophys. Acta, v. 173, p. 161-175.

Nordli, Erling, 1951, Resting spores of *Goniaulax polyedra* Stein: Nytt Mag. Naturvidenskapene, v. 88, p. 207-212.

Norris, Geoffrey, 1978a, Phylogeny and a revised supra-generic classification for Triassic-Quaternary organic-walled dinoflagellate cysts (Pyrrhophyta). Part I. Cyst terminology and assessment of previous classifications: Neues Jahrb. Geol. Paläontologie, Abh., v. 155, p. 300-327.

_______ 1978b, Phylogeny and a revised supra-generic classification for Triassic-Quaternary organic-walled dinoflagellate cysts (Pyrrhophyta). Part II. Families and suborders of fossil dinoflagellates: Neues Jahrb. Geol. Paläontologie, Abh., v. 156, p. 1-30.

Norris, Geoffrey, and Hedlund, R. W., 1972, Transapical sutures in dinoflagellate cysts: Geoscience and Man, v. 4, p. 49-56.

Norris, Geoffrey, and McAndrews, J. H., 1970, Dinoflagellate cysts from postglacial lake muds, Minnesota (USA): Rev. Palaeobotany Palynology, v. 10, p. 131-156.

Norris, Geoffrey, and Sargeant, W.A.S., 1965, A descriptive index of genera of fossil Dinophyceae and Acritarcha: New Zealand Geol. Survey Paleont. Bull. no 40, 72p.

Norvick, M. S., 1976, Mid-Cretaceous microplankton from Bathurst Island: p. 21-113 *in* Norvick, M. S., and Burger, Dennis, Palynology of the Cenomanian of Bathurst Island, Northern Territory, Australia: Bur. Min. Resources, Geology and Geophysics, Bull. 151.

Peters, Nicolaus, 1927, Das Wachstum des Peridiniumpanzers: Zoolog. Anzeiger, v. 73, p. 143-149.

_______ 1929a, Die Peridineenbevölkerung der Weddellsee mit besonderer Berücksichtigung der Wachstums- und Variations formen: Internat. Revue Gesamten Hydrobiologie Hydrographie 1928-1929, v. 21, p. 17-146.

_______ 1929b, Über Orts- und Geisselbewegung bei marinen Dinoflagellaten: Archiv Protistenkunde, v. 67, p. 291-321.

Pfiester, L. A., 1975-1977; 1975, Sexual reproduction of *Peridinium cinctum* f. *ovoplanum* (Dinophyceae: Jour. Phycology, v. 11, p. 259-265. 1976, Sexual reproduction of *Peridinium willei* (Dinophyceae): ibid., v. 12, p. 234-238. 1977, Sexual reproduction of *Peridinium gatunense* (Dinophyceae): ibid., v. 13, p. 92-95.

Pfiester, L. A., and Skvarla, J. J., 1980, Comparative ultrastructure of vegetative and sexual thecae of *Peridinium limbatum* and *Peridinium cinctum* (Dinophyceae): Amer. Jour. Botany, v. 67, p. 955-958.

Piel, K. M., and Evitt, W. R., 1980, Paratabulation in the Jurassic dinoflagellate genus *Nannoceratopsis* and a comparison with modern taxa: Palynology, v. 4, p. 79-104.

Pollingher, Utsa, 1981, The structure and dynamics of the phytoplankton assemblages in Lake Kinneret, Israel: Jour. Plankton Research, v. 13, p. 93-105.

Rahat, M., 1968, Observations on the life cycle of *Peridinium westii* in a mixed culture: Israel Jour. Botany, v. 17, p. 200-206.

Reid, P. C., 1974, Gonyaulacacean dinoflagellate cysts from the British Isles: Nova Hedwigia, Beihefte, v. 25, p. 579-637.

_______ 1977, Peridiniacean and glenodiniacean dinoflagellate cysts from the British Isles: Nova Hedwigia, v. 29, p. 429-456.

______ 1978, Dinoflagellate cysts in the plankton: New Phytology, v. 80, p. 219-229.

Reid, P. C., and Harland, Rex, 1977, Studies of Quaternary dinoflagellate cysts from the North Atlantic: p. 147-169 *in* Elsik, W. C. (ed.), Contributions of Stratigraphic Palynology, v. 1, Cenozoic Palynology: Amer. Assoc. Stratigraphic Palynologists, Contrib. Ser., no. 5A.

Reinsch, P. F., 1905, Die Palinosphärien, ein mikroscopischer vegetabiler Organismus in der Mukronatenkreide: Centralbl. Mineralogie, Geologie, Palaeontologie, Jahrg. 1905, p. 26-31.

Riegel, Walter, and Sarjeant, W. A. S., 1982, Dinoflagellate cysts from the Upper Cretaceous of southern Spain: new morphological and taxonomic observations: Neues Jahrb. Geol. Paläontologie, Abh., v. 162, p. 286-303.

Riley, L. A., and Sarjeant, W. A. S., 1972, Survey of the stratigraphic distribution of dinoflagellates, acritarchs, and tasmanitids in the Jurassic: Geophytology, v. 2, p. 1-40.

Rossignol, Martine, 1961, Analyse pollinique de sédiments marins Quarternaires en Israël. I: Sédiments Récents: Pollen et Spores, v. 3, p. 303-324.

______ 1964, Hystrichosphères du Quaternaire en Mediteranée orientale, dans les sédiments pléistocènes et les boues marines actuelles: Rev. Micropaléontologie, v. 7, p. 83-99.

Sarjeant, W. A. S., 1962, Upper Jurassic microplankton from Dorset, England: Micropaleontology, v. 8, p. 255-268.

______ 1963, Fossil dinoflagellates from Upper Triassic sediments: Nature (London), v. 199, p. 353-354.

______ 1967a, The genus *Palaeoperidinium* Deflandre 1934 (Dinophyceae): Grana Palynologica, v. 7, p. 241-258.

______ 1967b, The stratigraphical distribution of fossil dinoflagellates: Rev. Palaeobotany Palynology, v. 1, p. 323-343.

______ 1970, The genus *Spiniferites* Mantell (Dinophyceae): Grana Palynologica, v. 10, p. 74-78.

______ 1974, Fossil and living dinoflagellates: 182 p. Academic Press, London and New York.

______ 1978, *Arpylorus antiquus* Calandra, emend., a dinoflagellate cyst from the Upper Silurian: Palynology, v. 2, p. 167-180.

______ 1982a, Dinoflagellate cyst terminology, a discussion and proposals: Canadian Jour. Botany, v. 60, p. 922-945.

______ 1982b, The dinoflagellate cysts of the *Gonyaulacysta* group: a morphological and taxonomic restudy: Amer. Assoc. Stratigraphic Palynologists, Contrib. Ser. no. 9, 80 p.

Sarjeant, W. A. S., and Downie, Charles, 1966, The classification of dinoflagellate cysts above generic level: Grana Palynologica, v. 6, p. 503-527.

______ and ______ 1974, The classification of dinoflagellate cysts above generic level; a discussion and revision: Birbal Sahni Inst. Palaeobotany, Spec. Publ. No. 3, p. l9-32.

Schiller, Josef, 1931-1937, Dinoflagellatae (Peridineae), in monographischer Behandlung: p. 1-617 *in* Dr. L. Rabenhorst's Kryptogamenflora von Deutschland, Oesterreich, und der Schweiz, v. 10, Abt. 3. Leipzig: Akad. Verlags.

Schopf, J. W., 1968, Microflora of the Bitter Springs Formation, Late Precambrian, central Australia: Jour. Paleontology, v. 42, p. 651-688.

Schrank, F. vP., 1793, Mikroskopische Wahrnehmungen: Der Naturforscher, v. 27, p. 26-37.

Schütt, Franz, 1887, Ueber die Sporenbildung mariner Peridineen: Berichte Deutscher Geolog. Gesell., v. 5, p. 364-374.

______ 1895, Die Peridinien der Plankton-Expedition: Ergebnisse Plankton-Exped. Humboldt-Stiftung, v. 4, p. 1-170.

Sellberg, Björn, and Kjellström, Göran, 1974, Geometric characterization of the cavate dinoflagellate cysts of the genus *Deflandrea*: Neues Jahrb. Geol. Paläontologie, Mh. 1974, p. 315-320.

Sournia, Alain, 1973-1982, Catalogue des espèces et taxons infraspécifiques de Dinoflagellés marins actuels publiés depuis la revision de J. Schiller, I. Dinoflagellés libres: Nova Hedwigia, Beihefte, v. 48, p. i-xii, 1-92 (1973). III. (Complement): Revue algologique, n. sér., v. 13, p. 3-40, erratum, p. 186 (1978). IV. (Complement): Archiv Protistenkunde, v. 126, p. 151-168 (1982).

Sournia, Alain; Cachon, Jean; and Cachon, Monique, 1975, Catalogue des éspèces et taxons infraspécifiques de Dinoflagellés marins actuels publiés depuis la révision de

J. Schiller: II. Dinoflagélles parasites ou symbiotiques: Archiv Protistenkunde, v. 117, p. 1-19.

Steidinger, K. A., and Cox, E. R., 1980, Free-living dinoflagellates: p. 407-432 *in* Cox, E. R. (editor), Phytoflagellates. North Holland: Elsevier.

Steidinger, K. A., and Williams, Jean, 1970, Dinoflagellates: Memoirs of the Hourglass Cruises, v. 2, 251 p. St. Petersburg: Marine Research Lab., Florida Dept. Nat. Resources.

Steidinger, K. A., Tester, L. S., and Taylor, F. J. R., 1980, A redescription of *Pyrodinium bahamense* var. *compressa* (Bohm) stat. nov. from Pacific red tides: Phycologia, v. 19, p. 329-337.

Stein, J., 1983, Upper Maastrichtian dinoflagellate cysts from the western San Joaquin Valley, California: Stanford University, unpublished Ph.D. dissertation.

Stosch, H. A., von, see von Stosch, H. A.

Stover, L. E., 1975, Observations on some Australian Eocene dinoflagellates: Geoscience and Man, v. 11, p. 35-45.

Stover, L. E., and Evitt, W. R., 1978, Analyses of pre-Pleistocene organic-walled dinoflagellates: Stanford Univ. Publs., Geol. Sci., v. 15, iii + 300 p.

Tangen, Karl, Brand, L. E., Blackwelder, P. L., and Guillard, R. L., 1982, *Thoracosphaera heimii* (Lohmann) Kamptner is a dinophyte: observations on its morphology and life cycle: Marine Micropaleontology, v. 7, p. 193-212.

Tappan, Helen, 1971, Microplankton, ecological succession and evolution: Proc. North Amer. Paleont. Convention, Sept. 1969, pt. H, p. 1058-1103.

______ 1980, Paleobiology of plant protists: xvi + 1028 pages. San Francisco: W. H. Freeman.

Tappan, Helen, and Loeblich, A. R., Jr., 1971, Geobiologic implications of fossil phytoplankton evolution and time-space distribution: p. 247-340 *in* Kosanke, R., and Cross, A. T. (eds.), Symposium on palynology of the Late Cretaceous and Early Tertiary, Geol. Soc. America, Spec. Paper 127.

______ and ______ 1972, Fluctuating rates of protistan evolution, diversification, and extinction: 24th Internat. Geol. Congress (Montreal), Proc., Sec. 7 Paleontology, p. 205-213. Montreal.

______ and ______ 1973, Evolution of the ocean plankton: Earth Science Reviews, v. 9, p. 207-240.

______ and ______ 1977, Peridinialean cyst affinity, rather than gymnodinialean motile stage, of the Late Cretaceous dinoflagellate *Dinogymnium*: Trans. Amer. Microscop. Soc., v. 96, p. 497-505.

Tasch, Paul, 1963, Hystrichosphaerids and dinoflagellates from the Permian of Kansas: Micropaleontology, v. 9, p. 332-336.

Taylor, D. L., 1968, In situ studies on the cytochemistry and ultrastructure of a symbiotic marine dinoflagellate: Jour. Mar. Biol. Assoc. U.K., v. 48, p. 349-366.

Taylor, F. J. R., 1976a, Dinoflagellates from the International Indian Ocean Expedition. A report on material collected by the R. V. "Anton Brunn" 1963-1964: Bibliotheca Botanica, v. 132, 234 pages, 46 pls., 44 text-figs.

______ 1976b, Flagellate phylogeny: a study in conflicts: Jour. Protozoology, v. 23, p. 28-40.

______ 1978, Problems in the development of an explicit hypothetical phylogeny of the lower eukaryotes: BioSystems, v. 10, p. 67-89.

______ 1979, The toxic gonyaulacoid dinoflagellates: p. 47-56 *in* Taylor, D. L., and Seliger, H. H., (eds.), Toxic dinoflagellate blooms: Proc. Second Internat. Conf. on Toxic Dinoflagellate Blooms. New York: Elsevier.

______ 1980, On dinoflagellate evolution: BioSystems, v. 13, p. 65-108.

______ (editor), in press, The biology of dinoflagellates: Blackwell Botanical Monographs, v. 20.

Turpin, D. H., Dobell, P. E. R., and Taylor, F. J. R., 1978, Sexuality and cyst formation in Pacific strains of the toxic dinoflagellate *Gonyaulax tamarensis*: Jour. Phycology, v. 14, p. 235-238.

von Stosch, H. A. , 1964, Zum Problem der sexuellen Fortpflanzung in der Peridineengattung *Ceratium*: Helgoländer Wissensch. Meeresuntersuchung, v. 10, p. 140-152.

______ 1965, Sexualität bei *Ceratium cornutum* (Dinophyta): Naturwissenschaften, v. 52, p. 112-113.

______ 1972, La signification cytologique de la 'cyclose nucleaire' dans le cycle de vie des Dinoflagellés: Mém. Soc. Botanique France, 1972, p. 201-212.

______ 1973, Observations on vegetative reproduction and sexual life cycles of two freshwater dinoflagellates, *Gymnodinium pseudopalustre* Schiller and *Woloszynskia apiculata* sp. nov.: British Phycol. Jour., v. 8, p. 105-134.

Vozzhennikova, T. F., 1963, Tip Pyrrhophyta. Pirofitovye vodorosli: p. 171-195 *in* Orlov, Y. A. (ed.), Osnovy Paleontologii, v. 14. Moscow: Akad. Nauk.

______ 1967, Iskopaemye peridinei yurskikh, melovykh i paleogenovykh otlozhenii SSSR: 300 p. Moscow: Akad. Nauk, Sibirskoe Otdelenie, Inst. Geologii Geofisiki.

______ 1979, Dinotsisty i ikh stratigraficheskoe znachenie: 220 p. Novosibirsk: Nauka.

Walker, L. M., and Steidinger, K. A., 1979, Sexual reproduction in the toxic dinoflagellate *Gonyaulax monilata*: Jour. Phycology, v. 15, p. 312-315.

Wall, David, 1965, Modern hystrichospheres and dinoflagellate cysts from the Woods Hole region: Grana Palynologica, v. 6, p. 297-314.

______ 1967, Fossil microplankton in deep-sea cores from the Caribbean Sea: Palaeontology, v. 10, p. 95-123, pl. 14-16, 8 text-figs.

Wall, David, and Dale, Barrie, 1966, "Living fossils" in western Atlantic plankton: Nature (London), v. 211, p. 1025-1026.

______ and ______ 1967, The resting cysts of modern marine dinoflagellates and their palaeontological significance: Rev. Palaeobotany Palynology, v. 3, p. 349-354.

______ and ______ 1968a, Quaternary calcareous dinoflagellates (Calciodinellideae) and their natural affinities: Jour. Paleontology, v. 42, p. 1395-1408.

______ and ______ 1968b, Modern dinoflagellate cysts and the evolution of the Peridiniales: Micropaleontology, v. 14, p. 265-304.

______ and ______ 1969, The "hystrichosphaerid" resting spore of the dinoflagellate *Pyrodinium bahamense* Plate, 1906: Jour. Phycology, p. 140-149.

______ and ______ 1970, Living hystrichosphaerid dinoflagellate spores from Bermuda and Porto Rico: Micropaleontology, v. 16, p. 47-58.

______ and ______ 1971, A reconsideration of living and fossil *Pyrophacus* Stein, 1883 (Dinophyceae): Jour. Phycology, v. 7, p. 221-235.

Wall, David; Dale, Barrie; and Harada, Kenichi, 1973, Descriptions of new fossil dinoflagellates from the Late Quaternary of the Black Sea: Micropaleontology, v. 19, p. 18-31.

Wall, David; Dale, Barrie; Lohman, G. P.; and Smith, W. K., 1977, The environmental and climatic distribution of dinoflagellate cysts in modern marine sediments from regions in the North and South Atlantic Ocean and adjacent seas: Marine Micropaleontology, v. 2, p. 121-200.

Wall, David, and Evitt, W. R., 1975, A comparison of the modern genus *Ceratium* Schrank, 1793, with certain Cretaceous marine dinoflagellates: Micropaleontology, v. 21, p. 14-44.

Warren, J. S., 1973, Form and variation of the dinoflagellate *Sirmiodinium grossi* Alberti, from the Upper Jurassic and Lower Cretaceous of California: Jour. Paleontology, v. 47, p. 101-114.

West, G. S., 1909, A biological investigation of Peridiniae of Sutton Park, Warwickshire: New Phytologist, v. 8, p. 181-196.

Wetherbee, Richard, 1975, The fine structure of *Ceratium tripos*, a marine armored dinoflagellate: Jour. Ultrastructure Research, v. 50, p. 65-87.

Wetzel, Otto, 1933, Die in organischer Substanz erhaltenen Mikrofossilien des baltischen Kreidefeuersteins, mit einem sedimentpetrographischen und stratigraphischen Anhang: Palaeontographica, Abt. A, v. 77, p. 147-188; v. 78, p. 47-156.

Wetzel, Walther, 1952, Beitrag zur Kenntnis des danzeitlichen Meeresplanktons: Geol. Jahrb. v. 66, p. 391-419.

Wiggins, V. D., 1972, Two new Lower Cretaceous dinoflagellate genera from southern Alaska (U.S.A.): Rev. Palaeobotany Palynology, v. 14, p. 297-308.

______ 1975, The dinoflagellate family Pareodiniaceae—a discussion: Geoscience Man, v. 11, p. 95-115.

Wille, Wolfgang,1982, Evolution and ecology of Upper Liassic dinoflagellates from SW Germany: Neues Jahrb. Geol. Paläontologie, Abh., v. 164, p. 74-82.

Wille, Wolfgang, and Gocht, Hans, 1979, Dinoflagellaten aus dem Lias Südwestdeutschlands: Neues Jahrb. Geol. Paläontologie, Abh., v. 58, p. 221-258.

Williams, D. B., 1971a, The distribution of marine dinoflagellates in relation to physical and chemical conditions: p. 91-96 *in* Funnel, B. M., and Riedel, W. R., The micropaleontology of oceans. Cambridge: Univ. Press.

——— 1971b, The occurrence of dinoflagellates in marine sediments: p. 231-243 *in* Funnel, B. M., and Riedel, W. R., The micropaleontology of oceans. Cambridge: Univ. Press.

Williams, D. B., and Sarjeant, W. A. S., 1967, Organic-walled microfossils as depth and shoreline indicators: Marine Geology, v. 5, p. 389-412.

Williams, G. L., 1977, Dinocysts. Their classification, biostratigraphy and palaeoecology: p. 1231-1325 *in* Ramsay, A. T. S., (ed.), Oceanic micropaleontology, v. 2. New York: Academic Press.

Williams, G. L., Sarjeant, W. A. S., and Kidson, E. J., 1978, A glossary of the terminology applied to dinoflagellate amphiesmae and cysts and acritarchs: Amer. Assoc. Stratigraphic Palynologists, Contrib. Ser., no. 2a, p. 1-121.

Wilson, G. J., 1968, Palynology of some Lower Tertiary coal measures in the Waihao District, South Canterbury, New Zealand: New Zealand Jour. Botany, v. 6, p. 56-62.

——— 1971, Observations on European Late Cretaceous dinoflagellate cysts: p. 1259-1275 *in* Farinacci, A. (ed.), Proc. II Planktonic Conference, Rome 1970. Rome.

Wilson, G. J., and Clowes, C. D., 1980, A concise catalogue of organic-walled dinoflagellate genera: New Zealand Geol. Survey Rept. 92, 199 p.

——— and ——— 1982, *Arachnodinium*, a new dinoflagellate genus from the Lower Tertiary of Antarctica: Palynology, v. 6, p. 97-103.

Woloszynska, J., 1916, Polnische Süsswasser-Peridineen: Bull. Acad. Sci. Cracovie, Cl. Sci. Math. Nat., ser. B, v. 14.

Woollam, Ronald, and Riding, J. B., 1983, Dinoflagellate cyst zonation of the English Jurassic: Inst. Geol. Sciences, Rept. 83/2, p. 1-42.

TAXONOMIC INDEX

This index lists the names of all taxa mentioned in the text, tables, and figure captions. The author and date of valid publication are given after each name/combination and these are followed by references to all page (unmarked), figure (f - figure, front. - frontispiece), or table (t) numbers, except in the cases of four frequently cited genera (*Ceratium*, *Gonyaulax*, *Peridinium*, and *Protoperidinium*) for which only the more significant references are given.

Names of genera and species are marked as follows to indicate the nature of the material upon which the taxa were founded:

unmarked - Names of dinoflagellates proposed on the basis of geologically preservable material that surrounded (or, rarely, was contained by) the living cell, comprising mostly resting cysts which were probably zygotic. The preservable material is of organic composition (sporopollenin) except where the inorganic composition is identified in parentheses (cal. = calcareous; sil. = siliceous).

* - Names of dinoflagellates that were proposed on the basis of cells devoid of geologically preservable materials (i.e., mostly motile cells, with or without cellulosic thecae).

** - Names of non-dinoflagellates.

Abbreviations of author names:

Cook. - I. C. Cookson
DDSW - R. A. Davey, C. Downie, W. A. S. Sarjeant, and G. L. Williams
Def. - G. Deflandre
Ehr. - A. G. Ehrenberg
Eis. - A. Eisenack
Harl.- R. Harland
Lemm.- E. Lemmermann
Len. - J. K. Lentin
Loeb. Jr. - A. R. Loeblich, Jr.
Loeb. III - A. R. Loeblich, III
Morgen. - P. Morgenroth
Sarj. - W. A. S. Sarjeant
Voz. - T. F. Vozzhennikova
Will. - G. L. Williams

D

E

F

G

Q

R

SUBJECT INDEX

Bold face indentifies pages on which definitions of terms will be found; page numbers are unmarked, front. = frontispiece, f = figure, t = table.

Q

R